TRAVELS IN OMAN

The ultimate travel book about the Sultanate of Oman, *Travels in Oman: On the Track of the Early Explorers* is a massive compendium of all the significant travellers, explorers, adventurers, and wanderers, from ibn Battuta to Michael Loyd, from Engelbert Kaempfer to Bertram Thomas, who have left written records of their journeys in Oman, with an extended opening chapter on those, like Grattan Geary, who visited Muscat and environs, but never succeeded in reaching the mysterious Interior. Marco Polo and Palgrave probably never visited Oman, but many others did, including the influential Wellsted and his companion Whitelock, Cole, C. Ward, Haines, Pengelley, S. B. Miles, Arthur Stiffe and Percy Cox. Theodore Bent visited Muscat and Dhufar, Bertram Thomas Dhufar and Musandam, and Hamerton Buraimi. Their accounts have never been reprinted before from the learned journals in which they originally appeared, and their republication here constitutes a milestone in the literature on Oman. *Travels in Oman* has been sponsored by Muna Noor Incorporated, Muscat, Sultanate of Oman.

PHILIP WARD, F.R.G.S., F.R.S.A., A.L.A. has followed in the wake of these writer-travellers, adding his own narrative on Oman today, to indicate how far the land has changed, and how far it has remained recognisable from his predecessors' tales. He has also produced the classic documentary history of *Ha'il: Oasis City of Saudi Arabia, Japanese Capitals, Finnish Cities, Albania, Bangkok, Come with me to Ireland,* and *The Aeolian Islands.* Having lived and worked in the Middle East for nine years, he is an ideal companion to historic Oman and to the Renaissance that the nation has enjoyed since 1970.

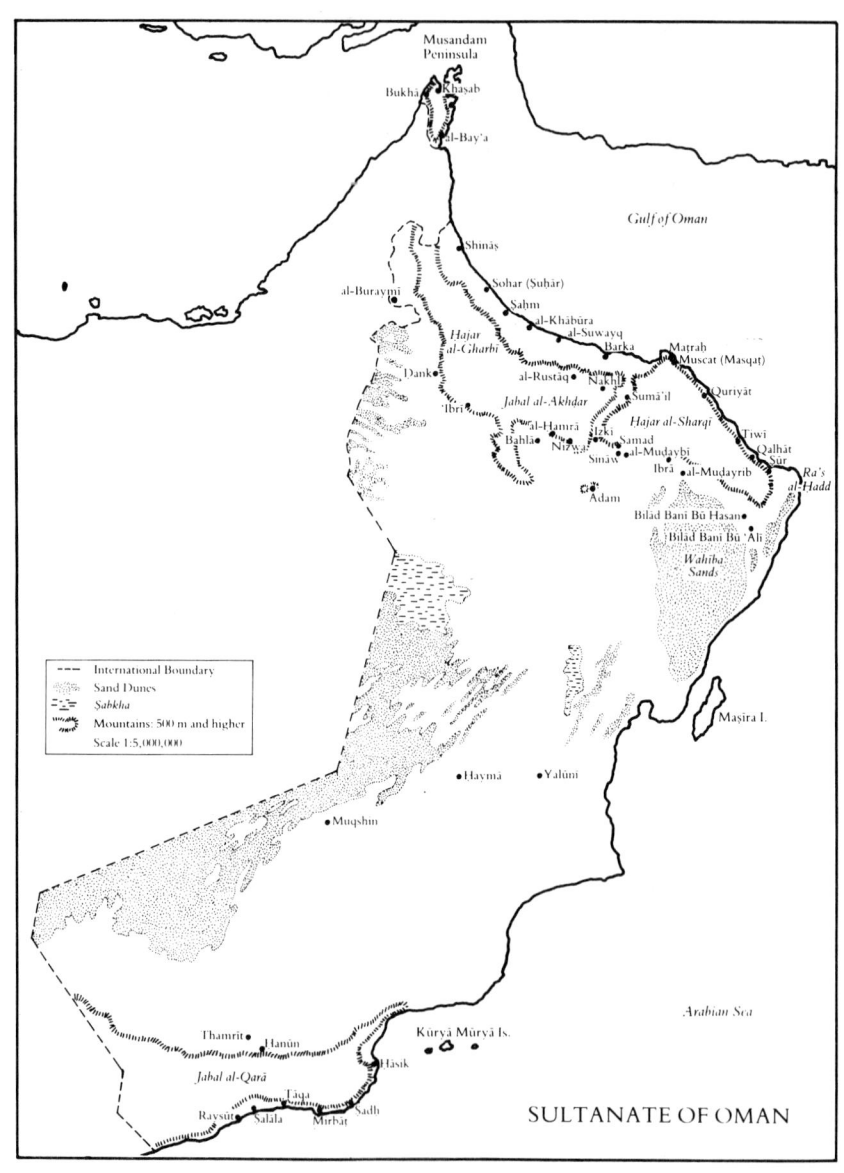

Musandam
Peninsula

Bukhā• •Khaṣab

al-Bay'a

Gulf of Oman

Shinaṣ•

Sohar (Ṣuḥār)•

al-Buraymī• Saḥm•
al-Khābūra•
al-Suwayq•
Barka• Maṭraḥ•
Hajar al-Rustāq• •Muscat (Masqat)
Danke• *al-Gharbī* Nakhl• •Quriyāt
•Ibrī *Jabal al-Akhdar* Sumā'il•
al-Hamrā• Izki• *Hajar al-Sharqī* Tiwi•
Bahlā• Niʒwā• Samad• •Qalhāt
Sināw• al-Muḍaybi• Ṣūr•
Ibrā• •al-Muḍayrib *Ra's al-Hadd*
•Adam
Bilād Banī Bū Hasan•
Bilād Banī Bū 'Alī•
Wāhiba Sands

International Boundary
Sand Dunes
Ṣaḥkha
Mountains: 500 m and higher
Scale 1:5,000,000

Maṣīra I.

•Ḥaymā •Yalūni

•Muqshin

Arabian Sea

•Thamrit •Ḥanūn Kūryā Mūryā Is.
Hāsik•
Jabal al-Qarā Ṭāqa•
Raysūt• •Sadḥ
Ṣalāla• Mirbāṭ•

SULTANATE OF OMAN

TRAVELS IN OMAN

on the Track of the Early Explorers

PHILIP WARD

Bent · Cole · Cox · Eccles · Geary
Haines · Hamdani · Hamerton
ibn Battuta · Kaempfer · Loyd
Miles · Pengelley · Stiffe
B. Thomas · C. Ward · Wellsted
Whitelock

THE OLEANDER PRESS LTD

The Oleander Press
17 Stansgate Avenue
Cambridge CB2 2QZ
England

The Oleander Press
210 Fifth Avenue
New York, N.Y. 10010
U.S.A.

British Library Cataloguing in Publication Data

Ward, Philip
  Travels in Oman.——(Arabia
  past and present; v. 21)
  1. Oman——Description and travel
  I. Title   II. Series
  915.3′53044   DS247.062

  ISBN 0-906672-51-1

Sponsored by Muna Noor Incorporated

Printed and bound in Great Britain

ACKNOWLEDGMENTS

This book owes its genesis, its progress, and its fulfilment to Muna Noor Incorporated, Muscat, Sultanate of Oman, with especial recognition to Ahmad Khalifah al-Balushi. In Muscat thanks are due to Ralph Daly, Adviser to the Royal Diwan on Conservation; Paolo Costa, former Archaeological Adviser to the Ministry of National Heritage and Culture; Col. Rashid bin Masoud bin Rashid Al Zaidi and Lt. Col. Peter Boxhall of the Sultan's Armed Forces Museum, Bait al-Falaj; Wing-Commander and Mrs Kirk; and James Lennox of the Ruwi Hotel; in Dariz, Shaikh Saif ibn 'Ali ibn Muhammad al-Ghafiri; in Bisya, Jocelyn and Jeffrey Orchard of the University of Birmingham Archaeological Expedition; in Bidbid, Khusayib bin Huwaishil; in Rustaq, Jum'a bin Muhammad bin Humaid; in Nizwa, Sulaiman Hamad 'Abdullah as-Sulaimani; in Suhar, Monique Kervran of the Mission Archéologique à Bahrain et Oman; in Zahab, 'Abdullah Salim Sa'id al-Mandhari; in Khasab, John Dymond; in Dhufar, Jum'a Ramadan Nasib; in Manah, Sulaiman bin Saif al-Masruri; in Suwaiq, 'Abdullah Khusaif 'Abdullah bin Khamis bin 'Abdullah al-Mandhari; in England, Charles and Jane Hepworth, and Michael Loyd; in Italy, Paolo Biagi of the University of Genoa and Renato Nisbet of the University of Turin.

Hundreds of Omanis helped to make this book. I refer not only to those who helped me, but also those who helped the early travellers in their epochmaking adventures in a land far too little known and appreciated overseas. If *Travels in Oman* helps to stir the blood – to excite wanderlust – then it will have achieved its objective.

CONTENTS

ILLUSTRATIONS

III: Sharqiyah

IV: Jabal Akhdhar and Dhahirah

1 THE CAPITAL AREA

It is with the arrival in Muscat of Afonso d'Alboquerque in 1507 that the West first made contact with Oman. Afonso set fire to the city, Quriyat, and Khor Fakkan, but left Qalhat unscathed (until 1508) because it submitted. The compiler of the *Commentaries* found that Muscat in 1507 was the main port for the whole coastal region, trading in dates, horses and corn, with a large population and a plain behind it, covered with saltpans, no smaller than the great square in Lisbon. The Portuguese destroyed a beautiful wooden mosque, and burned 34 ships. In the houses he found wooden water tanks, and hidden alcoves full of treasure.

Fantasies invented by the Portuguese of Muscat, in such rarities as Emanuel Faria y Sousa's *Portuguese Asia* (London, 1695) and Pedro Texeira's *Travels* (London, 1902) include a theory that the Three Wise Men had made rendezvous off the Muscat coast; 'it is so easy to catch fish that a hungry cat can come to the sea, put in its tail and when fish come to bite it, swish them ashore'; and 'at Muscat there are such sorcerers that they eat a thing inwardly only fixing their eyes upon it; with their sight draw the entrails of any human body and so kill many . . .'

The Portuguese did not attempt to interfere with the religious beliefs and practices of the Omanis; neither did they seek control of the hinterland, and this was another factor contributing to the separateness of the coastal people, outward-looking and familiar with foreigners, from the inward-looking, fiercely independent tribes of the interior. The Ottoman Empire, spreading its tentacles from Turkey and Egypt, took Muscat from the Portuguese several times in the 16th century, forcing the Portuguese temporarily towards the hinterland, but after the Turkish show of force in 1581, the Portuguese determined to build strongholds to defend Muscat harbour, and Matrah too, and both Mirani and Jalali were finished by 1588, standing to this day. In 1622 with the fall of the port of Hurmuz (on the opposite shore of the Gulf, to which Musandam points a knobbly finger), the Portuguese transformed Muscat into their major Gulf area naval base, but their hegemony began to wane. First, in 1624, Nasir ibn Murshid founded the Ya'ruba Dynasty from his redoubt of Rustaq, in the foothills of the Jabal Akhdhar, opening negotiations with the British East India Co in 1645; second, his successor Sultan bin Saif (1649–88) not only expelled the Portuguese from Muscat in 1650, but seized their boats and harried them, hoist with their own petard, to their strongholds in India and East Africa. It was Sultan bin Saif who created the massive fort at Nizwa over a period of a dozen years. Alexander Hamilton, in *A new account of the East Indies* (1727), tells of the courageous Omani onslaught against their invaders.

'The Portuguese flanked them, from their Forts on the mountains, with plenty of great and small shot; but the Arabs never looked back, nor minded the great numbers of their dead companions, but mounted the walls over the carcases of their slain . . . In the attack on the town (of Muscat), the Arabs lost between 4,000 and 5,000 of the best of their forces; and the Portuguese in their Forts were reduced to 60 or 70 . . . Those in small forts . . . were put to the sword except those who, to save their lives, promised to be circumcised and abjure the Christian Religion. Those in the great fort (Mirani) held out about six months, under great want and fatigues; and all hopes of relief being cut off, they resolved on a surrender, on which motion the impudent Governor, who was the sole cause of their calamity, leapt down a precipice into the sea where, the water being very shallow, he was dashed to pieces on the rocks . . .'

Sir Thomas Herbert, in 1617, noted that Muscat is 'seated on a plain, yet arm'd or propt with two rising advantageous Mountains; a ditche and parapet drawne from one hill to the other so environ her that she seems inaccessible: the Castle is large and defensive, fill'd with men, and stored with great Ordnance . . .'

The Italian traveller Pietro della Valle spent a fortnight in Muscat in January 1625, en route between Chaul (on the Malabar Coast) and Basra. From his *Travels* (London, 1665) we learn of the traditional *barasti* houses: 'sheds made of Palm-boughs'. The eastern wall which the Portuguese had started to construct is 'plain and weak, with a few Bastions, very distant from one another; which wall, drawn from Mountain to Mountain, incloses and secures their houses on that side, as the Sea doth on the opposite and inaccessible little Mountains on the other two sides.' Della Valle anticipated Stiffe's expedition to the hot springs at Ghala ('Kelhuh'), referring to 'sheds too small to stand up in' he found there.

In 1659, Mathew Arnold of Surat (an English factory north of the above-mentioned Chaul, in Gujarat) suggests a scheme for settling Muscat, sending to Sultan ibn Saif an embassy headed by Col. Henry Rainsford. 'If it pleases God to make him so happy as to compass the commaund of the Castle upon any reasonable tearms, wee are confident that there is no place upon these northern seas that can prove so profitable for to gaine your right of the Customs due to you in Persia, but to commaund all the Princes hereabouts to carry a faire correspondence with your people; otherwise you might right your selves upon their jounks as they enter the Gulfe of Persia.' The agreement was that the British should hold the fortifications with no more than a hundred soldiers. Colonel Rainsford died in May, but plans for settling Muscat continued to be worked out.

The Dutch traveller Jan Struys visited Muscat in July 1672 (*Voyages*, Amsterdam, 1681), and not in 1655 as Hogarth has it in *The Penetration of Arabia*. 'On the right hand as you enter the harbour is a

Muscat, 1672 (Jan Struys)

fort on a steep hill, which seems to be impregnable and sufficient to defend the whole harbour. It has also a private way leading to the harbour underground.' 'In the months of August and September it is so incredibly hot, that I am incapable of expressing the torment endured by foreigners, who feel as if thrust into boiling cauldrons or sweating tubs, so that I have known many unable to endure the heat throw themselves into the sea, and stay there until nightfall.'

In the 1660s, Jean-Baptiste Tavernier (*Six Voyages*, London, 1678) claims to have seen the most beautiful pearl in the world, owned by the Imam of Muscat, when the Imam showed it to the Khan of Hurmuz. The Khan offered 2,000 tomans for it, roughly £7,000. Later the Great Mughal offered 40,000 crowns for it, equivalent to £9,000, but Tavernier estimated the value of the pearl, which was nearly transparent, at about £30,000.

Engelbert Kaempfer on Muscat

The German traveller Engelbert Kaempfer arrived in Muscat on 16 July 1688 on a Dutch ship from Bandar Abbas, and had to leave for Ras al-Hadd on the following day, so enjoyed only a matter of hours ashore. In the circumstances, his drawings and notes are impressive, and deserve to be quoted in full.

'On 14 July we arrived at nightfall at the bay of Muscat. The harbour-mouth is hard to recognise as it is narrow, and like cliffs nearby which must be avoided because of the numerous rocky islands nearby. We finally found the entrance, and could have sailed into the bay, had the setting sun not compelled us to follow the Arabian custom of forbidding ships into the bay after sunset. So we had to anchor our ship in the roadstead outside the harbour. We could find no anchorage shallower than 30 fathoms; as we could not anchor at such a depth we drifted out more than half a mile, and had to anchor here to avoid drifting farther still. On 15 July we raised anchor early to recover, by tacking, the half mile lost the day before. We could not manage this, however, before nightfall so once again had to drop anchor in front of the bay, twenty paces from the rock at the right-hand side. We were warned by loud cannon to keep our distance, opposite the first bastion, so we gave up and stayed overnight at the harbour-mouth. On 16 July at first light we set out with our barge towards Muscat, leaving the ship to follow, and on shore acquired fresh stocks of water, our main requirement.

Schiffer, Müller the prebendary and I made the acquaintance of a rich old Armenian merchant called Hovhanessian who was paralysed, and had the misfortune of a failed voyage from Surat to Mocha in the Yemen, because of bad weather. He had made for Muscat with a Dutch captain, and under the Dutch flag, to replace his lost masts and repair the damage to his ship.

Hovhanessian invited us to his stone-built house, and treated us hospitably on the Friday and especially on the Saturday, when he was not required to fast, and ate meat with the fellow-Armenians and servants. We invited him back to our ship for the afternoon, together with two Dutch sailors, one of whom was captain of a Moroccan ship under the protection of the French flag. We honoured his friendship by firing two cannons when he departed.

Muscat Bay is, to my mind, the most wonderful that nature and art could ever devise. It is surrounded in deep semicircle by high precipitous rocks, which the Portuguese have topped with watchtowers, each equipped in the same way and painted with white lime. These fortifications seem no sturdier than paper, but must in fact be capable of withstanding any offensive from seaward. We could only marvel how guns could have been carried up such steep inclines. The bay has a breadth of about a hundred to a hundred and fifty fathoms from the extreme edge of the southernmost cliff to the northernmost counterpart, but a length of about 2,500 fathoms from harbour-mouth to town, in my judgement, supported by others'. The bay-floor is solid, between six and seven fathoms deep.

Muscat Town, surrounded by cliffs and mountains on every other side, is encircled by neat, elegant walls which, on the eastern side, washed by the waves except for a picturesque water-gate, run up the rocky hills, to make these ridges even less liable to successful attack. Entry to the northern side of Muscat is likewise closed by a wall winding over the steep hills. A small green valley along the beach is dotted with the small huts and boats of fisherfolk. In front of the southern side of Muscat there is a barren semicircle of rough ground with rocks, boulders, acacias, a rope-walk, wells to water the cattle, and a few small houses.

Since Muscat is entirely surrounded by mountains, it can be approached by land only on donkeyback. A watchtower is placed at the narrowest point between the rocks, about half a mile or a third of a mile from the town gate. At about a similar distance beyond the first hill, drinking water channels are led to the town, where two cisterns convey it to ships and other receptacles. This is particularly important, since the water obtainable on the opposite shore of the Gulf is brackish, except for the water of Kung (near Bandas Abbas) and of Basra. For the clear, sweet water of Muscat you pay one rupee for about 290 litres. Though we ordered the water early in the morning, it was not supplied until midday because, they said, both cisterns had been completely empty, so they had to await fresh supplies from the hills.

Two large bastions on the cliff along the bay, near the town on both sides of the bay serve to protect Muscat, one of them being crowned by a mosque. The northern side of the town, facing the harbour, is bounded by short, thick walls covered every few feet by heavy cannon. The pandemonium in the town resembles that in Bandar Abbas, being inhabited by native Arabs, Indian merchants, and Jews. An English resident stayed here, but was about to leave, due partly to poor business and dangers to shipping, returning home via Isfahan. His trade was incense (*olibanum*) and other types of gum. A small, vivacious man, he did nothing to reduce the respect I had always entertained for his compatriots.

The town houses are of three kinds. The *barasti* huts, covered with palm-tree leaves, are much roomier and cleaner than those on the Persian shore. Some homes are of mud and stone. But the majority are airy, spacious stone houses, one of them an Italianate building which served the Portuguese government as a Jesuit convent, and is now a warehouse, except when, in every third year, the Imam comes to Muscat and takes up his residence there. Another white European church is also now used as a warehouse. The bazaar, filling a large area, consists of covered, partly-vaulted streets, running at right angles to each other. One could buy there silk and linen, spices, incense, coffee, and among the foodstuffs the best and most delicious mangoes I ever tasted. I bought 100 mangoes for one *mamudi*, but when other foreigners arrived they soon put up their prices. As well as this year's small fresh dates, there were some from last year's crop. To ensure that everyone in our cabin could enjoy this beautiful fruit, I bought a whole bag of dates – as much as a native could carry to the beach – and filled a 150-litre barrel for a cost (including the native's pay) of eleven *mamudi*. In the market we found sour little lemons at one *mamudi*, later increased to 2 and more; onions and garlic – everything we wanted was to be had in abundance, such as black and white grapes, bananas and figs.

We found little butter, milk and chicken in the market, though there is no such scarcity in the country. Sheep, goats and cattle are roughly the same price as in Bandar Abbas. Produce of all kinds is brought to Muscat more usually by sea than by land.

The son of the former Governor was then in charge of administration. Blind in one eye, he was about 33 years old, of a pious disposition, and approached me because I was the only one speaking the Persian language. He asked me many anxious questions about the Portuguese ships which had recently arrived in Kangan, on the Persian shore of the Gulf: what was their number and their intention? He was afraid so much for Muscat town, where 200 soldiers paraded every evening, but for smaller villages nearby, which might be pillaged or destroyed by the invaders.

On the first occasion he was in the company of five men in a first-floor *majlis*, his sword at the ready, and treated guests to coffee, fruit, sweets, and the hubble-bubble. On the following day, however, only two little boys were with him, and a guardsman had to run off, coming back after more than an hour with our interpreter, an Indian merchant or *Banian*. Meanwhile we had gazed at each other like two dumb monkeys, ashamed of ourselves. On his part, he had only just woken up and was not yet properly dressed to receive. On my part, I was trying to recover two sailors who had fled the ship that morning, and he could merely offer to let them be sought in the streets, or in the houses under his jurisdiction. This did not help much, in view of impending dusk and our imminent departure.

The regent of the country calls himself Bil'arab and resides in Nizwa, four days' journey from Muscat. His overlordship extends seven days in the other direction, to Julfar [the modern Ras al-Khaimah], and he fears nobody but the Persians, according to his servants. He had at first captured in Muscat the son of the Great Mughal, Sultan Akbar, and had returned the fugitive to his father Aurangzeb. He had at first kept him in a distant village, without offering an audience, but as soon as a royal Persian messenger arrived and asked for him he was handed over.

His country, the ruler of which is always known as Imam, is not the worst part of 'Arabia Felix', but it is hilly, arid, and provided with but scant fresh water. The heat on this shore of the Gulf is no more tolerable than on the other, but the air seems lighter and healthier at such a heat, and fertile lowlands can be found not far from the Muscat mountains.

The natives are courteous, but not as restrained as the Persians; while conversing or trading they talk at the same time, and the young people shout everyone else down. But conversation in domestic privacy is more formal and civilised, if louder than among the Persians. When travelling through the Gulf, you can distinguish villages on the southern shore by the noise emanating from them even during the night. In Persia, on the other hand, you may travel at night without being able to tell inhabited parts from the uninhabited.

The people hereabouts are brown, slim, long in the face and thin in the cheek. Their noses are high, long and narrow, their beards long, thin and pointed, the hair black and stature average. Men dress in a long wide coat of linen, loose and untidy, with long wide sleeves and a belt round the waist. On this they wear a wide unlined linen coat. Their sandals have leather thongs. They wrap their head in long white towels, with their ends hanging down from their head. Their sword, long and usually straight, is carried on the back or shoulder, and their *khanjar* or dagger is worn by the side. Their muskets are of the same length as the Persians', and carried in the belt on top of their outer coats. All their weapons are old-fashioned.

The soldiers are pious, honourable elders or priests, with long beards, rather than the martial types seen in Europe, as we saw from their slow, dignified rifle-practice.

The coast of Arabia we had seen so far, from Julfar to that sector where it bends south-eastward, seems a uniform mass of rough brown hills, mostly bare, infertile, deserted. From there onward, the hills recede, making place for more pleasant lowlands, settled with villages and defended by forts, with some bushes and trees. Then the hills wind back towards the coast until here at Muscat the town has to find a place between them.

After leaving Muscat, we came upon nothing but beaches, except for some small villages and plains. We approached these beaches up to a thousand paces without being able to find a depth less than 27 fathoms as far as Ras al-Hadd. On Saturday 17 July we sailed before sunset, with an average wind, south-south-eastward.'

Engelbert Kaempfer's drawings are especially significant for their view of the town itself, as opposed to the fortifications shown in William Facey's important *Oman: a Seafaring Nation* (1979), which are drawn from 1610 (*Atlas de vinte folhas* by Manoel Godinho de Heredia); from 1635 (*Livro do Estado da India* (1646) by P. Barreto de Resende); and from 1680 (*Beschryving van Asie* by Olfert Dapper).

The population of Muscat in 1688 was under 10,000; J. L. Dubois believed it was 25,000 in 1793; and J. B. Fraser put the figure at 10,000–12,000 in 1821. Ida Pfeiffer (*A Woman's Journey Round the World*, London, 1852) estimated the population of Muscat at 4,000 in 1848 and Palgrave, in 1863, thought Muscat numbered 24,000 and Matrah 25,000.

The official British *Handbook of Arabia* (1916) reported: 'In

Muscat, 1688 (Engelbert Kaempfer)

number, the population of Muscat fluctuates, and is lowest in hot weather, when more than half the inhabitants of the suburbs seek a less trying climate in Sib, Barkah, and other places in Batinah. In winter, when the town is full, its total population may be set at 10,000, of whom some 3,000 reside within the walls.' By contrast, Matrah was in 1916 'the largest town in the Sultanate of Oman: total population 14,000, of whom 9,000 reside within the walls.'

By 1970, states W. D. Peyton in *Old Oman* (London, 1983), 'it would have been hard to find five thousand people there.'

Carsten Niebuhr, the Danish author of *Beschreibung von Arabien* (Copenhagen, 1772), saw Muscat in 1765 and noted that no Muslims he had met so far 'display so little ostentation and live so soberly as these Ibadhi; they never smoke tobacco, they drink little coffee and even less of strong liquors. Persons of distinction do not dress more splendidly than those of lesser estate, they do not let themselves be easily upset by violent passions, and they are polite towards foreigners . . .'

Abraham Parsons visited Muscat in August 1775 aboard H.M.S. *Seahorse* (with a certain Midshipman Horatio Nelson) and wrote of his experiences in *Travels in Asia and Africa* (London, 1808). The temperature reached 112° Fahrenheit. Goods so greatly exceeded in volume the warehouse space available that they lay in the streets, but were never stolen. Caravans overland brought ostrich-feathers, cattle-hides, sheep and skins, honey, and beeswax, taking back with them cutlery, toys, spices, rice, sugar, coffee and tobacco. The trade with Mocha was enormous: Muscat sent 20,000 bales of coffee every year to Basra, intended for Constantinople. Inward trade included Persian carpets and silks, pearls, *thalers* and Venetian *zecchini*. Again Parsons reported the mangoes better than those of India, though other writers have contradicted him. Water is taken on board ship by lighters loaded from skins, as the ground is too rough for casks to be rolled. Muscat, usually embroiled in war against Persia, was sending 34 warships to relieve the Persian siege of Basra.

Lt. John Porter writes of Muscat in *Remarks on the Bloachee, Brodia and Arabian Coasts* (London, 1781) and explored inland as far as 'Bushire' [Baushar] where he found water as hot as could be borne and 'reckoned Sovereign for all disorders of the Skin.' His account of obtaining water for ships is that a 'small channel made with Chinnam and Stones' runs from more than a mile beyond the walls to a reservoir from which it is fed by leather pipe into a boat. 'It seems strange', adds Porter, 'that though all the Rocks about this place are Lime Stone, and that the lime works out of the Rocks itself without the help of fire, there is very little made use of it, and the large houses in the Town are built with Mud instead of Chinnam.'

Dr John Griffiths describes, in *Travels in Europe, Asia Minor and Arabia* (London, 1805) how in 1786 'most of the merchants were nearly naked and each was cooled by an ingenious fan, for Muscat is the hottest place on earth. The stalls had an infinite variety of gums, grains, and medicines, and a very peculiar smell.' The interior seemed 'little known, for Europeans had no inducement to go far inland. The

French came from Mauritius for wheat and asses.' The following year, William Francklin found the police excellent. The Imam lived in splendour two days' journey inland but his Wakil, Shaikh Khalfan, made himself very attentive. Only a third of the population could be considered healthy.

Captain Matthew Jenour, in 1791, found the Muscatis 'handsome, brave and perfectly free from any *mauvaise honte*.' He notes in *The Route to India* (London, 1791) that 'they trade mainly in locally-grown coffee and pearls and sail in vessels with no deck except for a small part to cover the helmsman and perishable goods.'

In 1800 Sir John Malcolm met the Imam, simply dressed, without jewels or *khanjar*, who showed great courtesy. Malcolm presented the Imam with a diamond-set watch, a silver-ornamented clock, double-barrelled gun, a gold enamelled dagger or *kris*, a pair of pistols and telescope. The Governor at that time, Saif bin Muhammad, was reputed to have made sixteen voyages to Bombay, one to Calcutta, and eighteen others.

A British Residency was first established there in 1800, during the war with France. The fate of its early incumbents is put laconically by J. A. Saldanha in his *Précis of Correspondence regarding the Affairs of the Persian Gulf, 1801–1853* (Calcutta, 1906; reprinted by Archive in 1986):

'The treaty concluded by General Malcolm in 1800 with the Imam of Maskat provided for the appointment of an agent at that place. Mr Bogle was the first Agent appointed, but he died by the end of the year. Captain David Seton took his place, but was appointed to some other Agency in the Bombay Presidency in 1804. In 1805 he was sent again to Maskat in pursuance of the instructions of the Government of India. Lieutenant Watts was appointed to act for Captain Seton in 1808, but died in September the same year. A malignant fever carried off Captain Seton in August 1809 to the sincere regret of Government and all who knew his worth. There arrived at Maskat soon after His Majesty's ship *The Caroline*, and on Captain Gordon's being apprized of the sad event by the broker and consulted with respect to the Company's property, he appointed one of the officers of the ship, Isaac Roberts, to take temporary charge of the Residency. Mr Bunce was appointed by the Bombay Government to succeed the late Captain Seton, Mr Bunce died at Maskat, in December 1809. It was thought that English constitutions were unequal to withstand the baneful effects of the climate at Maskat, and Government had to decide to stop sending any more British officers as Residents at that place, and to place Maskat under the supervision of the Resident at Bushire, Mr Hankey Smith.'

A Frenchman, Cavaignac, stayed in Muscat harbour for ten days in 1803 without landing, for the Sultan was absent and nobody else could authorise giving him a house. The British influence was ubiquitous, since all the merchants supported them. The Sultan may have been able to raise up to 80,000 men, but Cavaignac did not consider him a useful ally and that 'as a minor Beduin chieftain', the Sultan deserved no more important an envoy than the 'lowest type of merchant.'

Views of the Sultan of Oman quickly changed in the West when Sayyid Sa'id ibn Sultan took over the rule in 1804. Our best source for the next fifteen years, Shaik Mansur's *History of Seyd Said* (London, 1819; reprinted by The Oleander Press, 1984), was in fact written by an Italian, Vincenzo Maurizi, who writes that the Muscatis dared not go overland to Jiddah, because they feared to find cannibals on the other side of the mountains.

Richard Blakeney visited Muscat in 1814; he tells us in a *Journal of an Oriental Voyage* (London, 1841) that the forts were then in a ruinous state, and that when a sailing boat capsized and the Royal Navy rescued thirty-two passengers, the eight women among them were put to death since they had been seen by Christians. Cattle were fed on a mixture of dried fish, dates, water and a soft kind of earth.

In November 1816, Lt. William Heude visited Muscat for six days. In his *A Voyage up the Persian Gulf, and a Journey Overland from India to England, in 1817* (London, 1819; reprinted by Gregg International, 1970), he expresses his excitement on coming ashore. 'The custom-house, the palace, and its vicinity, the bazars and principal streets, were crowded with Arabs of every description and tribe; with Jews, Hindoos, Belooches, Turks, and Africans. I was now among a race of whom it has been written, 'that every man's hand shall be against them, and their hand against every man;' and every man's hand was armed. The Arabs, each after the manner of his tribe, or his own convenience, with a curved asgailee, a matchlock, or a pike, the Beloochee soldiers, naked to the waist, with a crooked toffung, a knife, and a straight two-handed sword; the wild Bedooin might be distinguished from amongst the first, by a striped kerchief surrounded

Muscat, 1809

with lashes of whipcord, and flying loosely around his head; by a coarse shirt, a square striped cumlin over his shoulders, and a chubook; wild and uncontrolled; with a quick burning eye, an animated and restless countenance: he appeared the lord of the creation, and was even in his physiognomy the lawless robber of a desert land. The others beseemed, in truth, the condition which they filled; bare-headed, and with their black luxuriant hair floating to the wind, perhaps to increase the terror of their appearance; the deadly keenness of their look, seemed to indicate the savage servile instruments of that despot's will whose authority they served.'

James Silk Buckingham, author of *Travels in Assyria, Media and Persia* (London, 1830), visited Muscat the month after Heude: December 1816. He estimated the population at 10,000, 90% Arabs, and a further 3,000 (many of them Persians) in a suburb of *barasti* huts outside the city wall. The Customs House, he reported, was an open square with benches, where merchants met daily. Some twenty Arab ships between 300 and 600 tons trade dates, pearls from Bahrain and copper from Basra in exchange with India for muslin, spices, timber, rice, pepper and Chinese goods; with Mauritius for cotton and coffee; and with Zanzibar for gold, slaves, ostrich feathers and ivory. There is no standing army, but a few captains to command the forts and a hundred gunners. Some 20,000 fighting men could be called up if required. The Imam provides their ammunition and divides booty. The Arabs dress simply, with blue cotton turbans from Suhar, and even the Imam's clothes are simple. The people are so honest that goods are left unguarded.

Francis Erskine Loch, in Muscat during January and May 1819, is quoted by Sir Charles Belgrave in *The Pirate Coast* (London, 1966) as declaring that Sultan Sa'id had 'the most agreeable and polite manner of any Arabian or Persian I have ever met with.' He left 'to the inexpressible joy of all on board, for the weather had been more oppressively sultry than can be conceived.' In May 1819 Sayyid Sa'id also received in audience Capt. George Forster Sadleir, whose *Diary of a Journey across Arabia* (Byculla, 1866; reprinted by The Oleander Press in 1977) includes political negotiations.

Capt. R. A. Mignan's *A Winter Journey through Russia, the Caucasian Alps, and Georgia* (London, 1839) relates experiences in August 1820, when the author estimated the population to be 10,000. He found the people religious but not bigoted or averse to sharing their food with infidels. Sayyid Sa'id is a great soldier, who lends money to Omanis in financial straits. Six large boatloads of sardines can be taken in a day, and swordfish are so powerful that they can destroy a boat. The August heat reached 120°F in the daytime and the night dew seemed 'subtle and venomous as the cobra's sting.' Returning in April 1821 as a combatant against the recalcitrant Bani bu Ali, he found them more determined and courageous than any other force met by the British. in 1825, Mignan's wife was invited to visit the Imam's wife, and claimed to have been the first European lady thus honoured. The Sayyid first met her at the door, then provided sherbet and coffee, and

The Imam's Wife, 1820

finally led her to a padlocked door and upstairs to another padlocked door behind which they encountered the Imam's wife and two eunuchs. She spoke Hindustani, and wore gorgeous clothes, with an emerald bigger than a pigeon's egg. Her room, with a view over the harbour, had huge windows of plate-glass and coloured glass alternately. A bed stood in one corner, and round the room a divan barely three inches high, with fine carpets and a double row of cushions.

Another summer visitor exhausted by the heat was James Baillie Fraser. In his *Narrative of a Journey into Khorassan* (London, 1825) he found the nights of July 1821 suffocating, and the daytime temperature ranging from 80° to 102°. Muscat looked like a wretched Indian town, with a population of 10,000 to 12,000, of whom about a thousand were Indians. Arab women were dressed in a material like black silk, while negresses wore a blue shift from head to foot over their trousers and negroes were clad in ragged trousers, with a turban. Fish swarm like gnats around the boats; the oysters are excellent. The bazaar is strictly controlled. Any perishable goods not sold by brokers in the morning are auctioned in the evening. Everything is sold by the *Mahomedee* [the *mamudi* of Kaempfer's account] at the rate of 20 to the $1. A camel fetches $30–$300, a milch cow $16–$17, a common cow $1.50–$6, and a milch goat $4–$6. A hundred pieces of fruit cost: mangoes 7 *mamudi*, oranges 6, peaches 4, limes 2. The Imam, visiting the ship informally, claimed to be able to raise up to 100,000 in arms, but he never has more than 30,000 available. He said about 10,000 had died of cholera which had broken out near Ruwi. He collects duties of ½% on all merchandise passing up the Gulf in Arab ships (say $90,000–$120,000 a year) deriving other income from leasing land, working sulphur mines leased from Iran, and the manufacture and export of salt. He probably receives $40,000 a year from Zanzibar. Local produce includes the sweetmeat halwah, earthenware pottery jars called *murtuban*, turbans, *abba*s, girdles, cotton, canvas, and gunpowder.

A report in the July 1822 *Asiatic Journal* noted that the cholera outbreak in Muscat was so virulent that victims could not survive it for more than ten minutes. Corpses are sewn up in cloth or mats, and towed out to sea for burial with ropes round their necks.

Capt. W. F. W. Owen, commanding *H.M.S. Leven*, visited Muscat in December 1823. In his *Narrative of Voyages to Explore the Shores of Africa, Arabia, and Madagascar* (London, 1833), he describes how, when Sayyid Sa'id was due to come on board, the crew hung pigs brought for meat in nets over the side to avoid offending the Sayyid's Islamic sensibilities. Their squealing resounded all around the harbour, amusing everyone including the Sayyid, who courteously received from Owen a Bible in Arabic, offering in exchange a gold-hilted sword. He provided free firewood and water, and gave $50 each to pilot and interpreter.

Owen found Muscat *suq* knee-deep in mud after winter rains, a noxious pond, and *barasti* huts. The beef he reckoned the best they had eaten since leaving England.

In February 1824, Capt. Hon. George Thomas Keppel, who succeeded in 1851 as sixth Earl of Albemarle, visited Muscat, and left comments in his *A Personal Narrative of a Journey from India to England* (London, 1825). He estimated the total population within the walled city at 2,000, at least ten per cent being blind in one eye. 'They have a great regard for justice, and an universal toleration for other religions.' They do not smoke, according to Keppel, have no pomp in dress or buildings, no dervishes or monasteries. Keppel made the customary courtesy call on Sayyid Sa'id, who understood Persian, Hindustani, and possibly English, receiving all including beggars with truly patriarchal simplicity. He assisted Keppel to find horses at Matrah to take a party to the hot springs at Baushar. Each spring was protected by a fort.

The June 1825 *Asiatic Journal* reckons the population of the country at 460,000 men and rather more women. Every man carries arms, with a good grace for his own tribal shaikh but less willingly for the Imam, whose permanent paid bodyguard consists of three hundred Baluchis. His revenue is estimated at 522,000 Maria Theresia thalers, 120,000 of them from Zanzibar.

Joachim Hayward Stocqueler visited Sayyid Sa'id in March 1831, as he relates in *Fifteen Months' Pilgrimage through untrodden tracts of Khuzistan and Persia* (London, 1832). Described as 'a warrior and trader, just governor and chivalric lover', the Imam asked penetrating questions about the French Revolution. Sayyid Sa'id, surrounded by an impressive Arab guard armed with swords and spears, said he had 10,000 fighting men and displayed forty fine stallions and twenty mares from Kutch and Bahrain as evidence of his enthusiasm for breeding horses.

Testimony to the searing heat of a Muscat summer is provided by two travellers in 1833.

During the June visit of Major Thomas Skinner, as told in

14

Adventures during a Journey Overland to India, by way of Egypt, Syria and the Holy Land (London, 1836), the rocks of Muscat glowed like heated ovens, and the iron on the ship could not be touched long before midday, the shade temperature being 103°. Skinner returned to his vessel convinced that 'Muscat deserves its infernal reputation.' In September, 'perspiration poured from the body like rain,' wrote the U.S. Special Agent Edmund Roberts in his *Embassy to Eastern Courts* (New York, 1837), and 'no place presents a more forbidding aspect.' Weavers dig a hole for the feet, with their seat a step higher, and use a primitive loom with a palmleaf shelter. Blacksmiths use bellows with two skins so arranged that one fills while they blow with the other; their anvils are of stone; their fire and water are kept in holes in the ground. Any house not made of stone is demolished by rains in winter, pieces of it floating into waterlogged streets which become canals. Roberts found coppersmiths, sandalmakers, ropemakers and carpenters. Two fin-backed whales blow in the harbour: one is named Muscat Tom, visiting the harbour daily for the last twenty years. The circular fishnets, fifteen feet in diameter, are weighted to sink to a depth of ten feet. He estimates the population of Muscat at 12,000 within the walls and 5,000 in the suburbs, while Matrah has about 8,000 inhabitants.

The traditional tolerance of Sayyid Sa'id's Oman is emphasised by Jacob Samuel, whose visits in April and December 1835 are recorded in his *Journal of a Missionary Tour through the Deserts of Arabia to Bagdad* (Edinburgh, 1844). Samuel preached to twelve Jews assembled for the Passover, and found that about 350 Jewish families resided in the Batinah. Muscat had the only Islamic Government tolerating Banian temples.

Dr William Samuel Waithman Ruschenberger's *Narrative of a Voyage round the World* (London, 1838) included a visit to Sayyid Sa'id in the company of the aforementioned Edmund Roberts. A smartly turned-out guard some twenty strong presented arms, then the Sayyid received them, discussing French politics. His only jewel was a large ruby ring; he held a sword in a black scabbard with a gold mount. Dinner was given in a hall decorated with paintings of naval battles. Knives and forks were borrowed from the visitors. The meal, served cold, consisted of two sheep stuffed with nuts, dates and prunes. Education consists of studying the Qur'an up to a certain age, then a few sons of the wealthy study medicine in Persia, or others may go to India.

Tribes are distinguished by their turbans. Some women are dressed in loose yellow silk gauze, a short-sleeved jacket, with toe-rings. The population of Muscat was put at around 20,000, and Matrah about 18,000, the latter having a walled quarter of a thousand Luwatiyah (into which Arabs may not enter) and another quarter, with a gate opening on to the harbour, for some 2,000 Baluchis, who live in *barasti* huts and worship in a white mosque with two small turrets.

Muscat at night sounds romantic, sentries calling out every half hour, and boatmen singing. Only the three-storey Sultan's Palace and the Customs House can be seen from the sea. In the *suq*

Ruschenberger found swords from Persia and England, corn, pepper, senna-leaves, cloves and other spices, rosebuds, antimony pencils and glass beads.

Wellsted on Muscat

The next visitor to Muscat was one of the most important European travellers ever to set foot in Oman: the naval lieutenant James Raymond Wellsted, late of H.M.S *Palinurus*. Wellsted was commissioned by the East India Company, in the interest of their control of the Gulf, to explore the interior of Oman, to learn as much as they could about conditions there, and to assess the power and influence of Sayyid Sa'id, who had been pro-British since 1798. The French botanist Aucher-Eloy's forays to Nakhl, Saiq, Nizwa and Tanuf, returning to the coast from his base at Matrah via Wadi Suma'il, concentrated almost exclusively on trees, plants and shrubs, so it is to Wellsted that we look for the first detailed description of places and people in the interior. His credentials were impressive. He had engaged in surveys of the western and southern coasts of the Arabian Peninsula, had explored Socotra, and in the company of Cruttenden had travelled inland from Bal Haf, finding the ruins of Naqb al-Hajar.

These earlier journeys fulfilled the Company's objectives of assessing the suitability of coastal towns and villages for coaling stations. Now Wellsted was collecting data from the interior, such as the effect of the British expedition against the Wahhabite Bani bu Ali of 1821, undertaken by General Sir Lionel Smith with the consent of the Ibadhi Sultan. He also wanted to compile data on the life style of the badu of Inner Oman, to make the first accurate geographical maps and plans of the area, and to plot the various passes and mountains which had hitherto never been satisfactorily drawn.

Wellsted started by sea from Muscat, touching at Qalhat and Sur before travelling inland to al-Kamil, Bilad Bani bu Hasan, Bilad Bani bu Ali, Wadi Batha, Wadi Samad, Wadi Ithli, Manah, Nizwa, Jabal Akhdhar, Wadi Suma'il, Sib and back to Muscat. With Lieut. Whitelock (an Englishman he had met by chance at Samad), Wellsted started out in February 1835 for the Batinah, turning inland from Suwaiq to 'Ibri in the (vain) hope of obtaining access to Buraimi, vain because the Wahhabis encamped there had already begun to raid into southern Oman. Whitelock therefore made for Sharjah, and Wellsted for Makran and India.

Of Muscat, Wellsted wrote, the greater portion of the inhabitants 'are of a mixed race, the descendants of Arabs, Persians, Indians, Syrians, by the way of Baghdad and Basrah, Kurds, Afghans, Beluches, &c., who, attracted by the mildness of the government, have settled here, either for the purpose of commerce, or to avoid the despotism of the surrounding governments.' On the Sultan, Wellsted writes that he 'is very generous and every visiting Arab gets a present. He could raise 10,000 fighting men in three days and later another 20,000. He is the most respected prince in the East and has been called

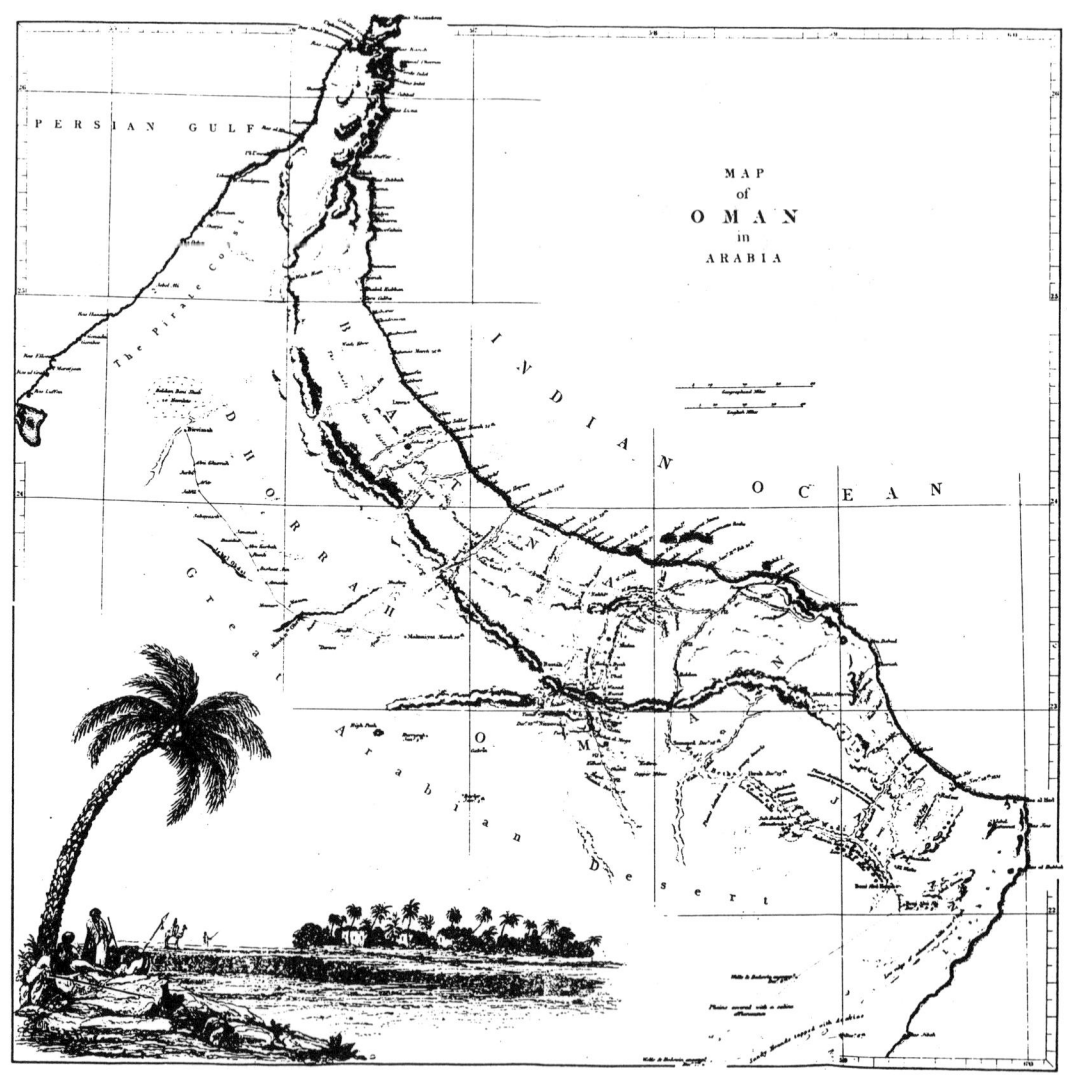

Map of Wellsted's Travels, 1834–5

the second Omar.' On commerce, the coffee trade is in the hands of the Banians, and is said to be very lucrative. The pearl fisheries in the Gulf are estimated at forty lakhs annually, and nearly two-thirds of that produce are brought hither in small boats, and thence conveyed to Bombay in ships or bagalas. They mostly arrive sealed at Maskat, and very few are disposed of there. In Bombay the Parsees are the principal purchasers, and a great many are sent by them to China. Maskat yields but few exports, and no duty is now levied on them. The principal are dates, taken to India, where large quantities are required to make the government arrack, or are sold at the different ports on the southern coast of Arabia; ruinos, or red dye, much valued in India; sharks' fins, shipped off to China, where they are used for making soup, and a variety of other purposes; and salt fish, much esteemed by the lower classes of natives in India. The returns for these articles are made principally in bullion and coffee.' Wellsted's highly important *Travels in Arabia* (London, 1838) have been reprinted by Akademische Druck u. Verlagsanstalt, Graz (1978).

In 1836, Pauline Helfer visited Sayyid Sa'id's wife, noting in the *Travels of Dr. and Mme. Helfer* (London, 1878) that the household comprised about a hundred women. The Imam's mother, simply dressed, was sewing, a recreation she said she had learned from European ladies. Mme. Helfer embarrassed her hostesses by being unveiled, since not even a mother sees a daughter's face after the age of 12, they said. The Imam's wife 'glittered in all the usual oriental splendour.'

We have three accounts of Muscat from travellers on the frigate *Columbia*, calling in October 1838. Rev. Fitch Taylor, in *A Voyage Round the World* (New York, 1843) relates that an American sailor died of exposure ten minutes after standing in the sun, was carried out beyond the town and buried outside the southern gate. In the absence of Sayyid Sa'id, the visitors were received by his son, aged 23. They were offered coffee with crystallized sugar by an aged eunuch. Joshua Henshaw, in *Around the World* (New York, 1840) describes a visit to a café outside the town walls, its stone counter covered in coffee boilers and tiny cups. He found *barasti* huts occupied by barbers, weavers, and bracelet-makers, and saw a juggler with vipers in his hair. In the Sultan's stud he saw some of the 1,000 horses, mainly of Kadishi breed, small, delicate and muscular. The third informant from the vessel, William Meacham Murrell, writes in *The Cruise of the Frigate Columbia* (Boston, 1840), that the boatswain's arm was numbed with fatigue in Muscat, after administering two hundred and forty lashes to the cooks for their dirty pots and pans.

In October 1850, Robert B. M. Binning landed at Muscat to take on coal. According to *A Journal of Two Years' Travel in Persia, Ceylon, &c.* (London, 1857) the then British Agent, Khoja Ezekiel, was 'an indolent and nonchalant rascal'. The population then was estimated at 12,000. Men rarely marry before the age of twenty or women before eighteen.

Thuwaini ibn Sa'id began his ten-year rule in 1856, the year when William Ashton Shepherd went ashore, as he tells us in *From Bombay to Bushire and Bussora* (London, 1857). He saw the whale called Muscat Tom, already described by Edmund Roberts in 1833, which was thought to keep sharks from basking in the harbour. Fishermen drew alongside Shepherd's boat, offering fish for a few pennies. The officers were entertained by Mahmud bin Khamis, whose home was decorated with coloured prints of racehorses and embellished with a library of French and English books. As the official interpreter, he had been despatched to London to congratulate Queen Victoria on her accession to the throne.

Stiffe on Muscat

Captain Arthur W. Stiffe surveyed the Muscat area in 1860, but his account was updated for publication in *The Geographical Journal* (1897).

ANCIENT TRADING CENTRES OF THE PERSIAN GULF.

By Captain ARTHUR W. STIFFE, R.I.M.

IV. MASKAT.*

THIS place is still an important place of call, and of trade with India, the Red Sea, and Zanzibar, chiefly only transhipment. It is the capital of the Arab country of Omán, and the residence of the prince, now designated Sultan or Seyyid, but formerly called the Imám of Maskat.† The country has long been in an unsettled state, owing to civil wars and dissensions, and has declined in importance, especially since the separation of the Zanzibar dependencies. It is now under British influence, and a Political Agency is established there. The town lies in a cove, one of a series close together on the north-east point of Arabia, which are all surrounded by precipitous rocky hills rising to several hundred feet abruptly from the water's edge. At the inner end of each cove is a small sandy beach, at the mouth of the little valley, or wadi, which forms the inland continuation of the

* The accent should be on the first syllable.

† This title has a religicus significance, and is not now assumed by the sovereign of Omán.

cove. The rocks are dark-coloured serpentine, here and there showing foliation well marked,* and it is part of an area of depression, the coves being submerged valleys, which had been excavated by subaërial agencies before the submergence took place. The bed of the sea sinks rapidly to a depth of upwards of 2000 fathoms at a distance of 10 or 12 miles from the coast.

The appearance of the coast from the sea is extremely picturesque, the rugged dark hills rising one range above another, until apparently joining the great back range, elevated about 6000 feet. Although from seaward the country appears utterly barren and desert, without any sign of vegetation, the valleys lying among the hills are more or less fertile where irrigated, which is done by means of wells, with so-called Persian water-wheels, and kanáts, or subterranean aqueducts. It produces fruits and vegetables, and many date-palms are grown. The coves abound with fish of excellent quality. The small cattle of the country are noted, and flocks of sheep or goats are numerous.

The map of the group of coves is from a survey by the author. Each one has a village or town at the head, built on the sandy beach close to the water's edge, so that the sea even washes the base of the houses, which extend back as far as the rugged ground will permit. They all may be considered suburbs of Maskat. The intervening hills are so rugged that, although there are some passes over them, much of the intercourse is carried on by water in large canoes. The houses fronting the water are mostly of two or three stories, and coated with white cement. The hills and passes are all crowned by small towers or forts for defence, with walls and gates across the passes. Maskat itself and Matrah, the place next in importance, are enclosed on the land side by walls, with towers at intervals. From Matrah only is there any pass into the interior of the country.

There are two large forts, built by the Portuguese, one on each side of the cove, on the summits of the hills overlooking the town, and two outer and less important, called Sírah, on the next projecting points of the cove; they are all in a very ruinous condition. The east side of Maskat cove is formed by three detached masses of rock; the outer, commonly called Maskat island, rises 350 feet above the sea, and can only be climbed with difficulty at a few points; it is much the largest, being 1400 yards in length, and is separated by a shallow strait, only a foot or two deep, from the second and smaller hill, and this again by a still shallower passage from the third, which is crowned by one of the great forts (Jaláli), and has a low sandy isthmus between it and the rocks surrounding the town, on which isthmus now stands the British Residency.

Outside the wall of Maskat there is a large suburb, occupying all the available ground in the valley, and consisting of huts of the usual material of Arab villages, viz. matting made of the stems and leaves of palm fronds.† Here there is a bazar, which is a busy scene in the morning. A curious kind of auction goes on constantly. Men walk about calling out the last bid made for some article they carry for sale, which seems to go on until some offer is made which they will accept.

Of the two principal forts, the eastern, Jaláli, already mentioned, occupies the whole of the top of the rock on which it stands. Its front is a long curtain wall with two tiers of embrasures, with a round tower with flagstaff at each end, the only access being by a flight of steps cut in the rock on the harbour side. This

---

* W. T. Blanford, 'Records of the Geological Survey of India.' 1872.
† Of the date tree, and largely of a dwarf palm called písh, brought from Makrán.

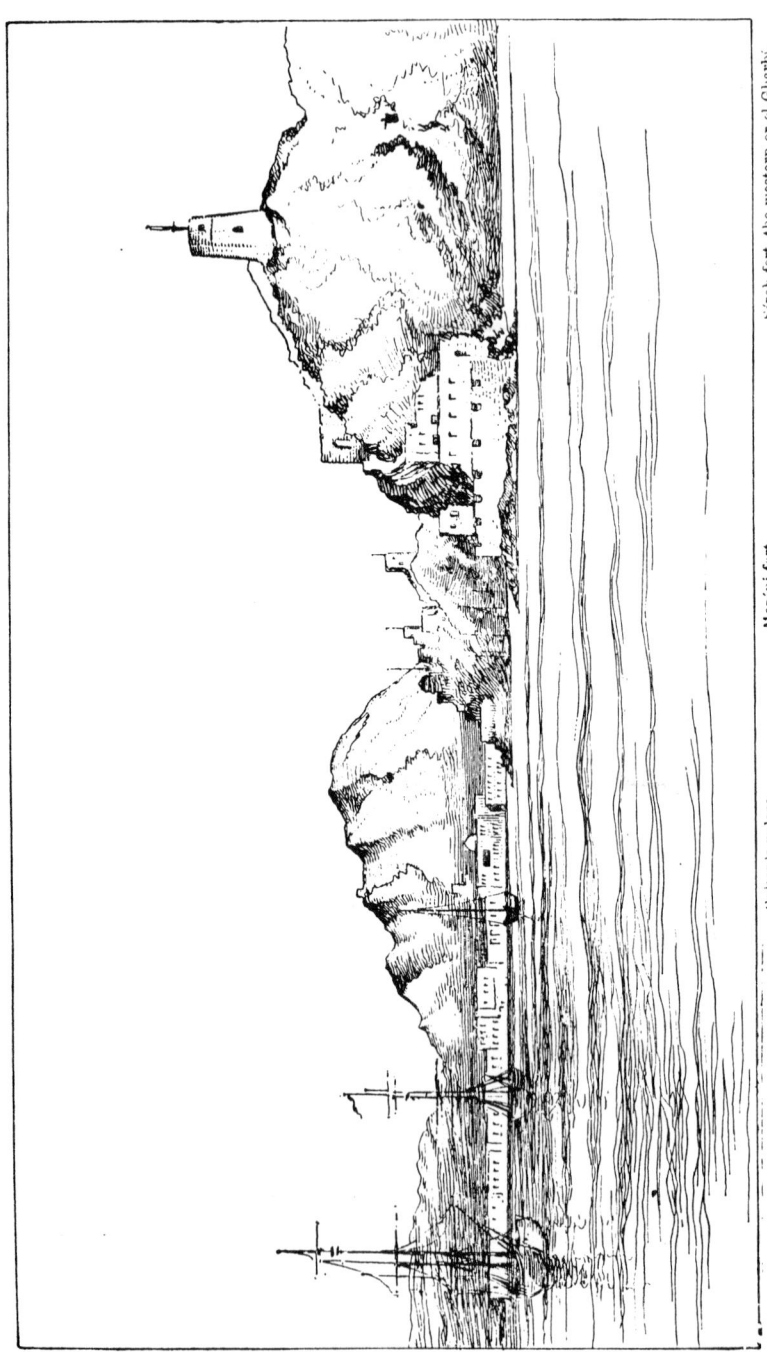

Sultan's palace.   Merâni fort.   Sirah fort, the western or el-Gharbi.

MASKAT, FROM THE ANCHORAGE OUTSIDE THE FORTS.

Muscat, 1860

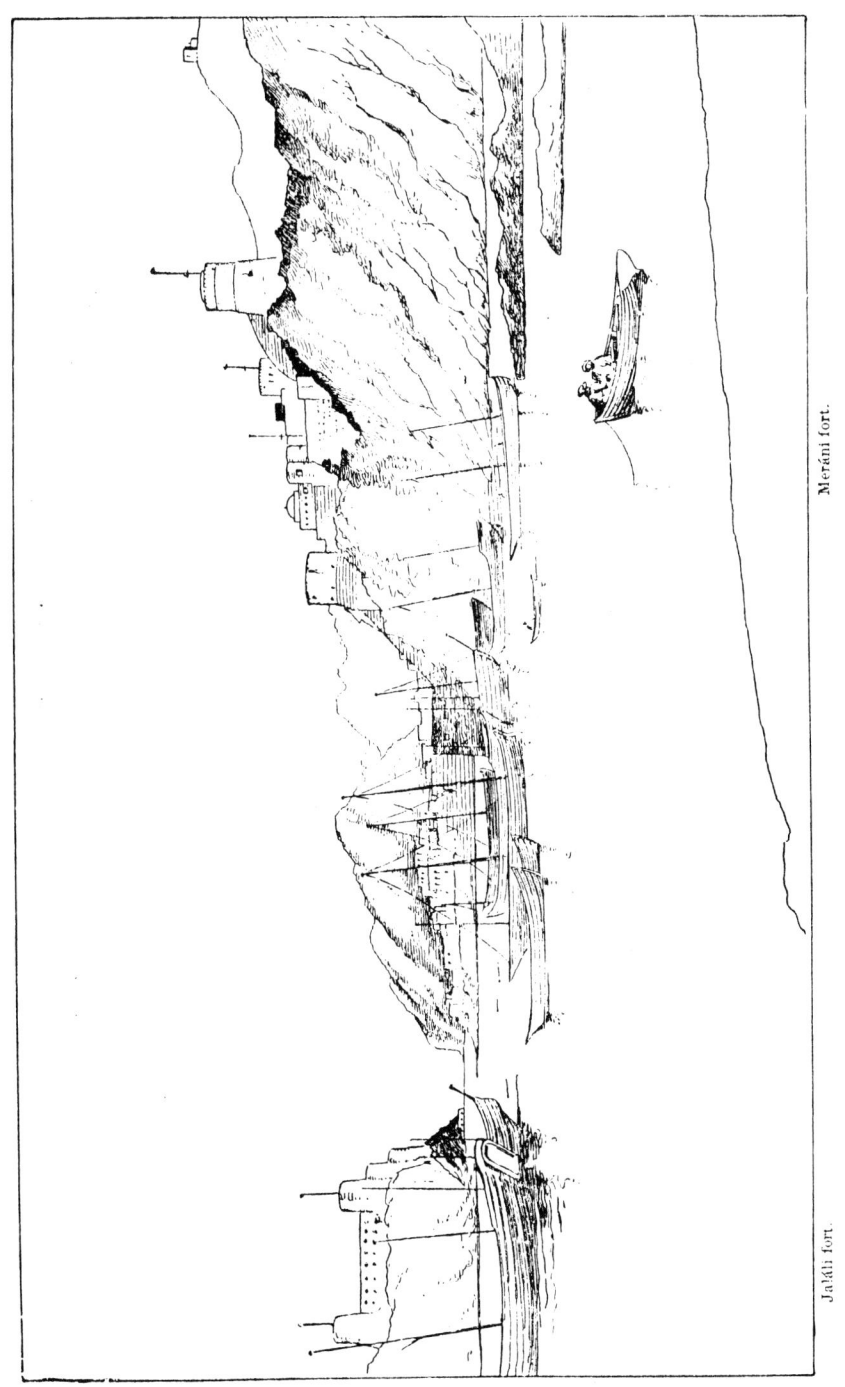

Jalâli fort.

Merâni fort.

Muscat from Makallah Bay, 1860

fort was much damaged by artillery fire during a civil war some years since (about twenty, I believe), and remains in a most dilapidated state, as the Arabs appear never to repair anything. We had given the Sultan of that time two guns of position—24-pounders, I think—and these were used with great effect by one side in the quarrel—I think, by the rebels, *i.e.* the unsuccessful party. This fort was built by the Portuguese in 1587, and called by them San Joao.

Opposite to this, on a hill overhanging the town, stands the other principal fort, now called Meráni. It has a ruined battery near the water, above which rises the body of the fort, occupying the top of the hill. It is an irregular structure adapted to the available space, and has two large round towers, with tall flagstaffs, as in all Arab forts, the higher tower being connected only by a double wall with the

REINAODHOMVAEFOHEPOED.
RORO·F·PRMEROEDS ENOME·R·
HE· S·NOSONOH OV KVOANOED·
SEVRÊNA DONACROADEPOR V
GMANDOVPOR DNDAREEDM
ENERE S SEWRO·R·DNDIAꝚE S EFI
RE S E·E SAFORKLE Ꝛ ÆꝚAE SLEICHR
Æ·S/PRMEROGPÏꝜOEFVNDDR 1588

INSCRIPTION OVER GATEWAY IN FORT MERANI.

main fortress. The guns mounted are very old, mostly Spanish or Portuguese, and the carriages dropping to pieces. The iron guns are all dangerously unserviceable. The brass guns are in better order; one bears the inscription, "Don Phelippe rey de España," and another the inscription, "Don Juan de Acuña de su conseio de cuera y su Capitan general de la artilleria año 1606." This fort was built in 1588, according to an inscription over the inner gateway in old Portuguese, which has been rendered for me as follows: "Reigning the most high and most faithful Henry, powerful and first of that name. King Henry our lord, in the eighth year of his reign on the crown of Portugal, ordered by Don Duarte de Menezes, his viceroy of India, that should be erected this fortress, of which Belchior Caleça was its first captain and founder, 1588." *

I cannot understand this date (unless it is the date of *completion*; it may well have taken many years to build, being so extensive), as Henry died in 1579, and did not reign eight years. It was named by them Fort Capitan.† In 1581 the news of the seizure of Portugal by the Spaniards reached India. The fort is only strong from the difficulty of access, and the old entrance gates, one within the other, are still carefully guarded. Over an arched window, 30 inches span, cut out of one stone, is carved, in letters 3½ inches long, an Ave Maria (copy attached). The two smaller outer forts, which had also batteries near water-level, are still more ruinous.

† A fort was commenced on this site by Da Lisboa thirty years previously, in 1552.

The Sultan's residence is a large three-storied building near the centre of the town, a quite plain, rectangular block. It is a relic of the Portuguese occupation, having comprised the governor's residence, factory, chapel, warehouses, and barracks. The Arabs call it El Jereza, a corruption of Igrezia (church). On an old wooden gate of the custom-house is cut " Anno 1624."

The wadi extends up behind the town for a mile or more, and is cultivated in patches, with vegetables and a few date trees. The wells, worked by bullocks, are about half a mile from the town wall, and are defended by a square tower or fort, loopholed for musketry. A small cemented aqueduct, generally out of repair, has

INSCRIBED OVER A WINDOW IN FORT MERANI.

been made to bring the water down to the landing-place for shipment. The water-course draining this valley passes through a culvert under the town wall into the sea. It is quite dry except after rain.

From Matrah there is a track or way winding through the hills into the interior, and, after following it about half a mile, you come on a plain among the hills with a small village, called El-Felej, where there is a castellated country residence of the sultan, very dilapidated, with a grove of date and other trees, and some cultivation. The water is brought from the upper part of the valley by an underground channel, or *kanát*. This is the only approach by land to Matrah, whence Maskat is generally reached by water.

As regards the actual productions of the place, they are unimportant; it being chiefly a port for transfer of trade. The speciality is the manufacture of *halwah*, a sweetmeat much in request, and of which large quantities are exported. It is made chiefly of the gluten of maize. Large quantities of dates brought from the coast of Bátinah are exported. It is a port of call for the Gulf mail steamers, and some English merchants are established here, also many Hindus (Banians), all traders.

The climate of Maskat is extremely hot, even in the winter, and there is but little rain, which falls in the winter. It is out of the track of the cooling south-west monsoon, which is cut off by Ras-el-Hadd ; but in that season light south-

24

easterly airs at times temper the heat. Abd-er-razzak,* 1442, says that in May "the heat of the sun was so intense that the sword in its scabbard melted like wax," etc. I can almost pardon him his exaggeration.

History.—Turning to the history of the place, it is only speculation to inquire whether the Moscha and Omana mentioned in the Periplus † are the Maskat and Omán of the present day. Dean Vincent argues that they are not, but it is possible they may be intended for these places, and misplaced in the itinerary, some confusion in the application of names having arisen. The description given of the places seems more applicable to Maskat, than any other part.

We have a brief glimpse of the place ‡ in the ninth century, indicating it as the last port of call for the Arab vessels proceeding to India, which is all I have been able to trace of its earliest history.

FORT OF FELEJ.

Colonel Taylor, formerly political agent in Turkish Arabia, and, I believe, a great Arabic scholar, gives a short account of the history § of Omán "from authentic sources of Arabian tradition," but does not specify any authorities. He says nothing about the period between the eighth and seventeenth centuries, and does not mention the long Portuguese occupation. His account says that the first native Arabian who entered Omán was one Malik bin Fakham of Nejd, four centuries before the Christian era, ‖ who, with some hundred followers of the Hiuávi tribe, settled at Jaalan or Bahla, two towns in the interior some 70 miles to south-westward of Maskat; and fortified Rastig, an ancient city in the mountainous district of Omán, 30 miles westward of Maskat. Successive additions to the numbers of these Arabs enabled them, after obstinate resistance, entirely to expel the Persians from the province. His successors continued in power until the mission of Mohammed.

* Hakluyt Society, vol. 22, 'India in the Fifteenth Century.'
† 'Periplus of the Erythrean Sea.' W. Vincent, D.D. 1800.
‡ *Geographical Journal*, vol. vi. No. 2, August, 1895, p. 169.
§ 'Bombay Government Records,' No. 24, New Series, 1856.
‖ Colonel Ross says the probability is that Malik bin Fahm entered Omán in the early part of the second century A.D., and that the part he played is probably exaggerated. The country up to his time was under the Persians, the date of whose conquest is not known.

In 571 A.D. a powerful prince of the dynasty, named Jalanda, equipped a fleet and seized Hormúz island, which was established as a rendezvous of a piratical fleet. They were converted to Islam in 621 A.D. by invitation of the Prophet.

Dr. Badger (Hakluyt Society, vol. 44, 1871) gives a translation of an Arab history of Omán, containing the period between 661 and 1856, by Salih-ibn Razik.

I propose here, however, to touch chiefly on the Portuguese occupation of the seaboard of Omán. The above Arab author gives only a short space to the period 1154 to 1557, and does not mention the Portuguese conquests; nor is there any record of the overlordship of Hormúz, which existed at the time of the appearance of the Christians. He admits a hiatus in the annals from 1154 to 1429, and these are also said to be obscure up to 1560.* His account of the recapture of the places from the Portuguese will be referred to later on.

Marco Polo (1260-95) does not mention Maskat *by name*, Kalatu † (Kalhát) is mentioned as frequented by numerous ships from India, and as "subject to Hormos." "Many good horses are exported to India ; the number from this and the other cities is something astonishing." The probable reference to Maskat is that "the Melic of Hormos has a castle which is still stronger than this city (Kalhát), and has a better command of the entrance of the gulf." Abulfeda (1273-1331) also does not mention Maskat by name. Ibn ‡ Batuta (1324-25) went by sea to Omán, and arrived at the city of Kalhát, "which is situated at the foot of a mountain. The inhabitants are Arabs and schismatics, which they keep secret, because they are subjects to the king of Hormúz, who is of the Súuni sect." He mentions the markets and a fine mosque, "whose walls are covered with coloured tiles."

Edrisi (1153) mentions in Omán, first Kalhat and Súr, and then Sohár, which he says is one of the most ancient cities of Omán, and of the richest. Maskat is mentioned as a populous town.

We may, I think, gather from all this, that in the thirteenth and fourteenth centuries Maskat was a less important place than Kalhát.

The first appearance of the Portuguese on the coast of Omán was that of D'Alboquerque,§ who with six ships left Socotra on August 10, 1507, and sailed up the coast of Arabia. He anchored, says the chronicle, at Calayate (Kalhát), an anchorage on the coast, about 25 miles to north-westward of Ras-al-Hadd, and there got some supplies. It was badly populated, with many old edifices, the sea beating against it; on the land side was a wall, about the height of a lance, reaching to the sea, not a single tree, and all supplies came from the interior. It was under Hormúz,‖ and they fished as far as Ras-al-Hadd, and it was the seat of the

---

* In Colonel Sir E. C. Ross's 'Annals of Omán' (translation of Kishf-al-Ghummeh. Calcutta : 1874), there is a similar hiatus from 1153-1406. Nor is there any mention of the Portuguese conquests, or of Hormúz.

† Colonel Yule's 'Marco Polo,' vol. ii. p. 448.

‡ 'Voyage d' Ibn Batouteh. Trad. de C. Depémery et B. R. Sanguinetti.' Paris : 1895. The translation of Rev. Sam Lee (1829) appears less accurate.

§ The Commentaries of the great Afonso D'Alboquerque, translated from the Portuguese edition of 1874 (Hakluyt Society, 1875-84), from which I abstract largely.

‖ Confirmed by Marco Polo, 1271-91 (see *ante*). Colonel S. B Miles says the supremacy existed since 1270. Colonel Miles was Political Agent in Maskat, and wrote a valuable report on the Portuguese in Eastern Arabia in the Administration report for 1884-85, printed in No. ccvii., 'Selections from the Records of the Government of India, Foreign Department' (Calcutta : 1885). He does not always quote the authorities he has consulted. I have quoted in one or two places from this report.

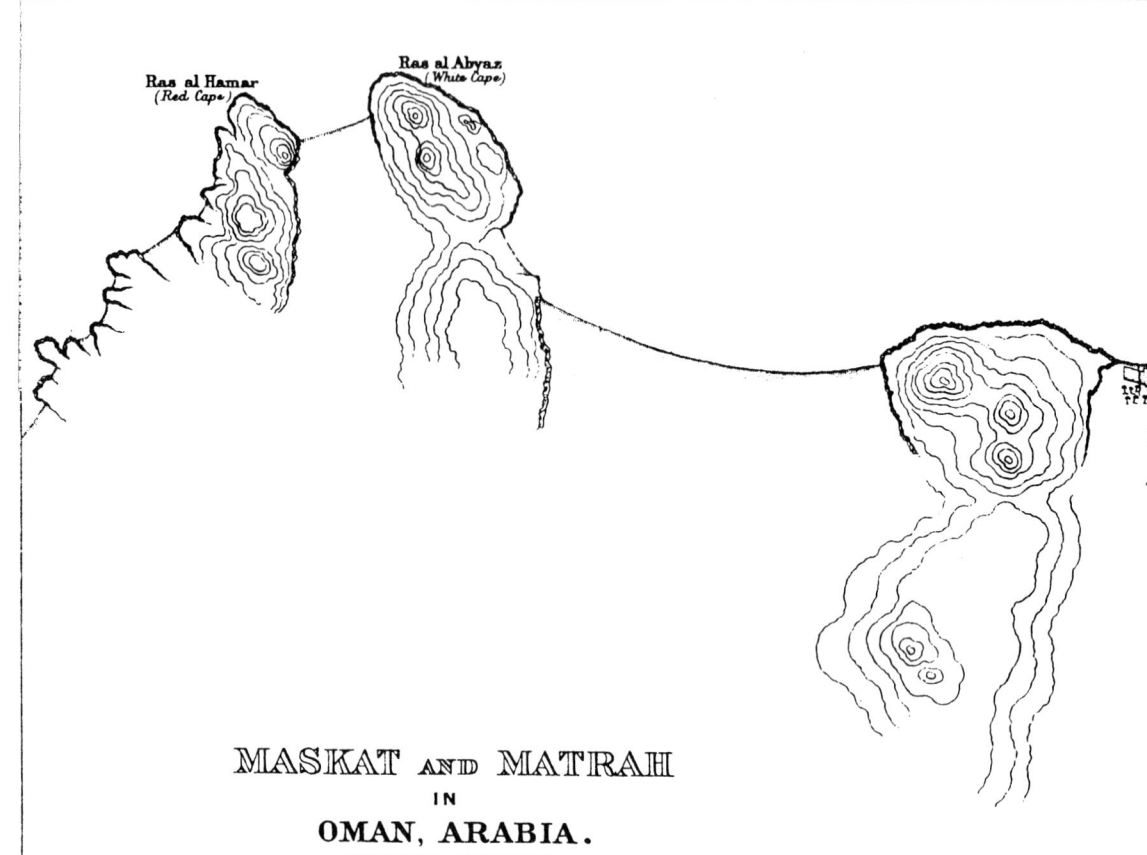

Ras al Hamar
(Red Cape)

Ras al Abyaz
(White Cape)

MASKAT AND MATRAH
IN
OMAN, ARABIA.

to accompany the paper by
Capt. A. W. Stiffe, late I.N.
Fishers Rock Lat. 23°37′55″N. Long. 58°35′68″E.

Scale of Miles.

0   ¼   ½   ¾   1   2

Natural Scale, 1: 30,000 or 1 inch = 0·47 miles.

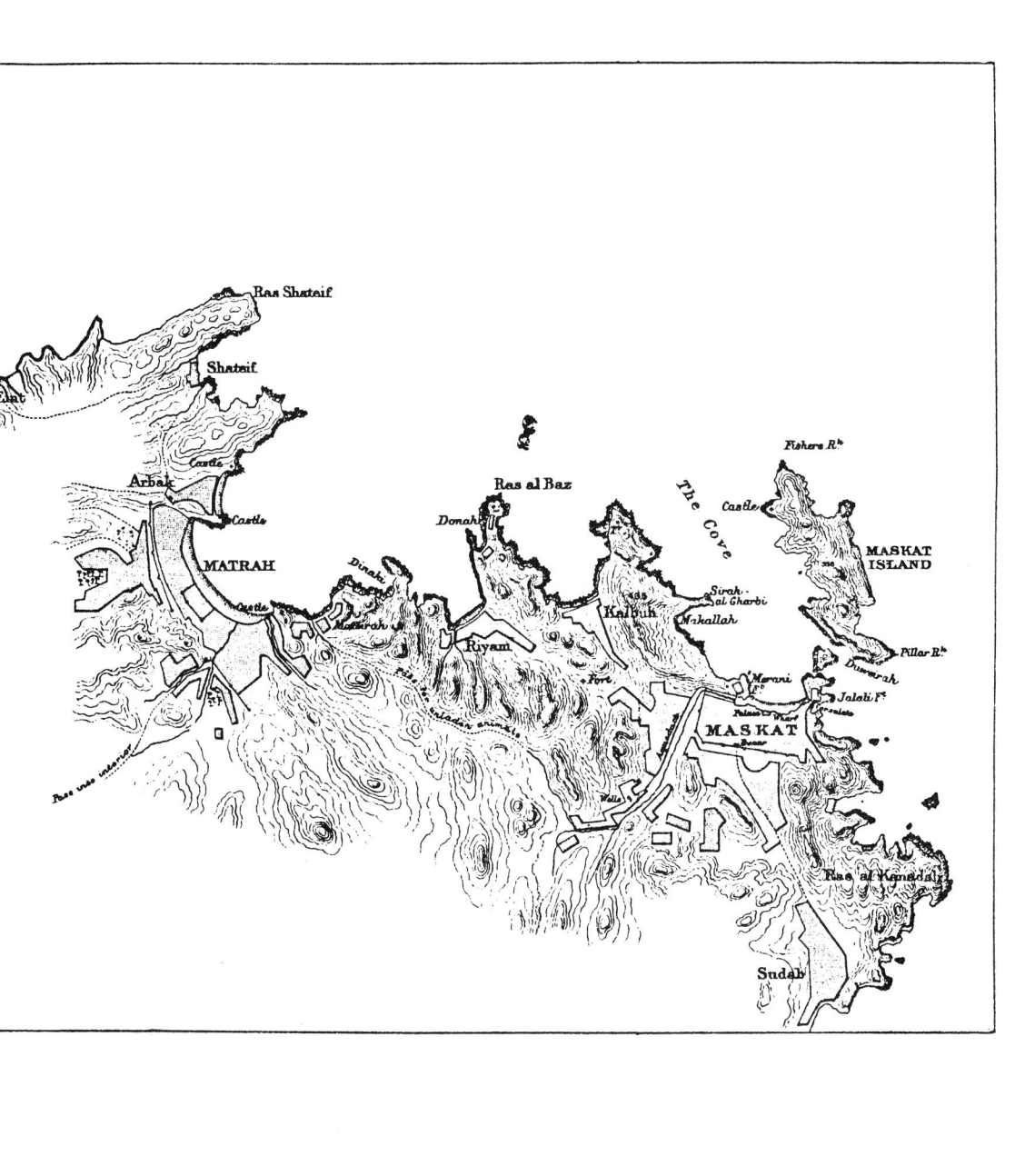

Ras Shateif

Shateif

Castle

Arbak

Castle

MATRAH

Castle

Matrah

Dinah

Riyam

Ras al Baz

Donah

Kalbuh

Sirah
al Gharbi

Nakallah

Fort

MASKAT

Bazar

Fishers R.

Castle

The Cove

MASKAT
ISLAND

Pillar R.

Duwarah

Merani

Jalali F.

Palace Wharf

Wells

Ras al Hamadah

Sudab

chief governor from Hormúz. They next anchored off Curiate (Karyát), stormed and took it, and put all to the sword who tried to escape, including women and children *; they plundered and burnt it, " so that not a house was left standing, not even the mosque, one of the most beautiful ever seen." They cut off the noses and ears of the prisoners, and sent them to Hormúz. Thirty-eight ships, great and small, were burned. It was a large town, and contained about 5000 to 6000 men, an entrepôt of ships which came to collect dates.

Thence the squadron went to Maskat, where the people submitted to be vassals of the King of Portugal, being aware of the destruction of Curiate, and agreed to pay tribute and furnish supplies. A "captain" having arrived with 10,000 men from the interior, hostilities ensued, and the town was taken after a stout resistance. D'Alboquerque put men, women, and children to the sword, sacked the town, and burnt it to the ground, including the large and beautiful mosque, and thirty-four "ships" in all. Some men and women who had been taken alive had their ears and noses cut off, and were then released. "It is," says the account, "a large and populous city, supplied from the interior with much wheat, maize, barley, and dates for lading ships. It is part of the kingdom of Ormus."

The unhappy "Moors" returned when the Portuguese embarked, to try and put out the flames. "The Moors call the interior the *island* of Arabia.† It is a very small land (!) governed by a king called the Benjabar"—this is the name of a tribe in the vicinity (Beni-jábar).

At Maskat the Portuguese got "Moorish" pilots, and, passing six desert islands—the Daimániyah group—came to Soor,‡ where they were at first defied by the "Alcaide;" but who, on the Portuguese preparing to attack, submitted to be vassals of the King of Portugal. They took possession of the fort, hoisted the Portuguese flag on it, and left the "Alcaide" in charge. The fort was of a square shape, with six towers round it, and two very large towers over the gate. There were about six thousand inhabitants and one hundred "cavaliers," the greater part "armed with steel armour: plates arranged after the manner of a roof tiled with slates. The fore quarters of the horses were similarly defended."

The last place in Omán they visited was Orfacào (Khor-Fakán), which was attacked and taken with the usual mutilation of captives and merciless slaughter, after which the place was burnt. It was a large town, with a wall on the land side, and lies at the foot of a very high mountain. Now it is a small fishing-village, and I saw no remains of the old town; it lies in a small cove at the northern end of the Bátinah district.

So much for the first visit of a Christian power to this country. The invaders thence sailed to Hormúz.

In August, 1508, Alboquerque returned to the coast *en route* to Hormúz, "intending to attack Calayate," which had been spared the previous year. It was taken after some fighting, and the town sacked and burnt, including the mosque, "which the Moors took much to heart, for it was a very large building with seven naves, all lined with tiles, and containing much porcelain hung upon the walls." It was burnt to the ground; twenty-seven ships, large and small,

* I must express my horror at the barbarous cruelty of the "great" commander, both here and throughout this cruise, towards people whose only offence seems to have been that they were of a different race and religion. The details are stated in the Commentaries in a matter-of-fact manner.
† As they do now, Jezíret-el-'Arab.
‡ Sahár, still the principal town on that part of the Omán littoral, called Bátinah.

were also burnt; and then Alboquerque ordered them to cut off the noses and ears of all the Moors whom they had captured, and left them on the shore and returned on board, "*giving many thanks to our Lord.*" They then apparently proceeded to Maskat. Faria y Souza * says Calayate was burned "to be revenged for some injuries done to some Portuguese." Alboquerque also touched at Maskat, on his way to Hormúz, in 1815.

In 1522 † a concerted rising took place simultaneously at Hormúz, Bahrain, Maskat, Karyát, and Sahár. Many Portuguese were killed; the number is given as one hundred and twenty. It was, however, suppressed, and Sahár destroyed with "fire and sword" by Dom Luis de Menezes, who was sent from Maskat with two ships to relieve Hormúz, then closely besieged. In 1526, Lope Vaz de Sampayo, on his way to Hormúz with five ships, reduced the "towns of Calayate and Muscate, which had revolted;" but no particulars are given. In 1550-51 "the great Turk," being offended at the proceedings of the Portuguese, fitted out a naval expedition ‡ consisting of sixteen thousand men, in "strong galleys" and other vessels, under the command of Pirbec (Pir Beg), who is described by our author as an "old pirate," but who was apparently the Turkish admiral, who attacked and took Maskat after a siege of a month; and, having failed in his siege of Hormúz fort, was beheaded after his return. He did not attempt to hold Maskat permanently, but sacked the place and removed all the ordnance. The garrison were made to work in the galleys, but were mostly released at Hormúz.

In 1581 § another Turkish expedition under Alibec (Ali Beg), "a Turk used to robbing," consisting of three galleys, was fitted out at Mocha, and surprised Maskat. He landed his main force at Siabo,‖ while "the galleys entered the port with those that remained, and began to 'play' their cannon furiously, so that he might come in on their backs; which succeeded, and he entered and plundered the town." His land force advanced through the narrow pass from Sudáb to Maskat, "so narrow that two men cannot pass it abreast; no one imagining he would attempt it." The Portuguese fled to Matero (Matrah), a town a league distant, and, not thinking themselves safe there, went to Bruxel,¶ a fort 4 leagues inland. They returned to Maskat after the departure of the Turks.

It was in 1588, according to the old inscription referred to already, that the fortress now called Meráni was completed, which strengthened the hold of the Portuguese on the place and country. Sahár, which had been taken and burnt by the Portuguese in 1522 (see *ante*), appears to have revived, for in 1616 its trade "much lessened the customs of Ormus and Mascate;" and an expedition was despatched from Maskat, which, with the aid of twelve hundred Arab auxiliaries, took and plundered the place, and left a garrison in the fort. It was retaken by the Imám in 1643.

After the loss of Hormúz in 1622, Maskat became the most important place held by the Portuguese, and was the headquarters of their fleet.

* 'The Portugues Asia.' Translated by Captain Jno. Stevens. In 3 vols. London: Printed for C. Brown, 1695, at the Sign of the Gun, at the west end of St. Paul's.

† 'Manuel de Faria y Souza,' *ante*.

‡ This expedition came from Egypt. Colonel Miles (*op. cit.*) calls him Pir Pasha, and says he was Capudan of Egypt, and that the expedition consisted of thirty sail

§ Faria y Souza, *op. cit.*

‖ This must be Sudáb, which answers to the description in the chronicle.

¶ This may be Bosher, a place visited by the author in December, 1859 (*Trans. Bombay Geog. Soc.*, vol. xv.).

In 1640, the garrison of Maskat, which had been much reduced, repulsed an attack on the place by the Imám's * forces. In 1648 Maskat was again besieged, and the Portuguese had to accept humiliating terms, being confined to Maskat, and giving up their other possessions in the country. Finally, in 1650, after another siege, Maskat also was surrendered, and the Portuguese finally expelled from Omán.

In the account given by Dr. Badger's author (already quoted), a detailed account is given, showing that the loss of the place was due, in part at least, to treachery on the part of a Hindu trader, "a worshipper of the cow," † whose daughter had attracted the attention of the Portuguese commandant. In a note to Colonel Ross's book (op. cit.), he says, "One story current is that the Arabs entered Maskat in the guise of peaceful peasants, hiding their arms in bundles of firewood, and that they took the opportunity of the Portuguese garrison being assembled without arms at chapel to attack and massacre them."

Captain Hamilton,‡ in his 'New Account,' gives a long and circumstantial account of the final scene, which he had "from a very old renegade who was at the tragedy, being then a soldier, who reckoned himself about a hundred years old, and by his aspect could not be much less." This story gives as the cause of the final hostilities a gratuitous insult offered by the commandant to the "king of that province." He says only those of the garrison were spared at the final surrender who "consented to embrace" Mohammedanism.§

During the reign of Nadir Shah, between 1736 and 1741, the Persians occupied Omán,‖ having, in the first-named year, gained a footing under the pretence of assisting one of two rival claimants to the Imámate, but they appear to have been finally driven out in the latter year.

The subsequent history of the country is not of sufficient general interest to relate at length. The fortunes of the country culminated under the great ruler Seyyid Said-bin-Sultan, 1804–56,¶ since whose death it has rapidly declined, owing to intestine wars and the loss of the African dominions (Zanzibar, etc.), which fell to another son, and has since remained a separate state.

About 1800 the French attempted to gain over the Imám in furtherance of their designs on India, but this was frustrated by the British. Seyyid Said continued throughout his reign our loyal ally, and co-operated with our forces in the expedition against the independent piratical ports in 1819, and in the disastrous Benu-bu-'Ali affair in 1820.

---

* Nassir bin Murshid, who appears also to have recovered Karyát, and all Omán except Maskat and Matrah, and was one of the strongest rulers Omán ever had.

† Colonel Miles (op. cit.) says that one of the several objections to this romantic story is that the Banians have never brought their wives to Arabia, much less their unmarried daughters.

‡ 'Pinkerton's Voyages,' vol. viii.

§ Wellsted gives the date of the final capture of Maskat as 1658, which is incorrect. He was not an accurate observer. The date has been the subject of controversy, but is now fixed by Mr. Danvers from Portuguese records ('The Portuguese in India,' by F. C. Danvers. 1894).

‖ As well as Bahrain and other islands in the gulf.

¶ Dr. Badger's author. Colonel Hamerton says 1807, and is more likely to be correct ('Bombay Government Records,' No. 24, New Series, 1856).

CORRESPONDENCE.
Maskat.

Lisbon, January 5, 1898.

THE inscription of the Fort Maskat, which Captain Arthur W. Stiffe publishes in the *Geographical Journal* of December, naturally not having been taken by an impression, is incorrectly copied and interpreted.

The consequence is, that Captain Stiffe cannot reconcile the date of 1588, clearly indicated in the inscription, with the reference which that gentleman attributes to it, to the eighth year of the reign of King Henry, our cardinal king, whose reign did not last eight years, and who had already been dead for a similar number of years.

Notwithstanding the imperfection of the copy, the Portuguese restitution of the inscription is easy, and evidently the following: "Reinando o mui alto e poderoso Filippe, primeiro deste nome Rei e Senhor nosso, no oitavo anno de seu reinado na Coroa de Portugal, mandou por Dom Duarte de Meneses, seu Viso Rei e Governador da India se fisesse esta fortalesa e a fez Belchior Calaça, seu primeiro capitaõ e fundador, 1588;" and its translation as follows: "Under the reign of the very high and mighty Felippe, first of this name our Lord and King, in the eighth year of his reign in the Crown of Portugal, he sent by Dom Duarte de Meneses, his Viceroy and Governor of India, that this Fortress be made, which was done by Belchior Calaça, its first captain and founder, 1588."

Not only Dom Henry died on January 31, 1580 (and not 1579, as stated by Captain Stiffe), but also Dom Duarte de Meneses was appointed Viceroy and Governor of India, not by that king, but by Felippe II. of Spain, *the first of this name in Portugal*, the crown of which he usurped in 1580. The date indicated corresponds exactly to the eighth year of this reign, and the initial F of the name of the king inserted is very clear in the copy, in spite of its imperfection.

Dom Duarte de Meneses left Lisbon to assume his post in India on April 10, 1584, arriving at Cochim on October 26.

Maskat, where the Portuguese had established a kind of co-dominion, more commercial than military, suffered frequent oppressions and depredations by the Arabs and the Mohammedans of India, that is, by the *Arabios* and *Mouros* as they were then called. The Turks, according to Captain Stiffe, when they dared to descend to the sea of Oman, assaulted and destroyed Maskat, and the people of Mogor—the *Mogores* as they were called—advanced through all the "kingdom of Sinde," treating better or worse the Portuguese, who were trading there, according as to whether they were nearer or further from our sovereignty and the resolution shown by them. Besides this, the struggle between the Dutch and English for the dominion of the Persian gulf was already advanced and bitter, and caused our expulsion from the same. The necessity of fortifying Maskat according to our custom or after the European fashion became manifest, as it was an important strategical point for the navigation and trade of Ormuz, and for the route and crossing of the Sea of Oman. The work of fortification had already commenced, advantage naturally being taken of the rudimental native defences. Dom Duarte de Meneses therefore received instructions and gave orders to consolidate in a more secure manner the occupation of Maskat. He appointed captain for this post Francisco Velho, a man of confidence, who, however, being implicated in a process, was soon afterwards substituted by Belchior Calaça (read *Calassa*, and not Caleça), whom it was that in 1588 was effectively in Maskat as captain. "Who is at present serving," says a royal letter of February 6, 1589, approving the substitution, but recommending that, as soon as Calaça would finish the time of his service and Velho be freed from guilt, the latter should succeed the former. When this letter arrived, the viceroy had already died on May 4, 1588, the same year in which Calaça placed the inscription on the Fort *Capitao*, or Merani, and when, in front of the same, and with it completing the guarding of Maskat, on the sea side, the fort of *Saõ Joao*, or Jaladi, was already erected, having been concluded in the previous year, probably also by Calaça.

The rightful successor to Dom Duarte de Meneses was Mathias de Albuquerque, who succeeded him shortly afterwards, and who, having embarked for Lisbon, left here on May 8, 1590, as viceroy, arriving at Goa in May, 1591.

The fortification of Maskat continued, as did also the development and commercial colonization of this place, as well as the perils and threats which threatened it and all our posts and factories of the Persian gulf.

Another captain, Dom Jeronymo Mascarenhas, commenced a new fort in Maskat, which was continued during the time of the new viceroy, Mathias de Albuquerque, and was recommended in the royal letter of February 5, 1597. Maskat was in a certain manner a conventional dependency of the Government of Ormuz, and the viceroy himself proposed to Lisbon that this dependency should become effective. A royal letter of February 18, 1595, adjourned, however, this resolution, determining that there should be only communication and constant accord between Ormuz and Maskat. Antonio de Sousa Falcao was at this time captain of the latter. The occupation became defined and organized in a military administration. The permanent garrison was fixed at thirty soldiers, who, as Viceroy D. Jeronymo de Asevedo said on November 29, 1613, "rarely became effective on account of the manners and frauds which they were accustomed to practise." By "Alvara" of this date, D. Luiz da Gama, who was then about to govern Ormuz, was ordered, that of the seven hundred soldiers who ought to be at this latter garrison, twenty should be made to reinforce the garrison of Maskat, increasing thereby the latter to fifty, "so long as there should be any fear of its being attacked by enemies from Europe." And in this same decree was announced the project of sending from Goa a further thirty soldiers as well as a captain, apparently for the purpose of the special garrisoning of the last fort constructed.

En passant, and in order that Captain Stiffe may see that our occupation was not so exacting and violent as one might suppose by his sentimental remark regarding the method we employed to secure the respect of the natives, a method absolutely justified by the means, time, and by the circumstances, I would state that still in 1609, and afterwards, we loyally divided the revenue of the custom-house of Maskat with the sheiks, with whom we contracted, or, to be more precise, with their descendants.

It was intended in the above year, and orders were sent from Lisbon, to negotiate with these participators the cession of the share which they had in the fiscal revenue, for a reasonable compensation. But since those persons did not agree to give up their rights, the right, which might have been disputed, was respected, and it was also resolved, in view of the fact that they were interested in the development and fiscalization of the trade, to maintain the traditional *regimen* decreed in the royal letter of January 23, 1612.

Many other interesting things might be indicated which must be reserved for a special work, which will fill up the gap generally found in the histories of Maskat which deal with the period of Portuguese domination. My present intention was simply to offer Captain Stiffe an opportunity of correcting what he states regarding the inscription which is published in his excellent and beautiful articles in any new edition which they may have, as I sincerely hope may be the case.

I will not conclude, however, without observing to the distinguished writer that it is not absolutely exact that in 1650, through the loss of Maskat, the Portuguese were finally expelled from Oman. This expulsion took place some time afterwards. In 1690 our flag was still hoisted in what we called the *Congo* of Persia, at a factory, of which João da Silva was manager, and Jorge de Freitas secretary; and one of our fleets, commanded by Antonio Machado de Brito, cruised in the Persian gulf, establishing treaties with Bassora, Maskat, etc., and in 1695 we seriously projected the recovery of Ormuz. Persia herself at this time implored our assistance against the Arabs, and offered it against the Imam of Maskat.

If we had not suffered in Europe the ominous union with Spain, surely would have been different the fate to our dominion in the East, which we were the first to open to Christian civilization.

Luciano Cordeiro, Hon. Corr. M.R.G.S.

London, January 15, 1898.

Referring to my paper on Maskat published in the December number of the *Journal* (p. 608), I have received a letter from Mr. Donald Ferguson, a gentleman evidently well acquainted with the language, criticizing the interpretation of the old Portuguese inscription, and suggesting the one given below, which is more probably the correct one, as it reconciles the discrepancy of data which had puzzled me.

I may say the reading I gave was furnished by a Portuguese gentleman, then employed under Mr. F. C. Danvers at the India Office in translating Portuguese records. I cannot ascertain his present address, or I should have referred it back to him.

It will be noted that the king's *name* is not given in full, only the *initial*, which has doubtless led to the difficulty. The letters in brackets are not in the original.

"REINAODO HO MVI A(LTO)E F(IDELISSIMO) HE * PODEROZO F(ELIPPE) (?) †
PRIMEIRO (?) DES(S)E NOME R(EI) HE ‡ S(ENHOR) § NOS(S)O NO HOUTAVO
AN(N)O DE SEV REINADO NA C(O)ROA DE PORTUGAL MANDOV POR DON
DVARTE DE MENEZES SEV VIZOR(EI) DA INDIA Q(U)E SE FIZES(S)E ESTA
FORTALEZA A QV(ELL)A FIS (?) ¶ BELCH(I)OR CALAÇA S(E)V PRIMEIRO
CAPITAO E FUNDADOR, 1588 "

In English—

"Reigning the most high, most faithful, and most mighty, Philip, first of this name, our king and lord, in the eighth year of his reign in the crown of Portugal, commanded by Don Duarte de Menezes, his viceroy of India, that should be made this fortress which (made?) Belchior Calaça, its first captain and founder, 1588."

* For E. My translator here suggested HENRIQUE.
† For Felippe. My translator amended this F to E for " and."
    ‡ For E.   § For Senhor.
¶ This word is obscure.

Of course, this was Philip the *second* of Spain, but first of Portugal, who was proclaimed king of the latter country in 1580 to 1581.

I hope, in view of the interest attaching to the subject, this explanation may not be deemed superfluous.

ARTHUR W. STIFFE.

P.S.—Perhaps I should have used the term "careful copy" instead of "facsimile" in the paper.

The leading annual exports from Muscat to Bushire in 1863 were listed by Col. Lewis Pelly in the *Transactions of the Bombay Geographical Society* (1863) as follows: mat bags (40,000 Bombay rupees), Janpoor indigo (Sind manufacture, 40,000 rupees), empty rice gunny bags (35,000), Bengal sugar (31,000), dry lemons (25,000), Cutch leather (12,000), coffee (10,000), lamp oil (Cutch manufacture, 6,500), tin (5,000), and pepper (5,000). The leading annual imports from Bushire to Muscat at the same epoch were raw silk (50,000 Bombay rupees), opium (30,000), wheat (12,000), ghee (6,000), rose-water (5,000), and cummin seeds (4,000).

Palgrave, if he visited Muscat at all, which Philby strongly doubted, saw the city in 1863. Readers are invited to refer to any edition of his *Narrative of a Year's Journey through Central and Eastern Arabia* (London, 1865), and then make up their own minds. Salim ibn Thuwaini governed Muscat from 1866 to 1868, when 'Azzan ibn Qais took over.

A. Germain reported a visit to Muscat in 1868 in the *Bulletin de la Société Géographique de Paris* (1868), indicating a drop in the population from 60,000 forty years previously to the current 30,000 within the walls and 4,000 beyond. Women now outnumber men by four to one. The city is half-ruined, many streets being impassable because of rubble which has never been cleared away. The forts stand in disrepair, and the city's four gates are guarded by sleepy *badu*. Only two Europeans reside in Muscat: the British Consul and the representative of British India Steamships.

Admiral Philip Howard Colomb noted the white flag of Azzan flying everywhere, and quoted his interpreter's warnings about the strict Islamic piety of Azzan: 'Plenty soldier–Plenty mosque. He look out for God. Bazaar nobody smoke. Azzan put him in chokee. Night smoke yes in house . . . no wear him silk, no drink him grog.' In Fort Jalali, Colomb, writing in *Slave Catching in the Indian Ocean* (London, 1873), found prisoners sitting apart in the half-ruined building on heaps of crumbling mortar, shackled to a pig of ballast about two feet long. Colomb reported that the slaves seemed very contented, and difficult to distinguish from free men. They cost roughly 20% more in Muscat than in Zanzibar, and 50% more in Basra or Bushire.

Sayyid Turki ibn Sa'id ruled from 1870 to 1888. Travellers recording their reminiscences of Muscat during his rule include Sir Bartle Frere, William Perry Fogg, Grattan Geary, Denis de Rivoyre, Edward Stack, Bishop Thomas Valpy French, and General F. T. Haig.

Sir Bartle Frere visited Muscat in April 1873. In John Martineau's

The Life and Correspondence of Sir Bartle Frere (London, 1895), he described the streets as 'more like passages in a rambling house than thoroughfares', and it is of Matrah he writes when noting that the *suq* was 'a very thriving labyrinth of fish, meat, cloth, grain and vegetable sellers, shoe-makers, cutlers and hardware sellers, and shops of beads and ornaments.' The Khojas had their own quarter and their mosques were decorated with dark blue flags with a star or scimitar in white or yellow, while the Arabs had a plain red flag, occasionally with a crescent or scimitar.

William Perry Fogg saw Muscat in 1874. Typical of all those visitors who went ashore during a long cruise for only a matter of hours, he noted in his *Travels and Adventures in Arabistan* (London, 1875) streets about four feet wide, full of 'savage-looking' *badu*. He estimated the population at 60,000, and reported that Sultan Turki went out seldom because he feared assassination.

Geary on Muscat

Geary, Editor of *The Times of India*, is by far the most sympathetic and informative of these writers, and his account of a visit in March 1878 is reprinted in full from *Through Asiatic Turkey* (London, 1878).

> Across the Sea of Oman to Muscat—The rock-bound harbour—The town and its narrow streets—Absence of beasts of burden and of carts—Large trade—Mixed population—British Resident—His highness the Sayyid—Household arrangements—The Bazaars—Arab swordsmen—Abyssinians—Beloochee mercenaries—His highness's army—The Politics of Oman—The gunboat and the Arabs—Visit to Sayyid Toorki—The lion at the palace gate—The royal staircase—His Highness on the Eastern Question—Inspection of Fort Mirani—How the Portuguese lost Muscat—The Worshipper of the Cow and his gentle daughter—The governor's love and unbounded simplicity—Fort Jalali—The Sayyid's early vicissitudes and present melancholy—The climate of Muscat.

WE left Kurrachee on the evening of the 18th March, and steaming across the Sea of Oman to the Arabian coast, arrived off Muscat, the capital of the Arabian kingdom of Oman, on the morning of the 21st, at six o'clock.

The little town of Muttra, situated on the coast about a mile and a half to the west of Muscat, is visible before anything can be seen of the capital itself. The cove of Muscat is entered from the north, and until the steamer rounds the point of the Gibraltar-like rock which forms one of the sides of the harbour, it is difficult to

imagine that there is a considerable city hidden somewhere in the midst of the inhospitable cliffs. The rugged
volcanic hills which surround the little harbour are as
bare of vegetation as Aden itself. They are four or five
hundred feet high, of irregular outline, and quite
precipitous. Two imposing forts occupy opposite points
commanding the harbour and the town, and along the
heights at intervals are towers of unusual height. The
place could be easily made impregnable; but the
harbour is not large, being little more than a quarter of
a mile from east to west, and less than half a mile from
north to south. It is completely open to the north.
During the north-west monsoon the sea comes in with
great fury, and though the anchorage is good, the
bottom rising gradually towards the rocky shore, ships
usually prefer to clear out rather than try the strength
of their cables. There is a depth of from fifteen
to thirty fathoms throughout, and the harbour is
perfectly sheltered from every wind except that from
the north or north-west. The amphitheatre of hills
which gives this shelter on three sides, allows a cramped
space at the head of the harbour for the town of
Muscat.

The town is surrounded on the land side by a wall,
strengthened with eight towers. The houses are built as
close to one another as it is possible for them to be;
there is not a street both sides of which you may not
touch with the hands as you walk along. Outside the
town there are suburbs, built of sticks and mats, in
which Beloochees principally reside, and there the streets

are not quite so narrow. But during the whole of my wanderings through this city of Muscat I saw neither horse, nor camel, nor mule, nor ass, and, of course, no cart of any kind. It would be impossible for four-footed animals, larger than dogs or cats, to move through the streets; there is scarcely room for two men to pass one another without jostling.

Nevertheless, Muscat is a place of importance, and possesses a considerable trade. The city population, estimated at 40,000, is perhaps as mixed as that of any of the ports in these seas. Besides the Arabs, there are numbers of Banians from Western India, of Beloochees from Mekran, of Abyssinians, Somalies, Nubians, and Persians. Of Europeans residing in the town there are two. It was an historical event when Captain Jourdan and myself landed, and doubled the European element for a whole day. The permanent British residents are the Political Agent, Colonel Miles, and Mr. Maguire, a merchant in a large way of business, the agent of the British India Company. Of course we visited both, and were most cordially welcomed. To Mr. Maguire, who is a very skilful amateur photographer, I was indebted for the photograph of Muscat, from which the engraving forming the frontispiece to the present volume has been taken. It is a very characteristic view. The large building on the left is the British Residency; in the centre is the Custom House; and on the right is Fort Mirani, of which more anon. Intimation of our arrival was sent to his Highness Sayyid Toorki, the Sovereign of Oman—usually, but erroneously, styled by

Europeans the " Imam of Muscat." His Highness very
courteously said he would be happy to give us an
audience at four o'clock, and sent word to the com-
mandants of the forts to allow us to go over them.
Meanwhile, we noted the architecture of the place, and
went through the bazaars. The houses are somewhat
Saracenic in general style. They have all two floors,
and some even three. The walls of the upper rooms are
divided into innumerable openings by long, narrow
windows, so that air may come freely in from all
quarters. No one lives on the ground-floors, which are
used as lumber or store-rooms. The roofs form the bed-
chambers; no one can sleep in-doors during the hot
weather on account of the stifling heat. When the
shumal, or hot wind from the desert, blows, the sleepers
are during the night watered, like plants, with a
watering-pot. This practice may account for the fact
that muscular rheumatism is by no means unknown in
Muscat.

The bazaars are different from anything of the sort
I had seen in India. All the business streets are roofed
in at a height of some eighteen or twenty feet. Poles
or beams go from side to side—they need not be long to
span the distance—and on them are placed layers of
mats, which are plastered over with three or four inches
of stiff mud. By this means the fierce heat of the sun
is effectually kept out. Every twenty or thirty yards a
hole is broken through the hard mud and the matting
to let in a little light; darkness is thus made visible.
The general effect is that of rows of shops, or rather of

stalls, in subterranean passages. The smell is sometimes heavy, but, strangely enough, the ventilation appears on the whole to be very fair. Doubtless, the great heat of the atmosphere above the close roofing insures an upward current through the light-hole. The city is tolerably clean throughout, far cleaner than any of the other cities we shall see before arriving in Europe. The dry-earth system is thoroughly understood in this Arabian capital; earth and ashes play a great part in keeping Muscat comparatively sweet and wholesome.

The goods on sale were the usual stock of Oriental bazaars. Manchester prints caught the eye at every turn. At one shop, padlocks made in Bombay were sold at two annas (threepence) each: they were coarsely put together, but strong and serviceable. Muscat is famous for its halwa, a compound of sugar, ghee, or clarified butter, and the gluten of sesame. It is usually supposed to be made of the milk of the camel; but that is a popular error. The halwa of Muscat is very different indeed from that which is sold in Bombay. It is quite palatable, and it is said to be highly nutritious and very fattening. Dates are to seen at every stall. Fruit and vegetables, from the cultivated spots in the interior, are abundant. The population is evidently well fed, and the traders generally have a comfortable, well-to-do air. We saw no beggars and no squalor. The bazaars were thronged with Bedouins fresh from the desert, who had left their camels and horses at Muttra, and come round by boat to Muscat, the narrow and tortuous defile between Muttra and the capital being almost impassable for animals. They were all armed to the teeth. Many had long, old-fashioned guns, highly ornamented, and all wore daggers or pistols in their girdles. A favourite weapon is a straight, broad, two-handed sword, the sweep of which would take off a man's thigh, or even cut him in two at the waist. The swordsmen carried over their shoulders small round shields of rhinoceros hide, eight or nine inches in diameter. The Arabs of these parts are very formidable swordsmen.

Half a century ago, a small number of the Beni Bou Ali being attacked by Captain, afterwards better known in England as Colonel Perronet Thompson, at the head of 350 British troops, rushed upon their badly-advised assailants, and cut down a couple of hundred of them in the twinkling of an eye. This led to a regular expedition of nearly 3000 men being sent against the tribe from Bombay. The Beni Bou Ali numbered scarcely a third of the avenging force, but they charged as before, sword in hand, and seemed to have a strong conviction that they ought to be again victorious. They were, however, defeated, and a number of prisoners were taken to Bombay. Mountstuart Elphinstone treated them very well, and sent them back to their own country, much to their surprise, for they imagined they would be all put to death by the infidels. Of course, the formidable swords of these lithe and active children of the desert will never again contest the victory with British bayonets, for the old Brown Bess has been superseded by a weapon which no swordsman can approach, but the skill and intrepidity of the Arabs of Oman are the same as ever. The perpetual feuds between the tribes keep the warlike virtues of the Arabs undimmed by time. Most of the Arabs we met in our visit to the bazaar were at least five feet ten inches in height; all were well formed and muscular, but so free from superfluous flesh, that they looked meagre and bony. The athletic negroes who do the rough work of the place, have the advantage of them in size and plumpness, and perhaps in muscular strength. The handsomer Abyssinians are manifestly inferior to them in energy and physical endurance. The Beloochees are a fine race, nearly as tall as the Arabs, and there is more flesh on their bones. They have a good-humoured smile, which contrasts pleasantly with the Arab gravity and taciturn air.

The troops of his Highness the Sayyid are composed of Arab and Beloochee mercenaries. The Beloochees are alone to be depended on in the struggles against pretenders or rebels, which the sovereign has almost always on hand. The Beloochees know nothing of the political factions which distract Oman, and they naturally obey the orders of the prince who gives them their pay. It is not so with the Arabs, who change sides with a facility which must be a source of considerable anxiety to the sovereign of the moment. The state troops are not drilled or armed on the European model. The long gun, the sword, and spear, appear to be the weapons most in use. Of drill there is evidently none. The Arabs wear the costume of the country, and unsoldier-like ringlets hang down their lank cheeks. There is nothing of the military swagger in their gait; they lounge along as if they suffered sadly from idleness and growing melancholy. The Beloochees are more alert in their air, but they are not a bit more soldierly. Doubtless, Sayyid Toorki knows his own business best, but it strikes one forcibly that if he put his troops through a course of military training, gave them smart uniforms, and armed them with modern weapons of precision, there would be better order in Oman, trade would flourish more, and his Highness's revenues would be considerably augmented.

The kingdom of Oman is about as large as that of England, and although it is not in all parts quite so fertile and productive, it boasts of many tracts which are well watered and fruitful. Its people have a certain turn for agriculture and trade. They have also some political cohesion, the best proof of which is to be found in the fact that Oman has preserved a sort of rude independence under its Imams from the eighth century until now, successfully resisting the efforts of the Persians on the one side, and the neighbouring Arabs on the other, to bring it into subjection. With a little vigour on the part of the central authority the predatory tribes could be kept in awe, and the incessant revolts of the local governors—for the most part " Sayyids," or princes of the prolific reigning house—put down with a strong hand. The British resident gives all the moral support possible to his Highness, and sometimes summons to his aid one of the British gunboats in these seas, and in that way some brief intervals of order are secured. Not very long since her Majesty's gunboat Teazer, lying in the harbour, was called upon to fire over the town at a crowd of Bedouins who had insolently taken up a position beyond the line of hills to the south, in the hope of finding their way into Muscat and sacking it. The shells created a great moral effect, the Arabs never having seen such missiles before. They said the shells had eyes and could see where to fall. They could not understand how otherwise the shells could come right into their midst, as they were of course out of the view of the gunners on board the Teazer, the hills intervening. The fire was really directed by signals from an old Portuguese tower on the summit of one of the hills. A shell fell in a field, and did not explode. It was immediately surrounded by a number of excited Bedouins, who determined at once to put out the "eye"—the copper percussion fuse—by which it guided its course. They struck the "eye" with their lances; the shell exploded, and eleven Bedouins were killed on the spot. Of course the marauders ultimately decamped. The support which the British power gives to the Sayyid is, however, of necessity confined to such diplomatic and naval assistance as can be rendered to him in his conflicts with enemies showing themselves near the coast. If his Highness wishes to protect his kingdom from troubles arising in other quarters he must reorganize his little army, and make himself formidable to all aggressors. As it is the country is declining, owing to a general feeling of insecurity which checks trade. The imports of Oman amount to about 300,000*l.* a year; the exports are considerably greater, amounting to 1,100,000*l.* Dates and cotton fabrics, fruits and

fish, are the principal articles of export. With settled
order, commerce would again flourish, as it has from
time to time flourished whenever a ruler of exceptional
vigour has borne sway at Muscat.

Having made the round of the bazaars we proceeded
to his Highness the Sayyid's palace, accompanied by a
very intelligent Arab interpreter provided for us by Mr.
Maguire. After going up and down a number of narrow
streets, or rather lanes, for there is not a genuine street
in all Muscat, we éntered one differing in nothing from
the others, in which was a crowd of armed men, Arabs
and Beloochees, his Highness's body-guard. The interpreter stopped at a ponderous gate, close shut, and told
us that was the entrance to the palace. He knocked,
and a little wicket was opened; he spoke a few words,
and we were invited to enter through a small doorway
in the great gate. We found ourselves in a sort of
court-yard, around which was built the palace, a very
unpretentious, two-storied edifice. To our left, close to
the gateway, was a good-sized room, in which reclined
a splendid African lion; the front of the lion's parlour
was formed of iron bars similar to those which protect
the plate-glass of jewellers' shops in London at night.
The royal brute looked at the strangers with considerable
curiosity. We returned his gaze, wondering what his
particular duty at the palace gate was. Doubtless he is
there to strike awe into the souls of people who may
seek an interview with the sovereign, so that they may
go into the presence duly impressed with an idea of his
Highness's power. In the centre of the court-yard a

leopard occupied a cage ; on our right were eight or ten
Arab mares, some of which were evidently of considerable
value, and the horse-keepers lay about on the pavement.

In a few minutes an intimation was sent that his
Highness was ready to receive us, and we were invited
to go forward. We were conducted to the foot of a
large ladder with a hand-rail, and we ascended. When
we got to the top we found ourselves in a sort of ante-
chamber with whitewashed walls, in which some half a
dozen Arabs were standing about. One of the simplest
of these Arabs advanced towards us and held out his
hand ; unlike the rest he had no turban ; he wore a small,
white skull-cap, and was very plainly dressed. This was
his Highness himself. We shook hands, and he very
politely ushered us into the next room, the hall of
audience. It was a plain apartment, with the long,
narrow windows characteristic of Muscat ; the walls
white and unadorned ; the furniture plain to a degree.
A few cane-bottomed chairs were ranged on one side of
the room, and in front of them was a small carpet. His
Highness motioned us to sit on the chairs, and then he
sat down on a sofa a little apart ; he had no carpet in
front of him ; he put his sandalled feet on a small bed
about four feet long by three feet wide, and made of
common bed-ticking—a most unprincely footstool.

His Highness is apparently about thirty-five years of
age, and is tall and spare, like most Arabs. His handsome countenance wears an expression of melancholy ;
he is grave and dignified, but perfectly simple and unassuming in manner. The conversation began by an

expression of satisfaction on the part of his Highness
that we had visited Muscat. He hoped we were well,
and said he had been informed I was going to Constantinople. Had I any news about the war? I told
him a treaty of peace had been signed between Russia
and Turkey, but that according to the latest accounts,
received when we were leaving Kurrachee, the Russians
were occupying some places near Constantinople in spite
of the protests of the Sultan. His Highness said that
since the Russians had got so near to Constantinople it
was not at all likely that they would go back. They
would no doubt stay there as long as they could. I
told him about the supposed understanding between
England and Austria to limit the results of the Russian
aggression, and his Highness paid great attention to
that point, putting repeated questions to elicit further
information. He remarked that England might have
interfered to advantage at an earlier period, and prevented the Russians from getting close to Constantinople, but that now they were at the very gates the
difficulty was much increased.

Sayyid Toorki, I afterwards learned, takes a great
and intelligent interest in the different phases of the
Eastern question, and his sympathies are strongly
aroused in favour of the Turks. The Arabs generally
hate the Turks with a cordial hatred, but in the
struggle against Russia the Osmanlis are regarded
as the defenders of Islam against the infidel. The
Sayyid appears to be a man of good sense, and to have a
fair knowledge of politics and geography. Being told

that Captain Jourdan was a French officer, his Highness asked whether he had come direct from Paris?
When informed that he was returning from Japan, and
intended to reach France through Persia and Russia, the
Sayyid remarked that the route was very long and
difficult. He appeared to know something about Japan
and the other countries mentioned. He asked eagerly
whether France would go to war with Russia if England
did so, and seemed rather disappointed when informed
that "a great bazaar" about to be held in Paris would
most probably induce that power to pursue a policy of
peace for several months to come. While the conversation was in progress, coffee was handed round; some
minutes afterwards sherbet, or rather orgeat, a preparation of almonds, and sugar, and water, was brought in.
We then rose to depart, and his Highness accompanied
us to the outer room, where the governor of the town
was in waiting, and he was formally presented to us
by the prince himself. We shook hands with Sayyid
Toorki, and so our long and interesting interview with
his Highness ended.

We then went to see Fort Mirani, one of the twin
forts which guard the harbour and the city on the
eastern and western sides. Mirani is the most interesting of the two, as it contains many relics of the
Portuguese, who built both. Albuquerque took Muscat
in 1507, and it was held by the Portuguese until 1652,
when it was recovered by the Omanis after a long
siege. A certain "worshipper of the cow," a Banian
from India, named Narrotum, had a good deal to do in

the delivery of these forts, and with them the town, into
the hands of the Arabs. The Banians appear to have
been always of some importance in Muscat. Narrotum
was a man of wealth and influence, and he had a beautiful
daughter, with whom the Portuguese governor, Pereira,
fell in love. His Excellency, greatly to his credit,
demanded her in lawful marriage, but Narrotum, far
from feeling honoured by the proposal, was horrified at
the notion of marrying his child to a Christian. Being
threatened, however, with the vengeance of the love-
sick governor if he refused his consent, he feigned
compliance, but asked for a year's time in which to
prepare the bride's *trousseau* upon a fitting scale.
Pereira, very foolishly, granted the delay. The city,
while all this was going on, was being besieged in due
form, but Sultan bin-Sief, the Imam, could make
little progress, so valiantly was the place defended.
Now Narrotum had, in the words of the Arab historian,
been given "the keys of the shops in the two forts, and
he was agent for the treasury and country;" and he
contrived a scheme for the discomfiture of the Portuguese,
so that the dreaded marriage might be averted. He
represented to the governor, who seems to have been a
simple-minded, honest fellow, that the water in the fort
tanks was bad, and should be renewed, and that the corn
was full of weavils, and ought to be removed, and good
corn put in its place. The gunpowder, too, he said, was
old, and not to be depended upon. Why not stock the
magazines with fresh gunpowder?

Pereira, suspecting no guile, gave Narrotum authority

to replenish the stores, so that Muscat might be in a condition to hold out indefinitely against the besiegers. The Portuguese had command of the sea, so that there was no difficulty in procuring supplies. When the wily Banian had emptied the water-tanks to clean them out, and removed the corn and the powder to make room for better, he sent word to Sultan bin-Sief to assault the city, and further advised him to assault it on a Sunday, when the Portuguese would be keeping their festival day, fiddling and drinking, and Sabbath-breaking generally. The Imam acted on the advice, and set out at the head of his troops, exclaiming, as he gave the order to march, "God is most great! O God, make the orthodox Moslems victorious over the beardless Portuguese!" The prayer was heard; the town wall was passed without difficulty, and the two forts were then taken by escalade. The guns would not go off, for the powder had been tampered with, and the garrison could make no effectual resistance. Two Portuguese ships still kept the mouth of the harbour, but "a hundred men to whom death was sweeter than wine to the wine-bibber," being well paid for the glorious enterprise, attacked them in small boats, and killed all their "polytheist crews."

Before the two great forts fell into the hands of the Arabs, one was named Fort Juan, and the other, the western, Fort Commandato. The former is now called Fort Jalali, or "the glorious," and the latter Mirani, after a former Beloochee governor. Mirani is placed on the summit of a hill about 400 feet high, and is itself very lofty, the walls in places rising some forty or fifty feet above the rock. It is built of sandstone, which in places crumbles to the touch, but is, on the whole, in a wonderful state of preservation. On the shore, near the flight of steps which leads to the entrance, are eight small bronze cannon, unmounted. They are kept there for salutes, and are in time of trouble a great temptation to the disaffected. One of her Majesty's cruisers has had on sundry occasions to make rocket practice into the angle of the rock in which they lie, to keep dishonest hands off them. Why they are not kept within the fort it is not easy to understand. Perhaps the Government of Muscat is afraid that such an increase to the armament of Fort Mirani would make it independent. At times the forts fall out, and they blaze away at each other across the harbour, and in front of the town, to the great interruption of business. We saw several marks on the face of Fort Jalali which testified to the occasional accuracy of the fire of Mirani. Fort Jalali not very long since fired on the town promiscuously, and the British Residency being within a hundred yards of the cannoneers, was pretty well riddled—it was difficult for them to miss a mark so conspicuous and so near. The British Resident, however, objected, and sent word that if the fire continued, it would be returned from a gunboat. Jalali ceased firing. But peace has prevailed, at Muscat now for a whole year—indeed ever since the Teazer, a twelvemonth since, aimed her shells over the town and the hills beyond at the Bedouins, as I have already mentioned.

When we entered Mirani the garrison consisted of
the governor and about half a dozen men, ununiformed
and unarmed. Possibly the bulk of the garrison was in
town off duty. The governor was a civil, inoffensive
old gentleman, with a cotton handkerchief wound round
his head by way of a turban. Most of the cannon
are iron, rusted apparently through and through;
their carriages are worm-eaten and falling to pieces,
and it is difficult to understand how they can be fired
with safety. Yet a salute of 101 guns was fired on the
1st of January from this fort in honour of the Empress
of India. As every shot reverberates in endless peals
amongst the hills which surround the harbour and town,
the effect must have been wonderfully fine. An
inscription cut in stone over one of the inner gateways
states that the fort was built in the reign of "Roro
Primero, 1588." A very fine bronze gun, some ten feet
in length, bears the Spanish arms, with the name of
Philip III., "Rey d'Espana," and the date, "1606." This
gun was therefore cast by the contemporaries of Guy
Fawkes, who little dreamed that it would for centuries
form part of the military strength of infidels, worse,
if possible, than heretics. It was doubtless placed in
this fort when the crowns of Spain and Portugal were
temporarily united. On the top of the main tower of the
fort is a small circular Portuguese chapel, with a pious
inscription to the Virgin, which the Arabs, much to
their credit, have not defaced. The holy water font
in stone is intact at the entrance; but no altar or cross,
or other mark of the original purpose of the little chapel

remains. It is at present used as a sleeping apartment by the commandant of the fort. When we had visited the chapel, we were asked to sit down under a canopy of reeds, whence we had a magnificent view over the harbour and town, and the valley behind the town. Coffee and sherbet were presented to us, and after the usual interchange of compliments, we took our leave of the courteous commandant.

It is not the forts nor his Highness's troops that protect Muscat from the Bedouins or save Sayyid Toorki from the endless machinations of his rivals. Colonel Miles, dwelling alone in the Residency, and the beautiful little gunboat at anchor in the harbour preserve Muscat and its prince from the ravenous prowlers who long to sack the one and depose the other. No wonder Sayyid Toorki is a prey to settled melancholy. Like Hamlet, he had a dear father—the Sayyid Thuwanney—murdered, not by an uncle, but by an ungrateful son, Sayyid Toorki's elder brother; and he would no doubt have been murdered himself, too, had not Colonel Pelly, then British Resident in the Gulf, partly by threats and partly by persuasion induced the parricide to let him out of prison. He spent a couple of years in exile in Bombay before the opportunity came for fighting his brother for the kingdom. During the desultory warfare which placed him on the throne he earned the reputation of being the most daring soldier in Arabia; and when he won, great hopes were entertained that he would prove a vigorous and capable ruler. He has certainly shown himself a very mild and well-intentioned prince, but he appears to lack the qualities necessary to organize Oman and control the turbulent. If it were not for the beneficent influence of the British Government, the " Kingdom of Security "—Oman means " security," or "settled peace "—would be in a state of complete anarchy.

During the day we spent at Muscat the temperature ranged from 81 to 83 degrees; nothing could be more delightful. The air was dry, clear, and exhilarating. For the previous three or four months the climate had been the same. Muscat has two seasons only—the hot and the cool—each lasting about six months. The hot season is something dreadful; for the black rocks all around give out during the night the heat they store up during the day. The place is a fiery furnace during the whole twenty-four hours; yet it is not regarded as particularly unhealthy. Possibly the six months of reasonably cool weather, and the complete absence of severe diurnal alternations of heat and cold at any time, keep the public health at par. About forty miles in the interior there is a range of tolerably high hills, which are described as perfect sanitaria; it is cool on their breezy tops when the plains are scorched as with fire. But under present circumstances these salubrious hills are practically inaccessible to the broiled denizens of Muscat.

Denis de Rivoyre was received in audience by the Sultan in 1880, and tells us in *Obock, Mascate, Bouchire, Bassorah* (Paris, 1883) that the Sultan presented him with locally-grown roses. The Sultan personally kept the keys of his arms and ammunition store; he kept 1,200 soldiers, many of them 'Kurds and Yemenis', who are billeted on villages in the interior as a means of punishment. The Sultan's palace had a great Persian carpet, and on the walls portraits of Queen Victoria, the Prince of Wales, and the Shah of Persia.

Edward Stack visited Muscat in February 1881. In *Six Months in Persia* (London, 1882), Stack refers to the rusty guns in the forts. He found Muscat flourishing, while Matrah reminded him of an Italian city, though he prudently does not say which one. The expectations of visitors vary wildly, for in March 1883 Bishop French reports Muscat 'an utter wreck of its past greatness and renown', exporting only donkeys and rock salt. Colonel Charles Grant, then British Agent, told the Bishop that the many slaves refused freedom whenever it was offered to them. Tribes from the interior were constantly raiding Muscat, and thirty-three men had fallen in a recent gunfight just outside the walls. French retired to Muscat in February 1891, practising as a Christian missionary until his death three months later. He preached in the *aswaq* of Muscat and Mutrah, lodging with a Goanese half-caste until he discovered that his landlord earned his living selling alcohol illegally. French then moved to Matrah, where the only other Christian was a retired Indian doctor. His biographer, the Rev. Herbert Birks, relates in *The Life and Correspondence of Thomas Valpy French* (London, 1895) how he found French's grave in 1894, with 35 other Christian tombs. General Haig, quoted in Birks' book, notes that the Arabs of Muscat are frank and tolerant and 'might well listen to the Gospel.'

Faisal ibn Turki ruled Muscat from 1888 to 1913. His first notable Western visitors were Theodore and Mabel Bent, who together wrote *Southern Arabia* (London, 1900). In 1889 they were received by Sultan Turki in a palace with a formidable door ornamented with brass spikes. Within stood two cages, one containing a lion and the other a prisoner. Theodore Bent climbed up to a gallery, where the Sultan sat in a red chair serving as a throne, and overlooked the sea. Fifty years ago the population had been three times greater, but the coming of the steamship had reduced the significance of Muscat to the status of a date-exporting harbour. The Sultan distributed two meals a day to the indigent.

Bent found the harbour full of brightly-painted canoes, with a corner set aside for seagoing dhows. The ice factory that the Sultan had established was now closed. The *suq* offered unusual iron locks with huge iron keys, wooden shields, sharkskins, and fine daggers ornamented with filigree silver. The cove of Shaikh Jabr, between Muscat and Sidab, is the site of the Christian cemetery, accessible only by boat. The richest merchants lived at Matrah, where Dr Atmaram Jayakar, an authority on wild life, and compiler of *Omani Proverbs* (The Oleander Press, 1986), had a house for the last twenty-five years. Bent returned to Muscat briefly in 1895.

Lord Curzon thought Muscat 'one of the most picturesque places in the world', a mixture of Aden and Corfu. He found the British Consulate, being rebuilt, 'the handsomest structure in the town', better than the Sultan's Palace, which was then merely 'plain' if 'substantial'. Curzon estimated the population within the walls at about 5,000. About 1,700 tons of coal for the Royal Navy was heaped in a hollow at the foot of the western rock, that is below Fort Mirani. The Customs were farmed out to a Banian for £17,000 a year, and Hindus monopolized the shops. Exports, then worth about £210,000 a year, included dates, limes, grapes, walnuts, and fish; imports, worth about £280,000 a year, included Bengal rice, cotton from Bombay and Manchester, sugar, coffee, opium, silk, oil, pearls, wheat and salt.

In December 1892, Edwin Lord Weeks was given an audience with Sultan Faisal, whom Weeks reported as 'a handsome young man interested in photography and Paris, whose reception room sounded to the tick of old clocks. The Postmaster accompanied him as interpreter. Weeks' experiences are recounted in *From the Black Sea through Persia* (London, 1896).

In 1898, Maurice Maindron contributed a detailed account to five issues in April and May of *La Revue Hebdomadaire*. He was received by Sultan Faisal in a gallery of the first floor of the Palace full of Indian furniture; the only thing hanging on the walls was a Remington rifle. Muscat seemed to the French journalist to be at the bottom of a giant's well, the whole city covered with dust, and many buildings in ruins. The royal horses are not stabled but maintained on an open square. The city wall, about fifteen feet high, has been recently repaired. The eastern gate is Bab as-Saghir, with a bastion called Burj al-Ghilah. The western gate, Bab al-Kabir, has acacia growing round it. Burj an-Naubah, to the east, is called 'Three Shots' because of the firing every night.

Major (later Colonel Sir) Percy Zachary Cox took up his duties as Political Agent in October 1899. In Philip Graves' *The Life of Sir Percy Cox* (London, 1941), he remarks on the unfailing source of interest: Muscat harbour. Socotran fishermen camp there in hot weather fishing from catamarans, and diving for jetsam. To catch mullet they use large rectangular fishing-baskets, anchoring them to the bottom with stones. Every well-wheel has its own particular wheezing sound; its owner can tell by a slight change in the sound if his workers are idling.

Major T. C. (later Lt. Col. Sir Trenchard) Fowle, Political Agent June 1930 – May 1931 and October 1931 – July 1932, described the port of Muscat in his *Travels in the Middle East:* 'The boom of the ship's signal gun woke me with a start, and by the grey dawn I saw that we were floating peacefully on the dark waters of a great lake – so it seemed – girdled with high cliffs still ringing with the echoes of the shot, while from in front a silent white-faced town stared gravely down on us with innumerable window-eyes. And this was Maskat – the Maskat of my first impression, at any rate. Later there were others. The lake was no lake, but a harbour: two castles flanked the town on either hand, and behind towered a peak-topped hill. Later still I went ashore. Maskat keeps its windows for the sea front. Once past the

Sir Percy Cox (Courtesy of H.B.M. Ambassador, Muscat)

British Consulate, and you find yourself in narrow byways, scarce room for two abreast, while on either side the blind walls on the houses rise cliff-like to a thin strip of sunny sky. But down below there is no sun, or heat, or sound – only the cool shade, the shuffle of one's feet in the sand, and the solitary figure of a cloaked Arab striding leisurely.'

Taimur ibn Faisal became Sultan in 1913, eleven years after medical missionary work had been begun by Dr Sharon John Thoms at a small clinic in a house on Matrah Corniche. The young Taimur gave Dr Thoms a great deal of encouragement even before he acceded, helping him to purchase the land for the Dutch Reformed Church, but Dr Thoms was killed in a fall in 1913 when he was only 42 and it was only in 1925 that Dr Paul W. Harrison (author of *The Arab at Home*, New York, 1924, and *Doctor in Arabia*, New York, 1940) resumed this work, and in 1948 Sultan Sa'id, who ruled since his father's abdication in 1932, granted land for the Sharon Thoms Memorial Hospital, now called ar-Rahma. William Wells Thoms, Sharon's son, said of the early years in Matrah, 'We lived in a two-storey native-built house on the sea shore near the private walled Khoja quarter. Father used the first floor as his clinic – here he examined and treated scores of patients in the mornings and did operations in the afternoons. Mother did her best to make the second floor comfortable and attractive. This was pretty difficult in those days when it was hot and the flies and mosquitoes and other insects buzzed around in swarms for we had no electric fans, refrigerators, modern plumbing; nor was it possible to screen more than two rooms against the insects.' Quoted in Wendell Phillips' *Unknown Oman* (London, 1966), Dr William Wells Thoms re-minisced about his first spell in Oman: 'We used to walk out to the date garden, which lay in the middle of the narrow valley beyond the city gate, sometimes in the evening to get away from the smells of the city and to rest our eyes on something green.' He had returned to Muscat in 1939, exactly 30 years after his parents had arrived and worked there until 1970, when he returned to America, where he died within a few months.

Sultan Taimur abdicated in 1932, in favour of his son Sa'id bin Taimur, who had attended Mayo College at Ajmer in Rajasthan. Taimur returned for five months in 1946, but then continued his self-imposed exile in Bombay until his death in 1965.

We shall choose six witnesses to life in Muscat during the reign of Sultan Sa'id, who spent much of his time at his palace in Salalah, and maintained a policy of isolationism and underdevelopment, rigid Islamic orthodoxy and rejection of Western ideas, including education and women's emancipation.

The first is the sailor Alan John Villiers, writing in *Sons of Sindbad* (London, 1940). 'The guns in the Portuguese fort faced the Sultan's palace. There was a mixed battery of ancient muzzle-loaders which included iron guns from India and from Yorkshire, and two beautifully carved old Portuguese guns. Their carriages were rotten and the platform on which they stood hardly looked strong enough to walk on. Muscat had fallen on evil days; there was no shipbuilding, and not much trade beyond a mean local trade from Mutrah across the Gulf to Baluchistan, or coastwise to Sur and Batina. Very few deep-sea dhows

Muscat Harbour, 1931

are now owned there, most of the long-voyage carrying of Oman being in the hands of the Suri and the ancient ships from Batina and the Trucial Coast [now the United Arab Emirates]. Beggars abound in the Mutrah *suq*, and there seem more Baluchi there than Omani. The only big ship sailing out of Mutrah at the time of our visit was a former Persian boom transferred there for the greater freedom of the Oman flag, the boom of Mohamed Kunji, whom we had met in Zanzibar at the mysterious establishment of Sayyid Sulaiman bin Sa'id.'

In December 1955, James Morris (now Jan Morris) left Salalah with the Sultan's entourage and travelled overland via Jaddat al-Harasis, Fahud, Adam, Firq, 'Ibri, Buraimi, Suhar and Sib to Muscat. The story is told in *Sultan in Oman* (London, 1957: reprinted by Century, 1983). At journey's end Morris wrote, 'They stood side by side, Muscat and Mattrah, in neighbouring ricky coves, like two old Arab merchantmen in adjoining berths. Only a protruding spit of land, crowned with fortifications, divided them, and they peered at each other over the intervening hills with a trace of crotchety jealousy. There were some cars in these cities, and a telephone system, and a little power house; a small English community sweated away the years there, and you could buy a camera film or a hair-net in the shops; once a week the steamer put in, on its way up the Gulf or back to India. Yet they remained the most leisurely, unspoilt and charming of the Arabian ports, untrammelled by noisy political aspirations, not yet smeared by the manicured finger of oil. They were gentle and friendly places, very different from the highlands we had just left, where violence was never far below the surface.'

Wendell Phillips made his fourth and last visit to Muscat in January 1958, on his way to head an archaeological expedition to Suhar, enthusiastically supported by Sultan Sa'id. In *Unknown Oman* (London, 1966) he writes: 'The Sultan speaks, reads and writes letter perfect English, is of modest stature, neat in appearance, of a retiring nature and extremely polite and courteous to visitors and guests. He is a devout Muslim, but not fanatical. Though completely Arab, the Sultan has great affection for Great Britain and America and listens to the B.B.C. and subscribes to American magazines. A man of conservative habits, the Sultan owns no Cadillacs, aeroplanes, or yachts but enjoys the simpler life of a man of action. He prefers to drive desert jeeps and trucks himself and is a keen sportsman, excellent shot and avid photographer. The Sultan has one wife, his second, who has borne him three children. The only son, the handsome and extremely intelligent Sayyid Qaboos, graduated from the Royal Military Academy at Sandhurst in September 1962. When Qaboos was a boy, his father the Sultan used to impress upon him that he was like all other Omanis; he must work or starve. He was to have no servants of his own.'

Barbara Wace, writing in *The Geographical Magazine* for May 1962, reached the airstrip of Bait al-Falaj by way of Bahrain, Qatar, and Sharjah. On the way to Muscat by jeep she 'passed strings of camels bringing produce from the mountains, and tribal Arabs, armed to the teeth astride diminutive donkeys, their white robes almost touching the ground, and their black-gowned and masked wives looking rather like dejected blackbirds walking behind. The men were usually crowned with brilliant head-dresses, with the ends hanging down rakishly. As we approached Matrah, we stopped for a fantastic procession. Perched high in a sort of sedan chair, wobbling on top of a camel, sat two young girls, veiled and dressed in orange and gold, and clinging desperately to each other. Around them were women with jewels in their noses and bangles on their arms and ankles, and long pantaloons peeping below gaudy, graceful gowns. It was a Baluchi wedding. One of the girls on the camel was the bride; the other was probably a younger sister. The women were her friends and relations, and they clapped and danced and chattered as they sped her on her way.'

When Barbara Wace left the British Consulate, after a dinner-party, 'it was pitch dark. The narrow alleys were fitfully lit. Going abroad at night without a lantern is forbidden and the few pedestrians cast ghoulish shadows on the white walls. There is another strict rule: you may not go through the town gates after sundown without a signed pass from the Wali (Governor).'

The two heavy-timbered panels of the swinging door of Bab al-Kabir were closed every night about three hours after sunset with a drumroll and three shots from a cannon. Thereafter doors were opened only for the Sultan's own vehicles or for others especially authorized. Pedestrians were able to enter the walled city through a small door, but only if armed with the prescribed lantern. Richard Burton noted in 1853 that 'there is a standing order in the chief cities of Egypt, that all who stir abroad after dark without a lantern shall pass the night in the station-house.' This custom was introduced into Aden by Stafford Bettesworth Haines.

Ian Skeet, author of *Muscat and Oman* (London, 1974; reprinted with an epilogue by Faber & Faber, 1985), lived in Muscat from 1966 to 1968. 'Our house was built, surprisingly, not more than, at the most, 150 years ago; it, and the others like it, look far more ancient, and one likes to feel the presence of the Portuguese in the massive walls, the arched doorways, the wide and tall balcony, the crenellated rim of the roof. It is not easy to establish any date with certainty, for the memories of Muscat are anything but precise, but I ended up with a general consensus of . . .1820 to 1840. The design of these houses is a mixture of Arab, Persian, Indian and African, and they are astonishingly small, in modern terms, for their enormous bulk. The ground floor of ours consisted of an entrance hall, a room for guards, and various store rooms, none really suitable for use as living quarters, all built round a central courtyard or well. The walls, even the inside ones, are of massive thickness, up to at least three feet, and this of course acts as a cooling device for the rooms within. The first, and only, floor was about twenty feet up, and the roof another twenty feet above that. The living rooms opened out on to a wide balcony, and although they were of a good size, there were only five of them. Each had a heavy wooden carved door, set beneath a square carved wooden lintel, into which the door could be chained. The tall wooden arches of the verandah were a reminder in silhouette of the Chehel Sutun in Isfahan.'

Col. David Smiley, on secondment, commanded the Sultan's Armed Forces at Muscat, and wrote in *Arabian Assignment* (London, 1975) that one of his duties was to visit the prisoners at Fort Jalali, 'a veritable hellhole', housing about a hundred inmates, political prisoners and common criminals herded indiscriminately together. Suzanne, Duchess of St Albans, visited the jail and in *Where Time Stood Still* (London, 1980), she described being 'considerably surprised by the unexpectedly open and cheerful look of the place when I went to visit it. The whitewashed cells, which are all shapes and sizes, are above ground and equipped with ceiling fans – looking more like aviaries awaiting the next batch of exotic tropical birds than prison cells. Built on top of the rock, the layout of the fort follows the formation of the outcrop, with the two great towers perched on the highest points, one facing out to sea, the other inland. Steps lead up from one level to the next in various directions: to the governor's tower, the battlements, the rooftops, and a long gallery overlooking the bay of Muscat.' Queen Elizabeth II and her consort Prince Philip, Duke of Edinburgh, paid a visit on the royal yacht *Britannia* during the stay of the Duchess of St Albans, who wrote of the occasion: 'The bay is thronged with boats of every shape and size, all bouncing about on the swell, with flags and yards of bunting fluttering in the wind. A large dhow chugs around ponderously, and an identical pair of green tugs, encrusted with barnacles and festooned with tyres and trailing seaweed, churn about spewing sulphureous, stinking fumes and making a terrible racket with their engines. Another boat draped in bunting is brimming with tiny boy scouts who cling unsteadily to the white rope of their gangway. The Muscat Divers are there in their fibreglass nutshell, and all the private launches are milling around, loaded to capacity with loyal, cheering Britons.'

Sultan Qaboos took over from his deposed father in July 1970, from which the Renaissance of Oman dates. W. D. Peyton states in *Old Oman* (London, 1983) that the old Sultan Sa'id had been 'not only extremely cautious about spending money not yet actually in his treasury but also sincerely concerned about how modernisation and education would affect the people of Oman.' Indeed, the old Sultan, who spent his few surviving years in England, had said, 'The people shall not have what they want, but what I think is good for them.' Having inherited nothing when he came Sultan in 1932, he had conserved some '£100 million, nearly all in oil revenues, in twenty-four bank accounts', to cite Peyton. This became suddenly available to the regime of Sultan Qaboos in 1970. Peyton himself, retiring from Petroleum Development (Oman) Ltd. in 1977, was asked to start a modern secondary boarding and day school, which opened that year as the Sultan's School at Sib.

John Townsend spent the years 1972–5 as Economic Adviser, and wrote the anonymous *Oman* (Muscat, 1972; 2nd ed., 1974) and *Oman: the Making of the Modern State* (London, 1977). 'Any foreigner who had visited the country prior to 1970, and then not again until 1976,' he writes in the latter, 'would be astonished by the differences. His first journey would have involved a perilous landing on a small airstrip surrounded by craggy mountain peaks; his trips around the country would have been by Land Rover on dusty and rough natural surface roads; his accommodation would have been as a guest in a private house. In 1976, he would have landed at a well-designed and constructed modern international airport; he would have been driven in an air-conditioned car on a well-made dual carriageway to a modern first-class hotel. He could see a great deal of the country on modern highways using his air-conditioned car. Before 1970, the traveller would have seen no sign of economic activity other than the ultra-modern oil installations and a pitiful, archaic agriculture. In 1976, especially in the Muscat area and in Salalah, he could be forgiven for thinking he found himself in the middle of a builder's yard, such is the level of construction activity going on on all sides. In his first visit, the traveller would have observed great poverty and need everywhere. In 1976 he sees, unless he is particularly observant and perceptive, only affluence.'

The American journalist Robert Azzi had visited Oman in 1970 and 1971 before the 1972 visit recorded in the February 1973 *National Geographical Magazine*.

The population of Muscat then was estimated at 6,500, compared with Matrah's 16,000. He stayed at the then new Al-Falaj Hotel next door to mud-brick houses and shops 'virtually unchanged from the 17th century.' Qasr al-'Alam had not then been finished, so Azzi spoke to Sultan Qaboos in a modest white house in the old walled city of Muscat. 'He wore an elegant brown aba, embroidered with gold thread, and a gold khanjar at his waist. His head bore the multicoloured turban that is worn only by the Al Bu Said.' "Since 1970 we have come far," the Sultan said as we sipped glasses of the popular iced lime drink called loomey. "A national feeling of working together has developed. This is generating great progress. I am encouraged by the spirit of the returning Omanis. Not only do they bring skills we need, but they make no complaints about the lives they have given up abroad".'

Tim Severin personified the spirit of swashbuckling adventure shown by mediaeval sailors in the Gulf, and by the modern leadership of Oman, when making the 7½–month sea–journey (1980–1) celebrated in *The Sindbad Voyage* (London, 1982) from Oman to China, in a vessel called the *Sohar* (at Sultan Qaboos' request) after Sindbad's legendary birthplace. 'We fitted out *Sohar* in one of the most spectacular harbours in the world, Muscat. On one side, high above her, loomed the walls of Fort Mirani, a soaring fortress originally built by the Portuguese to protect the Bay of Muscat against intruders. Now the claret-coloured standard of the Royal Guard fluttered from the flagpole in the light easterly breeze, as the fort was the barracks of the Sultan's bodyguard. Every morning on the way to work we drove past army lorries brimming with heavily armed guardsmen in berets, battledress and suede desert boots. On the Sultan's birthday the soldiers wheeled out a line of artillery guns, their brass and paintwork polished to a sparkling brilliance, and banged off a twenty-one-gun salute whose roar echoed off the hills around the bay. Aboard *Sohar* we got the full blast, holding our hands over our ears, for the guns were pointed directly at us and we could see the belch of flame from the muzzles and watch the flaming wads spit out. The object of the salute was the Sultan's Palace, an extraordinary creation of pale, umbrella-topped marble columns and huge glass windows, fringed by an immaculate emerald lawn that came down to the water's edge.'

Thomas J. Abercrombie noted in the *National Geographic Magazine* of September 1981 how the traffic had increased. 'Before 1970 there were only a few dozen private cars and trucks in the country; today, some 80,000. I was struck as much by the nature of the traffic as by the density. A wealthy sheikh darted by in a black Mercedes Benz, and a cabinet minister was flanked by a pair of wailing motorcycles. But most of it was workaday traffic: 20–ton dumpers, flatbeds loaded with cement and steel, rumbling graders, Land Rovers and pickups, busloads of Pakistani workers, a giant crane. Oman is, above all, a nation on the build.'

Abercrombie's 1981 interview with the Sultan took place at the Seeb Royal Palace. 'The Sultan wore traditional Omani dress: a black robe over a full-length white *dishdasha* cinched by a gleaming silver dagger, or *khanjar*. His flashing dark eyes were framed by a smartly trimmed beard, grayed at the cheek, and a powder-blue turban. The ramrod martial image I had watched on the parade ground gave way now to the quiet voice and manner of the private man. Yet as we explored the subject of Oman, the talk kept veering back to matters military.'

'We don't want to involve ourselves in the awesome conflict between the superpowers,' His Majesty said, shaking his head. 'Oman must look to its own defence, and we are capable of doing so. But we must expand the armed forces, especially our navy. We need minesweepers, better radar, antisubmarine planes. For this we count on backing from Europe and the United States.'

The Capital Area in the 1980s

The modern traveller in Oman will begin with the capital area, landing at Seeb International Airport, east of Seeb fishing village, and 40 km west of Muscat. A new triple-lane motorway (widened from a dual-lane motorway in 1985) takes you almost due east, past the Seeb Novotel and Ministry of Commerce and Industry exhibition centre, towards Ghubrah, al-Khuwair, Madinat Qaboos, Qurm, al-Watayah, Ruwi, Matrah and Muscat (properly Masqat). This street is named for the ruler, Sultan Qaboos, son of Sa'id, who ruled from 1932 to 1970, son of Taimur, who ruled from 1913–32, son of Faisal, who ruled from 1888 to 1913.

Al-Khuwair is best-known for its three embassies (United Arab Emirates, Egypt and Jordan) and the Ministries of Health, National Heritage and Culture (including the Natural History Museum), Petroleum, Interior, Labour and Social Affairs, and the Development Council. Madinat Qaboos has a modern residential quarter, and a splendid new sports complex. Beyond the Intercontinental Hotel, your eye carries up to Qurm Heights, beyond the nature reserve to the Gulf Hotel, behind which is the sprawling residential zone of Ras al-Hamra. The Oil Museum in the Petroleum Development (Oman) Exhibition Centre stands outside the gates leading to the oil refinery, and no pass is therefore needed to enter it. Al-Watayah is the site of two stadia for the national sport, Association football. Ruwi is the heart of the new business district which is gradually replacing Matrah, the historic commercial capital. Muscat is increasingly the domain of His Majesty the Sultan, with the Royal Palace (Qasr al-'Alam), the Royal Diwan, the Ministry of Finance, Bait Sayyid Nadir and Bait Graiza, Fort Mirani and Fort Jalali, and the British and American Embassies.

Bait Sayyid Nadir

On the corner of Sultan Faisal Street and Muhammad ibn Khalfan Street stands the carefully-restored eighteenth-century house named for its most celebrated owner, Sayyid Nadir bin Faisal, a great-uncle of the present Sultan. I arrived at eight in the morning, as the doors were being opened (it closes at 1 p.m. daily and all day on Fridays). From it you can imagine life in Muscat two centuries ago, when thick walls were necessary to keep out summer heat, and narrow windows allowed

64

House in Muscat, now owned by Municipality, with Bait Sayyid Nadir in background

sea breezes to circulate throughout. The ground floor was then used for servants, stores, and livestock, while the family of Sayyid Nadir lived inside the cool first floor during the day, and could sleep on the roof terrace during the hottest nights of the year. Whitewashed for a cool appearance, Bait Nadir is nowadays filled with birdsong, and the quiet absorption of visitors enjoying the atmosphere of an authentic old Muscati home. No entrance fee is charged; no postcards, guidebooks or souvenirs are touted. As you look down into the inner courtyard, with its waterwell, cannon, and woodcarvings, you can easily imagine yourself in the Muscat of 1860, when Captain Arthur W. Stiffe, R. I. M., surveyed Muscat.

The first case in the courtyard of Bait Sayyid Nadir comprises a range of firearms, such as an Omani gun once owned by Shaikh Yahya bin Imam Salim bin Rashid al-Kharusi and English guns of the last two centuries, together with seventeenth-century Omani swords. The second case shows a modest collection of ceramics from Europe and 18th–19th century China. The third case exhibits pottery types from Bahla, Bilad Bani bu Hasan, Saham on the Batinah coast, and Salalah.

Room 1 shows trunks typical of Oman, state mirrors, Chinese ceramics of the 14th–15th centuries, and beduin ornaments for special occasions. Here you can compare the typical coffee-pot of Oman with its counterparts in Eastern Arabia (centred on al-Hasa), and the northern Gulf. One case has decorations donated by the widow of the Sultan of Zanzibar, Sayyid Khalifa bin Harub (1911–60) and the other shows a range of jewellery, such as toe and finger rings, *kohl* pots, as well as Dhufari head-dress pendants, anklets, and bracelets.

Room 2 is adorned with a pair of Victorian corner chairs. In the first case you will find a camel's shoulder-blade used for writing as recently as the 19th century, even earlier market weights, a leather cosmetic box, and a spoon of 1904. Ornaments in the second include a bedu head-dress and tight necklace from the Sharqiyah, and in the third are necklaces with Maria Theresia coins (the *thalers* being local currency until the early seventies, and a selection of intriguing anklets, enough to weight down the lightest bedu damsel. The fourth case contains powder horns, a tiny silver pipe, and tweezers for taking thorns from the soles of your feet, a constant peril endured by barefoot bedu.

I found Room 3 especially interesting for the letters in the first case (English translations are provided), such as the fine calligraphy of the scribe responsible for a letter of 1886 from Sayyid Khalifa ibn Sa'id, Sultan of Zanzibar, to the Wali of Mombasa. Case 2 has a fine display of bangles, but your eye will be taken chiefly by dresses and silver that once belonged to Princess Salmah, daughter of Sayyid Sa'id ibn Sultan (1804–56), who was born in 1844–5. In Zanzibar she met Friedrich Ruete, a clerk in the German Embassy, left Zanzibar in August 1866, and married him in Aden the following year. They had one son and two daughters before the German died in 1870. She survived him for 52 years, living in Germany and writing her autobiography, translated as *Memoirs of an Arabian Princess* (Doubleday, New York, 1907) by Lionel Strachey. Her son Rudolph produced an interesting survey, necessarily at second hand, of his grandfather's life and times in *Said bin Sultan, 1791–1856, Ruler of Oman and Zanzibar* (Ouseley, London, 1929), which could be read in conjunction with the first-hand *History of Seyd Said* by 'Shaikh Mansur', actually the Italian Vincenzo Maurizi (Booth, London, 1819; reprinted by The Oleander Press, Cambridge, 1984).

From the terrace, the views are to the Royal Diwan, ahead, with Fort Mirani one way, and Citicorp at the side.

On the ground floor of Bait Sayyid Nadir a typical *majlis* or men's sitting-room has been assembled from traditional objects: a Qur'an on its special carved wooden reading-stand, rush matting and cushions on the floor. Before travelling to the Batinah and visiting the ruins of Bait Na'man, see the model of this splendid fortified palace built near Barka in the seventeenth century by Imam Bil'arab ibn Sultan ibn Saif al-Ya'arubi, and used by Imam Ahmad ibn Sa'id, founder of the Albusa'idi dynasty. I liked the nineteenth-century Oman wedding bed, the marble table from the former Qasr al-'Alam, and a great tray once owned by Hafidh ibn Muhammad ibn Ahmad Albusa'idi. Look for the window stucco from the eighteenth-century house of Shihhab ibn Turki al Sa'id rescued when the building was recently demolished. Specimens of woodcarving can be viewed in the open courtyard, and in the courtyard on the museum annex stand two cannon, with cannonballs.

If from Bait Sayyid Nadir you stroll down towards the sea, by the side of the Royal Palace (FISHING AND SWIMMING PROHIBITED), you can find little Yamaha-powered motor boats bobbing near the jetty, crabs scrambling out of your way. Here is a fine view of Fort Jalali, not open to visitors. You might, if lucky, secure a pass from an army officer to look inside Fort Mirani, though this is of course a privilege to be denied rather than a right to be expected. Look for Bait Graiza and the Khor Mosque near here, as you return from the parked cars below Mirani towards Bab al-Kabir, the main gate of Muscat until recently closed at night, after which nobody could enter or leave the capital freely. The old walls, dating to 1626, are now fully restored, like the gatehouse. You can now follow the wall's circumference outside the inner city of Muscat, passing the Kabritta Tower on the way to Bab as-Saghir, the new suq, and the Zawawi Mosque. Here is the other entrance to the city, with Waljat quarter below the massive Fort Jalali, best seen from the interior gardens of the British Embassy. Swivelling left, you can see Mirani above the round sweep of the palace lawns that jut out into the harbour.

British Embassy

A British Embassy of a sort (at first called an Agency) has existed in Muscat since the beginning of the nineteenth century, but we cannot be sure that it lay exactly on the site of the present Embassy, and indeed the earlier building on this site is not recorded before the 1820s. It had two storeys, the upper floor being the Agent's residence, and the lower his office, stores, cells for prisoners, and a post office.

The present Embassy, again necessarily in the Arab style to mitigate climatic excesses, was completed in 1890. Some walls are more than 66 cm thick, effectively keeping out the worst heat and humidity, lessening the effect of repeated shocks by cannon and other weapons, and helping the building to survive in a climate where houses are generally vulnerable. Lord Curzon, Viceroy of India, described the Agency in 1892 as the handsomest building in Muscat which, 'being situated close to a gap in the rocks where a side breeze comes in from the ocean, renders life less insupportable during the appalling heat of the summer months, when the sun's rays, refracted from the glowing rocks, seem literally to scorch, and the rocks themselves are like the walls of a brazen oven.'

I have been privileged to enter one of the traditional early houses of Muscat, built on to the living rock so that the inner walls on the easternmost side consist of sloping, jagged, elemental mountainside, as though one sat perennially at a cool grey picnic in the midday shade: where the rock is not scathed by the dazing sun, it becomes an ally, not a foe.

Fort Jalali, from the British Embassy

The Embassy's cool atmosphere is emphasised by the open courtyard, the veranda opening on to the evening breezes, and the shady staircase, decorated with portraits of Agents and Ambassadors, from the earliest at the top, to the latest, Duncan Slater, who left in Spring 1986. Three of these photographs are of (unwitting) contributors to this book: Lieut. W. M. Pengelley, Agent from May 1861 to January 1862; Samuel Barrett Miles, Political Agent from December 1872 to June 1877, January 1878 to June l879, October 1880 to August 1881, September 1883 to April 1886, and November 1886 to April 1887; and Colonel Sir Percy Zachary Cox (then Major), Agent from October 1899 to January 1904. Other distinguished Agents included Lt. Col. Herbert Disbrowe (Agent from January 1863 to February 1867), who contributed a great deal to the 'Selections from the Records of the Bombay Government,' n.s., no. XXIV (1856), compiled by R. Hughes Thomas and reprinted in 1985 by The Oleander Press of Cambridge as *Arabian Gulf Intelligence;* Colonel Sir Edward C. Ross (then Major), Agent from May 1871 to December 1872, whose translation of Sirhan's *Annals of Oman to 1728* (1874) was reprinted by Oleander in 1984; and Lt. W. H. I. Shakespear (officiating July–November 1906), whose biography has been written by Victor Winstone (Cape, 1976). The first Consul-General appointed, in September 1949, was Major F. C. L. Chauncy. The first Ambassador, Sir Donald Hawley (then Mr.), presented his credentials to His Majesty Sultan Qaboos ibn Sa'id on 22 July 1971.

The United States Embassy (established in 1972) can be seen along a narrow alley to the right, as you face the door of the British Embassy.

Bait Graiza was originally built for Ghaliah bint Salim ibn Sultan, niece of Sultan Sa'id ibn Sultan, about 1835, and later on became the Mission Hospital, being converted to a consular residence when M. Paul Ottavi was made first French Consul. It has since returned to private use, as the residence of the P. D. (O.) Government Liaison Officer, formerly Ian Skeet and W. D. Peyton and former offices of the British Bank of the Middle East, now headquartered in Ruwi. The B. B. M. E., part of the Hong Kong and Shanghai Banking Corporation, opened the first foreign banking facilities in Oman in 1948, remaining the sole foreign bank until 1968, when the Chartered Bank became established, a year before Grindlay's opened their first branch in Oman. The B. B. M. E. was instrumental in setting up the Oman Currency Board in 1972, two years before the Central Bank of Oman was established. Of the 22 commercial banks, eight are local, headed by the National Bank of Oman, and fourteen branches of foreign banks. There are three specialised banks to assist the Omani people with loans and credit: in order of establishment these are the Oman Housing Bank, the Oman Development Bank, and the Oman Bank for Agriculture and Fisheries. The last is particularly crucial in a country where the people depend for their livelihood on the resources of land and sea: it began lending in May 1982 and up to 1984 had made nearly 8,000 loans countrywide. Of these, 3,308 were to set up new farms, 1,874 to improve existing farms, 1,473 for farm mechanization, 878 for fisheries, 359 loans for agro-industries and marketing, and 105 for livestock, according to the 1984 *Statistical Yearbook* (1985).

Matrah and its Forbidden City

The poster outside my room at the Mina Hotel in Matrah blazed 'PEACE: WORLD WAR ENDS AS GERMANY SIGNS ARMISTICE'. It was a framed copy of the *Los Angeles Times* for 11 November 1918. The same quality of timelessness pervades the whole of Matrah. From left to right, out of my window, I saw the new port Mina Qaboos, the covered fish market, bus station, the picturesque *ghanjah*, forever beached here, little fishing-boats bobbing cheekily alongside it, the Corniche with seats, the lofty Matrah fort, and the headland separating us from Muscat bay.

'Atiyat ar-Rahman, the 130–ton *ghanjah*, will inevitably draw you down to the shore, between querulous, mewing seagulls and rusting anchors. It was built at Sur about 1930, and had a crew of nineteen, spending most of its years on the East African trade routes, and John

Jewell photographed it at Mombasa for his book *Dhows at Mombasa* (Nairobi, 2nd ed., 1976). The last *ghanjah* was built about 1952 at Sur, where you can see the other surviving example of this noble ship, distinguished by its square galleon-type stern associated by the British with the invading Spanish Armada.

Where Muscat is formal, neat and quiet as befits a nation's capital, its neighbour Matrah is informal, chaotic and noisy. My shyness responds to withdrawn Muscat; my expansive sense of life and humour responds to vivacious Matrah. It is in Matrah that you wave to friends, chat to strangers, chaff fishermen in the open fish *suq*. Here I turned round to scan the entire waterfront of Matrah, its skyline disastrously (and uncharacteristically in Oman) marred by 'National' and 'Citizen' neon signs. I explored the covered fish–market, with all the Gulf's fresh trove glinting silver in the early morning light: sardines, kingfish, snapper, mackerel, tuna, grouper, barracuda.

None of the fishermen reminded me of the last quarter of the twentieth century, with its computers and skyscrapers. Their dark brown skin has been tanned by a lifetime's spray and sun and wind. Their eyes gleam merrily, knowingly: their day's catch will be sold to this eager throng within a matter of an hour or so, but meantime they grin contentedly, rocking on their haunches, recommending their wares with wide, generous gestures to sceptical bargain-hunters such as blackveiled women or shrewd Bengalis. The fishermen know that the sea's bounty is as infinite as the compassion of Allah. A worried Japanese, short of breath and stature, is critically eying a kingfish. A laughing badu from the interior throws a bag of sardines into the back of his pick-up. An American woman in a grey skirt hands down a ten-rial note to a barefoot fisherman for change. The pandemonium in eleven languages, dominated by Omani Arabic, wanes and surges.

It continues next door, in the covered vegetable market, where the fruits of the earth are being prodded and poked by the same polyglot audience. The vendors call down prices from their podium above the vegetables and fruit to the passing customers. I buy grapes, eating them voluptuously in the presence of papayas, melons, pomegranates, chestnuts, bananas, limes, apples, oranges. A Zanzibari smiles and says 'Good morning, sir!' with a dazzle of white teeth. A Welshwoman steers her infant son in his push-chair, admonishing him in the lilt of the valleys to take care, though it is she who rams into others' ankles. 'Ya Muhammad,' yells an Omani, 'at-tuffah kam?' 'How much are the apples?' The gentle half–darkness gives way suddenly, as I emerge on to the Corniche again, to the blinding light Saul might well have taken for revelation. A ticket-booth advertises a night of Indian music, comedy, and dance at Seeb Novotel, arranged by Kalyanji and Anandji, tickets from 3 rials to 50. Unfortunately I already have a ticket that night for Tom Stoppard's *The Real Inspector Hound*.

Matrah corniche, lit up at night, offers scents and sounds that never pall on those who have known Strait Street in Damascus, the suq at Ta'if, or the Jam'a al-Fna Square in Marrakish. Frankincense, fresh leather cut swiftly into sandals, a bookshop crammed with religious texts and commentaries from Cairo and Arabia where I bought an Arabic edition of Sirhan's *Kashf al-ghumma* (published by The Oleander Press in the translation by E. C. Ross in 1985). Indian tin trunks, Japanese blankets, saffron from Spain. I tried on Omani pillbox caps (priced 1½, 2½ and 3½ rials) to the undisguised amusement of passers-by. Traditional money-changers like Khamis Salam and Sa'id al-Ghazali plied their trade opposite the Bank of Credit and Commerce. A Westerner would head unthinkingly for the bank; an Omani by contrast would relish the chance of a chat with Sa'id or Khamis, after greetings.

On the pavement, a Hindu was doing a roaring trade in prints of the Sacred Heart of Jesus and soft rabbits with floppy ears playing a tune, when pushed, that sounded very like 'Ye Banks and Braes of bonny Doon.' Poor scraggy cats side-stepped the unmindful feet of passers-by. Tiny coffee cups were being emptied at a gulp, in a roadside café. A butcher's shop proclaiming in English 'Fresh Mutton' undermined its pretension with a painting of a black goat. Everything

Matrah. Entrance to the suq

is closed on Friday, you may be told, but on Friday I found the fruiterers of Matrah *suq* as open as the goldsmiths, bookshops and fishmongers.

You can enter Matrah *suq* – one of my most enjoyable experiences in Oman – from Matrah High Street (Essa Saleh Trading Centre – Dealers in Business Gifts and Aquariums), or from the Corniche, by the cloth suq (Hassan Textile Store: English Curtains, Chinese Silk, Kuwaiti Chador).

If you choose the western end, off Matrah High Street, at the first street by the passenger roadbridge you will come across the old building which from 1971 until the mid 1980s housed the British Council building, its balcony still enclosed with wood, its open square still redolent of students with English grammars and dictionaries making their way to a good library with a cool verandah. The splendid new British Council building can be seen in Madinat Qaboos. It is planned to build the new Sultan Qaboos University Library into a National Library.

Following the alley southeastward from the British Council, you can wander through the open Hallat al-Hunud, or Indian Quarter, which ends at the Forbidden City, known as Hallat Sur al-Luwatiyah. In 1980 I explored parts of the imperial complex in Peking known as 'the Forbidden City': much is now, however, open to visitors of any nation on most days of the year, so 'forbidden' has now become a misnomer. The Moscow Kremlin, a name to strike awe, has opened its armoury, churches, and Palace of Congresses. Even the legendary Potala Palace in Lhasa has now entered the tourist circuit for anyone with the fare to Tibet. Such is not the case in the Luti enclave of Matrah. When I read, many years ago, in W. S. W. Ruschenberger (I thought his initials, as a traveller, stood for 'West South West'!), that Matrah's population of 18,000 (in the 1830s) included a thousand Luwatiyah who did not allow Arabs into their walled quarter, as the Luti women went unmasked about their daily business, I dreamed in my romantic youth of disguising myself, in the fashion of that hero Sir Richard Burton, and penetrating the fastness, if it still existed as such.

In 1986 I found that the prohibition against foreigners still existed. I could not go in, but prowled like a hungry wolf outside the thick walls. On the Corniche I spoke to a watchful guardian doing his stint on patrol. The big house towering above it, he told me, with a mosque behind it, used to belong to the Luti Hajji Ali Abdullatif. Now he is dead and the house belongs to his children. On this northern flank I passed in front of Bait Ali Abdullatif, past the Haideri cafeteria and a locked gate, then found the Gold Suq. By standing on the steps of a jeweller's I gazed into the historic ghetto, over a high wall blankly facing the Gold Suq with a walled-up door. Then another locked green door, another walled-up door on the eastern side. The southern wall faces the pedestrian alley which leads out westward back to the old British Council. There I found another locked gate, walls even higher, and then the main entrance on the southern side. I slipped through

Matrah. The Forbidden City (Harrat al-Luwatiyah), from the Corniche

with as much unconcern as I could muster, my pulse racing. My nonchalant wanderings, viewed with some anxiety by a few slow dowagers, took me to one cul-de-sac after another, like a rat in an experiment. To avoid being pelted with stones, manhandled, or hauled up before some Kafkaesque tribunal I returned the way I had come, deliberately losing myself twice and accidentally a third time before emerging (my camera concealed and unused) on the main alley. Many of the houses had no electricity, and there was clearly a water problem. Overcrowding had been solved by the voluntary exile to modern homes of the wealthier denizens. Around the perimeter I turned the southwest corner's watchtower-cum-bastion. Old balconies survive on this western flank, and must be protected by the authorities, for this forbidden city is a magnificent example of the old fortified *sur* (spelt with a *sin*, unlike the city of Sur, spelt with a *sad*) which allowed people and livestock to be protected as in a classical European motte-and-bailey. Turning the northwest corner, I found whitewashed houses and shops facing the harbour, with Haji Baba Oriental Carpets promising – out at the back – illicit entry to the quarters of the *sur* I had not yet seen.

Who are the Luwatiyah and how did they come to possess this residential quarter?

The Luwatiyah and Banians form two Indian communities in the Capital Area. The earlier settlers, concentrating in the walled city of Muscat, were the Banians, led in the late 19th century by the four families of Dowlatgirji Manrupgirji, a Gosain Brahman from Kutch nicknamed the 'Ace' who resided near Mandvi, and entrusted his Muscati business to a local manager; Ratansi Purshottam, a merchant nicknamed the 'King'; Virji Ratansi, a banker nicknamed the 'Queen'; and Damodar Dharamsi, nicknamed the 'Jack', who from time to time farmed the Customs. Other Hindus took their cue from these powerful family businesses, with the result that the names of Dayal Purshottam, Khimji Ramdas, Vallabdas Umarsi, Danji Murarji, and Gopalji Walji are very well known.

The later settlers, concentrating in the walled city of Matrah, were Khwajas (also spelt Khojas), Hindus converted to Islam, in the minority not deriving from Lohanas, but from Bhattias. One of the Bhattia *nookh*s or clans is the Panchlutiya ('Five Martyrs'), which probably gave the community of Luti (pl. Luwatiya) its name. Another theory, found in Muhammad Taqi Hasan al-'Umani's undated booklet *Dalil as-Sa'il*, suggests that the Luwatiyah descend from Hakam bin 'Awanat al-Lat, commanding the first Arab invasion of India, who became Governor of Sind. Hajji 'Ali Sultan, a Matrah resident, claims on the other hand that they were originally Hijazis of the Banu Lu'ayy, who went to India with Muhammad al-Qasim, there being converted to Shi'ism. A Luti I asked about the age of the community said they had been in Oman for four hundred years at least, showing a date on a gate to justify such antiquity. But Calvin H. Allen Jr. concludes 'while it is possible that Luti merchants came to 'Uman four centuries ago, there is no real documentary evidence to support this contention'; the inscription only tells us that the gate was built in the sixteenth century. Humaid ibn Muhammad ibn Ruzaiq's *Al-Fath al-mubin fi Sirat as-Sadah Al Bu Sa'idin* (1858, translated by G. P. Badger as *History of the Imams and Seyyids of Oman*, 1871) is the first manuscript asserting that Luwatiyah lived in Matrah: they were said to be its principal residents during the reign of Ahmad ibn Sa'id (1743–82). But ibn Ruzaiq was writing in the 1850s. Further, the Luwatiyah themselves claim to originate from the Haiderabad in Sind, but that city was founded as late as 1768, the Talpur dynasty which ruled there from 1783 destroyed the Hindu merchants and allowed Muslim merchants to take over trade. And as the Luwatiyah themselves, in a claim for privileges noted by Lorimer (in his *Gazetteer*, I, i, 535), state during a dispute of 1889 that their rights go back for a century (thus, no more), it is reasonable to assume that the Luti dominance of the walled *sur* in Matrah goes back no further than about 1789, probably within the first quinquennium of the Al Bu Sa'idi state founded in 1785. They suffered under Azzan ibn Qais, and it would only have been after the re-introduction of tolerant Al Bu Sa'idi rule in 1871 that a second wave of Luti immigrants reached Matrah. They spoke Khwajki (a compound of Sindhi and Kutchi) within their community, and Arabic outside it.

My circuit of the Forbidden City was completed by passing the Billah Trading Est. Foodstuffs Div. and the seaward main entrance to the *sur*. Two old men, chatting and gesticulating, lounged on guard to

74

prevent me from entering, despite my friendly wave and ingratiating smile. 'Is Muhsin here?' I enquired in Arabic. 'Muhsin who?' (Why did I have to think of a relatively uncommon name?) 'Muhsin 'Abdullah?' 'No.' 'Or his brother Ahmad?' (safer ground). 'No.' Their suspicions of me, well founded, may have been based on a report from within that I had been spotted roaming around earlier. I raised my right palm amicably and strolled off to take coffee in Haideri's. Dusk fell like an efficient confederate. I had noted two narrow alleyways breaching the seaward walls, one dark and one lit by a high ghostly lantern. I tried the first, barred by a youth in Western white shirt and trousers; the second was miraculously unguarded, except by a mangy cat. I slipped through the gap, and stood swallowed in a patch of darkness from watching women at high windows or men passing the well-lit street corner ahead. This is what Burton must have experienced in Makkah al-Mukarrimah; T. E. Lawrence skulking in Dira'a before capture by the Turks. Another cat sped yowling past me. In case a Luti entered by this alley I pressed myself into a doorway which seemed to be abandoned.

In the *suq* at Matrah I again listened as men examined watches in the gold *suq* while their swathed ladies considered bangles. Opposite Lal Buksh Muhammad I talked to the leather sandalmaker Arshad Mahmud from Islamabad, whose sandals fetch a princely 5 rials hereabouts: much more than they would ever cost in Pakistan. I sat at a pavement café not far from the discreet new multi-storey car park, and watched Fanja' and Ahli football teams attack better than they defended, amid 'Ya salaam!' and 'Allah!' from the crowded supporters as a shot whizzed past a goalpost, the 'keeper's acrobatics needless as a lorryload of sand in the Empty Quarter. I felt fortunate to have been born a man in this man's world, where the pallor of my skin did not make me unacceptable. I was taken for a Christian and a Briton, and as such tolerated in a land which has seen a long alliance with the British.

From the Mina Hotel in Matrah I observed the illuminated fort at night, the busy fish market at dawn, and the bustling harbour all day long. Just as close to the *suq* and the Luwatiyah quarter are the Sea View Hotel, or Funduq Shati al-Bahr, at the other end of the Corniche, Al Nahdha Hotel (the name means 'Renaissance', a key word in Oman connected with the period since Sultan Qaboos' accession), and the Khalil Hotel, which is just opposite the bus station. The Khalil is just the place for 'old Arab hands', who remember Oman as it used to be before modernisation. The building is ninety years old, and the hotel has been in existence since the time of Sultan Qaboos' father, Sultan Said bin Taimur (1932–70), whose reign is affectionately recorded in Neil McLeod Innes' memoirs *Minister in Oman* (1987). The Khalil will effortlessly waft you back in spirit to the times of Bertram Thomas or Hajji Abdullah Philby. I should not have been surprised to penetrate the disguise of a Richard Burton slipping through the portals on his way to the old *suq*.

The Mutrah Hotel itself is not on the Corniche, but located midway between the Muscat/Matrah conurbation and Ruwi's business district.

In front of my Corniche hotel, I listened contentedly to the babel of Malayalam, Arabic, Urdu, and watched searchlights flail the dark waters which would lap the disintegrating hulk of *'Atiyat ar-Rahman*, that bright anachronism, as I slept and dreamt of the disguises in which Caliph Harun would wander the warm streets of mediaeval Baghdad, and the galleons billowing over the waves between Suhar and Bombay, Sindbad at the helm.

Ruwi

'It has a watch-tower on a hill above it which at a particular moment of dusk looks like a great owl brooding over the village. Beside the road is a sort of Omani fish-and-chip shop, a large open roofed kitchen where they bake fish whole; you see the donkeys pattering along from Muttrah fish suq, a pannier bag slung each side with the forked fish-tails sticking out of the top, on their way to replenish the grill, and the smell of cooked fish pervades that stretch of road.'

Or it did. This description of Ruwi may seem antique, but it was penned by Ian Skeet as recently as 1967, when Ruwi consisted of a customs checkpoint between Muscat and Oman, a few houses with date-palms, and the military headquarters of the Sultan's Armed Forces, centred on the fort of Bait al-Falaj, close to the civil and military airport of Saih al-Harmil.

Ruwi today, between the Hamriyah flyover and roundabout in the south, and Darsait roundabout in the north, has become the premier business district of Oman.

Ruwi. Bus station

Ruwi. Communications Centre

Sultan Qaboos Street leads to the new airport. Near the shopping centre stands the Bus Station and taxi ranks. Here are the Family Bookshop and Falaj Hotel, the Kodak Shop and Ruwi Hotel, the Musandam Development Committee and the Sheraton Hotel.

Historically, we must see these novelties, tastefully designed and executed as they are in the conscientious traditions of Renaissance Oman, simply as a mirage that can be wafted away. The original Ruwi would have been protected by the fort of Bait al-Falaj, with a larger intent to protect Matrah and Muscat themselves. In the fort the new Sultan's Armed Forces Museum was opened by the Sultan himself in 1986. The Museum Director, Col. Rashid bin Masoud bin Rashid Al Zaidi, took me to see the Curator, Lt. Col. Peter Boxhall, whom I had known for a period of more than twenty years all over the Middle East, Cyprus, and the U.K. In his opinion there must have been forts here from time immemorial, because of the abundance of water, but the first recorded was destroyed in 1648 under the peace treaty between Imam Murshid and Dom Julião da Noronha, then Portuguese governor.

Ibn Ruzaiq mentions another fort there in 1743, when Saif ibn Himyar used it as headquarters from which to attack the Persians. Saif's fort was probably restored by Sayyid Hamid ibn Sa'id ibn Ahmad, who as Regent succeeded his father in 1779. It was Sayyid Hamid who moved the capital of the country from Rustaq to Muscat and he resided there most of the time in Bait Graiza. Bait al-Falaj formed the HQ from which Sultan Sa'id and Captain Perronet Thompson led a joint force of 2,000 Omani troops and 402 British soldiers and Indian sepoys against the Wahhabi Bani bu Ali tribe of the

Ja'alan. The forces were routed in a first engagement and retreated, to attack again in 1821 with larger troops under Sultan Sa'id and Major-General Lionel Smith. The historic fort saw many changes before it provided the stronghold of Sultan Faisal ibn Turki (1888–1913). From 1913 to 1978 Bait al-Falaj became the headquarters of the Sultan's Armed Forces; then Sultan Qaboos decreed that the old fort should become a permanent museum testifying to the loyalty of his armed forces. Members of the Oman rebel movement, opposed to the Sultan's rule, were captured in their dhow in 1962 and brought to Bait al-Falaj for interrogation. The north-east tower was prepared as a cell for two prisoners, with a microphone concealed in the ceiling in the hope that they would give away secrets which could be overheard. Brigadier Colin Maxwell, known as the 'Father of the Force' after more than three decades, ruefully notes that 'in the event, all our recorder picked up was a splendid dawn chorus of cocks crowing, the chirruping of sparrows, cooing of doves, and the general hum and noise of a busy military camp. Faintly, in the background, the prisoners whispering to each other could be discerned, but it was entirely unintelligible . . .'

The new museum retains (and in some cases restores) the nineteenth-century features of the fort, displaying within a selection of

Ruwi. Bait al-Falaj

78

weapons and photographs of the many other forts and fortified places that form so remarkable a feature of Oman's physical identity. The display of military might outside the museum pales into insignificance beside the nuclear arsenals held by the superpowers, but the aircraft and artillery constitute vivid testimony to the men who gave their lives for the country. Enjoy the *masbah*, or bath, on the western side of the fort; the charming restored mosque; fine original doors, and carefully-tended *shawisha* trees near the walls.

Ruwi was settling to its evening. Traffic on the roundabout had gradually evaporated with the end of the working day. The bus station stood stagnant with expectancy. Lights flicked on in the Sheraton and the Ruwi Hotels. The British 'pub' near Reception in the Ruwi was becoming boisterous with back-slapping and guffaws. An earnest white-uniformed electrician from Kerala was mending a light-switch near the restaurant. I waved to a Bombay-born friend (it was the day that his native city had officially renamed itself Mumbai to his hysterical merriment) and sauntered downstairs to the buffet, laid out in the Indigo Room, next to the tiny theatre where the Muscat Amateur Theatre was playing Tom Stoppard's *Real Inspector Hound* to the traditional packed house. All performances had been sold out before the run began.

'Has anyone seen Basil?' gushed a Baroness above the eager conversation.

'No,' instructed a lady with warts to her incipiently-bearded husband, 'we must have known them nearly ten years, and we dislike them more and more *all the time*. Now remember that.' I could have been in a diplomatic function at Vienna or Singapore: those intruders unlucky enough not to be British (like a bleating Dutchman, or a gigantic Finnish blonde pressed against her sweating escort) were more at home in current English slang than the old Muscat hands who said 'old thing' and 'jolly good.'

'Who's that woman with the red striped dress? It's June!' 'No, it's Judith.' 'It's June.' 'She's turning round – I don't recognise her at all.' 'Neither do I.' 'Funny, I thought I knew everybody.'

'Can't think where Basil's got to,' complained the Baroness.
'I like your hair. Where did you get it done?'

'Farasha's. I found out last year I was going bald, and suddenly it started to grow quite thickly, so I'm *revelling* in the fact that I've got hair.'

'I can't imagine where Basil is.'

'Well you see it's got this amplifier, this Bang and Olufsen thing, so the hi is *completely* fi.'

'Oh, *here's* Basil!'

All over the Capital Area, from Madinat Qaboos to Muscat, the evening divides. Quietly sumptuous dinner-parties; a riotous Indian evening at the Novotel; hurried telephone calls to Dubai, Abu Dhabi, Bahrain; squash at the Intercontinental; departures to Europe and Australia at the airport. Shops rattle their shutters and they clang the day to sleep. The muaddin calls the faithful to witness that there is no God but Allah, and that Muhammad is his Prophet.

Sultan Qaboos University

A World Bank report of 1973 stated that the Omani Ministry of Education had only two staff members with university degrees, and only five who had completed secondary education. It is some indicator of the enormous strides taken by Oman in recent years that a university could be planned, and completed for the start of the 1986–7 academic year. I received an invitation to visit the Sultan Qaboos University at al-Khawdh, south of Seeb, while the finishing touches were applied to the staff housing and main library. With its own farm and botanical garden, its own mosque and teaching hospital, the University has arisen from scrubland for an initial contract price to Cementation International in 1982 of £215 million, plus a further £65 million to a U.K. consortium led by Philip Harris International, for teaching aids and other equipment.

The world's newest university will cater to about 580 students in its first year, and by 1990 will have a capacity for nearly 3,000. With teaching, administrative and ancillary staff, the university town will soon number ten thousand, and will immediately become self-sufficient in water, gas and electricity.

All students will be required to take a foundation course in Arabic, English, and Islamic and Omani culture before taking up studies in the Colleges of Education and Islamic Sciences, Engineering, Science, Agriculture, and Medicine. The largest will be the College of Education and Islamic Sciences, as almost half of all students will be training as teachers to fill some of the places now taken up by teachers from Jordan, Egypt, Lebanon and many other Arabic-speaking nations. The new Teaching Hospital, opening in 1988, will avoid duplicating as far as necessary facilities at the Royal Hospital at Ghubrah, which opened in 1986.

From the point of view of the visitor, the most striking elements of the campus, apart from its sheer scale, will be the mosque, aligned like the whole of the central university district towards Makkah al-Mukarrimah, and the split-level walkways delicately reminiscent of the segregation of sexes throughout Islamic life: the men below and the women above. Eight men's hostels for 1,200 students stand north of the campus, and five women's hostels for 750 students south of the campus. Centrally dominating all faculties is the Main Library (there is a separate medical library), which will be the foremost academic library in Oman for several decades to come.

The University was announced by Sultan Qaboos in his National Day speech of 1980 as a gift to the nation: it was planned to open in six years. And it did.

Museum of Natural History

Just coastward from the main roundabout at al-Khuwair, on the Seeb-Muscat motorway known as Sultan Qaboos Street, you can take a slip road for the Ministry of National Heritage and Culture. In the main Ministry building a variety of handicrafts are offered for sale, such as pottery from Bahla or textiles from Suma'il, and Ministry shops are open during the mornings also in the Muscat Intercontinental (not in Muscat but on the beach close to al-Khuwair and Qurm), in the Gulf Hotel – owned by Gulf Air, with affiliates in Bahrain and Qatar, and in the al-Falaj Hotel in Ruwi. In the next building the Museum of Natural History is open during normal government hours, with captions in both Arabic and English. Interestingly, the emblem of the museum is the caracal lynx, adopted to symbolise the fearless courage and proud bearing of Oman. A lynx is found in a showcase with an adult male Arabian leopard. The Asiatic cheetah has probably disappeared from Oman, and the lion, *Panthera leo*, was last reported in Arabia in 1950, but if you explore desert regions you may one day glimpse the Arabian sand cat, *Felis margarita harrisoni*. The sociable white *Oryx leucoryx* roam freely in the flat, stony Harasis country of central Oman, having been reintroduced to the wild in 1982 from a base-camp at Yaluni: two herds of fourteen each roamed the desert in 1985. The Wadi Sarin Wildlife Reserve in the Jabal as-Aswad was established ten years earlier to protect the Arabian *tahr (Hemitragus jayakari)* and gazelle. A royal breeding centre west of Seeb concentrates on augmentation of Arabian leopard (found in Asir, Saudi Arabia), *tahr*, the Arabian wolf *(Canis lupus)* and the striped hyaena *(Hyaena hyaena sultana)*. Bird sanctuaries now exist at Daymaniyat islands, off the northern coast of Oman halfway between Seeb and Barka, out of bounds during the breeding season between May and October; near Suhar, on the northern Batinah coast; and close to the city of Salalah, in Dhufar. The Sultan Qaboos Public Park and Nature Reserve at Qurm (overlooked by the Gulf Hotel) consist of a park in which indigenous vegetation is preserved and representative trees and flowers from the rest of the country are planted, over one third of the territory, while the other two-thirds will be a protected forest of mangroves, *Avicennia marina*. The environment here, together with the nearby beach and cliffs of Ras al-Hamra', is affected not only by sea action, and a small spring-fed stream entering the sea by a shallow tidal channel, but also by the Wadi Adai, draining through the southern part of the reserve. The mangroves form part of a delicate ecosystem involving birds and marine creatures of numerous species.

Excavations at Qurm have centred on an oval-shaped mound called RH5 extending 90 metres by 45 metres on a north-south axis on the cape of Ras al-Hamra', a calcareous Tertiary terrace. Dr Paolo Biagi and Dr Renato Nisbet have written 'Some aspects of the 1983–5 excavations at the aceramic coastal settlement of RH5 (Qurm, Sultanate of Oman)', in *Orientalia Romana*, vol. 7 (1986). They note that, at least in the specific site excavated, settlements can be dated from 3530 B.C. to 2810 B.C., the dominant lifestyle being that of fishing and mollusc-collecting. Other activities include hunting gazelle, marine mammals and turtles, and rearing sheep and goats. Feline and equine bones occur, as does the earliest trace of sorghum in the Arabian Peninsula. The absence of polished stone implements such as chisels, adzes and axes leads to the conclusion that boats would have been elementary in design and construction.

Since many of the islands of Oman are barely accessible, take this chance to explore their wildlife vicariously: farthest north the Quoin Islands (in Arabic as-Salamah wa banatha), Salamah being the largest and highest, Didamar (Little Quoin) with its lighthouse, and Fanaka, autumn nesting-place of sooty falcons; the twelve Musandam islands, eleven of which are uninhabited and so attractive to seabirds; Suwadi and the Daymaniyat archipelago near Muscat; Fahal with its sooty falcons and pallid swifts; the turtles and seabirds of Masirah, 96 km. long and 16 km wide, Hamar an-Nafun; and the Kuria Muria islands with ten thousand pairs of masked boobies and, on Hasikiyah, a late summer colony of Socotra cormorants.

The Natural History Museum shows the amazing contrasts of life in Oman, from the crowned sand grouse *(Pterocles coronatus)* of the Interior to the nocturnal gecko *(Phyllodactylus elisae)* of Musandam. Here are the great junipers of Shuraijah in the Jabal al-Akhdhar; the four species of turtle found between Ras al-Hadd and Dhufar, the hyrax, ibex, leopards and eagles in wadis near Hasik, the large grasshopper identified in Jabal Qara in 1977, and named *Omania splendens*, and the ibis, flamingoes, grebes and osprey of Salalah's bird sanctuary.

By appointment you can inspect the growing Herbarium, and the shell collection, with over 6,000 specimens from more than 500 of the 570 known species in Oman. A botanical garden is planned.

The Oman Museum

At Qurm roundabout take the southern road and follow the signs to 'Information City' up a steep hill into a compound including the radio and television stations. A sign *Wizarat al-Alam* (Ministry of Information, in Arabic) is followed by a sign M.O.I. 0.5 km. on a kilometre-post.

Though smaller than the National Museum of Kuwait, for example, it is supremely interesting and must rank with the very different Bait Sayyid Nadir and Matrah's forbidden city as one of the three essential stages of any capital area tour.

Inaugurated in 1974, the museum consists of two floors: a general introduction to the land and people, with archaeology and prehistory; and on the upper floor the country's Islamic heritage, architecture, arts, crafts and weapons.

The earliest stone tools so far collected date to the Middle Palaeolithic, which in the case of Oman refers to the epoch from 70,000 to 30,000 B.C., but it was during the Bronze Age, in the 3rd millennium B.C., that Oman entered recorded history by participating in trade with Dilmun (Bahrain), Mesopotamia and the northern Gulf, and Meluhha (Indus Valley). The Harvard Archaeological Survey identified 17 settlements of this period in Oman, from the necropolis at Bat, near 'Ibri, to the copper region of Wadi Samad. Dhufar's frankincense exports created the conditions for the flourishing town of Sumhur (established as a colony of Shabwa about 100 B.C.), then the eastern border of Arabia Felix.

Suhar's dominance of the Gulf reached its apogee during the 7th to the 10th centuries, as we know from T'ang painted pottery found on the site. Qalhat replaced Suhar as the mercantile focus of the Gulf coast in the 14th–15th centuries. The Portuguese took Muscat in 1507 and constructed their forts Mirani, Jalali and Matrah to secure their hold on these distant strongholds. In 1650 the Portuguese were expelled by the Ya'arubah dynasty ruling from various capitals in Jabal Akhdhar (Rustaq, Jabrin, al-Hazm and Nizwa from time to time). The Imams, Sayyids and Sultans of the Albusa'idi dynasty have governed the country since 1741, at first from Rustaq, then from Muscat, beginning in 1779 up to the present day.

Oman has always been divided into two unequal parts: the long coastline, with its tradition of seamanship, boatbuilding, maritime trade, and small-scale fishing; and the introverted hinterland, Ibadhi in religion, agricultural and pastoral in activity. The Oman Museum has a splendid model of a *boom* in its section demonstrating ships and the sea in Omani life, and illustrations of the ships described by Will Facey in his *Oman: a seafaring nation* (Ministry of Information and Culture, 1979). Here are the *bajala* (no longer built), the fishing and cargo boats called *sambuq* and *shu'i*, and the smallest of all: the hollowed-teak *huri* and the ancient-style reed-boat called *shashah*.

When the Portuguese arrived in 1506 they introduced nails and the transom stern, giving rise to new types of boat, such as the *ghanjah* (like the Indian *kutiyyah*) and the *galeota*, arabicised as *jalibut*, which was used as a pearling vessel elsewhere in the Gulf, but in Oman primarily as a one-master deep-sea trader.

On the first floor, the Oman Museum displays important MSS. connected with Oman history, such as a part of Sirhan's *Kashf al-ghummah*, an autograph MS. of the compilation by Salil ibn Ruzaiq, and a copy of part of BM. Or. 6567 by the Omani twelfth-century poet Abubakr Ahmad ibn Sa'id as-Sitali. Architecturally, mosques in Oman cannot bear comparison with the great buildings of Qayrawan, Damascus, Cairo, or Fez, because Ibadhism favours simplicity rather than ostentation. So it is with the palaces, such as that of Jabrin, comparable with the Safavid palaces at Isfahan, perhaps, that Omani architecture reaches its zenith. Such fortified houses at al-Hamra', Rustaq and al-Hazm became power-centres from which local rulers could spread a limited degree of stability in troubled times against foes both internal and external. Here are the gems of military architecture, too, such as the castles at Bahla, Nizwa, and another near Sur. Here too are the seignorial mansions of Muscat.

As regards arts and crafts, the Oman Museum displays incense-burners from Dhufar, drums, pottery from Saham and Bahla, silver and gold made in Rustaq, bracelets from Sur, and copperware from Nizwa. In the weapons room you will find breechloaders and powder-horns; swords, daggers and knives with pre-eminence awarded to the *khanjar* which has come to symbolise the Sultanate; rifles and bullet-moulds.

Al-Bustan

Al-Bustan Palace Hotel, near the village of the same name on the shore south of Muscat,can be reached by the coastal road via Sidab or through Ruwi along Wadi Kabir. Created by a building consortium from Cyprus, France and Italy, it is owned by the Ministry of Commerce and Industry and managed by Intercontinental. In addition to the six public car parks, there is a helipad near the beach. Around the octagonal centre stand a ballroom, auditorium, swimming-pools for adults and children, and an open-air theatre. Levels 1, 2 and 3 comprise an international conference centre, with gymnasia, jacuzzi, sauna, squash courts and tennis courts. The lobby, on the fourth level, leads to a breathtaking atrium covered by a dome 38 metres high. One of the most luxurious and impressive hotels in the world, Al-Bustan Palace opened to the public on 1 December 1985, after inauguration as the residential quarters of visiting royal delegations during the Gulf Co-operation Council's summit conference.

On the roundabout near the hotel, the dhow *Sohar*, immortalised in Tim Severin's *The Sindbad Voyage*, moors at its last anchorage.

At Haramil we steered between cats, chickens, goats and barefoot urchins to park on the beach. Fishermen displayed their catch, and began with me that generous dialogue I found to my astonishment wherever I travelled in Oman. Though obviously of different race, colour, and creed to them, I was never made to feel an intruder into their working environment, requiring permits, appointments, and special arrangements, but a guest welcome at any time to share their tea and views, their interest in the world at home and outside. Previous travellers, especially perhaps those in the nineteenth century who felt that in some way Christianity, medicine, education, the British way of life, and the English language fused into an indissoluble amalgam clearly superior to all other societies, approached Oman much as we might think interplanetary intelligences would visit us today. Even the great Wellsted could be guilty of the unbelievable solecism of appointing a Persian *shi'a* as his interpreter among the Ibadhis of interior Oman! He at least possessed the saving grace of indefatigable curiosity, whereas most visitors from Britain goggled at the unfamiliar, and either laughed in embarrassed ignorance at what they could not comprehend; or felt it incumbent upon them to change the Omani's way of life to correspond more closely with that recognised by the Victorian English as 'civilised'.

Nowadays, with the gradual enlightenment provided by social anthropologists and comparative studies in politics, religion, sociology, and geography, we can see that each society adapts intelligently to its environment without any need for outside interference, except in the case of natural disasters, such as earthquakes or prolonged drought.

Hot Springs

On 23 November 1835, Lieutenant James Raymond Wellsted 'set out on a visit to the hot springs of Imam 'Ali, which are situated on the sea-shore, about seven hours to the westward of the town. Although the cool season was so far advanced the day proved excessively sultry, and when we pushed off in our boat for Matrah, notwithstanding a fresh breeze was prevailing outside, yet within the cove it was a perfect calm, and the heat thrown off from the sides of the mountains, along the base of which we were gliding, felt almost overpowering.'

The donkey track from Muscat to Matrah was so rugged that, though barely a mile as the crow flies, the towns were connected more frequently and easily by boat than by land. From Matrah Wellsted and his party proceeded by camel, noting some gardens and water-wells 'at the village of Rooah', or Ruwi.

'After sunset,' writes Wellsted in *Travels in Arabia* (London, 1838), 'we continued, occasionally, to pass groups of Bedowins, who had withdrawn some short distance from the road, and were seated with their camels round a fire. It seems customary on these occasions for neither party to proffer a salutation, which is contrary to the practice

Baushar. Fort

when they meet in the day, for then they exchange several sentences.'

The following morning Wellsted visited hot springs, though (despite his confidence) it is not clear whether these were the springs of Baushar, Ghala, or Imam 'Ali, the last also known as Hammam 'Ali. 'The water gushes with much violence from an aperture at the base of a hill of clay ironstone. Veins of a crystallized quartz run in a diagonal direction through the rock, and large fragments have been dislodged from it.' The local people he found would drink no other water. 'Neither its heat, nor any other quality which it may possess, prevents its nourishing the surrounding vegetation. As a cure for cutaneous and other local disorders, these waters enjoy a great reputation amongst the Bedowins and town Arabs, the former frequently undertaking long and painful journeys from a great distance in the interior, that they may remain and use them for several days. Although the temperature was so high, yet I witnessed the submersion of several of the patients, who were kept under the surface by force for some time. One, an old man of eighty, was so much exhausted by this rough treatment, that he

Ghala. Hot baths

appeared in a dying state: yet I was told that, if he lived, the operation would be repeated after an interval of two hours, for the natives believe if the waters fail in producing their desired effect, it is only because they have not been used sufficiently often. A few yards from the bath there is a small mosque, in which an old priest resides, who is ever ready to assist with his prayers those who may require them.'

My impression, from Wellsted's text, is that he visited just the hot springs of Baushar, where a mosque can be seen today. My photograph shows Baushar's fort; a modern house with *barasti* enclosure adjoining, date-palms and hill peak.

Lieut. Stiffe seems, from the description he left, to have visited not Baushar at all, as he states, but just Ghala, where the waters are hotter, and there are modern bath-buildings (one marked in Arabic and English ONLY FOR LADIES). This is what Stiffe writes about his visit to the hot springs, accompanied by my photograph of the pools and palms of Ghala in 1986 in the background behind a portrait of a local resident who had shown us the baths, having completed his own ablutions.

I could not resist a nagging suspicion that Wellsted's note about the 'hot springs of Imam Ali' related neither to the lukewarm waters above Baushar nor the warmer springs used by the people of Ghala: nobody had mentioned 'Imam 'Ali' or 'Hammam 'Ali' at either spot. So I retraced my steps to the Ghala roundabout and followed the line of the mountains away from Baushar, assuming that if the waters became significantly hotter as I wended my way westward, Imam 'Ali, with the hottest springs of all, must lie in the same direction. I came upon a group of Indians bathing in an enclosed building over a *falaj*. 'Is the water hot?' 'Yes,' they replied, splashing and beaming, 'come in!' 'Is this the bath they name for 'Ali?'

'No.'

'Do you know where those baths could be?'

'No.' Then a dark brown, supple, tiny figure pranced out before me like a classical Indian dancer, knees bent, teeth flashing, eyes gleaming black as coal, arms aristocratically posed for effect.

'I am Bangladeshi, sahib,' he announced with a mixture of deference and pride. 'I have two Bangladeshi friends.'

This seemed hardly relevant to my question, but as always I listened patiently for the woollen discourse to unravel.

'They are making new well,' my informant added simply, engaging another regal stance by a swift adjustment of shoulders, elbows and wrists. 'At place they call Hammam 'Ali.' I whooped with the lack of dignity that characterises my absorption in all Omani wonders. 'Where is it?'

He pointed down the road I had come. 'Turn off right, get out of your car, because no vehicle can traverse the rocks and boulders to Hammam 'Ali, and walk to the head of the valley. There, where you can walk no longer, is the hot spring of Hammam 'Ali.'

I followed the newly-cemented *falaj* that curved round the mountain, with a lantern now as night fell shiny black, pinpricked with innumerable stars. Steadily climbing, stumbling in my eagerness to reach the fastness that even the great Wellsted had never found, I picked my way among the massy fallen rocks, sometimes along the wadi bed, sometimes for safety on top of the level *falaj* heights. Then, after about twenty minutes, the *falaj* curved left as the wadi narrowed, and I saw an echoing lantern. It stood on the rim of a new well. 'Namaste,' I called out in Hindi. 'Greetings!'

Two bodies scurried up a ladder. Two eager faces smiled and ducked beside the still lantern. Nothing else was visible except a swathe of *falaj* illuminated by my own swinging lamp.

'Is this the hot spring of Hammam 'Ali?'

They nodded in the usual tolerant amazement of Asians on encountering a peculiar Englishman scrambling among rocks in the middle of the night. I touched the water, and withdrew my fingers immediately: they had been scalded. The water temperature must have been at or above boiling point! The springs gushing forth here were indubitably much hotter than those at Ghala experienced by Stiffe, which were again much hotter than those shown to Wellsted.

LEFT Muscat for Muttrah in a canoe, or bellum, as the Arabs call it, the usual conveyance in fine weather, the pass through the hills being so rugged and winding, that it is only used in blowy weather, when the canoes cannot ply, or by those too poor to be able to afford the luxury of a conveyance.

These canoes are very large, many of them carrying twenty or thirty people comfortably; they are made out of one tree, and all come from the Malabar Coast. The one I went in was three feet in breadth, and, perhaps, thirty-five feet long.

It being morning, we pass great numbers of canoes filled with well-dressed Arabs, most of the merchants of Muscat living at Muttrah, and going to Muscat every morning to carry on their business. As every body sits down in the bottom of the boat, with only the head above the gunwale, the bright-coloured head-dresses present a rather comical appearance.

The distance to Muttrah is about two miles, round a succession of bold rocky points, each with a fort on the top of the nearly perpendicular cliffs; and the boatmen, who are as usual Sídís, salute each point, as we pass it, with the formal Mussulman salutation.

In the little bays between these points are a series of villages, which come in sight round one corner, and disappear round the next, like a shifting scene. The sea being deep, carries its blue colour close up to the little white sandy beaches, on which all these villages are built. I imagine this spot is unequalled for wild, romantic effect, the hills rising suddenly from the sea to twelve or fifteen hundred feet, with the most irregular, fantastic outline imaginable.

Before starting, the British Agent at Muscat had supplied two letters from His Highness Syud Thaweyni (commonly called by the English the

MAP

To illustrate the Paper on the hot springs near Muskat,

prepared for the

BOMBAY GEOGRAPHICAL SOCIETY,

by

LIEUT. ARTHUR W. STIFFE, I.N.

From notes made during an excursion in December 1869.

Nautical or Geog.ᶜ miles.

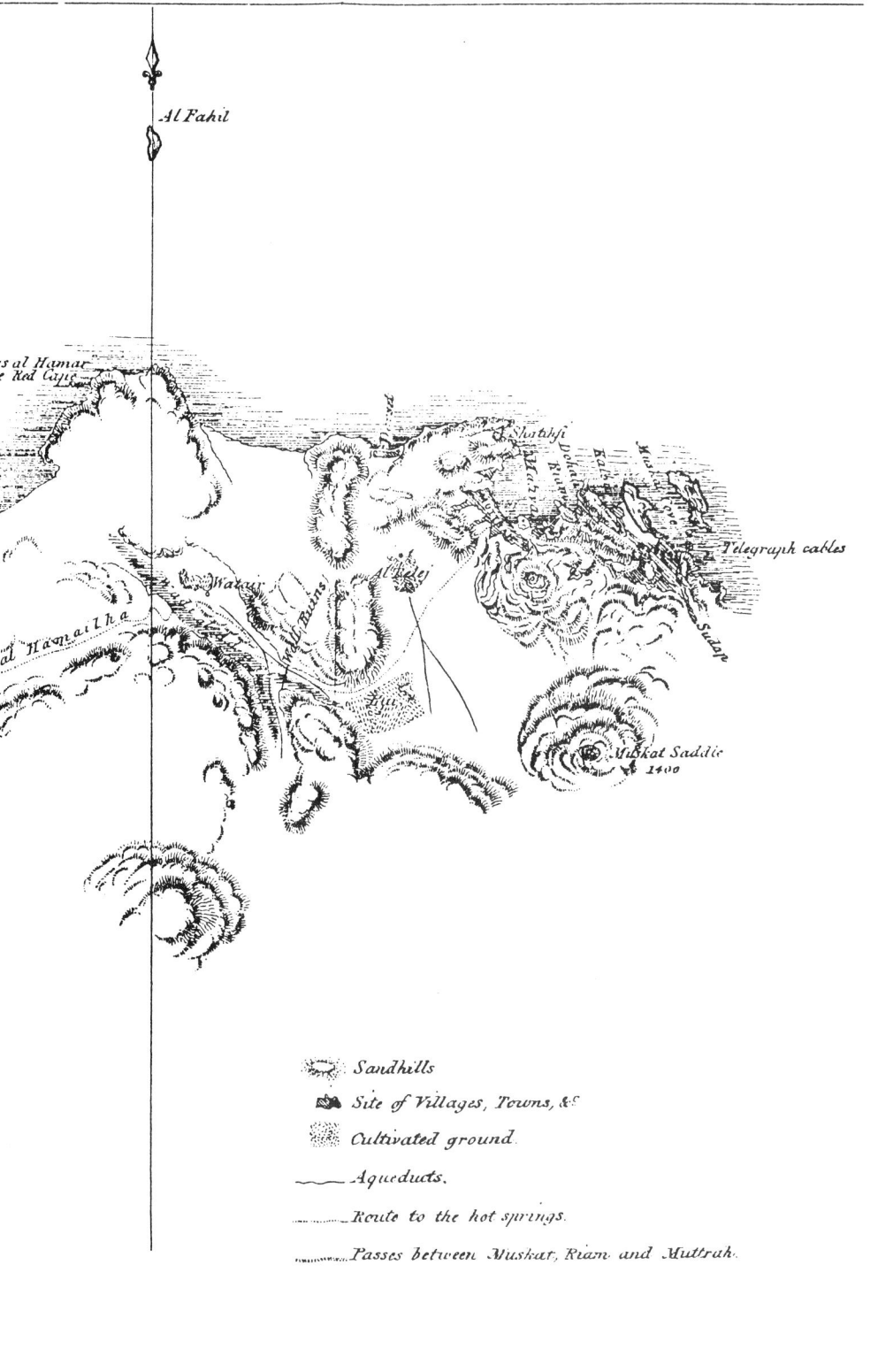

Al Fahil

as al Hamar
e Red Cape

Suttfi

Muhara River

Sudan

Telegraph cables

Watar

Wali Bains

al Hamaiha

Muskat Saddle
1400

Sandhills

Site of Villages, Towns, &c.

Cultivated ground.

_____ Aqueducts.

.... Route to the hot springs.

▬▬▬▬ Passes between Muskat, Riam and Muttrah.

Imám), one to the Wálí, or Governor of Muttrah, the other to the authorities in the village near the hot spring.

The Wálí of Muttrah, Rasás-bin-Bakhín, is a Sídí, and, by descent at least, a slave; but, as not unusual in these countries, has risen to the post of governor of a town of about thirty thousand inhabitants. He is a middle-aged, dignified man, and, I believe, much respected by the Arabs.

This gentleman is to provide me with a horse, and donkeys to carry kit, as well as an escort of three Arab soldiers. I am afraid the wishes of the owners of the donkeys are not consulted. I did endeavour to remunerate them on my return, but doubt if the money ever came to the hands of the proper people.

We get away from Muttrah at about 2 P.M., and wind through a pass in the hills behind the town, the only road, I believe, into the interior. It follows the bed of a watercourse, which must discharge a large body of water during the rains, that must pass through the gate of Muttrah, and on through the town to the sea. About half a mile of the pass, and we come on a small plain between the hills, across which the road passes.

Half a mile on the right is the village of Al Felej, so called from an extensive subterranean aqueduct. These are very common in Arabia and Persia, and are too well known to require description. This one extended, with its branches, for two miles or more. There is here an imposing looking fortified residence, belonging to some relation of the Sultan, with a prodigious flagstaff.

No residence of any one connected with a reigning family in Arabia is complete without a flagstaff, with topmast, and sometimes a third or top-gallant mast, and always unpleasantly out of the perpendicular. On approach, the fort is found to be dilapidated, and the cannon in a state of rust and decay, which European artillerists would hardly credit. A fine grove of trees, the result of the water from the aqueduct, has, however, a singular beauty in the midst of so much barrenness.

Across the plain to Rúí, a small hamlet, at a point where we enter a gorge in the hills, inhabited by tillers of the soil. Here is plenty of good water in wells, which are about fifty feet deep : the first twenty feet is walled round, the soil being loose, but below that depth they appear cut through a coarse conglomerate rock. (The depths are only estimated). At least half a dozen water-drawing machines are creaking away. Here are grown nearly all the vegetables which come to

Muscat—onions, lettuce, radishes, brinjalls, sugar-cane, maize, &c. There are some fine mango and tamarind trees, and many dates ; in fact, the valley for half a mile is a regular garden. Then the cultivation suddenly ceases, and we find ourselves in a barren watercourse, about two hundred yards wide, strewed with boulders and pebbles, while precipitous hills rise on either hand. Rúí lies at the entrance to this watercourse, which is called Wádí-al-Adai.

Following the Wádí to the north-westward, at a distance of half a mile from the cultivation, we come to the remains of a strong wall which has been built across the valley—a distance of about two hundred yards : nothing now remains beyond the foundation, the piers of a bridge which allowed the waters to pass, and an arched gate, which latter is in tolerable repair, the wall ends at the hills at each side being built as far up their side only as they are accessible.

On down the Wádí to north-westward, about two miles, to Watair. Having received several accessions, the valley is here about a quarter of a mile wide ; the hills also on each side are much higher than at Rúí, where they were only from two to three hundred feet. It must be a grand stream in the rains, judging from the size of some of the boulders : now that it is dry, it much resembles a gigantic railway cutting.

At Watair is a fine building in the eastern style, with a date-grove. It is a summer residence of the Sultan, but out of repair and very dirty inside, with the usual carved-wood trellis work, stained glass-windows, &c. An aqueduct is in course of construction to this place, all the way from Rúí, with a branch extending to a great distance up another Wádí. This watercourse goes on in a north-west direction to the sea, but I could not make out its mouth. I imagine it to be somewhere in the bay westward of Ras Hamar.

Here we leave the valley, and up the hills on the west side, through a pass of no great height, called Akabat-ar-Hamaitha. Here the escort ought certainly to have come under "Martin's Act," for they insisted on riding up the hill two on one donkey, in heavy fighting order, with matchlocks, swords, and shields, &c., one of the proprietors of these useful animals having contrived to abscond at Rúí while watering.

The road now winds along the north face of the mountains, which here take an east and west direction, having low, rocky hills on the right hand, just sufficiently high to intercept the view of the sea. After about three miles of rough ground, we emerge from the pass, and get a view of the sea, distant about two miles, with the picturesque Island Al Fahil, and the great bay called Jubet Hail. Here we halt, while the escort perform their afternoon devotions: they do not appear particularly attentive, stopping now and again to talk, or look after the donkeys, and then going on again with their prostrations, in ludicrous alternation.

From this point the road bends round to the south-westward, keeping close to the hills, which recede from the coast and increase in height as we proceed; the ground also becomes sandy, and high sand hills shut out the view of the sea.

Three miles of this, and we reach a cultivated plain, dotted with forts and date-groves. This is the district of Bósher, which it is satisfactory to hear, as the evening is closing in. A great deal of drumming and firing of matchlocks is going on at one of these forts : some great man is being married, and there is a great concourse of guests. As we pass, they are very anxious we should partake of the marriage feast, which is embarrassing, as presents would, of course, be expected, and I am unprovided for such a contingency.

We excuse ourselves on the plea of its being late, and go on across several watercourses. Most of these villages seem to be called Al Felej, or, at any rate, the escort say so.

About three miles through this valley, and over a low ridge of hills, and we reach the village of Ghulla, where the springs are. By this time it is nearly dark : we pass through the village, knock up the owner of the letter of credence, and are forthwith accommodated in a house belonging to the Sultan. It is an agreeable place for hot weather, without doors or windows; but the nights are now chilly—thermometer at 64°; and the musketry and drums in the next village are distinctly audible all night. A wood fire is found objectionable, there being no arrangements about smoke.

The hot springs, five in number, issue from fissures in the rock at the foot of a mountain about 3,000 feet high, chiefly of volcanic formation, all within a space of about two hundred yards. Four of them are not much larger than can be stopped with the hand, and at these there is no perceptible deposit; the fifth, however, which is considerably larger, deposits near its source a white earthy substance in a stalactitic form. (I have shown a piece to Dr. Carter, who considers it to be sulphate of magnesia.) The natives call this spring the " cow " from this circumstance. This spring discharges about as much water as a five-inch pipe would. The temperature of the water at this spring was 100° Fahrenheit.

The spring most frequented, and the one considered to have such marvellous properties, is another and smaller one, but which has a temperature of 115°. It is drunk in large quantities by the sick and lame, who abound at the spring, and great quantities are carried away. The people have great faith in its virtue, but I could not detect any specific taste : it appeared simply hot water, and it is very limpid. I have, however, brought a bottle well sealed to Bombay, should the Society consider it desirable to have it analysed. It did not appear to contain any fixed air.

The other three springs range from 107° to 110° in temperature, and are not in any way remarkable.

The Aneroid barometer indicated an elevation of 380 feet above the sea.

The water runs from the springs into baths, enclosed within small buildings, of which some are roofed. The two springs described above have two baths each ; the others one. These baths were full of people all the time I was there, alternately parties of men and of women. Bathing seems the favourite form of benefiting by the waters.

After leaving the baths, the water is allowed to run into shallow reservoirs to cool, and is then used in irrigation.

The village is small, but is in a beautiful clump of trees and gardens ; it is the most fertile little spot conceivable—dates, mangoes, limes, pomegranates, plantains, and many other trees, and plenty of corn and vegetables.

The mountain rises quite over the village. On going a little way up, there is a fine view, over the trees, of the sea (distant about four miles), with the numerous groups of islets, called Jezirat Dehmányeh ; and nearer, the villages and forts seen through the date-groves along the shore, for near this place begins the fertile coast of Batneh, where, for near two hundred miles, the coast is one continuous date-grove. On the left, distant about 30 miles, rises the stupendous chain of mountains called Jebel Nakhl, of which some peaks are near 8,000 feet.

2 EAST FROM MUSCAT

Yiti

The Batinah road westward from Muscat leads all the way to Musandam and the United Arab Emirates. But the road eastward properly so called does not exist at all. There is a new asphalt road as far as Quriyat, which touches the Gulf for the first and last time at Quriyat, and the long road looping south through Bidbid, 'Ibra, and al-Kamil which touches the Gulf for the first and last time at Sur. Apart from Quriyat and Sur, none of the fishing villages of the coastline between Sidab and Ras al-Hadd possesses an asphalt road link.

Curiouser and curiouser, when one recalls the feats of civil engineering in Musandam, which are infinitely more difficult and costly than any stretch of road near Dibab or Tiwi. But then, Musandam is strategically significant to oil tankers passing through the Strait of Hurmuz, and to a political system vulnerable to fanatical Shi'a fundamentalists in Iran or to Soviet adventures in the Middle East. And on the positive side, inconvenience for the farmers and fisherfolk of Fins or Qalhat is no novelty. Travellers can see their countryside much as it might have been seen two centuries ago. But Wellsted and the rest took ship at Matrah for Qalhat and Sur, missing the charms of cliff and oasis, beach and wadi of this delightfully varied coastline.

One trip, of half a day from the Capital, takes you off the asphalt road at al-Hajar to Yiti and as-Sifa. A weekend of two whole days will take you along the fast road to Quriyat, then a graded road to Wadi Shab and Tiwi, allowing time for a walk inland, along Wadi Shab. Another weekend, if you start early enough on the first morning, will take you along the Eastern Hajar asphalt road to Sur, then on a graded road to Qalhat. There is no road joining Tiwi to Qalhat at present, a stretch which may be covered in any one of the local boats plying for hire. The route from Sur to Ras al-Hadd is similarly best enjoyed by boat (take a hat against sunstroke at all seasons). Leaving Ruwi roundabout, we arrived in 2 km at the Wadi Adai roundabout, signed 86 km to Wadi Hatat, and signed also to Quriyat. Km 8–9 flew by between high mountains with a little melancholy vegetation, supporting no agriculture. Km 11 sees al-Amirat village 2 km off right, the wadi spreading out again in a wide arc. The photograph at km 16 shows part of the main road below modern bungalow-style housing (each room with an air-conditioner clearly visible), and the barren hills

Km 16 from Ruwi, just before Yiti turning

under a midday sky. At km 17 you see the Ibn Sina Psychiatric Hospital to the left, and at km 20 the turn off left to Yiti (so far asphalted to km 34) along Wadi al-Maih. At km 27, still on the asphalt road, al-'Atkirah is signed just off the road. At km 28, the road is signed straight on to Saih Teman and Quriyat, and off left at 25 km to Yiti and Wadi al-Maih. We explored the lovely village of al-Hajar, with its abundant palms, at km 29, then returned to the Yiti road at km 34, where the graded road starts and signs are: Wadi al-Maih 20 km; Yiti 22 km; Yankit 27 km; al-Khiran 30 km; and as-Saifah 47 km.

The rugged wadi and mountainside here, where Wadi Hatat meets Wadi al-Maih, is illustrated to show how close in time and space romantic desert conditions are situated to the sophistication of international hotels. At km 40 we found the rough wadi bed snaking between palms at both sides, then jagged rocks and pebbles crudely flattened were marked in Arabic 'Nadi Wadi al-Maih ar-Riadi': Wadi al-Maih Sporting Club. It was a football pitch, of a sort, showing the

Wadi al-Maih. The mountains close in

Wadi al-Maih village

tenacity of the world game in the most spectacularly unlikely wadi beds. Even the most accurate pass could only succeed by chance on these rough stones and a fierce tackle would graze tackled and tackler out of the game for medical treatment. Several times a season the entire pitch will be swept away in torrential floods. At km 56 a signpost showed no observable track 3 km right to al-Hilw (the name means 'Sweet'), and 6 km left towards Hanshaft.

Instead of taking the track straight ahead, direct to Yiti, the spirit of adventure drove us on the track where signs pointed to al-Khiran 9 km. and as-Saifah, along the coast, the Nissan Patrol bucketing like a roller-coaster on the exuberant gradients. An alluvial valley spread before us and below, with a few houses nestling in its depths, 5 km north of Yankit.

Back in the elongated fishing village of Yiti, we sped and splashed the car through the silky, sunlit, lazily spreading waves. Windswept

Yiti. Village with barasti huts

palms on the wadi floor, below the low cliffs, glimmered green beside *barasti* huts. Is the absence of a road to Sidab and the Capital really such a disadvantage? Commercially, perhaps, for there is no restaurant (though you can buy food and ice-cold drinks in a shop) and little for the people of Yiti to do apart from scrape a living from the harvest of the sea and a few date-palms. But remember that here, on a beach unpolluted, unpopulated, and in every sense unspoilt, you are a mere 55 km from Ruwi's hectic roundabout. Remember, and enjoy the loveliness of Yiti before it falls victim to the commercialism of our time.

Quriyat

One glorious Friday morning, while denizens of the Capital Area were still taking their rest, prior to the day's holiday, Ahmad Khalifah al-Balushi and I sped out through the illuminated streets from Ruwi roundabout to the Wadi Adai roundabout, to the strains of Omani music on the cassette-player, past al-Hajar at km 20, the Yiti turn off left at km 25, al-Haji 1 km left at km 30, Tuyan just left and Hindi 5 km off right at km 33, al-Mundhiriah 1 km off right at km 37, as-Sa'adi 2 km off right at km 39, and Dawad off right at km 40.

The road to 'Arqi off right at km 41 leads to Siyah and Wadi Sarin. Before you start, if you are interested in nature reserves or the habitat of the Arabian *tahr*, you may request a permit to visit the Wadi Sarin reserve set up in 1979 by the Sultanate's Adviser on Conservation, Ralph Daly, whom I met in his office in the Capital.

Siyah is an isolated village now well-known as a small centre for naturalists. Local tribesmen act as game wardens to protect the *tahr*, a member of the goat family which maintains its precarious existence on the high *jabal* here in the Sharqiyah, and would soon have been hunted to extinction without conscientious governmental intervention.

Continuing on the main road to Quriyat, completed as recently as 1978, as-Salil is signed 1 km off left at 48 km, then you enter a broad wadi between the mountains, entering Wilayat Quriyat at km 52. Signs indicate Hifadh 2 km off left at km 61, al-Fayad off left at km 62, at-Tarif 5 km off right at km 62, and al-Habubiyah 2 km off right at km 65.

Few hamlets dot the barren, broken foothills at km 76–7, but then vast views spread out at km 80 as you approach ever closer to the sea, and at km 85 we took the asphalt road all the way to Hail al-Ghaf (7½ km), the alternative being 7 km straight on to Quriyat. If you want to explore the Devil's Gap (Wadi Daiqah), you will have to do it on foot between Mazara and Ghubrat at-Tam. One method would be for two parties to set out in interchangeable four-wheel-drive vehicles. The first would leave two or three hours ahead for Wadi Tayin off the Bidbid–'Ibra road and park at Ghubrat at-Tam. The second would follow the Quriyat road as far as the Hail al-Ghaf turn, then take the sign to the lime groves of al-Misfah at km 88 on our Quriyat route, then park at Mazara. The walk between Mazara and Ghubrat at-Tam would

Al-Habubiyah, 67 km from Ruwi

take four to six hours, and should be planned to end well before midday in winter, to avoid sunstroke and exhaustion. Remember to take adequate back-packs of liquid and first-aid, headgear, and sunglasses, with strong footwear and light but strong clothes. The two parties could then exchange car keys en route within the Devil's Gap, to avoid following the same itinerary back.

Hail al-Ghaf is one of those sunsoaked villages lost in time, off the main roads, which remains inviolate despite changes in architecture or development. A shady avenue of mango trees wafts its heavenly scent towards you, so you enter on foot beside the parallel *falaj*. The road ends abruptly in the main square, where two lone cannon point at right angles to each other against a foe long departed. Instead of such a foe, I found a village elder, crouched by a wall, in oblique shadows guillotining his knees to ankles, under a dishdasha brilliantly white. As still as granite, he did not want to be disturbed in his thoughts: he looked as though dusk would find him there, and dawn, serene, untouched by transience.

In Quriyat, Ahmad and I took a belated breakfast in a buzzing Indian restaurant near the fort, which was being restored, with thin scaffolding labouring up its sides like lianas impressed by military discipline. A Shell station outside the town is an obligatory call before driving on to Tiwi: a full tank now can get you back to Quriyat without mishap.

A bus station with a timetable in Arabic and English announced daily departures to Matrah at 4.55 a.m. and 4.30 p.m., fares being 500

100

baiza to Wadi Hatat, and 600 baiza to Ruwi or Matrah. A few taxis waited in the dusty square, without much hope of custom. The gold-domed mosque on Quriyat's main street is architecturally timid: it could be in Benghazi or Ajman. At the cinema 'Tonight' at 9 p.m. they were giving a Bengali film called *The Life Partner*. Goats browsed inquisitively between parked cars, in the gutters, securing scraps of gift-wrapping and newspaper for their insatiable appetites.

To reach the sea, we left the asphalt road at the right-hand side of the fort, and cruised across the soft wet sands cluttered with tiny crabs to the village of Sahil. I waded out to the islet in the sea pinnacled by a ruined fort, which is cut off by high tide.

Tiwi

By 9.30 we had left Quriyat behind us at km 116; you have a choice between the old track to Daghmar marked 17 km, or the new asphalt road (not marked on any map I had) which forked south at km 118, with Daghmar marked 12 km, choosing the latter because of its

Jabal Tayin from as-Sinsilah near at-Tawiyah

novelty. Turnings include ar-Ramlah 2 km off left at km 120, al-Karid 1 km off right at km 121, al-Fala' left at km 127, al-Khuba left at km 129, before reaching Daghmar at 9.45, km 131, in 1986 the end of the asphalt road. The new eight-columned fish *suq* is both simple and characteristic of Oman: there is no otiose gesture, when the function is fulfilled. Such is the religious architecture of the country, from Batinah to Jabal Akhdhar; such is the unaffected hospitality of a villager whose first instinct is to offer you a drink, a meal, and rest from the heat of the sun. At Daghmar I found simple wooden panels resting on simple fishing-boats. Without a road sign to follow, we headed south until we came to a wooden stick pointing to the right, its peeling paint just legible: 'Wadi Arba'in 70 km' at km 147. That adventure would have to await us another day.

The northern foothills of Jabal Aswad ('Black' Mountain here implies 'Leafless') seem as desolate as any moon landscape we have ever examined in a photograph. Except – if you look very closely at the photograph I took near km 150 – two hardy trees can just be discerned in the middle distance. At km 160 we hoved into sight of the fishing

Daghmar. The beach, looking south

Barren landscape between Daghmar and Dibab

village called Dibab, and made our jolting way over stones, sand, and
pebbles to the shore, where seabirds strutted, unconcerned by our
ephemeral intrusion.

We resumed the Tiwi track at km 162, encountering in 3 km a sign
'Block Factory', where building blocks were being cut for a waiting
lorry. At km 179 we entered the coastal hamlet of Bimah, its shore
scattered with rusting anchors, a *huri*, lobster-pots, and a cool fish
market, empty except for a short, spare elder with henna-coloured
nails, who pressed us to share his tea. We expressed gratitude for his
kindness, but continued our way along the fast track, arriving in Fins at
km 192.

Here a dozen fishermen stood or sat lounging in their covered,
open-air *sablah* with its strategic views of track and sea. Every Omani
village has a more or less formal meeting-place for the local notables,
where business can be transacted by the consensus method that has
almost invariably dominated Arab decision-making. Foreigners are
often misled into thinking that because an Arab country possesses a
ruler, whether Sultan or Amir, King or President, and it possesses no
parliamentary system of democracy, like the House of Commons or
the U.S. Senate, then the democratic process cannot exist in Arab
countries. This misconception arises from the difficulty faced by
non-Arabs when faced by the Arab mind or the Arabic language. In
fact, because advisers, tribal leaders, elders, experts, ministers, and
others are not subject to pressure groups lobbying them in Parliament,
they often take a much wider view than party politicians whose first
objective is to be returned at the next election, sacrificing the
long-term objective in favour of expediency, and the floating voter.

Virtually anybody may gain access to an Arab ruler at well-publicised times of audience, and every sector of society feels itself to have a voice, which is heard. Grievances about lack of proper housing, education, and schools often arose from a nation's poverty, and these have been overcome in most Arab countries recently as a result of oil income; the poorer Arab countries have benefited by interest-free loans and credits from the richer neighbours.

So I joined in, diffidently, at the Fins *sablah* which was debating the purchase of new Yamaha-powered fishing boats: who would benefit from a goverment grant, and who was considered an outsider, having arrived in Fins too recently to benefit from local expansion. Ahmad and I listened, while the boatman Juma'a ibn Said Salmani of the Ghadani tribe handed round bottles of soft drinks to all of us. Most of the other elders were of the Husaini tribe, and it is clear that Juma'a felt himself isolated. He offered to take me in his *huri* from Fins to Tiwi for ten rials. 'It takes half an hour. Much quicker than car.'

'How long does it take by car?'

Fins. Shell, with Salim Muhammad Khumayis at-Kuwaili

'Thirty minutes.'

As we settled on five rials for the one-way boat trip, I arranged with Ahmad to meet him by the shore at Tiwi. A charming, toothless elder in his sixties brought me a beautiful shell to admire. He introduced himself as Salim Muhammad Khumayis al-Kuwaili, and the shell – sprawling on my desk as I write these words – passed into my possession for 500 baiza in memory of a splendid day.

Juma'a pushed me off in the fragile reed *huri*, then lightly leapt in and started the motor.

'Sit perfectly still,' he admonished me, 'otherwise we shall both fall overboard.' Visions of hungry, lithe and rapidly streaking sharks ensured that I performed my passive part without error or objection. And who more enviable than I? Impalpable yet enslaving, the air of that extraordinary day again filled my lungs with that passion for the East which has overtaken many a head more level than mine: Hogarth and Cox, Pelly and Miles, Sadleir and Wallin, Musil and Niebuhr. A wild wind soared, vanished, ruffled my hair and vanished again like an invisible lord. Low cliffs visited by herons and terns cut into sandy beaches, the barren mountain range lolloping behind the whole way. Cormorants stood aloof on rocks, expecting a suitable fishy victim. Half an hour of blissfully riding the easy waves, and then the call 'Tiwi', and we glided up to a fishing village beyond the town, huddled like a tawny puppy up against the bitch fortress, of the same desert tinge. The upper part of the Tiwi ravine is called Wadi Bair, and there nestle nine villages: apart from Tiwi itself these are Hallat ibn 'Isa, Hasan, Hallat Bida', al-'Aqr, Saimah, Mibam, Sawi, and 'Amq Kubair.

Ahmad and his vehicle stood alongside the unmarked restaurant, which was still closed for Friday prayers. We gazed straight out across the bluest sea the world had ever seen, waving to the retreating figure of Juma'a, who was already on his way back to Fins. Not advertised as such, the restaurant can be found next door to the 'Muqeem Trad. and Cont. Est. Carpentry.' The shutters were removed, and the worshippers emerged from the mosque, some heading homeward, a few joining us in the restaurant. As always, no woman or girl could be glimpsed. Four South Indians from Kerala were served first, facing each other as if in a fast rice-eating contest. They stuffed balls of rice and curried chicken into their mouths while speaking Malayalam, the combination being inimitably incomprehensible.

After lunch we sat on the beach, watching the swoop and birds above the skitter of crabs. A boy came up and chatted with us as if he had known us all his life. Young men are treated as adults early on in Oman, with duties expected to include earning money, helping in the date plantations or in the fishing boats, and looking after younger siblings. These responsibilities weigh lightly on their shoulders, insofar as they are expected of everyone alike.

'Ali bin Ya'qub bin 'Ali told us he was of the Ghadani, like the boatman Juma'a, but the dominant tribe hereabouts is the Salti. Others include Muqaimi, Sa'di and Qalhati. Tiwi is now connected to the Capital by radio and television. Muhammad ibn 'Abdullah ibn Battuta, one of the two greatest of all Arab travellers, visited Tiwi in the fourteenth century, calling it 'one of the most strikingly beautiful of all villages, with flowing streams and plentiful orchards.' His book, bearing one of those characteristically ornate rhyming titles beloved of Arab authors in every period, was called *Tuhfat an-nuzzar fi ghara'ib al-amsar wa 'aja'ib al-asfar* which might best be rendered 'A gift for those interested in the curiosities of cities and the wonders on the roads.' H. A. R. Gibb (and latterly C. F. Beckingham) have translated the book as *Travels in Asia and Africa, 1325–54: o lingua frigida!*

'Ali bin Ya'qub told us that there is still no motorable track from Tiwi along the coast to Qalhat. By boat it takes about four hours, and a more pleasant way of spending the time would be difficult to conceive. The fare will be 10 rials for an Omani, but – he looked at me critically – he did not know how much a foreigner might have to pay.

Wadi Shab

A pleasant weekend can be spent camping at the wide entrance to Wadi Shab, north of Tiwi, or along its steep sides. During the winter the whole of the wadi may be filled with rainwater pools, and one heavy thunderstorm may sweep your tents away, so be sure to camp on high ground, or in a shelter. Precipitous rocks tower over the shady date groves beyond the sand bar, then deep freshwater pools refresh you. Birdsong sparkles in the clear air. Park and lock your vehicle at the wadi head, then roll up your trousers and remove your boots or sandals to cross the wadi. Occasionally ferry-boats will be drifting and lolling: you stand in one and haul yourself across the wadi by a rope; but these boats are often ankle-deep in water. If no ferries are around, you can still wade across if you ask which is the shallowest point: here you will sink only up to your knees.

About 5 km from the fort you will ascend a series of charming waterfalls, and hamlets with hanging gardens. Do not try to walk along the slippery, uneven, and often steep paths in the dark, even with a torch: allow ample time for the return walk by daylight.

Wadi Tayin

A flavour of the timeless Oman that we can accept as intrinsically changeless, despite the spread of cars and television, radio and air conditioning, can be found along Wadi Tayin, a valley running northwest to southeast below Jabal Abyad (White Mountain), itself the foothills of Jabal Aswad (Black Mountain).

The road west from Ruwi roundabout (km 0) reaches the New Military Hospital at km 41 and Rusail Industrial Estate at km 42, entering Wilayat Bidbid at km 44, near the oasis and watchtower of Rusail. The ascent towards Jabal Akhdhar really begins at Fanja' (km 64), with Bidbid village signed off left at km 69. By the filling station on the main road we filled the Nissan Patrol with petrol for the desert drive ahead.

At km 72 we took a left fork ('Ibra and Sur), where the Jabal Akhdhar road would have taken us to the right. The new highway bypasses Sarur (left at km 79), Luzugh (left at km 83), Dasir (left at km 92) and Wadi as-Saijani (right at km 96). At km 120, Somrah is signed off 11 km, 220 km short of Sur. Wadi Andam is signed 56 km off right at km 124, and Wadi Tayin 75 km off left at km 126.

Miles on Wadi Tayin

Our best classic account of the trek through Wadi Tayin, dating from February 1884, is by Colonel Samuel Barrett Miles, and is here reprinted from *The Geographical Journal* of 1896, with a map of his route, which shows the circuit Ruwi, Sibal, Ghubrat at-Tam, Devil's Gap, Hail al-Ghaf, Quriyat, al-Hajar, Ruwi. Asphalted stretches nowadays run from Ruwi to Quriyat and Hail al-Ghaf, and the Ruwi-Jabal Akhdhar and 'Ibra road turn-off. Four-wheel-drive

Lt.-Col. S. B. Miles (Courtesy of H.B.M. Ambassador, Muscat)

vehicles can cope with the graded roads on the other stretches, except
that camel, donkey, or Shanks' Pony are still necessary for the Devil's
Gap, that wild, romantic ravine that would have aroused the ardour of

such Romantic painters as Loutherbourg or a Caspar David Friedrich.

Samuel Barrett Miles might almost be considered, with Wellsted perhaps, the presiding genius of this anthology of travellers' accounts. His biography has never yet been written, and even the *Dictionary of National Biography* excludes him. Yet he was a great man, whose monuments are *The Countries and Tribes of the Persian Gulf* (London, 1919; reprinted by Cass with an introduction by J. B. Kelly, London, 1966), and the articles scattered about journals and here collected in volume form for the first time.

His destiny was clear. The son of Major-General William Miles, late of the 9th Regiment, Bombay Native Infantry, S. B. Miles was born in 1838, educated at Harrow and, in 1857, he entered the military service of the East India Company as an ensign in the 7th Regiment, Bombay Native Infantry. After serving nine years in India, Miles was posted with his Regiment to Aden in 1866, and a year later entered the political service as Cantonment Magistrate and Assistant Resident at Aden, where he stayed till March 1869. Life back in India no longer attracted him, and he applied for a political position, obtaining the post of assistant Political Agent on the Makran Coast in 1871, and that of Political Agent and Consul in Muscat in 1872. Here, despite the notorious climate, he remained from 1872 to 1886, with leave periods away and temporary posts elsewhere.

His toughness can be judged from the records of the early years of the British presence in Muscat.

All the early British agents and residents at Muscat died so rapidly that the post was abolished for thirty years. Miles outstayed all the others, with brief intervals in Baghdad (1879–80), Zanzibar (1881–2), Northwest Province and Oudh (1885) and as acting Political Resident, Persian Gulf (1886). He was promoted Colonel in India, staying at Meywar from 1887 to 1893. During his early retirement years he spent much of his energies in writing, but he gradually went blind (like the traveller to Yanbu', Ta'if and Jiddah Charles Didier, whose *Sojourn with the Grand Sharif of Makkah (1854)* has recently been translated and published by The Oleander Press, 1985) and died, before completing his major book, in 1914. Many of his excellent writings are to be found in *The Persian Gulf Administration Reports,* collected in ten volumes by Archive Editions (1986).

So that when we traverse Wadi Tayin, we can evoke memories of Miles in the company of Sayyid Nasir ibn Muhammad Al bu Sa'idi, and the guide Shaikh Nasir ibn Ghurayib al-Jabri, on 11 February 1884.

JOURNAL OF AN EXCURSION IN OMAN, IN SOUTH-EAST ARABIA

By Colonel S. B. MILES.

It is, perhaps, needless to remark that our present geographical knowledge of Oman, in Eastern Arabia, and especially of its orographical system, is derived almost exclusively from the map of Lieut. Wellsted, I.N., whose 'Travels in Arabia' was published in 1838. The value of the work accomplished by Wellsted has been universally acknowledged, and, considering the difficulties he had to contend with as the pioneer explorer, he deserves the greatest credit and commendation for the light he has thrown on the country.

Among the regions Wellsted did not personally visit, and consequently did not describe, is that portion of the great mountain chain which forms the backbone of Oman, lying between Maskat and Ras Al Had.

The system of mountains hereabouts is somewhat complicated, but may be said to consist broadly of two parallel ranges forming a continuation of the Jebel Akhdar chain, and embracing between them as far as Kuryat a rich and thickly peopled valley known as the Wadi Tyin.

This valley, which rises a little to the north-east of Semmed, terminates at Ghubra el Tam, where the torrent has excavated its way through the hills, forming a very remarkable cañon or channel, the seaward side of which is known to navigators as the Devil's Gap.

The exterior of this gap I had seen in 1874, in company with an officer of H.M.S. *Rifleman*—Lieut. Black, who afterwards unfortunately perished in the ill-fated *Eurydice*; but the opportunity of fully exploring the gorge and the Tyin valley did not occur until ten years later. In the month of February, 1884, however, I was able to make arrangements for the journey, and the Sultan H. H. Seyyid Turki appointed Seyyid Nasir bin Mohammed Al Bu Saidi to lead the escort, our guide and kefir being Shaikh Nasir bin Gharayib el Jabri, who was directed to join our party the next day at Natat.

It was on the morning of the 11th that we mounted our camels and rode out of Mattrah into the broad and shallow Wadi Harmal, or "Vale of Rue." Passing Felej castle and the village of Ruwi, one of the market gardens of Maskat, we strike off to the left up the Wadi Adi, a winding gorge or defile about 4 miles in length cut through the hills, which leads into and drains part of the small plain called Seh Hatat, a basin or opening in the lofty and precipitous hills behind Maskat. The Wadi Adi, which has a rough stony bed of many colours, and the mural surfaces of which disclose a singular variety of geological strata, bifurcates at its outlet from the hills into two streams, the torrent after heavy rain pouring not only into the Wadi Harmal, but also into the watercourse that runs by El Wateyeh.

At 10 miles from Mattrah, we arrive at Al Birain, a hamlet of the Beni Wahaib tribe, so called from two copious fountains issuing from the rocks hard by. From these springs two new felejes, or underground streams, have been conducted for the purposes of irrigation. The water being sweet and unfailing, orchards and gardens have been planted, and are thriving famously. Palms and various fruit trees, grain, and lucerne luxuriate here. A lofty tower, that indispensable and ubiquitous adjunct to an Arab settlement, without which the source of water-supply for the use of the inhabitants would lie at the mercy of their enemies, is being built on a gentle eminence, and will soon be completed. I had heard something about these felejes at Maskat, and was therefore much interested at seeing them. One of them was the property of a joint-stock company there, of which the Sultan's wazeer was the chief shareholder and promoter. The other shares were held by Indian merchants. The spring had been purchased from the Arab owners for a consideration, and I found hat the shafts of the felej had been already sunk, and that the underground connections were in progress. The shareholders receive the water in their gardens in proportion, of course, to their interest in the company, the allotment being made every ten days. We halted here for the night. The elevation above the sea is 420 feet.

In the mountainous parts of Oman the roads run almost invariably along the beds of the hill torrents or wadies, which form the natural highways into the interior, and are sometimes sandy watercourses, sometimes deep rocky ravines, and sometimes broad fertile valleys.

Our second march was to lead us over the northern and more elevated of the two mountain ranges I have mentioned above, by the Kahza pass, which forms one of the main channels of communication between Maskat and the Sharkiyah, or eastern district of Oman; and as our day's journey was likely to be a long and toilsome one, we started early in the morning, and travelled in a southerly direction to where the Wadi Kahza enters the plain. This expanse is occupied not only with the usual stunted trees and shrubs of the wilderness, which are in this land never too abundant, and which have been here sadly diminished by the race of woodcutters who infest the neighbouring valleys to supply Maskat with firewood, but also with many singular natural pillars of considerable size, their surfaces indicating, with a precision that would be very interesting to a geologist, the composition of the surrounding hills.

We soon enter the Wadi Kahza, and commence the ascent up a gentle gradient along a good gravelly bed, treeless and waterless, cut through a deep bed of coarse conglomerate.

By-and-by we pass on our left the outlet of the Wadi Amda, which, scoring the northern side of this range in a direction almost parallel with the Wadi Kahza, forms a shorter and more direct route to the Tyin valley. It is, however, a very rugged and difficult pass, the defile, I was told, being only a foot wide at one part, with a wall of rock on one hand and a precipice on the other. After a short rest under the grateful shade of a clump of large trees, we pursued our journey up the ravine for an hour and a half, the acclivity gradually increasing in sharpness as we proceed. Above us here tower two lofty peaks, one on either side, that to our right being a cap-shaped point called Jebel Sell. And now commenced the real struggle of the ascent, the zigzag path of which was so frightfully steep, and the footing so rugged and insecure, that the camels, though helped and encouraged with the utmost endeavours of the Arab drivers, only climbed it with extreme difficulty.

By the time we had reached the summit at 3 p.m. I was able to realize the truth of the warnings given me beforehand by the Arabs as to the perilous nature of this pass for beasts of burden, for the way-side was strewn with the whitening bones and skeletons of camels that had fallen over the edge and been left to perish. Fortunately, we had no fatal accident, though five of the animals fell during the ascent, and bruised their legs. The elevation of the Akabat el Kahza I found to be 3900 feet; but Jebel Tyin, which stands in front of us, is a giant by comparison, and rears its head to an altitude of 5250 feet. Just as we mounted the top we met Shaikh Saud bin Hamad of the Rehbiyin tribe, who was on his way to Maskat to solicit pardon from the Sultan for his perfidious conduct in opening the road to the rebel army on their way to besiege that town in September, 1883, on which occasion they were completely defeated. The Rehbiyin tribe have for many centuries possessed and occupied the Kahza pass, and have always been in receipt of an annual subsidy for holding it closed against the Sultan's enemies. The rencontre with our party did not appear very pleasing to the shaikh, who had always been profuse in his protestations of loyalty to the government, and after a hasty salutation he commenced to descend the path by which we had ascended.

With our faces still towards the south, we crossed the ridge and began to move down the Wadi Mugheira, a ravine which I found, to my surprise, to present not only a more severe gradient, but even a more rugged and formidable path than the Kahza. The banks are as steep as if artificially scarped, for the stratification is vertical or at very obtuse angles, and runs parallel to the direction of the wadi. Shapeless blocks of blue limestone, large enough to form a serious obstacle, are piled up in profusion in the rocky bed, rendering our progress very slow. Patient and docile as the camels were, it seemed impossible sometimes that they could extricate themselves from the confused masses of rock among which they appeared to be entangled, and keep

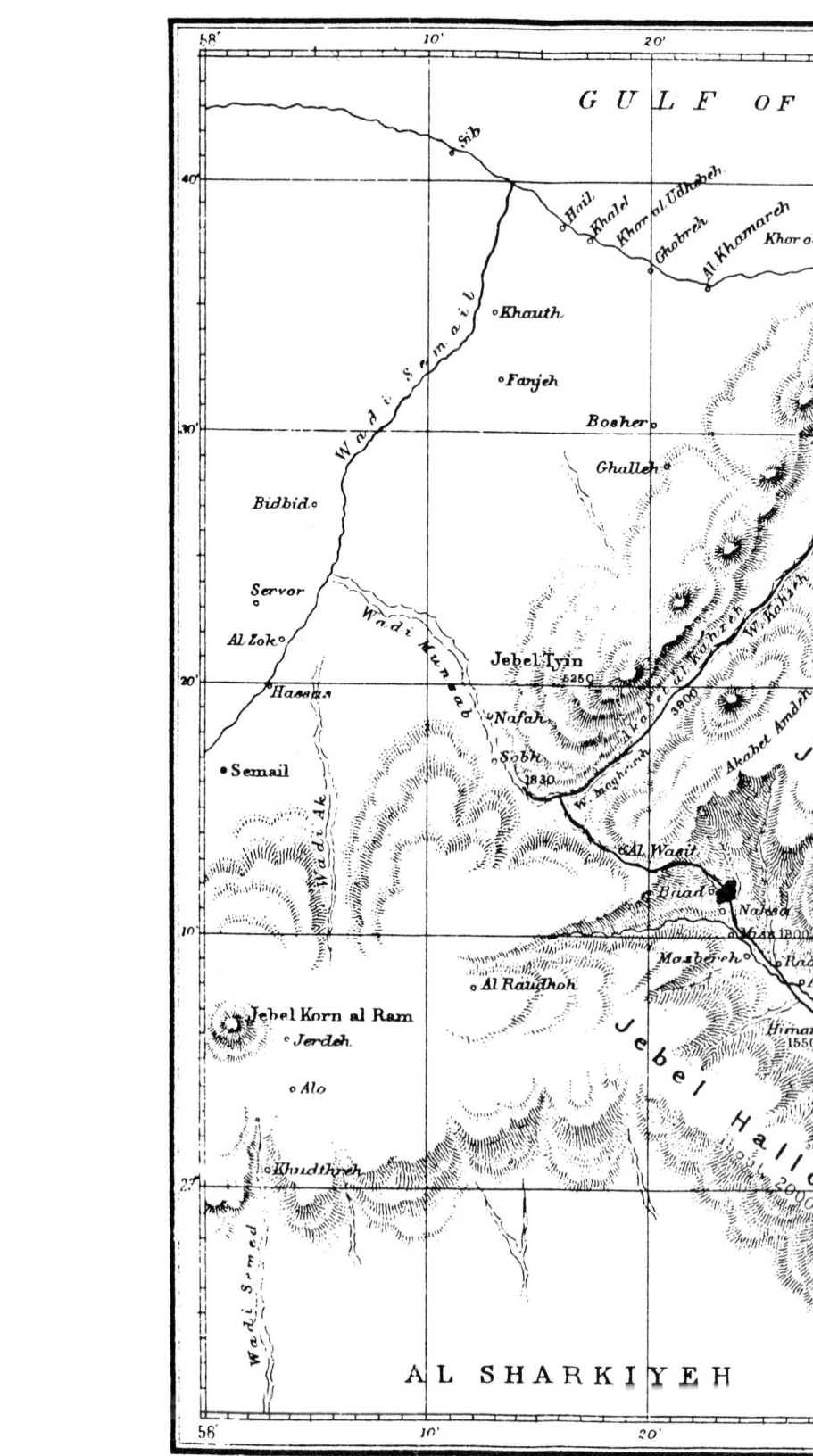

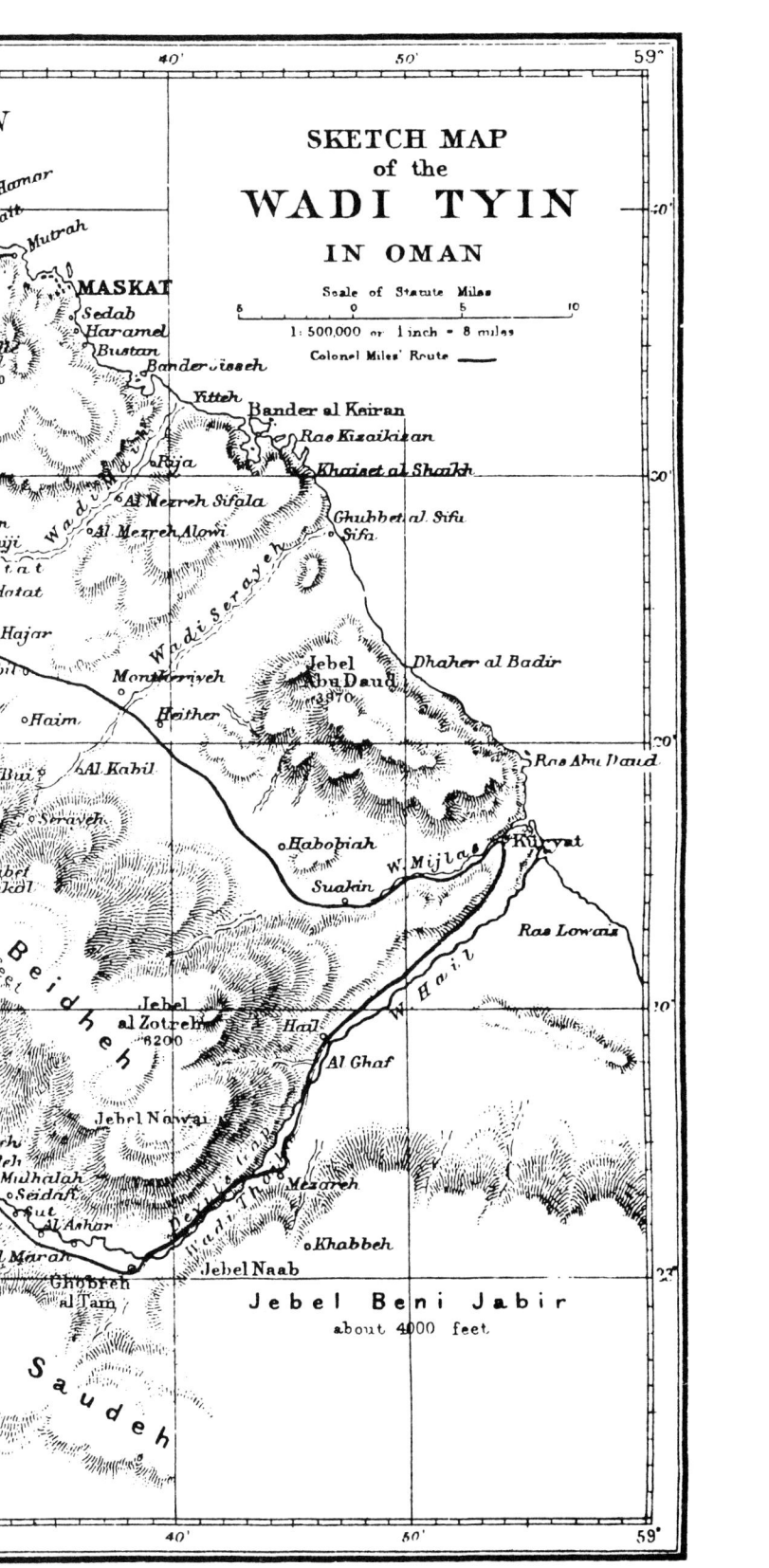

SKETCH MAP
of the
WADI TYIN
IN OMAN

Scale of Statute Miles

5   0   5   10

1 : 500,000 or 1 inch = 8 miles

Colonel Miles' Route ——

N

damar
att
Mutrah
MASKAT
Sedab
Haramel
Bustan
Bander Jisseh
Yitteh
Bander al Keiran
Raja
Ras Kizaikizan
Khaiset al Shaikh
Al Nezreh Sifala
Ghubbet al Sifu
Al Mezreh Alowi
Sifa
Wadi Serayeh
n.
uji
tat
lotat
Hajar
Jebel
Dhaher al Badir
Monakerriyeh
Abu Daud
Haim
3970
Heither
Bui
Al Kabil
Ras Abu Daud
Seraveh
Habobiah
Kuryat
bel
W. Mijlas
Kol
Suakin
Ras Lowai
B
e
i
d
h
e
h
W. Hail
Jebel
al Zotreh
Hail
6200
Al Ghaf
Jebel Nawai
rh
rh
Mulhalah
Mezareh
Seidah
ut
Al Ashar
Wadi Tyin
Khabbeh
Marah
Ghobreh
Jebel Naab
al Tam
Jebel Beni Jabir
about 4000 feet

S
a
u
d
e
h

40'   50'   59°

their feet on the slippery water-worn stones; but their drivers managed to get them through without mishap.

Owing, perhaps, to the protection afforded by the rugged nature of this torrent bed, and also to the presence of pools of water among the huge rock fragments, vegetation is tolerably abundant—tamarisks, oleanders, kafas, euphorbias, the tirucalla or milkbush, rhamnus, and acacias are the most common and characteristic, but many others I did not know are to be seen.

It was nearly sunset before we were able to remount our camels, and quite dark before we reached the foot of the Mugheira ravine, where we were to camp. Here we found water, but no shelter, and as it was too late to look about, we lighted fires and bivouacked for the night. Our camping-ground was 2070 feet below the summit of the Kahza pass, and the air was cold and humid here, owing to the quantity of vegetation and the numerous pools of water around.

On the following day we retraced our steps for a short distance, and then, turning to the south-east, began to ascend the Wadi Mansab, a broad highway with a slight acclivity and smooth, sandy bed, offering an excellent road. This wadi flows into the Wadi Semail near Surur, and forms, indeed, the principal route between the Tyin and Semail valleys. There are two large villages, Subh and Nafaah, in it, besides several hamlets, and it is fringed along the greater part of its course with fine date groves. I had been advised at Maskat before starting to travel by the Semail-Mansab route as the more easy and convenient way, but I had decided on taking the shorter though more troublesome route, as I was anxious to examine the Kahza pass, which I had never seen, while the Semail valley was very familiar to me. The nejd or upland of the Wadi Mansab was reached in two hours, and we soon espied in front of us the village of Al Wasit, belonging to the Rehbiyin, where I intended to halt.

The quarters assigned to me were in a pretty orchard composed of numerous fruit trees interspersed with flowering shrubs and plants, and here I soon made the acquaintance of the whole community, a very small one certainly, who crowded round to see their first English visitor. Our march this day had been a very short one, only 6 miles, as the camels wanted rest after the fatigues of the Kahza pass. Al Wasit is a picturesque little place in itself, but its surroundings are bleak and desolate in the extreme. Situated on the barren slope of the mountain range, nothing meets the eye but dark masses of hills, exhibiting no signs of human habitation, and decked with but scanty verdure. The head shaikh of the tribe resides here with his attendants, but there is no village, as the paucity of water does not admit of a large population. One little rill trickling from a neighbouring glen is allowed to collect in a reservoir, from which it is drawn economically for domestic use and irrigation. Notwithstanding its apparent insignificance, however, Al

Wasit is much frequented by native wayfarers and traders, as well as by the chiefs of other tribes, who come to visit the influential Shaikh of the Rehbiyin, Salim bin Hassam bin Mohammed, who at this time happened to be away.

On occasions when the martial tribes of Al Sharkiya contemplate rising and marching on Maskat with the object of extorting concessions from the Sultan, or of ousting him in favour of some rival, it becomes a matter of importance to gain over the tribes holding command of the various mountain passes through which a passage for the invading army must be purchased or forced. Most of the tribes holding such passes as Akk and Kahza are in receipt of a regular stipend from the Maskat government to keep them closed against the tribal coalitions on the warpath, but this precaution does not always avail. In September, 1883, the Rehbiyin tribe had played false to the Sultan, and, though in receipt of a good subsidy, had listened to the cajolements of Seyyid Abdul Aziz and Shaikh Salih bin Ali, and had given passage to the enemy, whereby they had been able to pour down the Kahza pass with such secrecy and dexterity as to take his Highness by surprise. The attack, however, owing in great measure to the support afforded to the Sultan by H.M.S. *Philomel*, which shelled the rebels from their position, an action fully approved by our Government, had signally failed. Some of the rebels had been already chastised, and the Rehbiyin, among others, were still afraid of reprisals against themselves.

February 14.—The road hence to the Wadi Tyin led us in a south-easterly direction down the Wadi Wasit, a well-wooded shallow watercourse with a few palms. Leaving this nullah through a small gap between two vertical cliffs of white limestone, we arrive, after a short ride, at Naksa, a hamlet lying at the angle formed by the junction of the two wadis, and, passing this, we find ourselves at once in the stony bed of the Wadi Tyin. To our right, some little distance up the ravine, is a village called Baad, which is the highest settlement belonging to the Tyin tribes; but the source of this wadi is at Rautha, further south.

Beneath us now lay stretched, in all its picturesque beauty, one of the largest, most beautiful, and most populous valleys in Oman, the Wadi Tyin, a broad, straight vale lying between two mountain ranges, and extending north-west and south-east for a distance of nearly 25 miles. This rich oasis contains twenty-nine villages belonging to the Rehbiyin, Beni Arâba, Siâbiyin, Nahaya, and Beni Battâsh tribes, embosomed in dense palm groves, with orchards and fields of varied cultivation. Many of these settlements are concealed in the secluded and secure ravines which deeply gash the mountain slope on either side of the valley, but many others extend along the fertile bed, their orchards and plantations fringing the oasis at intervals for miles. It is watered by no less than 360 springs according to the Arabs, with whom this number is a conventional one, and though it does not possess a broad continuous river, the supply of water in the bed is perennial and abundant, flowing sometimes on the surface, sometimes disappearing in the porous soil. The northern of the two ranges holding Wadi Tyin between them is known in the valley by the name of Jebel Beida, or the White Mountain. It is tabular, and appears to be mainly composed of limestone in horizontal stratification, the average height being perhaps 3000 feet. Arid and sterile as these hills look, they are nevertheless inhabited by a considerable population of shepherds and goatherds, who rear large flocks. Their dwellings are mere oval shanties constructed of loose stones, and they subsist on the flesh and milk of their animals, cultivating only a few vegetables and indigo, which is in extensive demand in the valley. The crevices and hollows in the hills produce an abundance of thorny undergrowth and coarse herbage, from which the goats manage to extract food and nutriment in the most inaccessible spots. But the botany here and throughout the hilly districts is decidedly limited in character, and there is no great variety of species; indeed, the French botanist, Aucher Eloy, reckons that the total number of species in Oman does not extend five hundred. The usual vegetation here consists of colocynth, rue, acacia, vera and arabica, aloes, calotropis, senna, liquorice, euphorbia, brambles, sidr or rhamnus, and others. The wild animals are the ibex, called wail of the Arabs; the wild goat recently identified by Mr. Oldfield Thomas, and named after its discoverer, Dr. Jayakar; hares, foxes, hyenas, etc.

The range on the southern side of the Wadi Tyin is known to some as Jabel Hallowi, to others as Jebel Sauda. This range forms a minor spur from the great chain striking off near Zikki. Its mean height appears to be about 2000 feet, and it has no peak of any great altitude. On the southern flank, which merges gradually into the desert, it throws out several large shallow watercourses, such as the Wadi Andam, Wadi Beni Khalid, and Wadi Halfain, all flowing to the sea south of Ras al Had. The dwellers in this favoured vale, the Wadi Tyin, have made the most of the natural advantages and capabilities of their rocky home, and have, with the indomitable energy and perse-verance of the Arab character, industriously laboured to produce all that the fertility of the soil, conjoined with warmth of climate, is capable of. In addition to the usual grains and vegetables raised in Oman, they cultivate kumkum or turmeric, sugar-cane, bastard saffron, and henna. But it is to fruit culture that the Arab specially devotes his attention. The date, the most characteristic tree of Eastern Arabia, ranks, of course, first, and the vast assemblage of palms in this valley is very striking. Vines, peaches, apricots, custard apples, guavas, figs, pomegranates, plums, limes, sweet limes, quinces, oranges, bananas, citrons, mangoes, melons, and mulberries are also cultivated, and with considerable success.

We now pass, at a rapid trot, several villages in succession, each embowered in an umbrageous date grove, the water here perennially flowing in a copious stream. The first village is Al Bir, with a watchtower perched on an eminence commanding an extensive view down the valley, and having an aqueduct raised on stone pillars crossing the road. At this point the valley is 1900 feet above sea-level. Nestled in a corner is a town called Miss, inhabited by the Beni Jabir, with beautiful orchards, and enshrined in a little forest of date-palms, which rear their graceful crests over it as if to shade it from the scorching rays of the sun. As our guide, Shaikh Nasir, belonged to this tribe, we were heartily welcomed by the people, and invited to stop, but this I was obliged to decline.

Almost adjoining it is another town with a good cluster of houses, and possessing a masonry aqueduct built up 20 feet above the bed of the wadi to feed the palms and cultivation, a work of considerable pretension for such a place. We halted at Sibal, a fine, large village some miles further on, with an abundant supply of water, and every kind of fruit growing in the gardens. The venerable and courteous chief of this place, Shaikh Mesud, who had met us on arrival, furnished me with a comfortable hut to lodge in, and after we had sat together for a little time to drink coffee, took me for a walk round the town. There are some substantial houses in it, four mosques, and several *sablahs*, or meeting-places. The shaikh pointed out Akabat Amda, which bears a little to the north-east, and informed me that another pass called Akabat Mankal lay almost opposite Sibal and eastward of Amda; he said it was steeper than the Kahza, and was formed by two ravines, viz. Wadi Tima, running into Wadi Tyin, and Wadi Sarreya, on the Kuryat side. The shaikh was acquainted with three roads leading from Wadi Tyin to Al Sharkiya, two of which branch off from Malhal hah, and one from Gubra el Tam. There is much traffic between Al Sharkiya and Maskat by these passes, the roads converging at the Akabat Amda on the northern range, which forms the most direct passage.

In other parts of Oman I had found that certain tribes, particularly those who possessed large herds of camels and droves of asses, had obtained, in the course of ages by length of usage, peculiar privileges throughout the country as carriers of merchandize, but in this valley I could hear of no tribe that could boast of this privilege; all the traders, pedlars, and peripatetic vendors of goods being Mekrani Beluches, who act as agents or travellers for the Hindu and Khoja merchants of Maskat and Kuryat, the latter never venturing to penetrate into the interior of Oman themselves. Collecting from the various settlements in the valley and adjacent parts, the fruits and other produce of the soil, these Beluches bring back in return the sugar, oil, cloth, metals, and other commodities required by the people, though the large Arab proprietors have, of course, direct transactions with the Banians.

The name of this place is suggestive. In the Omani dialect, Sibal means an ape or monkey, and it was the name of a famous idol worshipped by the Arabs in pagan times before Islam. As there are no monkeys in Oman, it is possible that here in ancient days stood a temple dedicated to that image.

Sibal is 1500 feet above the sea, and is distant from Al Wasit about 13 miles. The journey hence to Ghubreh el Tam took us five hours, the distance being about the same, and the aneroid showing a fall of 500 feet in elevation.

From Sibal to Ghubra el Tam the valley continues to present the same character as in its upper part, occasionally contracting and expanding, but on the whole widening considerably as we proceed eastwards, while the hills on either side maintain their altitude. Water is abundant everywhere, in pools and in motion, and many streams pour down from the hillside to swell the volume of the main torrent in floodtimes. We pass many villages on our way, cultivation and palm groves alternating with desolation and arid rock. Two of the settlements of the Beni Battash, Hidda and Akdah, are so close together that they form but one large town extending for upwards of a mile, picturesquely flanked by two lofty watch-towers, which command the approaches and guard the water-supply. We were riding slowly and peacefully past Akdah, when our advance was suddenly and unexpectedly challenged by the inhabitants, who, either from panic or some other cause, gathered on the road in front of us in great excitement and offered to bar our progress, firing their matchlocks in the air and brandishing their spears and swords frantically, as if we deserved instant annihilation. This insult on the part of the people greatly provoked Seyyid Nasir, who abused them roundly, and could not conceal his vexation. However, it soon appeared that the noisy demonstration was not of an alarming nature, and as the shaikhs and elders of Akdah used their efforts to quiet the mob, the storm was soon over, and we proceeded quietly on our way. I did not mention the incident to the Temima of the tribe, Shaikh Shamas, when we met next day, but he heard of it from others, and was greatly ashamed of his people's conduct. He apologized to us for it, and announced his intention of punishing those concerned, but I doubt if he possessed the power of doing so.

The town of Ghubra el Tam is very picturesquely situated on the skirt of an eminence, which, lying at the end of the valley and thus forming a barrier against the onward progress of the stream, has caused it to swerve to the northward and cut its way through the mountain range down to the sea. It has some good houses and a population of over a thousand of the Siâbiyin tribe, and is protected by a strong fort of oblong shape perched on the western extremity of the hill.

At this time there was very little water in the wadi, the unusual dryness of its bed being due to the severe and long-continued drought, from which this part of Oman had been suffering, and our party were congratulating themselves on having arrived at such an opportune time for passing through the gorge, when their joy was suddenly turned into dismay by a slight shower of rain which fell in the evening. The clouds now began to gather so ominously in the sky, that if it had not been so late I should have pushed on at once without halting. It had, however, already become too dark to permit of this, and with some foreboding—for the intensity of the heat seemed to threaten a thunder-storm—we took up our quarters for the night in the habitation our hosts the shaikhs of the town had allotted to us. Had it rained heavily, as many of us fully expected, I should have had to wait here until the torrent had subsided sufficiently to allow of our proceeding through the gap, which would undoubtedly have entailed a delay of several days.

The exploration of this cañon had been one of the main objects of my journey, as it had not before been traversed by a European, so I was resolved to seize the present chance of visiting it at all risks. Fortunately, the night passed without the expected downpour, and though the morning of the 16th broke gloomily and lowery, the rain still held off, and the stream flowing at our feet had risen but slightly. After a consultation, we deemed it best to face the peril of a sudden rush of water through the gorge, and hazard the passage before the storm, which now appeared inevitable, could burst upon us and unite the rills and streamlets of the valley into a swift and overwhelming torrent. Having hastily loaded the camels, therefore, we started early, and crossed the bed of the wadi, in which the water was running a little over 2 feet deep, just opposite the town. We then found ourselves at once at the entrance of the great cleft, which is as sharp and abrupt as if we were entering the portals of some monstrous castle and stood immured within its massive walls. Towering loftily, sheer and perpendicular above the narrow floor, the huge walls of rock give the appearance as if the mountain range had been suddenly split in twain from the base to the summit by some convulsion of nature, exhibiting a singular illustration of impressive grandeur. The breadth of the passage here is about 100 yards, but it varies throughout its length from 500 to 150 yards, while the cliffs rise to an altitude of from 1000 to 1500 feet, as near as I could judge. The stream appeared to flow 4 or 5 miles an hour, and gradually increases in volume as we progress, being fed by the springs of water which burst from the crevices in the walls. Throughout the chasm the camels were wading nearly up to their knees.

After riding along this grand and curious gallery for a quarter of a mile, we are told to dismount, having arrived at a sort of deep step or waterfall called the Akaba. Here the camels are relieved of their baggage and saddles, and are taken along a ledge of the precipice on the left bank which leads circuitously to the bed further on, while the men of our party are let down by a rope over the projection on to the floor of the wadi below. This remarkable step or fall in the rock offered a very serious impediment, as it was of considerable depth, while huge blocks and fragments of blue and white limestone, that had fallen from above, added to the difficulty, and presented an obstacle which was absolutely insuperable to the camels, even when freed of their loads. The path leading to the fall, along which we had to scramble, was so rugged and slippery, and the cliff was so smooth and waterworn, that even the Arabs, who are as nimble as cats, did not find it easy work.

The solicitude evinced for my safety, not only by my own party, but also by the Siâbiyin who had accompanied us from Ghubreh, was almost touching, though the descent could not in fact be called perilous. Indeed, throughout my excursions in Oman, I always had reason to be grateful to the Arabs of my escort, and not unfrequently to the local Arab shaikhs, for their zeal and self-sacrifice on my behalf. They never resented the inconvenience and fatigue I often caused them, but deferred without question to my wishes as to the when and the whither; while on any occasion of unusual toil or danger, they seemed to regard my safety and comfort as a main point of consideration.

At the bottom of this pass, called Al Makuba by our Siâbiyin guides, we waited an hour for the camels, which, though carefully led by the drivers, did not traverse the narrow and dangerous ledge on the other bank without serious difficulty and hazard. Fortunately, however, they arrived at last in safety, and the baggage, which had in the mean time been lowered down by the Arabs, having been replaced, we mounted and resumed our journey.

The channel is here at its broadest, but it narrows further on, and becomes gloomy and cavernous, the mountain frowning above to a height of about 1500 feet. The cliff on the right bank at this part is known as Hail el Kebir, that on the left as Hail el Harim. Winding along this stupendous chasm, we occasionally have to encounter immense fragments of rock, piled in confusion on the floor, and obstructing the road, while above us are to be seen curious crags, overarching rocks, and other peculiar features of natural architecture. There is no lateral opening throughout the entire length, and only one small ravine falls into it, this being on the left bank. The geological structure of this range, as disclosed by the walls of this chasm, is mainly limestone, superimposed, probably, on the plutonic formation of which the rocks at Maskat are an outcrop. The lowest stratum to be seen is conglomerate, the upper layer of which is arenaceous. Overlying this with a horizontal stratification are courses of limestone, white or blueish, the upper rocks appearing to be of a reddish colour.

As may readily be supposed, the heavy and tumultuous torrents that frequently sweep the bed preclude the possibility of trees and plants surviving the rush of water, and we consequently find here no vegetation whatever. Even the long period of three years that had elapsed since the last flood had not produced any sign of bush or reed that I could see.

After heavy rain, the volume of water flowing through this chasm must be enormous, and the surging and raging torrent must then be a magnificent sight. It not unfrequently happens that travellers and caravans coming from Kuryat are engulfed and overwhelmed by the sudden rise and rush of the stream, as the innumerable tributaries and affluents in a drainage-area of some 200 square miles, swelling after rain, would concentrate at the gorge with marvellous rapidity and force, and form a mighty and irresistible wave, destroying everything in its path.

This effort of nature to provide an outflow for the pent-up waters of the Tyin valley through a mountain range is the most singular specimen of earth-sculpture I have seen in Arabia, and consists, in short, of a narrow, winding, vertical-sided gallery or cañon, extending for about 6 miles in a north-east and south-west direction, excavated through the solid limestone rock by the erosive action of water in a period of countless ages.

The peculiar character of this chasm, and the grand and picturesque scenery of its surroundings, create an impression on the mind which is not easily effaced. The Arab name for it is the Wadi Thaika, meaning the "Strait or Narrow Torrent."

It was one o'clock before we emerged from the cañon, our rate of progress in it being necessarily slow, and we found the opening at this end less abrupt than at the other, the walls gradually receding on each side and declining in altitude as we proceed. The high point of the range known as Kuryat peak to navigators, and to Arabs as Jebel al Zatri, now lies to our left, and raises its head 6200 feet above us, falling in terraces to the plain, while the mountain cliff to our right over Dagmar has been reckoned at 4000 feet.

Winding round a low hill, we come all at once upon the town of Mezára, the chief settlement of the powerful Beni Battash tribe, surrounded by thousands of date-palms, rearing their tufted heads in a dense grove; and so sudden and unexpected is our appearance, that no little commotion is caused among the inhabitants, who fly to arms, and rouse themselves into an absurd fit of excitement. Much firing and shouting ensues, but the hubbub evaporates on the appearance of Shaikh Mohammed Adi, who holds this part of the town, and who is most pressing for us to be his guests and remain the night.

But Seyyid Nasir whispers to me that the two shaikhs are not on good terms, and suggests the expediency of moving on to the castle. We accordingly politely decline Shaikh Mohammed's invitation, and ride up to the spot where Shaikh Shamas bin Mohammed, the temima or paramount chief of the tribe, is awaiting us.

I had known this venerable and noble-looking shaikh for many years, as he often came to Maskat, and I had learned to like and respect his character. His reception now was most cordial, and I was much gratified at it, as it greatly increased the pleasure I felt at being able to pay him a visit in his own home. As we halted and dismounted on the bank of a small but rapid stream that intersects the plain and winds among the palm groves and settlements, Shaikh Shamas came down from his castle at the head of a long procession of his people, and gave Seyyid Nasir and myself a hearty welcome. With a levity and humour uncommon among Arab shaikhs, who are usually grave and dignified, he put his hands on my throat and declared he would throttle me if I did not promise then and there to spend a day with him and accept his hospitality. There was no disputing with him on this point, so, having been carried over the river by his men, I walked up the eminence with the shaikh and Seyyid Nasir, followed by a dense crowd of Arabs to the fort. Just as we approached it, an old twelve-pounder gun lying unmounted on the ground in front of the gateway was fired off in honour of the occasion, the report reverberating finely among the surrounding hills. From this gateway, and in a higher degree from the roof of the castle, the landscape presented to the view is one of exceeding beauty. The town is situated in a small circular plain, the low hills encircling which, with their dark background of lofty peaks and tabular mountains, render it invisible from the sea, and constitute a natural fortification, giving an assurance of peaceful security. The lower part of the plain is filled up by habitations, plantations, orchards, and cultivation, while standing high on the east side is the castle, square, substantial, and imposing.

The temima took us into his reception-hall, where we sat down to talk while coffee was being roasted, pounded, and boiled at the other end of the room. The interior of the castle does not accord with the pretension of its external appearance, for it is but scantily furnished and decorated, and can boast of but little comfort. Chairs, sofas, and furniture, as we understand it, are not to be found in the shaikh's residence, for the Oman Arab is a plain man, simple in his habits, and free from ostentation; his wants are few, and, however well off he may be, he does not indulge in luxurious sloth, or surround himself with many articles of needless luxury. Even the women's apartments are bare and empty; a carpet, a box of clothes, and articles of domestic use are the only things to be seen. The temima, however, was extremely courteous, and treated us very hospitably.

He informed me that the abutment of the range on the north-west side of the Thaika gap was called Jebel Nuwai, and that on the south-east side was Jebel Naab; the Beni Nuwai and Beni Naab being two of the pastoral tribes who occupy the tableland of these hills, and subsist on their flocks of goats and sheep. These shepherds are probably of mixed Arab and aboriginal descent, and form the main following in war of the settled tribes in the neighbourhood, of which they are, in fact, the Bedouin portion.

Mezara greatly charmed me by its quiet beauty. The rich orchards and gardens and the rippling brook made the locality very delightful and attractive, and I almost envied Shaikh Shamas his residence in so favoured a spot; but, shut in as it is by hills, I should imagine it to be excessively hot and oppressive in the summer months. Just below the town is an aqueduct on five arches, constructed a few years ago for the purpose of leading water from the stream to irrigate the cultivation. Running into the Thaika valley beyond Mezara is a ravine called Wadi Khabba, with a large village of the same name; and under the lofty cliffs to the south-east of Hail are Wallja and Sunt, the latter belonging to the Sâbiyin tribe.

Good fortune had befriended us in allowing our passage through the dreaded Thaika without being overtaken and submerged by a flood. We had, however, only just escaped in time, as the heavy masses of clouds which had been brooding over us since yestermorn began now to descend in a deluge, and a grand thunderstorm broke upon the hills and valleys.

The rain continued all the afternoon, and was most joyfully welcomed by the Arabs, who naturally looked on our arrival in their town as a propitious omen.

Our departure from Mezara the next day was accompanied by the same demonstrations of friendship as our arrival, and, after taking leave of our hosts, we mounted our camels in the presence of the whole community. For 5 or 6 miles our course led along the bed of the river, which, known above as Wadi Thaika, receives here the appellation of Wadi Hail. The banks near Mezara are about 1000 feet high, and perpendicular, but they continue to decrease in elevation and to broaden out as we advance seaward. Along the left bank runs a fine masonry felej, with shafts to raise the water at intervals, recently constructed to replace an old one built by the sultans of the Yaareba dynasty, two centuries before, and now fallen into decay. It soon leads off northwards to supply the gardens and groves of Hail el Ghâf.

On arrival at the settlement, we were met by Seyyid Hilal bin Said bin Hamad and his two brothers, and were escorted to a building forming a single spacious reception-room, situated in a pleasant garden. Here we sat and rested for several hours, enjoying the fragrance of the flowers and the balmy atmosphere of this salubrious spot.

The Hail el Ghaf settlement is said to have been founded by Seyyid Khalfan Al Bu Saidi, a notable man in Oman in the early years of this century, who took a conspicuous part in cementing the friendship of the English with the Maskat government, he being at that time wali or governor of that town.

Prior to the selection of this locality for building and planting by

Seyyid Khalfan and his family, Hail el Ghaf was merely a patch of elevated ground, partly surrounded by the wadi, and covered with a thicket of acacia trees (ghaf), from which circumstance it derives its name. The land was purchased from the Beni Battash tribe, who, however, retained a portion of it, and still exercise a sort of protection over the place. The benign sway of the Al Bu Saidis, and the liberal manner in which money has been expended, have caused the village to flourish and the population to increase to its present number, about 1200 souls.

Hail abounds in orchards and cultivation, and a vast quantity of fruit and vegetables are sent to the Maskat and other markets. The Arabs are as passionately fond of flowers as they are of strong perfumes, and take great delight, when they can afford it, in horticulture. The number and variety of fruit and flower trees, imported at various periods from Persia and India into Oman, testify to the appreciation by the Arab of these plants, and to the care bestowed on them. The "bostans" or gardens, therefore, as may be supposed, absorb most of the time of the aristocracy of Hail, who seem to take the keenest enjoyment in tending their roses and jasmine, and in spending the livelong day in the shade of their sablahs or summer-houses. The most beautiful feature in the vicinity of Hail is a broad, straight avenue of superb mango trees over two miles in length.

At 4 p.m. we started for Kuryat, where I intended to pass the night, and on the way we encountered another thunderstorm and heavy downpour of rain, which drenched us through. It was hailed as a blessing by the Arabs to the parched and thirsty soil, and their loud exclamations of gratitude sufficed to reconcile us to our discomfort. We rode fast to escape the deluge, and covered the 12 miles between Hail and Kuryat in an hour and a half.

I took up my quarters in Seyyid Hamad bin Khalfan's large house, which is fortified with two towers, and here I was detained the whole of the following day by the rain, which fell in torrents and without intermission, much to the delight of the inhabitants, but somewhat to my annoyance, for I had visited Kuryat on many previous occasions, and knew it too well to desire its further acquaintance. This town lies 31 miles south-east of Maskat, and is situated about a mile from the shore, on the maritime plain lying between the great range and the sea. The plain contains about a dozen hamlets, and is intersected by two wadis. It forms an extensive pasture-ground, on which in former days an excellent breed of horses was reared by the inhabitants, who exported them from Kuryat in considerable numbers to Surat to supply the Indian market. Chiefly owing, I believe, to internal dissensions, this trade became very precarious, and ultimately came to an end about two centuries ago. The population of Kuryat is 3000, and it has a good bazaar with several shops belonging to Hindu traders, who supply the settlements in the Wadi Tyin and in the hilly districts east of Maskat with foreign merchandise, the Wadi Tyin being the main artery through which the traffic between Kuryat and the interior passes. This port and Sur are the Karteia and Tsor, the Carthage and Tyre, of the race whom we know as the Phœnicians, and who, as far back as the time of Solomon, or earlier, had trading-stations along the southern coast of Arabia. They are undoubtedly of great antiquity, and retain their primitive names to this day. Their convenient and important position on the Arab coast just opposite India, must, like Kilhat and Khor Jerama (Corodamon), have led to their early occupation as trading depots by the merchants of those times who were engaged in exchanging the productions of the East and West.

The sun rose on the 19th in a clear sky, and we were soon beyond the outskirts of Kuryat, ascending the Wadi Mijlas, a deep and narrow ravine which leads in a tortuous fashion and in a generally south-west direction to Sawákin, whither the Wali of Kuryat accompanied us on horseback. Sawákin is a small and pretty hamlet, and forms a triangle with Hail el Ghaf and Kuryat, from which latter town it is 9 miles distant. Here, enshrined in a fine plantation of palms, is a large house built by Seyyid Said bin Khalfan, who used this charming and peaceful retreat as his country house in the hot weather. After an hour's halt at Sawákin, we leave the Wadi Mijlas, and, striking off to the west, traverse a rough, desolate, and very broken country, a confused mass of ridges and hillocks of limestone, the strata of which appear to slope generally southwards. We twist and turn along the watercourses, which, adorned with dwarf acacias, thorny shrubs, and jungle herbage, intersect the country.

We pass several villages on the road, Heither, Muntheriya, and others, all belonging to the Beni Wahaib and Beni Hassan; and about halfway to Maskat, Sarraya, a small town of cloth-weavers, is pointed out in a well-watered and fertile ravine 6 or 7 miles away to our left. In the evening we reached Al Hajar, a village in the Wadi Hatat, where we camped for the night. It is a comparatively new settlement, for when I first visited the spot in 1876, the ground was being prepared to receive Busreh date-palms, and a felij, half a mile long, had been projected and commenced by Seyyid Hilal bin Ahmed, who had purchased the fountain from its proprietors, the Beni Wahaib. Though water is very plentiful and the soil tolerably productive, the plantations did not appear to have thriven as well as might have been expected. The felij was destroyed, it seems, by a flood soon after it was completed, and had to be rebuilt, and other causes have combined to retard progress; but, like its neighbour Al Birain, which I have mentioned before, it struggles on, and is fairly profitable to its enterprising founder. Tobacco is one of the chief articles cultivated about here, and is grown for the Maskat market.

The Wadi Hatat, in which we now stand, extends, under its various names of Wadi Kahza, Wadi Hatat, and Wadi Maih, from Jebel Kahza to Yiti, being joined on the way by the Wadi Amda and numerous other ravines. At the foot of Wadi Kahza the hills open out, and a small plain is formed 9 or 10 miles long called Seh Hatat, which has been the scene of many a sanguinary battle in Oman history. It contains several villages and much cultivation, and is possessed by five tribes, viz. the Beni Wahaib, Beni Hassan, Beni Jabir, Beni Battash, and Al Maashera.

The range to our left, on the other flank of which runs the Wadi Tyin, and to which the Rehbiyin and Siâbiyin tribes gave the general name of Jebel Beida, is not known by this name to the Beni Wahaib, who, indeed, could give me no appellation for it.

The Wadi Maih, which extends from Al Hajar to the sea at Yiti, is a narrow vale about 9 miles in length, with a rough stony bed and a fast-flowing stream of water, in which the fragrant and ubiquitous oleander is extremely abundant. It is in general barren, with occasional patches of cultivation. The hamlets in it are Al Mezra Alowi, where the orchards are walled or revetted up on the banks to preserve them from the encroachment and rush of the torrent; then El Mezra Sifala, then Rijaa, and then Yiti.

The geology of the Hatat valley is extremely curious and interesting, as it exhibits on the one hand the nature and stratification of the sedimentary rocks, of which the great mountain chain of Oman appears to be principally composed; and on the other, the metamorphic or igneous structures forming the dark group of hills at Maskat. The bed of the Wadi is throughout of limestone pebbles, underneath which is a coarse conglomerate. The hills are most varied in colour, and the strata lie at all angles.

In the Seh Hatat there are many curious natural pillars about 25 or 30 feet in height, standing some distance apart, and apparently the effect of denudation, the general aspect of this plain suggesting the idea that it may have been in remote times the basin holding the drainage of the surrounding hills in the form of a lake, until the eroding action of the water had excavated its way through the Wadis Maih and Adi down to the sea.

Near Al Hajar the limestone rocks were of a reddish tinge, and appeared to be mixed with layers of blue mud.

In the Wadi Maih the mural section shows the plutonic action in a most unmistakable form, the rocks being all confused and crumpled up, and the strata lying in folds or arches, as it were, over huge blocks of limestone.

We left Al Hajar the following morning, and pursued a course varying from north-west to north. After passing Al Birain, we enter abruptly the defile of Wadi Adi, and then, turning down the Seh Harmal, we soon arrive at Muttrah.

Wadi Tayin Today

Since the time of Miles the track through Wadi Tayin will have
changed but relatively little, confined as it must be to the relatively
shallow watercourse below the limestone of Jabal Abyad. The level of
the water table, according to local informants, seems also to have
declined scarcely at all, since the population has probably remained
much the same, a small increase in the local birth-rate being offset by
migration to the Capital Area.

At km 132 we racketed into al-Jirdah on a corrugated switchback of
a track, and I stopped to chat with Khalfan bin Hamid bin 'Ali of the
Awlad Mahriz. In the rare Admiralty *Handbook of Arabia* (1916) I
found his tribe listed: '800 souls; Ibadhi in religion; settled at Ba'ad in
Wadi Tayin'. But Miles on his map shows 'Jerdeh' nearly 40 km
southwest of Ba'ad. With a magnificent beard and an elegant turban
around his head, Khalfan embraced me cordially and invited me to
take coffee. I equally affably declined his courtesy, but asked him if he
could spare a few minutes to show me his gardens. He pointed to pipes
bringing water from a shallow well to his date-palms. 'It would cost me

Al-Jirdah, with Khalfan bin Hamid bin 'Ali of the Mahrazi

128

Mazbar. Open-air sablah

three thousand rials to have the well dug deeper, and that is more than I can afford,' he shrugged. The Government has spent millions of rials all over the Sultanate on improving open and underground *aflaj* and digging new wells, but the cry is always 'more water' in a desert nation which never receives enough rainfall.

Ba'ad, 13 km. east of al-Jirdah, is a large oasis still occupied exclusively by Awlad Mahriz. Of the population of 2,000 or so, 300 attend primary school. I was shown over the village by the army officer Mubarak bin Ahmad bin 'Ali. I met his bearded father, Ahmad bin 'Ali bin Hamud. Ba'ad's prosperity is due to the large number of abundant wells. At km 148 a sign showed Naqsa and its Block Factory, producing limestone for building, off to the left (whereas Miles shows Naqsa to the right off his route), and at km 151 the wadi widened, and the road with it, permitting a more comfortable ride on a better choice of surface. Next we encountered the idyllic village of Miss, with its new co-educational twelve-class primary school. At km 156 magnificent views of mountains on both sides of the wadi made a stop compulsory. We breathed the pure desert air, and watched children playing in a fresh water pool and others in a palm grove below the track. Then the road descends into the wadi at Mazbar, and we wandered dreamily among the fertile palms until accosted by Salim bin Khamis bin Said, who introduced us to his uncle Sulaiman bin Rashid bin Sulaiman, of the Rahbiyin.

'Have you all the water you need here, Shaikh Sulaiman?', I asked the old man, 'Al hamdu lillah, we have enough for all of us. While the rest of the Sharqiyah complains of drought, the wells of Mazbar never dry up. Why should we leave to go to a government job in the Capital, or work in the oilfields at Fahud while we have the blessings of Allah in

Sibal. Wall and watchtower

our own gardens?' I saw these fruitful gardens, echoing to the peals of children's laughter; the plain, stylish watchtowers protecting these Ghafiri Ibadhis from envious invaders; and was invited to join the men in their discussions at their morning coffee (it was ten o'clock on a Thursday, eve of the holy day).

They had brought flasks of coffee and bowls of dates and oranges to a mat in the open air. Old men, the middle-aged, and youths past adolescence gathered in a circle to discuss matters of common concern in the place chosen as *sablah*.

No architectural or sociological study has yet been made of the Omani *sablah*, yet there may be grounds for considering it as crucial an aspect of Omani social life as the mosque is to Omani religious life. A *sablah* achieves by informal conversation and discussion what a more complex and divisive community must achieve by parliamentary means. For the typical Omani community is not only virtually self-sufficient, it is cohesive and mutually supportive. Members of a coastal community tend to unite in fishing or, at Sur, boatbuilding, and trading by boat. Members of an inland community tend to unite in agricultural pursuits, sharing *falaj* irrigation time, and responding in similar ways to common problems. Whereas three neighbours in a western city suburb might be engaged in computers, insurance, or secretarial work, and thus have nothing in common but the street where they live, three neighbours in Mazbar or Nizwa or Suhar will probably be engaged in a similar occupation, and will probably tend date gardens and own animals in addition to their main occupation. A taxi-driver will (outside the Capital Area) normally also be a farmer or a fisherman; so will a shopkeeper. The shortage of vocational training for specialization means that most men can claim experience in most skills, from house building to car-repairs. A calamity affecting one, whether fowl disease, or drought, will probably affect all, so the concept of community becomes an effective reality.

A *sablah* may be held in the open air as at Mazbar, at a spot conventionally agreed as convenient to all, hence near the centre of a village; or it may be a special building constructed by the *shaikh* for use at determined times; or it may be a covered building with an open courtyard, suitable for any time of the year, as at Ihda'. At Fins, on the coast, we found it a simple concrete floor, columns holding up a rush roof, open on all four sides to allow breezes to waft in, whatever their direction.

At km 160 al-Badi'a was signed off left and at km 167 we entered Ghayadah (the Rodaidah of Miles' map), with its wide, simple cemetery, like a field strewn with stones; at km 168 al-Mudairah (presumably the Al Madeira of Miles' map) was signed straight on and ar-Rihani off right (both in Arabic only) and at km 173 we entered the large village of al-Badi'a. A massive new walled enclosure, with an iron ladder up the side (as if up the side of a ship) protected many hectares of date plantations with new concrete *aflaj*, the whole belonging to 'Abdullah Musa of al-Badi'a. Shat was indicated 1 km off right at km 174, and Sibal off left at km 175. At Sibal, particularly, I was impressed by many new houses of the Bani 'Arabah, once a great tribe decimated by cholera. One never uses the singular of their name to describe a tribesman because *'arabi* means 'hill donkey.'

The extraordinary individuality of these villages needs to be stressed: here at Sibal, for example, a great long wall, defended at intervals by large watchtowers, rises on a natural outcrop like a miniature Great Wall of China against the pallid Jabal Abyad behind it.

We passed al-Hammam at km 176, Qar (Miles' Kan) off right at km 179, and al-'Alia off right at km 180, indicating by their proximity how fertile the wadi bed must be.

Now at km 181 I encountered one of the many unforeseen delights of Wadi Tayin: the Bani Battash of Ihda'. Readers will recall the alarm of Miles at 'Hidda and Akdah' of the text and 'Hiddeh and Akdeh' of the map. The inhabitants 'either from panic or some other cause, gathered on the road in front of us in great excitement and offered to bar our progress, firing their matchlocks in the air and brandishing their spears and swords frantically, as if we deserved instant annihilation.' I must admit that the inhabitants of Ihda' offered to bar *our* progress as well, but their motive was to offer us hospitality in the form of coffee, fruit, and shady rest in their *sablah*.

'We don't have enough water in the summer,' said the elderly shaikh Zaid Saif Zahir Bil'arab, sipping his coffee delicately to avoid spilling a drop on his luxuriant black beard. 'We have to use *aflaj* because there are no springs hereabouts.'

Three members of the Battashi tribe posed for a photograph in the shade of a tree in the latest communal *sablah* of Ihda', the previous one having a mosque attached. Salim Hamud Adi al-Battashi accompanied

Ihda'. The sablah, with men of the Bani Battash

me around the village, as we talked of books and education. New classrooms are burgeoning throughout the Sultanate, but the greatest impact seems to be in districts distant from the Capital, where people have never traditionally considered schools and literacy a part of their lives. Everybody benefits, but the remotest oases seem to benefit most. Sulaiman Habib Shuwaini al-Battashi told me that the *shaikh* of the Bani Battash was the most important of all the *shuyukh* in Wadi Tayin. The sections of the Bani Battash, who are Yemeni in origin, Hinawi in politics, and Ibadhi in religion, are the Bani Dhakar, Awlad Faris, Bani Ghasain, Awlad Hazam, Awlad Jum'a, Ma'ashirah, Awlad Malik, Awlad Salt, Bani 'Umr, and the Awlad Ward. I showed my passport to Sulaiman: 'Ward' I spelled out in Arabic: 'waw, ra, dal,' and he grinned in delight as we recognised each other again. 'Our neighbours are the Awlad 'Arifah at al-Ajif, 3 km away, the Na'bi at al-Madighat, 3 km away, and 'Aqdah 3 km farther along the wadi.' I was privileged to be shown the private *sablah* of paramount Shaikh Saif Hamud Hamid al-Battashi, newly rebuilt with concrete walls, a wooden roof, and low wooden windows allowing light on the floor level, where visitors would recline against pillows, barefoot on a carpet placed on rush mats. Here was the only substantial private library I saw in Oman. As founder of the Private Libraries Association in 1956, I always hope to see examples of interesting personal libraries wherever I go, but I admit that Ihda' in Wadi Tayin had not at first sight struck me as a promising prospect.

Sulaiman gestured from the dusty alleyways of paradisiac Ihda' up to the slopes of Jabal Abyad. 'Seven hours away by donkey, you will come to where the *badu* roam,' he said casually. There is always

Ihda'. The private library of Shaikh Saif Hamud al-Battashi

someone more remote than yourself, as you recall from Dino Buzzati's
Il deserto dei Tartari or *Robinson Crusoe*.

Yet only a few minutes drive along the bumpy track you come to a
football pitch, gravel it is true, but with regulation nets, for boys who
are taught by Cairenes and have seen on television the Brazilian stars
of Rio de Janeiro or São Paulo.

And then, 6 km from Ihda', there is the administrative centre of
Wadi Tayin, with a car workshop, filling station, and a restaurant for
chicken biryani and ice-cold drinks. This is Malahlah, unmentioned in
Miles' text, and misspelt Mulhalah on his map. Signs abound: to the
Town Hall, to the Wali's Office, to the Hospital, to the School, 15 km
straight on to Ghubrat at-Tam for the Devil's Gap or Wadi Khabbah,
15 km left to Badi'a, 15 km right to Wadi Nam and the asphalt road at
'Ibra.

It sounds like a town, but Malahlah was evidently chosen for its
location as a crossroads rather than for its intrinsic importance or
fertility. Or perhaps for its neutrality in a sea of fractious tribesmen
jealous of their own importance. Malahlah is scarcely a village, but
more of a speckle of the wadi like spots on a leopard, its tiny white
one-storey buildings dwarfed by the shallow tawny plain and the rising
umber mountains.

After lunch we maintained our progress towards the end of Wadi
Tayin: Rakakiyah at km 196, al-Baydh at km 197, an Indian file of
palms marching left in the lee of the rocks left of Sut (km 201), the
widening valley at al-Murrah, and finally Ghubrat at-Tam (km 208),
home of the Hanadhilah tribe, Ghafiri in politics and Ibadhi in
religion.

I greeted the smiling Said Nasir Saif al-Hanadhilah, who showed me over his gardens, padding ahead of me barefoot, his fine silver *khanjar* glinting in the afternoon sun.

'If you want to get through Wadi Dayqah to Mazarah, one of the *badu* will take you by donkey for 15 rials', said Said Nasir, except during periods of rainfall, when sudden torrents reaching the chasm simultaneously would, as Miles wrote, sweep away everything in their path. As well as the Hanadhilah, our informant Amr Ahmad said that the leading tribes in Ghubrat nowadays were the Siyabiyin (Ghafiri and Ibadhi), whose main centre of influence is at Nafa'ah in Wadi Mansah; and the Rahbiyin (Ghafiri and Ibadhi likewise), concentrated at Wasit, between Tayin and Mansah.

The Siq at Petra in Jordan bears the nearest physical resemblance to the Devil's Gap that I can recall, except that at the Nabataean city one is transfixed on release from the defile by the apparition of the rose-red Treasury, or Khaznah. The Omani ravine rears natural as the day earth sundered it apart, victim of geological circumstances, yet determining the fate of men who trust their destiny within it for a few short hours. I feared, looking up into the craggy darkness high above, to see the Minotaur of Dante's *Inferno* XII, as portrayed in Doré's engraving:

> 'che da cima del monte, onde si mosse,
> al piano è sí la roccia discoscesa,
> ch' alcuna via darebbe a chi su fosse;
> cotal di quel burrato era la scesa.'

(Just as the rock, down from the height hell-bent down to the plain lies spread, that one might climb from summit to the depths by that feared descent, such was that dread precipice.)

Returning from Wadi Tayin to the 'Ibra road by another route, one has the choice between Wadi Dima, accessible from Ghayadah, or Wadi Khabbah, accessible from Ghubrat at-Tam. Acacias, those hardy clingers to the most arid rock, survive in the Wadi Khabbah where little else can. It is only at al-Waljah, at km 215, that terraced fields defiantly spread their richness below barren limestone cliffs. Then the valley narrows until, at km 220, walled palm groves give livelihood to a small population, houses built up the wadi sides to avoid flash flooding. We came to ar-Rafi'ah at km 230, then saw a sign off left to Maqta (16 km) at km 235, where the 'Ibra road is marked 64 km distant. Settlements cease with the drying of the wadi bed, which widens at km 247 and becomes faster, as bumpy corrugation gives way to smoother tracks. Unadvised by signposts, we took one or two wrong turnings until we found someone to ask, at one point a suspicious *badu* herdsman near a cluster of tents. This is Wadi Nam now, sparsely populated with very few hamlets, such as an-Nabi (surmounted by a watchtower) at km 279 and al-Mu'tarid (km 283), with its new school. Finally, at km 297, we came to the Motor Workshop just off the main road 1 km east of 'Alayat 'Ibra, rejoining Route 23 after taking a welcome cup of tea at al-Walid Restaurant. Other weary travellers

134

were following a televised football match between an-Nasr of Salalah and al-'Aruba of Sur. At km 301 a sign off right to Wadi Dima showed 78 km.

Wadi Bani Khalid

One day I breakfasted at 6 a.m. off papaya, toast with marmalade, and coffee at the Mina Hotel in Matrah, and sat there worrying whether dawn would ever emerge across the fort and rocky headlands: if not, I should never see Sur and Ja'alan.

At last night's curtain disintegrated into shreds, like rice-paper on the tongue, and the first buses drew up outside. Schoolgirls crossed the pedestrian overpass, then an endless seep of cars circling Corniche Roundabout like worry beads in a nervous hand, and small fishing boats glided through the bay's glinting waters. Taxi-drivers touted for business to keep warm, for they were dressed for midday, not for this shivering chill: 'Masqat, Masqat, Masqat!' Cocks were clearing their morning throats, while the first street cleaners rocked brooms from side to side as if lulling the rubbish to sleep. Early customers looked over the night's haul of fish laid out on the cold bright morgue of stone. Musa's taxi eventually took off for Muscat with two passengers crammed into the front seat, and three behind.

At seven Ahmad Khalifah drew up, and we were away on the long road to Sur, through the Sharqiyah, or Eastern Region. The first-class road between the Capital and Sur can lead to any number of side-tracking adventures: to the Jabal Akhdhar, to Samad and Sanaw, to Wadis Dima and Tayin, 'Amq and Muqal at the end of Wadi Bani Khalid, Ras al-Hadd and Ja'alan from Bilad Bani bu Hasan, al-Ashkharah, and the Wahibah Sands.

We chose the Wadi Bani Khalid route to begin with, which is taken from al-Mintirib, known to the early explorers as al-Badi'ah or some such name (Bedi'ah in Wellsted, Bedeyer in Cole, Bediyeh in Miles). The sign shows 39 km to Wadi Bani Khalid, off left, approaching from Matrah, but that is merely the entrance to the wadi, and your drive will take you much farther, and road conditions will worsen as terrains become much rougher, and traffic decreases in density. A four-wheel-drive vehicle is absolutely necessary to reach Muqal, unless an asphalt road is laid there. There is nearby accommodation in the motel called al-Qabil, though it is actually situated in Dariz, and was given its name (according to James Lennox of the Ruwi Hotel) only at the insistence of a local *shaikh*.

At km 9 from al-Mintirib my notes are already epileptic from the attempt to write on board the internal combustion engine's equivalent of a bucking bronco, as the corrugated track juddered and threw us. Nobody and no living plant disturbed the windswept valley, baked brown in the interminable stone-cracking dry heat of the Sharqiyah. Finally, after 14 km we came across one old woman, standing with a glance of eternity in her eyes, one arm akimbo and the other concealed by her side in black. Had she walked to the track from her home in the mountains? But surely nobody would dare to give a lift to a woman by herself, for fear of being arrested on a charge of seduction? She made no sign, so to this day I cannot imagine what she can have wanted. Slightly farther on I saw a spectacular geological formation, contorted behind a few stubborn bushes, and at km 19 we began to climb steeply, arriving at the top of the mountain at km 21 and then bearing down again into Wadi Bani Khalid proper, and its fertile oasis of the same name dominated by the tribe of Sa'adiyin, Ghafiri and Ibadhi. The Wali's office can be distinguished by its radio mast and flagstaff: turn left there and make for 'Amq, with the buildings, new school, and palm plantations on your right. Here we parked our poor labouring vehicle: if it had been a donkey I should have had it put down in sympathy. Village *shuyukh* came out to welcome us, and we shook hands with everyone separately, murmuring the usual courtesies.

'Ahlan wa sahlan.'	Greetings!
'Marhaba!'	Welcome!
'Kaif haluk?'	How are you?
'Zain, kaif al-hal?'	Well, how are you?
'Al-hamdu lillah, zain.'	Thank God, alright.
'Allah yu'afik.'	God grant you health.
'Kaif halukum?'	How are you all?
'Tafaddalu.'	Come and join us.
'Shukran.'	Thank you.

Even the smallest boys ceremonially lined up to shake our hands, excitedly chattering among themselves to see a funny Englishman in an even funnier sunhat. 'I'm going to Muqal,' Muhsin Nasir Bukhit told

'Amq. The wadi ravine seen from the falaj track to Muqal

me after a refreshing cup of tea at their open-air *sablah:* 'would you like to come?'

I agreed with the kind of boyish alacrity that would make any cynic smile, but it was entirely genuine, and the magnificent views from the precipice we followed from 'Amq to Muqal were worth any jewel. Our path lay along a *falaj* newly cemented (according to a date marked in it) in December 1980. A sheer drop on the right made it imperative for me to keep my attention on the *falaj* to avoid falling down and rolling several hundred feet to the valley floor, but as the *falaj* snakes inventively close to the mountainside, one is also in permanent danger of stumbling over tree-trunks and bushes, or being stunned by protruding rock-edges, and *falaj* twists give the chamois-footed a distinct advantage. Muhsin, long accustomed to the quick ramble, skipped nimbly barefoot ahead of me. To him the new 'road' to Muqal remained of purely theoretical interest. Whenever he wanted to visit his relatives and friends in Muqal he took the quiet path alive with startled birds. He took my elbow: 'Shuf–Abu 'l-husain!'

A quick brown fox it may have been but there, jumping over a rivulet, I managed a glimpse of its taunting brush before it disappeared into a clump of bushes. Great wings of cloud fanned my face upward. Muhsin had turned another bend in the sheer rock-face, gliding like a circus tightrope artist on the thin brink of heaven.

He told me of an older *falaj* track running parallel around the swollen head of the mountain, higher up and even more dangerous as it splinters and cracks in its senescence. We saw a watchtower built into the rock, surrealistically perfect as a beautiful object but useless defensively as the rock around it has eroded away almost completely, leaving the tower stranded precariously, like a stone diver on a high board.

At a certain point you will see below you an inverted syphon, luring ground water from the other side of the wadi by means of a pipe; here you can choose between scrambling to Muqal along an ever more slithery *falaj*, or the wadi's *terra firma*. In either case, following the mountain's curve, you finally arrive at Muqal. In the best possible light head for the dammed lakes, whose waters flow into the *falaj*, creating a permanent oasis of green serenity, pierced by arcs of birdsong and sunlight. Keep left of these lakes: it is here that the athletic will find their stamina and agility useful, for there is no path at all, but a geological rage of toppled boulders. Little boys race you to the head of the wadi, and gleefully win. You ascend to peaceful pools, whose tiny fishy denizens nip your naked toes. Five minutes farther on, you find along a dry wadi bed which bends right a tall metal ladder recently placed to assist entrance to the caves.

Take a flashlight and you will be able to explore a cramped low distance of twenty-five metres until you come to a cave about 13 metres high, the outermost of a series of infernally hot caverns occupied by families of bats. From this first cavern you can explore either left or right, with adequate local guiding. Left is a complex appealing to the potholer, fossilised trees deep within. Right leads to an underground lake, cold as fear. Muhsin is a jovial raconteur whose fund of stories about the caves, treasure, and lands of paradise seem to echo the *Alf layla wa layla*, or 'Thousand and One Nights'. In Arabic 'Open sesame!' becomes 'Sālim bin Salīm Salām', or 'Greetings, Sālim the son of Salīm!', and anyone reaching the caves of Muqal and intoning these words will find the sheer rock opening before them, and be able to enter a land bewitched, full of eternal fountains in endless gardens filled with birdsong. Muhsin recounted the cautionary tale of the adventurer who had come from afar to test the truth of the legend. He spoke the magic words, entered as promised, and spent so many hours listening entranced to the contented buzz of bees and song of larks that he lost his way. He came upon a man who, hearing of his plight, offered to lend him a magic ring that would lead him back to the entrance; once there, he had to take off the ring and leave it inside. The stranger to Muqal gave his word but, on reaching the entrance his avarice overcame his honour and he stepped out from the cave into the world with the ring still on his finger, only to be trapped as the rock snapped across his struggling body. The man who had given him the ring suddenly appeared and offered to release him from agony if he would surrender the ring of return. The stranger did so, the rock parted again, and he fled as fast as his legs would take him.

Back in the village of Muqal, Ahmad Khalifah and I were greeted by a zigzag chorus of young boys, some of whose hands I shook twice, and a more sedate line of elders. We were the only visitors, and gladly accepted the offer of an accompanied tour of the village, which rode easily on the spur of mountain rock like an Omani *badu* on the back of a camel. Ahmad had driven up the boulder-strewn track, the vehicle

'Amq. From the mountain track to Muqal

crazy like a goat in his quaking hands. He needed a rest, and we sat in
the shade, crosslegged, swigging Pepsi Cola straight out of warm
bottles. The villagers of Muqal were as proud as new fathers of their
new 'road', which made it possible for the first time to bring heavy
machinery, building blocks, and other great loads up from 'Amq. It
was made only six months ago, and will be a talking point for many
more. But this is still donkey country: there are two thousand
sedentary inhabitants in the zone, but a further 2,500 *badu*, who rely
on donkeys for the carrying of women, children, and burdens such as
animal fodder. As elsewhere, we found intricately carved doors
testifying to the skill of local woodcarvers, even if housebuilding
materials seem rudimentary. I photographed Yaqub Said Mas'ud, the
nephew of Muhsin Nasir Bukhit, with 'Abdullah Hilal Mas'ud and
Amr Muhammad Amr beside one of these doors.

We crashed rock by rock back along the track, downhill now,
slithering on falling sand, and holding tight on to the steering-wheel,
door-handles, and indeed anything fixed to avoid bruising ourselves
more than necessary on the helter-skelter down to 'Amq, the worst
road in Oman.

Sur

From Wadi Bani Khalid we returned to al-Mintirib, there resuming the
fast and excellent asphalt road to al-Kamil ('The Perfect One'), an
oddly-named village on crossroads north-east to Sur, south-east to
al-Ashkharah, north-west to al-Mintirib, and south to the Wahibah
Sands. There is a filling-station here, a modest restaurant with smiling
Indians, and a shop selling chicken curry tins from Malaysia, and a
range of sardine tins from Australia and Japan.

At eight in the evening we struck north along the wide valley called
Wadi Fulaij ('The Little Canal'), finishing after sixty swift kilometres
in Sur. A newly-opened hotel surprised us: we expected to sleep on the
beach because in Muscat we had been told that there was no hotel in
Sur. A ceiling fan proved necessary in the small rooms even in winter;
as it lazily circled I read Wellsted on Sur:

'Sur, the port of the district of Jailan, is situated on a low sandy
shore, utterly destitute of vegetation or trees.' (This would have come
as a distinct shock to the cartographers of the British Airways
magazine *High Life*, who show Sur positioned deep into the Wahibah
Sands.) 'It is merely a large collection of huts erected on either side of a
deep lagoon,' noted Wellsted in 1835, 'which are separately occupied
by different tribes.' He did not name the two districts and the two
tribes, respectively 'Aiqa and Sur as-Sahil; and the Bani bu Ali and
Janaba. Sur as-Sahil is about thrice as large as 'Aiqa, with a
correspondingly larger boatbuilding industry and silk weaving industry. The huts, in Wellsted's epoch, were 'very compactly constructed
with the branches of the date palm, airy and spacious and, as the streets
are kept very clean, the whole wears a neat and pleasing appearance.
There are no shops here, the bazar being situated about a mile and a

Sur. Fish market

half from the beach, where a considerable number of the inhabitants reside. Thither, accompanied by its Sheikh, who had been sent for directly intelligence was received of my arrival, I set out, and was much gratified at finding when I reached the village that my tent was pitched in a delightful spot, and that guards had been placed, and every precaution taken for the safety of my baggage. Here I was to remain until camels and guides could be collected for my journey.'

Wellsted says of the inland *suq* at Bilad Sur: 'A daily market is held on this spot, at which grain, fruit, and vegetables are exposed for sale. The houses, though small, are strongly built of stone and cement, and the largest and best are occupied by the Banians and people from Cutch, who monopolise a considerable share of the trade.' This old *suq* is some little distance closer to the beach than the bright modern market, with streets at right angles to each other, in which the new hotel is situated. I was rudely awakened at 5.30 a.m. when I had asked for a 6 a.m. call, then again at 5.50, but there was no call at 6 so, bleary-eyed and befogged with sleep, I stirred downstairs to breakfast at 6.30, but there was nobody in the restaurant, kitchens, or even at Reception. Breakfast consisted of fresh orange juice, tomato omelette, thin unleavened bread with cheese spread, jam, and a pot of tea. Spots of last night's rain still wobbled iridescently on the dust outside. The old market must have looked much the same in Wellsted's time a hundred and fifty years ago. Wooden kiosks, *khanajir* on sale (though now they fetch a thousand pounds sterling in Sur!), traditional turbans, white *dishdashas*, woven baskets, the blind eyes, and their architectural similes, drawn shutters, patient donkeys waiting for their carts to be filled, fresh fish in their silvered armour, limes, carrots and oranges in sprawling piles. Then you see a line of pickups and vans to catapult your fancy into thinking yourself back into a twentieth century of 'Typing and Photostat', as a notice announces. Here are still signs of Indian traders: JETHMAND TULSIDAS ASHAR, MULCHAND UDHARAM AND BROTHERS. Next to the Al-Afrah Restaurant stands the gleaming National Bank of Oman Ltd.

Ahmad and I spoke to a taxi-driver who had bought his car from high wages earned higher in the Gulf, working for an oil company. He had never contemplated staying away once he had earned enough, and indeed it proved that, throughout Oman, family ties and self-respect alike decree that one should return to one's original home when affluent. Unemployment is in fact quite a problem in Sur, and no new long-term industries have been established, though there is short-term construction work on the new jetty (in progress in 1986), and on new schools and a new hospital. Fishing is a small-scale activity, involving a few dozen families who live at the beach, drawing up their boats near their houses, and selling their catch fresh in the market or dried and salted inland. Fish meal is used as agricultural fertilizer.

Fishing and boatbuilding may derive from the Levantine Sur (which we call Tyre and is spelt in the same way with a *sad*), but then so must the great traditions of seafaring and navigation, in which Sur vies with Suhar. Tim Severin, in his *Sindbad Voyage*, undertaken in 1980, describes a retired sea-captain renowned in Sur and farther afield: Salah Khamis, 'a gaunt-faced sprightly old man of at least seventy. He walked with a very bad limp, but this did not stop him from clambering up the scaffolding, which was now twelve feet high, and prodding at the plank joints with his camel stick. Hudaid, the head shipwright, told me . . . that no-one could tell me more about the old trading days. So I tracked Salah Khamis in his house in the middle of Sur and passed a delightful afternoon with him. He was the Omani equivalent of an old sea dog, as full of yarns and vigour and gusto as a man half his age.' (My experience of yarn-spinning Omani elders is that they generally *are* half their age.) 'He told me how he had taken command of his first ship, when he was only twelve years old, on the return passage from India when his father, the owner of the vessel, had died at sea. The twelve-year-old boy had brought the ship home, and for the next forty years had sailed regularly two or three times a year across to India. Salah Khamis waved his arms excitedly, and his eyes gleamed as he spoke of storms and shipwrecks, of men rescued from the water, of record passages, and how eventually his own son, who took over his ship when Salah was too old, ran her aground on the south Oman coast. She was a total loss. No other ships was built to replace her, because the trade was dead.'

A dozen wrecks scattered forlornly across Sur harbour reminded me of a sunny Scapa Flow. Only about twenty working vessels rode the shimmering waves, but a new corniche is being built and a renaissance of Sur may be imminent, for I counted ten *sambuqs* in the making, apart from a variety of lesser craft. I talked to carpenters from Kerala, bare-armed and bronzed, crouching by seasoned timbers. They were fashioning a new *sambuq* for Sabait al-'Araimi of Sur, and proudly showed me over its massive teak hull; they would spend five months on it and then it would sail.

Don't be misled by the so-called 'Hotel Suhriya' near the shipyard: this is merely a restaurant, cheery with the racket of Malayalam conversation and rapid laughter.

We drove back through narrow, dusty streets of mainly single-storey white homes, blind to the street, their windows looking into the central courtyard. An ancient castle, tawny as a old lion, straddled a hill next to a new telecommunications saucer beside the main road. At Sabakh village, we turned through an oasis signed to Hassi, and there we found another mighty fort, with a carpenter's shop leaning eccentrically against one wall, below a round tower where the flag of Oman waved in the morning breeze. The shop was declared to be that of Khamis bin Ahmad bin Khamis al-Ya'arub (or 'Al-Earobe' as he chooses to spell it in English). In the quiet suq we found a shop selling cassettes of Omani and Indian music, and I bought one featuring the dark-voiced Abu

Sur. Ghanjah in the harbour

Sur. A new sambuq in the making

Hassi, Sur. Fort, with carpenter's shop

Fa'iz of Sur. Ahmad and I played this cassette many times, to the extent that the throbbing African-style percussion rhythms, and the low male chorus became our signature tune; we tapped our fingers in time to Abu Fa'iz across the wide Wadi Fulaij and down to the Ja'alan sea below al-Ashkharah. As I write this I set the cassette on my tape deck and back comes the atavistic love song, of desolation and longing that effortlessly links Abu Fa'iz to the *fado* singers accompanying Portuguese ships to the Gulf, to Goa and Macao.

Qalhat

To the vibrant chant of Abu Fa'iz and his circle we headed north along the coast to Qalhat on a road rough, but not too discommoding. Seventeen km out of Sur, we encountered Oman Sea Farms off to the right, then we came in another 7 km to a ruined archaeological site called by the local people Mqabba, but actually ancient Qalhat, destroyed by an earthquake some time before the 14th century A.D., when it gave way to Muscat as the leading trading centre on the coast of Oman. Alboquerque visited Qalhat in 1507, then devastated the city in 1508. The Portuguese decided on Quriyat as their southernmost stronghold, and by the end of the Portuguese period of domination, in the later 16th century, Qalhat subsided to become a mere outpost of the rising Sur.

Marco Polo wrote of the place in the 13th century:

'Calatu is a great city, within a gulf which bears the name of the gulf of Calatu. It is a noble city, and lies 600 miles from Dhufar towards the northwest, upon the seashore. The people are Saracens, and are subject to Hormos. And whenever the Melic [Ar. *malik*, king] of Hormos is at war with some prince more potent than himself, he betakes himself to this city of Calatu, because it is very strong, both from its position and its fortifications. They grow no corn here, but get it from abroad, for every merchant vessel that comes brings some. The haven is very large and good, and is frequented by numerous ships with goods from India, and from this city the spices and other merchandise are distributed among the cities and towns of the interior. They also export many good Arab horses to India. For, as I have told you before, the number of horses exported from this and the other cities to India is something astonishing.'

The Book of Duarte Barbosa, completed about 1518, translated from the Portuguese by Mansel Longworth Dames (Hakluyt Society, 1918) contains a succinct comment on Qalhat: 'Clarate is a large Moorish town, of fine, well-built houses, wherein dwell many merchants, wholesale dealers and other gentlemen', an observation taken before the city's destruction under Alboquerque. The reason for the Portuguese action was aggression against the princes of Hurmuz who, according to the chronicle of Turan Shah, claimed to have come originally from Qalhat (see the *Travels of Pedro Texeira*, edited by Sinclair and Ferguson, Hakluyt Society, 1902). Dames states that the name Qalhat 'is simply the Arabic *qala'at*, a fort, and is applied to

144

Qalhat. Mosque of Bibi Maryam

many places in Persia, Afghanistan, etc.' but all Arabic maps show the
consonant 'h' instead of "ain' and I suggest relationship with the
consonant cluster 'qlh' meaning 'yellowing of the teeth' according to
Hans Wehr. For yellow is how the rocky teeth of the mountains above
Qalhat seem in many a light between dawn and twilight.

Just imagine a city with a complex system of *aflaj* running down from
the mountains behind, and a deepwater anchorage rivalling that of
Muscat or Matrah. Sitting on the ground, on a hillside near Bibi
Maryam's Mosque, your gaze cannot fail to light on countless
potsherds from centuries past; but they must be left intact, exactly
where they are, to provide the abundant surface clues that
archaeologists require to piece together the past.

In the early nineteenth century, Wellsted found 'only one building'
remaining 'in a state of tolerable preservation. This is a small mosque
which, judging from the writing on various parts of it, has been
frequented by Indian Mussulmans. Its interior is covered with
party-coloured glazed tiles, on which are inscribed, in rilievo,
sentences from the Koran.' This is the Mosque of Bibi Maryam, still
standing today, though stripped of its tiles.

The *Handbook of Arabia* of 1916 found Qalhat 'an unwalled fishing
village . . . consisting of about 125 mud houses and a small bazaar of a
dozen shops; there are a few wells, but no dates or cultivation.' The
inhabitants were then recorded as possessing 12 *badan* boats between
them, trading between Muscat and Sur, as well presumably as Quriyat,
Tiwi and Fins.

Today, Qalhat is dominated by a brilliantly white new school, on the
track leading to the shore, at the head of the Wadi 'Isa. Three boats

Qalhat. Village, from the shore

were beached up on the pebbles, and I wandered zigzag, as the spirit moved me, picking up cowries and other shells. I spoke to a passer-by as I explored the village, just by the little Quranic school, and asked about transport by boat to Tiwi, the land journey being impossible. About a dozen craft are available for the half-hour trip, which costs RO 11 or 12. He complained that Qalhat still enjoys no fresh water of its own, all drinking water coming from Sur.

Ja'alan

More than a hundred and forty years earlier, C. G. Ward, Acting Master of the *Palinurus*, noted his impressions of the journey from Sur to Ruwais (his 'Soor' to 'Roves' on the coast of Ja'alan), and this is how my namesake spent four days, concluding that the Omanis had been 'kind and attentive to us in the extreme'.

Account of a Journey from Soor to Jahlan, and thence to Ras Roves. By Messrs WARD, SYLVESTER, and JAMES, of the Hon'ble Company's Surveying Brig *Palinurus*.

[Presented by Government.]

FRIDAY, *November 6th*, 1845.—Landed at Soor with Mr Cole, C. in C., the interpreter, who accompanied us for the purpose of arranging with the Sheik to accommodate us with camels and guides for our journey.

On arriving at the town, which is situated on the beach, we found that the Sheik was absent on business at the other town of Soor, about two miles inland in a north-westerly direction. Said, the brother of the Sheik, took us to his house, and refreshed us with coffee and sweetmeats ; whilst the inhabitants of the place completely filled the room in order to satisfy their curiosity by staring us out of countenance. We afterwards, accompanied by several natives, walked to the other town, which is very prettily situated in a grove of date trees, and contains a capital bazaar, and a fort, mounting three guns, built of mud. We were here conducted to the market place, where a seat was placed for us that we might wait the arrival of the Sheik, and underwent a scrutiny from the curious eyes of the natives. Some Banyans, of whom there were several, presented us with plaintains and sweet limes, which were gratefully accepted after our walk.

On the arrival of the Sheik, we returned to the town, followed by a numerous concourse of men, fully armed with swords, matchlocks, spears, &c., and having mounted us on camels, they performed a species of war-dance, accompanied with singing and firing their matchlocks, in honor of our visit. On arriving at the town, our interpreter, having agreed with the Sheik that he should provide us with camels and guides, returned to the vessel, and left us to enjoy the hospitality of the Sheik, who took us to his house, and served up to us an excellent fish curry, with dates and hulwa. We afterwards smoked a pipe with our host, and retired to rest in the house which he very politely placed at our disposal for the night.

On the following morning a considerable dispute arose as to the number of camels and men we should take, but it was at last decided that four camels, and the like number of men, would be sufficient. Before starting, an unfortunate accident occurred to the only watch we had amongst us. One of the slaves attending us, happening to see it wound up, imagined he could do it equally well, and taking it up unobserved, turned the key the wrong way, and so injured it that we lost the use of it during our subsequent journey, and were obliged in consequence to judge of the time by the sun. We took leave of the Sheik and started on our journey, each having a camel, and the guides, riding in turn, on the fourth, first proceeding to the inland town to procure food for ourselves, attendants, and camels, consisting of dates, hulwa, and small cakes made of coarse flour.

Our route for the first six miles lay in a south-easterly direction, when we came to a well of fresh water in the midst of a date grove: having filled our water-skins, and allowed the camels to drink, we continued our journey, now in a south-westerly direction, over a hilly and very barren country. Our road at this part skirted a ravine which had apparently been a water-course, continuous with the Khore at Soor. On arriving at its termination we took the following bearings:—Jibel Menkab, S. E. ½ E.; Jibel Minarf, E. by S. At about 8 P. M. we took up our quarters for the night in a small enclosure of trees, within a short distance of the camel-track. We were visited by two men of the Beni-boo-Ali tribe, who were passing on their way to Soor. A heavy dew fell during the night, wetting our bedding completely through.

At daylight we were awakened by the guides, who had lighted a fire from the dried wood about us, it being exceedingly cold, and having partaken of breakfast, consisting of dates &c., spread on a greasy sheepskin for a tablecloth, we mounted the camels and pursued our route in a south-westerly direction. At about 8 o'clock we reached a well, where we replenished our skins, and performed our oblutions in the best way we could. The country hereabouts presented nothing of particular interest, being very hilly, and here and there interspersed with stunted trees. We passed several flocks of sheep, tended by women, and attempted to strike a bargain, but in vain, as they would sell none but the very worst of the flock at a very unreasonable price.

At about 11 o'clock we halted in Wadee Talif, and here our guides procured a sheep, which they killed and dressed in a very original style, first cutting the meat from the bones, and grilling it, with the intestines, upon stones previously heated in the fire. Three men of the Beni-boo-Ali tribe joined us, and partook of the good cheer. At about 3. P. M. we resumed our route in the same direction, and over the same description of country, and halted for the night in Wadee Biah, the greasy sheepskin being again in requisition.

On the following morning we pursued our way in a W. S. Westerly direction. The country, as we approached the district of Jahlan, presented a more fertile appearance, the trees becoming more numerous, and of a larger growth, and the soil of a finer description. Several wild asses came within shot of us, and after staring for some time, took to their heels, running with considerable swiftness. As we gained the summit of a hill overlooking the valley of Jahlan, we descried the town of the Beni-boo-Hassan tribe about three miles distant, surrounded by groves of date trees, these being the first we had seen since our departure from Soor. The great hill of Jahlan is about seven miles distant from the town, although, from its immense size, it does not appear so distinct.

On approaching the mud wall which encloses a portion of the town, we were met by several of the tribe, who conducted us into the town, where we were cordially received by Ali ben Saleen, the Sheik, a middle aged man with an intelligent cast of countenance, and standing upwards of six feet. A carpet was spread under the shade of one of the houses, and crowds of Arabs assembled to satisfy their curiosity: amongst them were two natives of Cutch, dyers by trade, who understood Hindoostani, and with their help we were enabled to carry on a conversation, and learned that on that day a peace had been concluded between themselves and the Beni-boo-Ali, and that a party, headed by Hamood ben Sabin, had proceeded to their town for the purpose of ratifying it. In the evening we were called to witness the return of the deputation, consisting of about fifty men mounted on camels, fully equipped for war. We were then introduced to Hamood ben Salin, the Sheik's brother, and leader of the expedition,—a most pleasing and intelligent young man, very superior to the generality of the Arabs we had seen before.

The town of the Beni-boo-Hassan is of considerable size, spreading over a large extent of ground. It is defended on the northern and eastern sides with a wall of mud, and here and there a fort of the same material: the houses are very irregularly built, and composed some of mud others of cadjan. That of the Sheik is much superior to the rest, being neatly built of cadjan, and nicely matted and carpetted in the interior. The walls were hung with glass bottles, and many articles of English manufacture, while a plentiful supply of earthenware was displayed on a species of shelf occupying one side of the apartment, and a double barrelled gun—a present from the Imaum of Muskat—hung in a conspicuous situation as being an article not a little prized by its owner.

On the southern and western sides of the town there are numerous date groves, on which much care is apparently bestowed: streams of beautiful water intersect them in various directions. In many of the groves we observed maize and cotton growing, which, by their beautiful green tint, presented a delightful contrast to the sandy appearance of the soil. There is a very good bazaar, in which the principal articles sold are dates, bread, fish, poultry, and meat (chiefly mutton); together with cotton cloth, which they manufacture with a species of hand-loom.

This tribe is possessed of several horses, some of which we saw: they are of a small compact make. Most of them have nice Turkish saddles, and they ride without stirrups.

We received a very pressing invitation from Kallipheea ben Ali, Sheik of a Bedouin section of the tribe,—who had come here as mediator between the two tribes, on account of the ensuing month being Dulbajja,—that they might perform the pilgrimage &c. unmolested: when that time has elapsed, the war will be renewed as before ; but our time being limited, we could not accept the invitation, his territory being three days' journey distant.

A slight incident occurred during the first day of our visit, which will serve in some measure to display the curiosity of the Arabs. We had signified our wish to enjoy the luxury of a bath after our journey, and were conducted to a small stream outside the town. Such was the curiosity of the people that they crowded the walls overlooking the stream, to such an extent, that, on our leaving the water to dress, one of the mud banks more laden than the rest, unable to sustain the weight imposed on it, gave way, and the fragments, together with a fat Arab, rolling in the direction where we were standing, completely buried our clothes, and we only escaped the same fate by jumping again into the stream. The natives appeared much concerned at our misfortune, and immediately assisted in removing the mud and stones, and recovering our effects.

After staying with the Beni-boo-Hassan tribe two days, we left to pay a visit to the Beni-boo-Ali tribe, from whom an invitation had been sent the day previous. We were accompanied by the Sheik, and four of his tribe, on horseback, as far as the boundary of his territory, which is indicated by a mud fort erected close to the road or camel-track : at this part a quantity of wells were being dug. The road for some distance lay, before reaching the Beni-boo-Ali territory, through thick date groves interspersed with streams of delightfully clear water, and neatly fenced in. The distance between the two towns does not exceed six or seven miles.

This town presented a very different appearance to that we had just left, the houses being more scattered, and many, which had originally been much superior to the present ones, partially knocked down, which they informed us had been dismantled by the British troops in 1821.

The Ameer, Mahomed bin Ali, received us very cordially, and, on entering his house, or rather mud hut, which was very imperfectly covered with mats, entertained us with coffee, &c. He is a man of very venerable appearance, about 70 years of age, and appears to be very much respected by his people. He showed us the letter he had received from the Bombay Government, by the *Palinurus*, and was apparently much surprised when we told him we had not brought any other for him, and were on no political mission whatever.

There is a mud wall being built round the town, but they are not at present proceeding with it. We were shewn the ground (which is just without the walls,) where the action was fought between the British troops and themselves: there were one or two broken pieces of artillery still remaining, and in the Sheik's compound there are two entire pieces. The town is very thinly inhabited, and the people seem to have no occupation, as in the other town, where all was business: they also appear to be wretchedly poor in comparison with their neighbours. We did not find them nearly so curious as in the other town, but left us much more to ourselves. We visited the bazaar, but found it a very poor one.

On the morning of the third day we took leave of our entertainers, and commenced our journey to Roves, where the Ameer's eldest son (Hamed) was to meet us, it being his intention to visit the ship. Our route for the first few miles lay in a S.S.Wly. direction, passing several date groves: we then struck off to the westward, where nothing of any interest was presented to our view, it being a sandy desert, and here and there a stunted shrub. We halted for the night in Wadee Sahal, and resumed our journey before sunrise the next morning, travelling still in the same direction, and over the same sandy desert. At about 9 A. M. we arrived at Roves, a small fishing village composed of cadjan huts. On enquiry we found that the ship had not arrived, but was still to the northward. We met here a man who had acted as pilot to the *Palinurus* the year previous, who very kindly proffered his boat, but previously took us to his hut, (where we met the Ameer's son, who had arrived a few hours after us,) and regaled us with rice, fish, dates, &c. We then returned to the beach, and, after much delay and deliberation, the boat was launched. We embarked with Hamed and suit, and reached the ship at about 8 P. M., very much pleased with our journey, the natives throughout having been kind and attentive to us in the extreme.

(Signed.)　C. G. WARD, *Acting Master.*
　　,,　　C. J. SYLVESTER, M. D., *Assistant-Surgeon.*
(True Copy.)　A. MALET, *Chief Secretary.*

Wadi Fulaij is a wide, mainly barren valley with few hamlets or grazing grounds. Starting from Qalhat we reached Hassi on the outskirts of Sur at km 27, with the Sultan's Naval Training Centre, then at km 45 a turn showed 8 km to Taima and 15 km to Ifta', shown on the map as Fita (the usual vowel/consonant variation which reappears with Zikki/Izki and Manha/Manah in the literature). The mountains are low, with sparse feeding for flocks, and we saw very few *badu* near the good asphalt road. At km 53 we entered the Wilayat al-Kamil and al-Wafi, reaching the Shell station in al-Kamil at km 81. After a lunch of chicken omelette and yoghourt, I enjoyed a stroll in the main square of al-Kamil, examining the squat keep, and trudging the walls under the relentless midday sun. Centuries of tribal conflict in this district have caused families to live in the same area as their ancestors, and ally themselves with their neighbours against those more distant. Each tribe must control its own water, its own land, and its own livestock: to that end it used to have to man watchtowers and occupy forts, though with the Omani renaissance since 1970 old feuds have died, tribalism has diminished somewhat in favour of nationalism, and migration has to some extent dissolved old rivalries. What is left is the archaeology of hatred: these castles, keeps, and dungeons. The walls around the palm groves in al-Kamil are untypically high, but the separate watchtowers are as characteristic of old Oman as is the locked wooden gate in the sleeping castle.

Five km away stands the pugnacious fort of al-Wafi, with its tall keep. Nothing stirred in the sandblown square except for my steps, and the slow wag of a dog's tail in its unceasing toil against flies. The same graded road, churning up sand as we drove along it, brought us to Bilad Bani bu Hasan. Wellsted, in 1835, found the *badu* dwellings here to be 'mostly huts, erected beneath their date groves. They are very straggling, and we were three quarters of an hour passing from one extremity to the other. As soon as the intelligence of our arrival had spread, they crowded around us in great numbers. Their curiosity was unbounded, and they expressed their astonishment at all they saw in the most boisterous manner, leaping and yelling as if they were half crazy.'

Many of the homes are now single-storey stone dwellings, of the open courtyard type, with room for animals to shelter from the sun. The village is just as straggling as it was during Wellsted's time, and the evocation 'from an antique land' is strengthened by the absence of a modern asphalt road, though new primary schools are being erected. You can fill up with petrol at a Shell station here, and the presence of a

Al-Wafi. Fort

152

Bilad Bani bu Hasan. Al-Balushi boys in the renovated mosque

Malaria Control Centre denotes that the Sultanate is still aware of the
need to monitor levels of mosquito breeding, though malaria has been
eradicated almost totally. Bilad Bani bu Hasan ('Bilad' means 'region'
and Bani bu Hasan is the name of the tribe) has a very large fort, now in
such a ruinous state that you can drive straight through a gap in the
walls. The main keep, however, has been restored and closed. A
cannon dated 1790 was mounted with an Arabic commemorative
plaque in 1984. I wandered through the prosperous date plantations,
and at the Hamid Amr al-Kasbi Restaurant asked a vivacious lad
called Khalid 'Ali Salim Matar al-Balushi to show me a local mosque.
He, with Muhammad Rashid Matar and Ahmad Rashid Matar, took
me to a newly renovated Friday Mosque that in any other part of the
Islamic world would have passed unnoticed – from Fez to Mashhad.
The keynote is plainness: a simple niche for a mihrab, two windows in
the mihrab wall and two little niches for copies of the Qur'an al-Karim,
a roof of palm trunks, undecorated arches but, as a concession to the
modern age, a loudspeaker attached to the low minaret, reached by
simple wooden rungs fixed in a corner wall.

Another 7 km and you are in Bilad Bani bu ʻAli, which you might think almost indistinguishable from Bilad Bani bu Hasan. But the people here are Ghafiri and Wahhabi, not Hinawi and Ibadhi; their feuds go back to time immemorial, and ʻAlawi is still considered the most 'hard-line' sect in the relatively liberal religious climate of the Sultanate. On 9 November 1820 they defeated a force of British Indian Sepoys on their home territory, and were in their turn badly defeated there on 2 March 1821, when the town was gutted, many lost their lives, and many prisoners taken. But their date groves were spared in a singular act of clemency which enabled the village to survive, however tenuously, after retribution had been exacted.

Wellsted understandably felt apprehensive about travelling among these warlike tribesmen within two decades of these bloody battles, yet found nothing but goodwill there. After consuming a meal of camel, mutton, and rice, Wellsted encountered 'the whole of the tribe . . . of about two hundred and fifty men assembled for the purpose of exhibiting their war dance. They had formed a circle, within which five or six of their number now entered. After walking leisurely round for some time, each challenged one of the spectators by striking him gently with the flat of his sword. His adversary immediately leaped forth, and feigned combat ensued. They have but two cuts, one directly downwards at the head, and the other horizontally across the legs. They parry neither with the sword nor shield, but avoid the blows by leaping or bounding backwards. The blade of their sword is three feet in length, straight, thin, double-edged, and as sharp as a razor. As they carry it upright before them, by a peculiar motion of their wrist they cause it to vibrate in a very remarkable manner, which has a singularly

Bilad Bani bu Hasan. Mosque

striking effect when they are assembled in any considerable number. The shield is attached to the sword by a leathern thong; it measures about fourteen inches in diameter, and is generally used to parry the thrust of the spear, or jambeer. It was part of the entertainment to fire off their matchlocks under the legs of some one of the spectators, who appeared too intent on watching the game to observe their approach, and any signs of alarm which incautiously escaped the individual added greatly to their mirth.'

The huge castle of Bilad Bani bu 'Ali could accommodate thousands of refugees from tribal warfare if required, but pacification since 1970 has led to enhanced sentiments of unity between groups formerly factious and troublesome. The great fort has been completely restored on all four sides, as well as within; we found a tiny wooden door and a larger door both imperturbably shut. The empty square felt radiant with summer, though the date was mid-February. I wandered amid scatters of single-storey courtyard houses; occasionally a bearded goat eyed me anxiously, then as I made no threatening move it continued the search for scraps of paper on the sandy tracks between houses. The oasis spreads for miles in every direction, a kind of agricultural nomadism to parallel the wanderings of the *badu*. Yet here is a 'Toyota Spare Parts' shop, next to 'Bridgestone and Yokohama Tyres', and a diminutive 'supermarket' no more defined by its name than the 'Hotel' Jasmin.

I returned to the historic fort, now silent, recalling the 'thousand welcoming villagers' that had greeted Bertram Thomas in 1927, as he narrates in *Alarms and Excursions in Arabia* (London, 1931):

'Drums were beating, and the crowd swayed left and right to their rhythm: quivering sword blades flashed in the sun as sword dancers leapt hither and thither, and low chanting grew loud as we approached. Swinging round to form a corridor for us the tribesmen, holding their rifle butts to their hips for the *feu de joie,* sent a hail of friendly bullets pinging over our heads. We went to where the Amir [Muhammad of the Bani bu 'Ali] stood for my reception, before the fort, on a carpet placed in the large open square, a favourable position for witnessing the horsemanship and camel-racing that now took place; for, as I have said, such a display is the inevitable feature of an Arab welcome in Oman. A dozen horsemen galloped past, now in this formation, now in that, curveting and firing their rifles at the same time, or racing in pairs down the straight, one rider standing upright on his stirrupless saddle, gripping only with his toes, and maintaining a parlous equilibrium by placing an outstretched arm on the neck of his more comfortably seated fellow-rider. Reforming, the party would move past in close formation at a jog-trot, chanting heroic verse, an ancient Badawin custom deriving from the 'Antar of antiquity. The leader gabbled his lines and, at the end of each couplet, the rest of the party shouted in chorus "Allahu akbar!"'

Al-Ashkharah

The flat plain now gives way to duneland, and the track is subject to sand drifts. We met a *badu* camp 14 km from Bilad Bani bu 'Ali, and in another ten km sped along smooth sands, having left the hills of Ja'alan behind, a small sign (red on white) indicating al-Ashkharah in Arabic only. Ten km more and the sea appeared before us, riding even and majestic in contrast to the bumpy, corrugated track. In 4 km we reached a Fish Refrigeration Plant, then a Shell petrol station and the flat-roofed houses of al-Ashkharah appeared, like a crowd of brown squares flung on to a beach. We drew up before the Tawfiq Restaurant of Abu Arif Trading, and enjoyed cups of hot sweet tea. Young Omanis jostled to shake our hands and hospitably gave up their seats in front of the television set, operating off its own generator like the lighting, for electricity mains had not reached al-Ashkharah by early 1986.

C. S. D. Cole on Ja'alan

C. S. D. Cole, of the brig *Palinurus*, has left us this account of his journey in 1845 from al-Ashkharah (the 'Leskkairee' of his title and 'Leshkairee' of his text) to Bilad Bani bu Hasan, al-Wafi, al-Mintirib ('Bedeyer'), Sanaw, Muscat, Nizwa ('Nuswee'), Tanuf, Firq ('Fark'), Suma'il, Matrah, and Muscat.

An account of an Overland Journey from Leskkairee to Meskat and the " Green Mountains" of Oman. By Mr COLE, of the E. I. C. S. B. *Palinurus.*

[Presented by Government.]

ON the 22d of November, having obtained leave of absence to travel overland to Meskat, landed at the Village of Leshkairee in the Beni-boo-Ali territory. About a week previous, when off Roves, a small fishing town, I made an arrangement with one of the Bedouins in attendance on the Sheik of the Beni-boo-Ali's son, who was then on board the vessel, to meet me at this place, and accordingly I found him on the beach in readiness to start. Our arrangements were soon completed, and in less than half an hour from landing, mounted behind my guide, as he had only provided one camel, commenced the journey by proceeding towards Jahlan, where we arrived at 9 A. M., and went direct to the Ameer's house. When about six miles from the town my guide was alarmed by the report of fire arms, expressing great fear that some foul play was going on, but after the lapse of about a quarter of an hour, a noise of an approaching party was heard. The guide was immediately with his head on the ground, and after listening for a few seconds, he jumped up and moved on merrily, pronouncing the approaching individuals to be friends : and shortly after we met, and learnt that the firing had proceeded from them. On entering the town, my guide, Akmar, having dismounted, the camel suddenly took fright at a stray bullock, and threw me. I was somewhat stunned at first, but fortunately not injured. The Bedouin shewed the greatest solicitude, and it was some time before I could convince him that no damage was done. The distance from Leshkairee is about sixteen miles : the first part of the road was very broken and sandy, the latter generally level and more firm, with the usual tree of the desert, the Acacia, called by the natives " Gharf," but not at all plentiful. As soon as the sun was up next morning, I received several visitors, all those who had seen me last year expressing great delight at my paying them a second visit, and anxiously enquiring whether I would return again next year ; to which I constantly replied with the usual answer, " Insh Allah." The old Ameer, I was sorry to observe, was much altered for the worse. He appears very infirm now, and is constantly ailing, and I was much surprised to find that the fortifications he was so intent on repairing last year, were entirely abandoned, being exactly in the same state as at that time. Immediately after the noonday meal, took leave of the old Ameer, and set out in company with five or six companions for the adjacent town of the Beni-boo-Hassan's, where I purposed passing the night with my friend Ali bin Salam, who received me so courteously on my first visit to this country.

2. Two of my companions belonged to the tribe of " Beni Rasim," whose chief town, " El Wafee," I had been informed was the most flourishing in the district of Jahlan. They invited me to visit them, which I promised to do, and then parted. Having arrived in the Beni-boo-Hassan town, I proceeded direct to Ali bin Salam's house : he was out at the time, but a hundred messengers were soon in search of him with the news of Salam's (my travelling name) arrival, and in a few minutes he made his appearance. All the town soon heard the report, and, as usual, an im-

mense crowd assembled—one fellow, a slave, looking as proud as a peacock in a white jacket, and trowsers, which had been given to him by my shipmates (Messrs. Sylvester, Ward, and James,) when here. Many enquiries were made after these gentlemen, and all were greatly chagrined to find that they had actually been visited by an English doctor without their knowing it. This ignorance was extremely fortunate for the gentleman concerned (Dr. Sylvester,) for had his character been once known, the whole population would have beseiged him with an account of all sorts of imaginary diseases. Here I obtained a second guide, a man of the Hanaweger tribe, my other being a Junebi of the Beni Gharferee branch, —individuals of these tribes forming a mutual protection for each other in any part of Oman.

3. At 10 a. m. next morning took leave of Ali bin Salam, and started for "El Wafee," two hours' travel (five or six miles) bringing me to that place, and was kindly received by a young man (Abdalla bin Ali,) a cousin of the Sheik who was then absent. "El Wafee" is a walled town, —the houses larger and of a more respectable appearance, than those in the Beni Hassan country, but not so numerous. Lime trees are very plentiful; oranges, plantains, pomegranates, tamarinds, olives, and grapes, in moderate quantities; together with the usual articles, viz. git and mysably, being a kind of grass given to cattle, and dokhan, which produces a very small seed used for making bread. Barley and a little cotton is also grown, but no wheat. The Beni Rasim is not a very large tribe, amounting to no more than five hundred warriors, who are mostly armed with the matchlock, while some only carry spears, but the "jambeah" is universally worn. I found them busily engaged constructing a wall to protect their date groves from the Beni Hassan, a powerful tribe residing principally in the range of mountains above "Soor," with whom they were then at war. The rate of pay for labourers employed on this work was at the rate of one dollar for every twenty-four feet in length, the height being ten feet, and two in breadth.

4. At sunset took leave of my kind entertainer, and after two hours' travelling reached "Felaij el Maskayekh," a small Bedouin place, consisting simply of a date grove with a few miserable huts. The whole of the road from "El Wafee" was thickly covered with the Gharf tree of a much larger size than any I had hitherto seen.

5. The family of one of my guides lived here, and having expressed a wish for a drink of milk, I was nearly suffocated with the great quantities which their ideas of hospitality compelled me to swallow. Slept in the open

air, and next morning my blanket was completely drenched from the excessively heavy dew that had fallen during the night. I was anxious to make an early start to-day, as Bedeyer, my next station, the natives called a day's journey; I therefore concluded it to be distant about twenty-five miles: but with all my endeavours, it was noon before I could prevail on my guide to take leave of his family. Our road now was over a complete desert, a few bushes, and those very parched, only meeting the eye. From having set out so late, night closed in long before reaching Bedeyer, and my guides pretended to be very apprehensive, constantly stopping and looking anxiously around, and enjoining the strictest silence for fear of attracting any passing marauders: this however I felt pretty confident was merely assumed to enhance their services. As they were continually endeavouring to impress upon me the great dangers of the road, my constant reply was, that " having no feuds with the people of the country, none of them would molest an Englishman." This opinion, some time after, when they found that I was not to be imposed upon by such idle reports, they fully concurred in, admitting that in no part of Oman was there the slightest danger,—that on the contrary, wherever I went, every attention would be paid me, particularly as I was a " Serkali"—a term almost as generally understood in these parts as in India. At 8 P. M. arrived at Bedeyer, having travelled a good twenty-five miles. Passed the night in a small hut. The first thing next morning I was conducted to a silversmith's shop, where I found a numerous circle, who all rose to receive me, and coffee was immediately served round. My entertainer the silversmith was a very respectable-looking man, and had never seen an Englishman since our troops were at Jahlan. All present were very cleanly dressed.

6. Bedeyer is not a regular built town, but comprises some eight or nine date groves, which have sprung up around so many different springs: they are all separate, some three and four miles apart, the intermediate spaces sandy and barren, and the whole are situated in one large plain; the Soor mountains to the Nd. and Ed., with a range of sand-hills to the Sd. and Wd., beyond which all is desert.

7. This is the place from whence the dry date eaten in India is exported: it is the most valuable description grown, and is called mybsalee. Caravans of five and six hundred camels, laden solely with this article, leave daily for Soor in the date season, from whence they are mostly shipped for India. This trade renders Bedeyer the most wealthy place in Oman, Meskat excepted: the dates grown elsewhere being of a less valuable quality, are entirely used for home consumption. A single date tree of the mybsalee kind is worth on the average from ten to fifteen

dollars, whereas the various other descriptions never exceed five. There is not a single respectable house to be seen : the only thing like a building is a large fort with five towers, standing in the principal plantation, and inside of which is the market. The inhabitants are of the tribe of El Haiyarien, but have far more of the townspeople in their appearance than that of the Bedouin. There are several Sheiks, but none possessing any particular authority. This being the month of the pilgrimage, everything is gaiety, the principal amusements being horse and camel races ; and the whole population, dressed in their best, are strolling about from morning till eve. The women, who do not cover their faces, and are not at all good looking, generally wear a profusion of silver ornaments,—the ears in particular, which are perforated in numerous places, being completely studded with large rings : the weight is supported by a cord passed through the whole and tied over the head. A large ring round the neck, with dollars pendant, is universally worn. All eyes were of course directed on me, but more especially on the part of the ladies, who collected round me in crowds, and I was not a little surprised at their strain of conversation. The most indecent questions were put to me, and the whiteness of my skin greatly astonished them,—one woman exposing my breast, to the admiration of her surrounding friends, comparing her own with it: the colour of my hair, red, was also a matter of great curiosity, and when I asserted it to be the natural colour, the exclamations of surprise by the fair sex were unbounded.

8. On the second day of my stay here, observed a large cavalcade entering the open space in front of the fort, where the races were going on, and immediately recognised one of the party to be " Said bin Abdalla," cousin to the Sheik of Soor, who I had met at that place a short time previous. I then learnt that this was the escort of the Sheik of the " Al Wahabies," (Khalifeen bin Ali,) who was en route to his own country from Soor, where he had been on a visit to the Sheik Mahomed. His arrival was extremely opportune, as I was anxious to pay him a visit (my shipmates having met him at Jahlan, and been much pressed to accompany him home, which circumstances prevented), while my guides were as desirous to avoid it, representing that part of the country to be extremely dangerous for strangers. Meeting the Sheik himself, therefore, was very fortunate, the matter being settled by my joining his party ; and accordingly at 4-30 P. M. that day, riding by the side of the Sheik at the head of some forty of his followers, set out from Bedeyer and travelled till about 8 P. M., when we encamped in a small valley thickly covered with the gharf tree, for the night. A large fire was soon enlivening our

bivouac. A sheep purchased from some Bedouins living in the valley, was slain, cooked, and devoured in less than an hour : the method of cooking was by spreading the flesh on large stones previously heated in the fire. It was extremely good, and as I had tasted nothing since the morning, I might have competed with my wild companions in doing justice to it, but unfortunately there was nothing to take with it—not even a bit of salt. Before sunrise next morning we resumed our journey, and at noon halted at a small date grove, from whence the Sheik's house is distant about one mile. The first hour or two after our arrival was wholly taken up in receiving and returning salutations, every soul apparently in the place flocking towards us ; and as on the approach of every new comer all present arose and remained standing till he had saluted every one of our party, I was never more pleased in my life than when a slave desired me to follow him for the purpose of taking some refreshment. I was greatly surprised to observe that even a slave on his approach to salute the Sheik, was treated with the same respect by the company rising to receive him : the slaves, however, invariably kissed the Sheik's hand, and most of them made the motion of doing so to all of us. At sunset set out for the Sheik's residence, which we soon reached, and found it a mere hut erected under the shelter of a tree. Carpets were immediately spread under some adjoining trees for the accommodation of myself, Said bin Abdalla, and our attendants, and supper served up. The Sheik did not join us, as it is not the custom for the host to do so, but made his appearance with the coffee when the edibles were removed, and remained till near midnight. In the morning, when about two miles from the date grove where we halted for the day, met a party of three Arabs and with them a slave. Some conversation took place, when I observed the Sheik order one of his followers (a nephew) to dismount and give his camel to the slave, who returned in our company, as also his companions, the three Bedouins. From observing no signs of anger at the time, I fancied they were a party purposely come out to meet the Sheik : I was therefore greatly surprised to learn, in the course of conversation, that these men had kidnapped the slave from a neighbouring tribe, and when met were on their way to Bedeyer to sell him. One of the thieves was present when the Sheik told me this, and instead of shewing the slightest annoyance at having failed in his design (the slave was to be returned,) he was one of the loudest in the laugh raised at my expense when I explained how theft was punished in my country. The slave trade was likewise discussed, and the utmost astonishment was expressed at the efforts of the English to put it down, observing that it was an indisputable fact that " negroes were especially created to be the servants of the white man."

The Al Wahabies are a powerful tribe, nearly as numerous as the Junelie, numbering about one thousand warriors, most of whom are armed with the matchlock. Their principal wealth consists in camels, possessing more than any other tribe in Oman with the exception of the El Awakman, the most powerful tribe in this part of the country, and who entirely live in the desert, subsisting on their flocks of goats, as they do not possess one single date grove. On the 30th of November, being the third day of my stay with the Al Wahabies, resumed my journey, and after three or four hours travel reached the town of Senow in company with Khalifeen bin Ali, who was about paying a visit to Meskat, where he repairs twice a year to receive " Buksheesh" from the Emaum. " Senow," a small town belonging to the Al Wahabies, and distant about nine miles from the residence of the Sheik, is situated on a hill with date groves stretching beneath. The houses, two stories high, and built of mud, are tolerably respectable ; and the inhabitants number from three to four hundred. A little before midnight set out from Senow, still in company with Khalifeen bin Ali, and continued at a quick trot till 1.30 A. M., when we encamped in a small valley sheltered by the acacia, having travelled about eleven miles. Next morning I was surprised to find the valley completely crowded with men and camels, upwards of a hundred men having assembled during the night to follow the Sheik to Meskat, as none on such an occasion come away empty-handed. I was greatly pleased with the scene thus presented, as, from the morning being chilly, fires were kindled in every direction, surrounded by different parties, who, as the sun rose, approached our fire to salute their Chief. So great a number, however, was anything but pleasing to that personage, who only desired some twenty followers to accompany him ; I was therefore not at all surprised, on arriving at Meskat, to hear that he had returned to his own country on the plea of being unwell, this being the only means of ridding himself of the incumbrance, for had he expressed any open disapprobation, a quarrel in all probability would have ensued as to who should be the lucky individuals to form the escort. The hint, I was informed by some of the tribe, will have the desired effect, and after a few days' delay the Sheik will be enabled to resume his journey with whatever number of attendants he may please to take. At 9 A. M. Khalifeen and the whole party resumed their journey, when we parted company, as his rate of travelling was by far too rapid and fatiguing for me. They have to make a wide circuit, in consequence of being at war with the people who occupy the direct route, which is no more than eighty miles, whereas the " Al Wahabies" will have to traverse at least twice that distance. Being very stiff from last night's journey, my progress to-day was very slow. Passed several small date groves, but all so very poor that we could not procure a little grass for our camels : two of them were entirely destroyed, the trees being all dead from the spring in their vicinity having dried up. At 4-30 P. M. reached "Minnik," where I was kindly received by the Governor of the place, who immediately conducted me to a house,—the same that Lieutenant Wellstead occupied during his stay here. The distance travelled from the encampment of last night was about sixteen miles: the first part of the road had a range of hills on either side, and the latter environed by detached hills. " Minnik" is a walled town with narrow streets. The houses, two stories high, are built of mud and stone. The population is from four to five hundred : in Wellstead's time, (of whom particular enquiries were made) it exceeded double that number. This place has suffered more than any hitherto seen from the want of rain; no less than seven out of the eight springs with which the town was supplied are now quite dry, and, in consequence, large tracts of once cultivated land in the vicinity now lay perfectly waste, and whole date groves are entirely destroyed. The only flourishing plantation at present is to the N. and Ed. of the town : it is walled in, and formed into regular roads, the largest leading from the N. Eastern gate, being about five or six hundred yards in length. Here I saw a little sugar-cane growing : at one time great quantities were produced, for the manufacture of sugar, which they clarified with eggs.

9.   On the second day of December, a little after noon, set out for "Nuswee," distant six or seven miles, the surrounding country being very hilly : passed several date plantations, many of them ruined and deserted. At 3-30 P. M. reached the town, and proceeded direct to the fort, where I found the Governor and several of the principal inhabitants, dressed quite gaily, making my appearance, which was anything but clean, still more conspicuous. I was received very kindly by the Governor, Seafe bin Ahmar, an old gentleman with a very pleasing appearance ; and an apartment just outside the fort, where he resided, was immediately prepared for my accommodation. Shortly after my arrival a large party, all armed, and preceded by drums and horns creating a most discordant noise, approached the fort, when those present with the Governor joined them, and invited me to do the same. The whole party then formed into procession, the drums and horns leading the way, headed by a man who was continually crying out " a cheer for the Imaum, a cheer for his son, a cheer for the Englishman." This was responded to by the whole multitude yelling with all their might, until arriving at the grand square, when a circle was formed, and sham fights with the spear, sword, and matchlock,

took place. The tops of the houses on either side were crowded with well-dressed women with faces uncovered, but I certainly did not observe more than half a dozen pretty ones amongst the whole of them. The fight continued for about a quarter of an hour, when the party moved on to the middle of the square and commenced again. A third movement was then made, bringing us to the end, where we remained till a gun was fired from the fort, which is situated at one extremity of the square, and at the other extreme are two small towers, placed one on either side; when the word was given to face about, and we returned exactly in the same manner as in the advance, the ceremony having occupied nearly two hours. It is observed every year, at the conclusion of the pilgrimage, and continued for three days, during which time the market is closed, and business of every kind suspended. After the evening meal, the Governor, accompanied by half a dozen friends, came up to my quarters, when the old gentleman enquired if I was fond of singing; and being answered in the affirmative, he immediately dispatched a messenger to summon the " Bard of Nuswee," who he assured me had frequently performed in the presence of the Imaum's son, Said Khwainee, Governor of Meskat. The individual in question soon made his appearance, and after being regaled with coffee, commenced in true Arab style. He was rapturously applauded by the company, with the exception of my guides, who complained of not understanding a word he said : the performance, however, was tolerable, but by no means to be compared with the public singers at Cairo.

10. Nuswee, the largest town in Oman (Meskat excepted,) is situated about five miles from the base of the Green Mountains, and is entirely surrounded by hills : the heat in consequence is far greater than in other towns of the province, which are generally in more open situations. The town is built with regularity : the houses two stories high (stone) and of a respectable appearance, with narrow streets. The fort, which is round and built of stone, is about eighty feet high. I was not allowed to enter it, the Governor having orders to admit no stranger without an order from Meskat. I only saw two or three old rusty guns projecting from the embrasures. Date groves extend in every direction, but, as at Minnik, several are withered up, and a great deal of land laying waste for want of water.

11. The drought has continued now for five years, and it is feared that, unless a change speedily takes place, the whole province will be ruined. Notwithstanding all the philosophy of the Mahomedan character, the

greatest anxiety prevails throughout all classes, and I was continually applied to to know if any rain was to fall this year. The population at present does not exceed three thousand, as many families, from the ruin of their property, have been compelled to leave, generally repairing to the towns on the sea coast. It contains, I was credibly informed, nearly three hundred mosques: none of them, however, are of any considerable size—on the contrary, the majority are very small, and all built of mud, and, as they have no spires, the appearance of these buildings is very mean. The market is small but neat, and covered in: its length is about seventy-four paces, and twenty-five broad, containing four rows of shops, one on either side, with a centre building of two more. Copper vessels are manufactured here; and Nuswee was once celebrated for its sweetmeats, but none good are procurable now, from most of the confectioners having left. At noon on the third day of my stay here, set out for the Green Mountains, and a two hours' ride brought me to the small town of Tenoof, distant six or seven miles from Nuswee. This is a pretty little place, surrounded by a wall pierced with loopholes, and a castle stands at the eastern extreme, with a fine flourishing date grove to the westward. The houses, built of mud and stone, are small, and the population about three hundred. I had been given to understand that this town was actually in the mountains, but I now found its position to be immediately at the base of them. The Sheik received me very kindly, and undertook to provide me with a donkey and guide for the ascent; and at 7 o'clock next morning, having sent my camel back to Nuswee, I was again in the saddle. In five hours I reached the summit. The ascent was very steep, and the road dreadfully bad, being covered with large loose stones, so that at times I was quite at a loss how to proceed, as I could not discern the vestige of a path,—but not so the donkey, for, after taking a little rest, he would start on again, and carry me over places that I should have thought impracticable for any beast of burden. We sat down for an hour or two here, my guide being rather fagged, and then proceeded by going down a very deep descent on foot, after which the road led over a continued succession of ridges, but none too steep for riding,—four hours' travel bringing us to the first inhabited valley (Waddy bin Abeeb,) where I passed the night. This valley is narrow and precipitous, in most places not more than thirty yards across. The houses are small, and stand on one side about half way down, and both sides of the valley are laid out in terraces rising one above the other, where most of the cultivation is carried on, consisting almost entirely of fruit trees, viz. pomegranates, figs, peaches, apricots, walnuts, and sweet limes: the pomegranates are very plentiful, and a little wheat and barley is also grown ; while the hills on either side are completely covered with the vine. About noon set out for another valley, (Waddy Seek,) distant about four miles. This valley is much larger than bin Abeeb, and not at all precipitous, the approach to it on either side being by a gentle descent. The same fruit trees abound here as at the former valley, with the addition of the mulberry, which is very much prized by the natives. This was the limit of my excursion, as only five days remained to the expiration of my leave. At 10 A. M., therefore, on the 7th of December, commenced retracing my steps, and by 8 P. M. was once more partaking of the hospitality of the Sheik of Tenoof. I found the descent far more fatiguing than the ascent, and had I not been barefooted, (my shoes having been cut to pieces during my short stay in the mountains,) I would never have ventured on riding, for one false step would have been destruction to both rider and donkey.

12. The Green Mountains of Oman are merely a small part of the great range that traverse throughout this part of Arabia. The whole of the cultivated valleys might easily be seen in a couple of days, and their appearance is very tame, the hills being quite bare, and no trees of any size to be seen, with the exception of the walnut, which is cultivated with the other fruit trees in the valley. I was received everywhere with the greatest kindness, and here again frequent enquiries were made after Lieut. Wellstead. The inhabitants of these valleys indulge in wine of their own making ; it is like the common black wine of Europe, only much inferior. Their excuse for doing so is on account of, as they consider it, the excessive cold of the mountains. I found it rather cold in the mornings about daybreak, but I do not think the thermometer could ever have been lower than seventy during the three nights of my stay here. I saw but few cattle, and altogether the natives appear to be exceedingly poor. Such a thing as coffee was not to be procured. At noon on the 8th of December, accompanied by two horsemen supplied by the Sheik as an escort, started on my return to Nuswee. At the moment of leaving the town a large party arrived, one of whom was the Sheik of a district about twenty miles distant : he pressed me very much to remain and accompany him home, and nothing would have afforded me greater pleasure, had I had time to do so, but as my leave expired on the 11th, I was obliged to decline. On nearing Nuswee my escort took leave, the Sheik of Tenoof being at variance with the Governor of that place. My guides were delighted to see me return on the day specified, as they were anxious to reach Meskat, for which place I desired them to be ready to start the next morning. Accordingly, at 8.15 A. M., accompanied by the

Governor and three of his friends, set out from Nuswee, and at 10.30 A. M. stopped at a small village (Fark) to breakfast, which was prepared in the mosque, when we again proceeded on our journey, and about two hours after reached a large village (Berkee) situated between hills, with a large building surrounded by a wall to the westward, and to the eastward a date grove, distant about ten miles from Nuswee : here we came to a halt, a carpet being spread under a tree for our accommodation. As this was the place where I was to part with my kind host, he insisted that I should spend the remainder of the day with him : it was therefore arranged, that, as I had no time to spare, my delay during the day should be made up by setting off immediately after the evening meal. The inhabitants of this village appeared to be in very good circumstances, and all were desirous that I should prolong my visit. A short time after sunset sat down to smoking dishes of rice and goat flesh, with liberal supplies of milk for drinking. I was placed by the side of the Governor, an extra dish of five wheaten cakes, soaked in ghee and sugar, being prepared for our particular gratification. Directly after coffee took leave of the worthy old Governor, who expressed great hopes of seeing me again : his friends, however, still accompanied me (one was his brother-in-law,) on account of the natives of this part of the country being very suspicious of strangers, particularly of Bedouins. Travelled till midnight, when we halted for the night at a travellers' bungalow in a village named Mettee. Most places in Oman have a building set apart solely for the use of travellers. On awaking the next morning found several of the inhabitants seated around a blazing fire with my companions, and soon after an excellent breakfast was provided by the Sheik of the village ; after which the journey was again resumed, one hour and three quarters bringing us to a valley inhabited by the "Beni Ruwahi" and "Beni Yaber." At 11 A. M., one hour after entering the valley, alighted at the Sheik's house. Here my friends from Nuswee left me, a son of the Sheik's taking their place. After coffee, remounted, and at 4.45 P. M. reached " Sumaiyel," having travelled since the morning about twenty miles. This valley, known by the name of " Waddy Beni Ruwahie," is about fourteen miles in length, and generally very stony. The first part is narrow, but opens out gradually, with date groves throughout, protected by towers on the neighbouring hills on either side of the valley, and is in consequence considered almost unassailable by the Bedouins.

13. On arriving at Sumaiyel, took up my quarters in a large room just outside the principal fort, where the Governor (a Beloochee) resides. This I found to be a place of considerable extent, and the most flourish-

ing of any that I had seen in Oman. Water was plentiful; the date groves, which are very extensive, were in the best condition; and everything about looked green and healthy, forming a great contrast to the other parts of the country that I had passed through. The scenery, too, was far superior, the surrounding country being very hilly, with small fortifications on different heights; but all of them are in a sad state of dilapidation, and none possess any artillery, except the citadel, and that only boasted of two or three rust-eaten six pounders. Early the next morning left Sumaiyel. My guides were not at all pleased with the Beloochee Governor, because he had not furnished a very sumptuous supper the preceding evening, though, considering his means,—his pay not being more than twenty dollars a month,—I thought our entertainment very good. I was sorry to find that my guides on all occasions were very greedy, as, when anything superior, such as sweetmeats, was supplied us by the people through whose district we passed, they were not satisfied with what they eat at the time, but invariably, when an opportunity of doing so without detection offered, pocketed as much as they could, notwithstanding my remonstrances against such conduct. Throughout the day's journey the country was very hilly, with date groves here and there. A little before sunset put up for the night in a small village about nine miles from Meskat: this was the only place where the slightest inhospitality was shewn me during the journey, the natives declaring that they could neither give or sell us any kind of food—not so much as a little rice, or a cup of coffee, but were anxious that we should proceed, by declaring that there was a large village a little further on, where everything in abundance might be procured; but, in consequence of a fall from my camel in the afternoon, which had stunned me, I was in no condition to travel more that day, so that I was determined to abide here for the night. My guides did not approve of this resolution by any means, as they did not at all relish the idea of going to bed with empty stomachs. Being very much knocked up I soon fell asleep, and on awaking about two hours after, I was greatly surprised to find a goodly supper of rice and meat, with a large bowl of milk, placed before me. On enquiring how this came to pass, in a place where at first things had appeared so unfavorable, I learnt that my guides, hearing that a woman, whose husband was in Government employ at Meskat, resided in the village, repaired to her with the information that a sick "Ingleez" had arrived, and was very much in want of some refreshment, for which liberal payment would be given. The appeal was not in vain, for she set about preparing supper forthwith, but rejected all offers of money. For-

168

tunately I had some few things remaining from the presents supplied me on leaving the vessel, though none of any value, with which I was enabled to make some slight return for her extreme kindness. Before sunrise next morning, 12th December, mounted my camel for the last time, and at 9.30 A. M. reached Matrah, and was rather disappointed to find that the vessel had not yet arrived, as I had been informed, when at Sumaiyel, that she had arrived three days previously, which induced me to hurry on with all speed. At Matrah I immediately took boat for Meskat, where I was soon comfortably lodged in the Agent's house. Being much in want of clean linen, and having none of my own, I was glad to put on the flowing robes of the East till the arrival of the ship, five days after, enabled me to obtain a fresh supply of my own costume. The only animals I could learn of as inhabiting Oman, were the following, viz. a large animal the size of a bullock, with straight horns, and perfectly white, called Aboo Sota, and the Ibex, as also hares, gazelles, wolves, foxes, and leopards, the whole of which are eaten by the Bedouin. Rock salt is very plentiful, and likewise brimstone.

The natives of Oman do not smoke, and none of their mosques have spires,—in which they resemble the Wahabies, for whom, however, they have a great dislike, and are rather apprehensive that this sect will again become troublesome to their neighbours.

(Signed.)   C. S. D. Cole.

(True Copy.)   A. Malet, *Chief Secretary.*

Miles on al-Ashkharah

S. B. Miles visited al-Ashkharah in September 1874, travelling there by sea, on board H.M.S. Philomel, with Sa'id bin Sulayyin. His account is drawn from *The Countries and Tribes of the Persian Gulf* (London, 1919). His purpose was 'to enquire into a piracy case', which in British colonial terminology of the time meant punishing raiders with a fine, levied by Miles at 600 dollars. The whole narrative is well worth reading for its pungent evocation of al-Ashkharah, Qalhat, Yiti, Ras al-Hadd, and especially the individuals concerned, judged with a kind of rough objectivity within the British Victorian world view.

VISIT TO AL-ASKHARA IN 1874

August 31st.—Left Muscat at 4 p.m. for Al-Askhara with Saeed bin Seleyyim on board H.M.S. *Philomel* to enquire into piracy case.

September 1st.—Saw Kilhat at 10 a.m. and reached Soor at 2 p.m. Sent Dervish and Saeed on shore to call on Wali and landed at 5 p.m. and found Wali just arrived. Held a conference in Jethoo's house and arranged that Majid and Abdulla should be written to, to meet us at Al-Askhara. Was told that a collision had occurred the day before, in which two or three men were killed. The Wali Nasir Bin Mohammed al-Bu Saeedi arrived at Soor eight days ago overland, but he has no power and is not likely to stay for he gets no revenue and has had to pay 60 dollars already to Beni Bu Hassan Bedouins. Visited town and Khor afterwards and saw a good many bughlas still in creek getting ready for sea. Returned on board in evening. Dervish stayed.

September 2nd.—Dervish and Jethoo came off. Wrote to Majid and Hamod. Jethoo says no danger in travelling about the country provided you have khafeers Ghafiri and Hinawi with you; these protectors are necessary to ensure safety. If the Bedouins find you alone, even in the temporary absence of these men, they will certainly plunder you.

People in Jaalan are haramis and not to be trusted when they scent money. Jethoo has been to Kilhat, where there is a ruined palace or fort under which, the people say, is a buried treasure, but they are afraid to dig on account of the Jenebeh and Beni Bu Ali, who would come down and plunder them of all they had and perhaps drive them out of the place altogether. Seedees sometimes go and look for gold coins there after rain, which they sell to the Soonais at

Soor for a dollar or so ; these coins, from the descriptions, seem to be cufics.

At 4 p.m. landed to go to Belad. Road leads north over stony and rising ground. Outside the town the khor appears to go some distance up but is here shallow. The forts of Senaisila and Shenoo are passed on the right of us ; forts have guns but are not garrisoned.

Road then turns to left until we reach Eis, which is a small defile with a mined fort on top, destroyed by Thowaini. It belonged to the Jenebeh. Belad is now visible; it is in a low plain or valley and is a mass of date-trees, few houses being visible. On reaching the fort, which is in a tumbledown condition, I was received by the Wali with a salute. Sat in the gateway for some time and then took a walk over the place. Outside the fort is the ruined Sook or bazaar, which was destroyed by Sedeyree the Wahabee. The Banians' houses are still standing. Not much cultivation here. Melons and a little grain seem to be all. I was invited by the Beni Bu Hassan Shaikhs to visit their country, which is beyond Wafee, they said. Wali seems an intelligent and determined man but has no sinecure in dealing with these Jenebeh scoundrels.

Returned to Sahil about 7.30 p.m. ; quite dark, obtained a pilot, Ali bin Mobarek, who came on board with Dervish at 10 p.m. By the inhabitants Soor is called Al-Sahil, and the upper town is known as Al-Belad.

September 3rd.—Started at 5.30 a.m. and reached Ras al-Had at 9. The actual point of this cape is very low, merely a sandy spit, but the hills in the background are visible from a great distance. The residence of the Shaikh is a small fort with three towers, and the hamlet is called Belad Ras al-Had. There are, perhaps, 600 people under the Shaikh, called Beni-wa-Beni, and consist of three divisions, the Al-Mowalik, Beni Ghozal, and Beni Mahra, the last tracing its descent from a Mahra ancestor. The coast is bold but the rocks very bare to Ras Khabba. I saw a few Bedouins on the hills of the Beni Harbee, a small sect of thirty men. Between Khabba and Ruwais the coast is called Saweih Beni Bu Ali. It strikes me to-day that Kilhat is the Acilla of Pliny. Forster has missed this identification, but the article in Al-Kilhat is incorporated with the name.

Arrived at Al-Askhara at 4 p.m. Much less windy than I expected round the cape. The town appears to be large, of mat huts mostly, with a few stone ones, and lies among the sandhills a few miles from the beach, but on account of the hillocks of sand only part of it is visible from the sea. The fishing boats are observed all drawn up in line and covered with matting, about forty of them. There is no fort, but two ruined towers stand on the beach. There is a projecting line of reefs to the south of the town, which serves to distinguish it and acts as a shelter to the boats from the surf. We see only one boat still in the water, and no signs of anyone coming off.

September 4th.—Shaikhs arrived this morning and are pitching a tent on the beach made out of an old sail apparently. At 10 o'clock a canoe with two Seedees came off with letters from Majid and Abdulla. Majid says he is ill, and Abdulla wants some one sent on shore. I sent Saeed bin Musallim and the pilot to them with letter requesting them to come on board (not too many) and promising safety. Seedee says no slaves have arrived this year from Zanzibar, there are few slaves in Al-Askhara but numerous freedmen. At 12 o'clock the boat returned with Shaikh Abdulla, Yusof bin Ali, the Jenebeh Shaikh, Ali bin Khalfan, and the pilot explained to them what I had come for and heard their explanations of the matter which was that the Lashkhareh boats, seeing a disturbance among the Kitia men on the island, went to interfere and stop the quarrel, but that the Kitia men, seeing them coming, became afraid and ran away. Told them I had come to demand restitution of the plunder and the surrender of the offenders and their boats. They admitted the plunder to have been about 300 dollars and agreed to pay it up and said they submitted at once to the demands of Government whatever they were ; I gave them until to-morrow morning to consider and reply. Ali Mubarek says he considers the limit of Oman to be the same with furthest land of Al-Jenebeh at Ras   *

* Name omitted by author, probably he meant Ras Noos.

opposite Kooria Mooria. The Jenebeh extend to Jibsh, then come Hikman, then Bar al-Hadhei, then Al-Waheebeh, then Jenebeh again to the Garas tribe. Al-Hadhramaut commences from east at Shihr. The Mahra country is not in it. There is no general name for the country between Jaalan and Al-Gara. Mesineh belongs to Jenebeh. It has no cultivation and the inhabitants are Bedoos, who live on milk and flesh. The limit of Oman towards the desert is Adam. The Awamir at Ras Noos are genuine Bedoos and wear Hadhramaut clothes and speak a different dialect to Oman; it is hardly intelligible but is better Arabic.

Some Awamir have lately arrived at Muscat and joined Turkee, they came overland all the way on camels; in some places they were five or six days without water. They pay no road tax but are obliged to take kefeer. If protégé of kefeer be murdered, blood money cannot be taken, he must be avenged by blood. The same with the kefeer; the custom of kefeer is punctiliously adhered to. If the traveller or kefeer be killed, and, if both the kefeers are Ghafiri and Hinawi, the tribes rise and unite to avenge him. Camels in some districts are only watered once or twice a month. In passing waterless tracts, camels go without, and men drink the milk only.

A schooner was wrecked on this shore in Rassam's time and the crew taken overland to Muscat by Bedoos, who fed them on camels' milk and dates, which gave them diarrhœa, as a remedy for which the Bedoos seared them all over with hot irons.

September 5th.—At 10 a.m. I sent a boat for the Shaikhs, who came off and said that during the night the offenders had bolted from the town into the interior, but they had sent some of the Lashkhareh people after them and expected they would be brought back. The night before they had agreed to surrender themselves, on condition of the other Shaikhs accompanying them to Muscat, to which they had consented. They said the Jaaferees had never committed themselves before and were very repentant, and that if the offenders did not surrender they would hand over the six boats belonging to them. I said we remembered good actions as well as bad ones and their conduct in shipwrecks was not forgotten, that they had misbehaved once and might do so twice, etc. I then said I would take 600 dollars on account of the property plundered and as a fine, but as I was going this evening I wanted it at once. The Shaikhs said they could not collect it for three or four days but said they would deposit jewels with me until paid. They said they would come off in the evening with Jaafera headmen to conclude business. I gave Abdulla a gun and some cloth. At 5 p.m. I sent Saeed on shore to fetch them, but he returned alone at 5 p.m. and said the Shaikhs had not been able to collect anything yet and had therefore not come. The six men concerned had returned to the town.

September 6th.—Sent off Saeed at 9 a.m. to the shore; he returned at 2 p.m. with the Shaikhs and about half a dozen of the Jaafera elders, dirty looking men. I admonished them, and they assured me they were not with the six men in this matter and had not approved of their proceedings at all, and promised it should not occur again. Yusof said they had endeavoured to collect the money and jewels but had got little. In the whole place they had collected less than 200 dollars and the ornaments were so paltry that they were ashamed to show them. They promised to come round to Soor with me, Yusof and Abdulla, and borrow the money from Jethoo, and hand it over there. In the meanwhile, prohibiting the Jaafera from going to sea and seizing their date-trees until the money was paid up, I required that some of the Jaafera should also come round to Soor, and it was so arranged accordingly, namely, that eight of them should accompany us. I told them to be ready at daylight to-morrow.

September 7th.—I sent a boat at daylight, but it not coming off I sent Saeed in whaler to hasten them. They came off at 9.30, Abdulla, Yusof, and eight or ten others. The reason of the delay appears to have been that on sending for the six culprits, the latter had taken up arms and refused to come without a fight, a row ensued and matters were looking serious. Saeed, however, ended the matter by letting them off. The Shaikhs have bound the Jaafera over not to put to sea this season until the fine is paid, the dates and gardens belonging to them have also been pledged to Abdulla until released.

The Beni Bu Ali and Jenebeh are Wahabees and will not eat meat killed by others. The Shaikhs hold 200 dollars and have to produce 400 more to make up the account. Abdulla, it appears, has dealings with Jethoo, but Yusof's credit is not good apparently. I arrived at Soor at 6 p.m. and sent them on shore to do what they can with Jethoo. The boat did not return till 8 o'clock.

September 8th.—Seyyid Nasir, the Wali, and the Shaikhs with Jethoo came off at 11 a.m. and said the arrangement was concluded, as they had obtained the loan. Gave them some more warnings and advice, especially about the slave trade. Told them they would not be interfered with unless they carried slaves in their vessels. Jethoo said he was willing to lend them the money on the terms accepted by them. They have handed him over 200 dollars and are to stay in Soor until the balance is produced. The Wali said there was no fresh news from Muscat or about Sultan. The Shaikhs returned on shore at 1 o'clock. At 4 p.m. I landed with Garforth and went to Jethoo's, where we saw the Shaikhs and Wali. Garforth and others then went with Wali to Al-Belad, while I went on camel to Senaisila to examine the fort there. It is untenanted and is in a state of disrepair but is still serviceable, the gate is the strongest part and is towards the sea, but has a wall in front. This fort is completely commanded by the hills near.

The Senaisila are Ghafiri and number 200. I saw a moolla there, apparently a Pinjabu or Hindustanee man. There is a large musjid and several decent houses. I returned at dusk and had a conversation with Abdulla and Yusof about Turkee having cast off Ghafiris and taken up Hinawis. They said that Saleh bin Ali looted Muttrah and had since had all his own way in the country and they were afraid he would induce Turkee to attack the Beni Bu Ali, as Saleh was bitterly hostile to them on account of their having killed Azzan and put Turkee on the throne. I reminded them that when Saleh came to Muttrah not one Ghafiri in the country had raised a finger to help

Turkee and that he had not been supported by the Ghafiris as he ought to have been, but that the Beni Bu Ali were friends of the E   ,* and if any split occurred between them, I should be ready to use my good offices to prevent hostilities and to maintain a good understanding and peace in the country, etc.

They produced the bag of dollars afterwards, and it was formally accepted of them. Dervish then distributed his presents to them all. The pilot got fifteen dollars. I promised Shaikh Abdulla bin Salim a gun for his son Mohammed, who is a promising lad and has influence. Yusof is to have a gun when he next comes to Muscat.

The Wali returned with Garforth and party at 8 p.m. and said he had to go back immediately as some of the Beni Bu Hassan Bedouins were there and were coming on to Muscat to get nafa. The Wali receives a handsome reward as he has been most attentive and obliging all through. Returned on board at 8.30 p.m.

September 9th.—Started at 6.30 a.m. for Kilhat, where we arrived at about 9 a.m. Landed at 10 and visited ruins. They are situated between the shore and the hills, which rise about a mile up and are of very uneven ground, covered with loose cobble stones, boulders, and coral.

The southern limit is a wall, stretching from the sea to the hills towards Soor, in which were three towers, the outlines of which are clearly traceable; one of which fell down only last year. The foundations of many houses are visible and the walls of a few are still partly standing. There is also a tomb, once domed, and lined inside with coloured tiles, it is about fourteen feet square and is now minus the domed roof; there is a vault underground which probably held the coffin but is now empty. Close by is a rectangular and deep reservoir for water, partly filled with rubbish, but with a fine wide arch over it; this was evidently to catch rainwater, being in a water-course.

To the north of this is the Wady Asin, which comes out of the Wady Khalid, and has a flow of good water, sufficiently plentiful to

---

* Name omitted in original MS.

irrigate some fields at Wady Sakrat, a valley five or six miles up, where there are kanats, and there is a large pool of water at Meida, a short way up, where the women go to wash, bathe, and fetch water; the outlet of the torrent is wide and shows that a considerable body of water comes down it. On the north of this is the present Kilhat, now merely a fishing village, inhabited by the Shaaban, a petty tribe of 200 souls; these people draw their water from shallow wells or pits in the torrent bed and possess goats and sheep. To the north of this is another smaller torrent, which debouches into the sea near the village.

North of this again are some more ruins, and this appears to have been the chief part of the old town; there was a tower and perhaps a wall here too, and here I got a fragment of an inscription. I have not, however, inspected this part properly. Here are said to be the remains of an aqueduct leading from the interior and is in good preservation, the name of which is Ghala. Coins are found here with Oman mintage, probably struck here or at Nezwa in the time of the Bowide Dynasty. This place is not much like Aden except in miniature. The view from the Front Bay, where the harbour is under Seera Island, would present something of the same appearance, only Aden is on a larger scale and the grandeur is wanting in this. Kilhat looks a poor place and most unsuitable for a town at first sight, but has advantages on further inspection, having plenty of good water, which is scarce at Soor. This was probably the cause of its being chosen in preference to Soor. The houses and buildings were all of coral and cobbles, no hewn or shaped stones anywhere being visible. Received from the Shaikh some tiles from a tomb brought from Persia and an inscription in coloured stucco done by a Persian artist. The water is very deep close in shore. To the south of the south wall is a large iron ring in the rock over the sea, probably used for mooring. The coast all along here is steep, ten to fifteen fathoms being found within 100 yards. There is a low headland or reef sticking out some way at Kilhat south of the khor, which forms a good anchorage for bughlas. Soor is certainly more suited for a large port, but I can easily understand why Kilhat was chosen in preference in former days ; the bar and the entire absence of sweet water at Soor militate strongly against it.

At 12 o'clock I went on board, after giving the Shaikh a few dollars and some powder, in return for the inscription which I carried off. At 2 p.m. the *Philomel* anchored off Taiwee, which is situated at the mouth of an apparently deep valley with very precipitous sides, but there is a patch of palm and a streak of water visible. The village is to the north of the wady and close to the beach. At 5 p.m. the Shaikh came on board. Afterwards I landed and walked up the valley and found the town clean with many stone houses, having 500 or 600 inhabitants, all Beni Jabir. Khutba, until fifteen years ago, read for Sultan of Turkey ; Sunnis and Wahabees in Jaalan still do so.

The Shaikh considered that the Sultan of Turkey was Khalifa of Islam. Up the valley is the reputed tomb of Ibn al-Mukaire, a celebrated poet of Al-Hasar, who retired here and built his own tomb on the hill. To the south of Taiwee is Wady Fakk, which is quite distinct from Taiwee valley and shorter but with a greater flow of water. Just south of Taiwee are the ruins of Jereyf, an ancient town destroyed at a very early date, but there is a tower still standing and some remains.

178

Sand dunes 10 km before al-Ashkharah

Al-Ashkharah Today

By 1916 the *Handbook of Arabia* estimated the number of houses in
al-Ashkharah at some two hundred, 'a few of stone by mostly huts,
inhabited by the Ja'afirah section of the Bani bu 'Ali, owning about
fifty fishing-boats and one large *sambuq*. The place is the port for all
the Ghafiri tribes subject to the Tamimah of the Bani bu 'Ali.'

In 1986 I was assured that the population of al-Ashkharah was about
five thousand, but no census has ever been taken in Oman, the
inhabitants spend part of the year here and part inland, some are
fishermen out for weeks or months of the year, and many *badu* come
and go, depending on the weather and grazing conditions. Our
particular friend in al-Ashkharah proved the epitome of courtesy:
Muhammad Rashid bin 'Ali al-'Alawi, thus a member of the Bani bu
'Ali tribe who had fought the British. Bearded and moustached, he
wore a plain white *dishdasha* and turban, and felt as comfortable with
me as with his compatriots. A mine of information, he exuded that
all-encompassing wave of charm and hospitality that all Omanis
demonstrate when one speaks their language, however imperfectly,
and enthuses about their magnificent landscapes.

'As there is no hotel nearer than Sur or al-Mudairib, could I stay
anywhere in or near al-Ashkharah?'

'Why not?'

'How could I find a place to spend the night?'

'Leave it all to me.'

Muhammad Rashid drove me in his taxi to an old mud house, with
new cement additions. He called out from the car, 'Ya, Musbah!' and

after a few moments, an old man emerged through a small wooden door. After greetings, they discussed the matter of accommodation, and Muhammad Rashid explained that for a matter of RO 2–3 I could have the use of a house of a relative of Musbah who had been appointed to a frontier customs post, and had taken along his whole family, as is usual. The place, a few doors away, with its barred windows, and thick layers of sand, induced a sparkle in my eye like that of a Jan Steen genre scene, or a coloratura aria by Donizetti: an 'isness' before which the ordinary fades like ice in sun. No degree of dull comfort can make up for these 'epiphanies', as Joyce called them, when our lives take on that elusive rich texture one detects in a Turkish *kilim* or the Mughal gardens of Lahore.

'It is very cool here even in summer,' said Muhammad Rashid kindly, mistaking my silence for disappointment. A black goat peered round the door we had left open, but nothing had been left save buzzing flies and memories by the family that had vanished to the border. Sea-tang lifted to my nostrils; another goat joined the party. Muhammad Rashid understood without being told. 'You'll take it,' he said. I nodded, enjoying the peace that had come to al-Ashkharah since the turbulent times of Cole and Miles.

He took me to meet Salim Muhammad al-Ghafiri, owner of the fort. We entered through the tiny door, so manufactured as to permit only one person to enter at a time. The graceful Iranian-style arches within had been recently restored. Outside, a damaged cannon rested on rocks, pensioned off after an existence of long waiting for an enemy that rarely if ever came. Even then the cannon might have been unprepared, without ammunition. Houses have been suffocated by the

Al-Ashkharah. Fort, with its owner, Salim Muhammad al-Ghafiri

rising sands near the little fort: one of them was owned by Rashid bin 'Ali bin Salim until he could exist there no longer and built his new mud house opposite. There was no ancient *suq* to prowl in, so we walked down to the sea below al-Barahima (the other sector of the town is called al-Majbali) and scarred the empty beach with our footsteps. Swimming for pleasure is not a custom here, any more than it is elsewhere in the Arabian Peninsula, and only a few little boys were splashing in the afternoon waves. The school has about 500 pupils, and they had dashed out with *dishdashas* and satchels flying, shortly after I arrived. Between ninety and a hundred boats, according to Muhammad Rashid, sail out for trade and fishing to Masirah, Ruwais, and Ras al-Hadd.

We drank hot sweet tea and spoke of the excellent government of Sultan Qaboos ibn Sa'id, of the understanding that men can achieve without a common language if they merely exercise their natural goodwill, and of the need for more schools, more learning, and more tolerance, exemplified by the surge of new schools all over the Sultanate and of the opening of the first University, which was imminent.

Muhammad Rashid told me about his travels in Sur and Muscat, his plans to go to England, his joy in reading and football. He belonged to al-Ashkharah and always came back, but he sought an equilibrium between the nomadic life of his ancestors which brings adventure, and the sedentary life which brings stability. 'And you,' he ended softly, 'you too are very happy in al-Ashkharah, as though you were born here and lived here all your life.'

'Yes,' I said simply, 'this too is the paradise like that revealed to the Prophet Muhammad (salla Allahu 'alayhi wa sallam).

Al-Ashkharah. Cannon outside the fort, with Muhammad Rashid bin 'Ali al-'Alawi (right) and Jamil 'Abdullah al-Ja'afari

Here I am, like the 'pure fool' in *Parsifal* an innocent between heaven and earth, neither rich nor poor, neither employer nor employed. I am midway between stupid and intelligent, between weak and strong. I have no illusions about God or gods; I dare to believe ardently in the love of men and women, despite all our imperfections: in trust, hope, duty, diligence, altruism, even in a kind of foolish 'love for its own sake.' Unlike a pauper, I can travel; unlike a king, I can journey anonymously, passing off others' business cards as my own to affect a different identity with each slight acquaintance: I am a short-story writer weaving fantasies and ironies about slips of paste-board that identify me as an Italian geologist, an American lawyer, a French banker, a Scottish wool-manufacturer. It is my magic; my vanishing trick.

3 SHARQIYAH

The so-called 'Eastern' part of the interior is a predominantly flat area of gravel plains and shallow *widian* bordered to the north by the Eastern Hajar, to the south by barren desert plains inhabited only by *badu*, to the east by the isolated Wahibah Sands, and to the west by the oilfields of Natih and Fahud on the arid plains west of Adam. On the new asphalt road from Ruwi to Sur lie the Sharqiyah towns of 'Ibra, al-Mudairib, al-Qabil, and al-Mintirib. South of this major artery there is a well-populated graded road leading to Samad and al-Khadr, al-Mudaibi, al-Musalla, and Sanaw. From Sanaw you can take a track north and west into the Jabal Akhdhar, through Humaidiyin to Manah and Firq, or to Izki.

To explore Sharqiyah, you can stay at the comfortable new Qabil Motel. 'Cawbil', wrote Wellsted in 1835, 'is walled round and has several forts.' It was originally the region's capital, and the principal fort is to be found today, unnecessary in a pacified land and therefore merely picturesque, with its stone towers either still intact or cleverly restored. This is a centre of the Harth (or Hirth) tribe, one of the great elite tribes of Omani society (like the Nabahinah, for instance, and the Riqaish), which achieved special renown for its colonial activities in Zanzibar. Its capital is traditionally 'Ibra, though it was strong also in al-Mudairib, al-Mudaibi, Nizwa and Samad. Its eighteenth-century capital was al-Qabil, and it was only the persistence of drought conditions that led them to migrate to Zanzibar where they became leading merchants and influential citizens. The Harthi elite, Awlad Salih bin 'Ali and Awlad Humaid, originated from Dariz, and the other sedentary groups came from 'Ibra, except the Daghaishi, who come from Bani Riyam.

Al-Mintirib

Al-Mintirib sleeps, forgotten in the sun, off the main road, its squat old fort contrasting with the bright new mosque. Domestic buildings are probably more attractive than the public faces of the town, like its surrounding district called Bedi'ah by Wellsted. During his time, long before the road was completed in 1977, Wellsted found it a relatively prospering place, 'a collection of seven hamlets, situated in as many oases, each containing from two to three hundred houses. The suk or market is held on that which is the most centrical. . . One striking feature in the appearance of these towns is their low situation. They are erected in artificial hollows, which have been excavated to the depth of six or eight feet, and the soil thus removed is left in hillocks around their margins.' He goes on to describe the manner of construction of the *aflaj* or artificial water-courses, which have best been described by J. C. Wilkinson in his magisterial account, *Water and tribal settlement in south-east Arabia: a study of the aflaj of Oman* (1977). The *falaj* system of irrigation in Oman most probably dates to the Persian occupation of Oman in pre-Islamic times, and many of the systems have fallen into disuse, partly because of migration, changes in agricultural and settlement patterns, effective digging of additional wells, and the evolution in very recent times of other trades and industries. Yet during my travels in 1986 I saw wide evidence of restored and repaired *aflaj* wherever settlements remained active. It is also interesting to note that Sirhan ibn Sa'id ibn Sirhan (if indeed he *is* the author of the *Kashf al-ghummah* translated by E. C. Ross as *Annals of Oman*, Cambridge, 1984) states that the Persians destroyed many of their *aflaj* in Oman in the 2nd century A.D. as a ploy to weaken their enemy during a truce with the leader of the Azd, Malik bin Fahm, when Malik had withdrawn to Qalhat and the Persians to Suhar. *Kashf al-ghummah* reports that Sulaiman bin Da'ud, known to us as King Solomon, son of David, had constructed 10,000 such 'aqueducts' in Oman: this is of course a purely conventional number, like the Persian 'thirty' meaning 'a lot'.

The population of Mintirib was reported to be 6,500 by the *Handbook of Arabia* (volume 2, 1917), with the most valuable dates in Oman (of the 'mibsali' variety) which are exported to Bombay via Sur.

Al-Mintirib. Fort

All the villages possess camels, donkeys, cattle, sheep, and goats.'
Today the population of cattle and sheep has dropped significantly,
and Japanese pick-ups replace camels and asses as load-carriers, but
the goats have never interrupted their determined depredations on the
vegetation.

Wahibah Sands

South of al-Mintirib lie the extraordinary Wahibah Sands, currently
the subject of extensive research by an Anglo-Omani Expedition
promoted by the Royal Geographical Society. In an area the size of
Wales, Wahibah constitutes a range of dune-types and desert
environments as varied as may be found elsewhere only in much larger
environments, such as the Sahara. More than twenty different
sand-dune formations have been identified, so that Wahibah is a
significant microcosm for study purposes, especially notable for large
Prosopis forests on its eastern fringes. Below the ancient dunes
scientists have identified eroded remains of a sand sea cemented by
calcium carbonate to produce aeolianite, exemplified by the escarpment at Ras Ruwais. The high density of calcium carbonate indicates
that the sands derived from shell fragments blown onshore from the
Arabian Sea.

The woodlands are crucial for the human, animal and vegetable life
in the zone. Shelter for tribesmen and their herds, they offer grazing
for goats and sheep, firewood, and building materials. The white-
tailed mongoose has been found there, in addition to previously-
known habitats in north and south Oman. Zoologists have found there
not only the Arabian gazelle, as expected, but also the much rarer sand
gazelle. The long-term project includes studies of the history of the
sands, including the geological background and human occupation;
plant-distribution, soil moisture and sand movement, assisting a
programme for sands control; biological resources with special
reference to *Prosopis* conservation and range management; economic
interactions between such activities as fishing, farming, and pastoralism, with proposals to assist the local people; and the effect of the local
oil industry and co-ordinated development. The team, operating from
the Taylor Woodrow-Towell Research Base, a specially-designed
project camp at al-Mintirib, has benefited enormously from logistical
support by the Sultan's Armed Forces, including a full convoy from the
Coastal Security Force.

Loyd on the Wahibah Sands

Michael Loyd's secondment to the Sultan's Armed Forces was due to
end in 1964, but he had already seen the remains of three oryx that had
been shot while travelling from Zauliyah to the southern fringe of
Umm as-Samim, and was keen to seek evidence on the basis of reports
from the Jiddat al-Harasis by two Harasis soldiers in the Muscat
Regiment.

Sa'id and Muhammad of the al-'Amr, Central Wahibah Sands (Courtesy of Charles Hepworth)

He passed information to the Zoological Society of London and received a reply from them and from the Fauna Preservation Society, requesting more information and the habits and habitat of the Arabian oryx. So little was then known that almost any information would be of great value. The Force decided to disband its camel platoons and, with the support of the Force Commander, Colonel Hugh Oldman, Colonel Loyd was permitted to purchase the camels neccessary and to spend his terminal leave gathering information on the Arabian oryx within the Sultanate. The tale is taken up by Michael Loyd, whose account is reprinted here by courtesy of the author and the editor of the *Sultan's Armed Forces Association Newsletter* (September 1985):

'Sir Hugh also obtained for me permission from the Sultan, Said bin Taimur, to travel in the Wahibah Sands and Jiddat al-Harasis, also letters of authority from Muhammad bin Ibrahim, the Minister of the Interior, to tribal *shuyukh* of the areas I would visit. The opportunity to travel through the Wahibah Sands was particularly exciting because the area had not been surveyed by an oil company and was believed to be impassable to motor vehicles and consequently had not been entered by a European since Wilfred Thesiger traversed it in 1949.

I chose to take with me the two Harasis soldiers in the Muscat Regiment, Ibrahim bin Saqar and Salim bin Khamis and also an Al bu Shams soldier from Sunnainah, Khalud bin Muhammad and an Al bu Shams civilian, Said bin Salim, who came along with his Saker falcon for the sport. I had met Said bin Salim with a hawk near Sunnainah the previous year and he had been a camp follower of my reconnaisance platoon ever since, finding hawking from a Land Rover more productive than from a camel.

Falconry in Oman is confined to the Al Bu Shams and Na'im tribes whose territories border with Buraimi. Peregrine and Saker falcons are passage migrants and winter visitors thoughout the Sharqiyah and Batinah and the Houbara (Macqueen's Bustard), the falconer's favourite quarry, was at that time common throughout Oman. On one occasion when travelling with me from Bidbid to Muscat, Said flew his hawk and killed a bustard at Rusail, close to the area now enclosed by Muaskar al-Murtafa.

I left the selection of camels and equipment to these bedu, and purchased seven camels which allowed two to carry our food and water. For food we started with 200 lbs of dates, several pounds of coffee, beans, flour, rice and some samn (butter). All our supplies and water were carried in goatskins.

We set out from the camp at Izki on 21 February 1964 and went first to Mudhaibi where I was well received by the Wali, Shaikh Ali bin Zaher, who produced a Wahibah badu to take me to where their shaikh was encamped. We found Muhammad bin Hamed, senior shaikh of the Wahibah, near Tawi Barrud on the edge of the Sands. At first he refused to provide a guide to take me to Tawi Harian in the Sands and then to Nafi in the Wadi Andam and arranged a meeting with the other Wahibah Shuyukh who were encamped in the same neighbourhood. The Wahibah were suspicious of my reasons for entering the Sands and agreed to let me do so only if I obtained authority from Shaikh Ahmad bin Muhammad, Tamima of the Harth and paramount shaikh in the Sharqiyah. Shaikh Ahmad lived at the fort at Qabil, a return trip of fifty miles, but there seemed no alternative to seeing him if I was to enter the Wahibah Sands. I set off with Ibrahim, leaving the rest of my party with the Wahibah. Shaikh Ahmad read my letter of authority from the Minister of the Interior and questioned me closely on my reasons for wishing to enter the Sands, and then gave me a letter to the Wahibah Shuyukh telling them to provide me with a guide to take me anywhere I wished to go.

We departed from Tawi Barrud on 27 February and rode that day across a pleasant sandy plain which held plenty of bustard and hare. We enjoyed some excellent hawking and saw a lot of gazelle. We made south east gradually closing on the line of dunes and stopped for the night at its foot. We reached Tawi Harian two days later. Our camels, being unused to the sand, were finding the going difficult and we often had to divert from the direct course in order to cross a line of dunes at a low point. There was ample grazing for the camels but the heat of the sand and the lack of shade made the middle hours of the day oppressive. Tawi Harian is situated in a depression between the dunes with three ghaff trees growing nearby. We were joined at the well by an old badu who provided a long rope and bucket so that we could draw water for our camels. Later my companions asked him whether he had ever seen a Christian, he replied that he may have done fifteen years

Tawi Haryan, Wahibah Sands, 1971 (Courtesy of Charles Hepworth) Note gypsum strengthening compared with brushwood surround seen by Michael Loyd, 1964

previously when a tall man with big feet passed there with a party travelling from Nafi to Badi'ah. He explained how this man had an instrument in his hand and asked him to point to other wells in the Sands. The man he was referring to was probably Wilfred Thesiger. The map I used was compiled by the Royal Geographical Society from Thesiger's traverses and remained the only map with detail of the region.

We reached Nafi two days later and there we parted with the Wahibah guide. My enquiries of the badu confirmed that oryx are never seen in or near the Wahibah Sands. One of my bull camels had become badly lame in the Sands and had to be left with the badu. From Nafi we made the short trip to Uruk in the Wadi Halfayn where we rested a day with the family of Ibrahim bin Saqar. There were several families of Harasis in the Wadi Halfayn including Shaikh Shargi bin Argis, camped near the well at Farai, but most of the tribe were much further south beyond Duqm.

After Farai we joined the main camel track from Salalah to Oman. This track follows the course of the Wadi Halfayn and then a line of wells close to the coast. We passed a caravan of Januba taking dried fish from Duqm to Adam.

The route often crossed salt flats and the wells were very saline. One of our camels was in milk (a *murri*) which we could mix with the brackish water to make it more drinkable; the badu call this mixture Shanin. I was impressed by the badu's sense of direction. I would sometimes pass the time by taking compass bearings off the map of other wells and more distant points such as towns in Oman and ask a badu to point towards it, he always did so with complete accuracy. Similarly they thought I was quite good but they did not know about my compass. I did less well when I could not cheat; for instance, each morning our first task was to find our camels which, though hobbled, frequently wandered out of sight in search of grazing. The badu thought me stupid not to be able to distinguish the tracks of my own camel from the others.

It was my intention to go to Bowi, a Harasis well near the Huqf inland from Duqm but before reaching it we came across two Harasis families near a well called Thath. There I met Said bin Fox who had just come off the Jiddat where he claimed to have shot an oryx and thought he could lead us to more. He agreed to come with us as our guide to Haima.

I was worried by a second bull camel that showed signs of exhaustion so left it and two other weak camels with Salim bin Khamis, at Thath where the grazing was good, with instructions that he was to wait for us at the well at Mugaibara to the North. We knew from experience that our camels could not be relied upon to go more than three days without water, but there were eighty waterless miles across the Jiddat to the oil company's pump at Haima.

After leaving Thath we watered finally at Mugatib where the hawk killed a bustard. The well had fallen in and took some hours to dig out; we then climbed the Huqf escarpment onto the Jiddat, seeing four wild

Lake near al-Huqf, Wahibah Sands, 1971 (Courtesy of Charles Hepworth)

goat on the way. We lost the hawk on our first day on the Jiddat. She had been flown at a hare but was not sharp set so ignored her quarry and did not return, and because of the need to reach water we could not afford to spend time the following day to look for her. On the second day we came across the tracks of oryx and that afternoon, when approaching a place of thick vegetation in an area called Yaluni, we came across two adult oryx and a calf. At first I was misled by the animals' white silhouette seen against the dark gravel background into thinking that it was canvas sheeting left by bedu but the calf was a fawn colour similar to a gazelle's. We had approached downwind and I was able to observe them quite closely before they sensed us and trotted away. Said bin Fox who possessed an ancient Martini-Henri carbine explained that when hunting oryx he would approach walking directly behind his camel and when within range would fire along the camel's flank. Said bin Fox collected samples of the plants he claimed were favoured by oryx which I brought back to be identified at the Royal Botanic Gardens at Kew.

We reached Haima the next day and found a number of Harasis employed at the seismic camp. There was no satisfactory grazing nearby so my badu took our camels some hours distant whilst I remained a day as guest of the six European oil company employees who were amazed to see me. We saw no oryx on our return across the Jiddat to Raqi where we descended from Huqf and found Salim bin Khamis waiting for us at Mugaibara. The bull camel had not recovered, so we left it with some Harasis we had met the previous day collecting their camels from pasture.

Although I had seen only three oryx, I had found tracks of others and scrapings the animals had made under acacia trees. From this evidence and from talking with the Harasis who claimed shooting five in the past three months, I concluded that there must be at least one hundred oryx in the area but that they were greatly threatened. The Harasis admitted that their numbers had declined, particularly between 1957–59 when a previous oil company had surveyed the area and the Harasis employees had been able to chase and shoot them from company vehicles.

After leaving Mugaibara we travelled north for fifty miles across barren land which alternated between sand dunes and salt flats. It was here that we came close to losing Said bin Salim. One morning we crossed the tracks of gazelle and as we had no meat Said dismounted and went in pursuit of it on foot, intending to catch up with us at our midday halt and take a camel back to collect it. The rest of us continued for rather longer than usual in search of grazing before halting when a strong wind then got up which completely obliterated our tracks. Said bin Salim had not rejoined us by sunset and our search had failed to find him. We were worried because, being a stranger, he had no local knowledge. By dawn the wind had dropped and the Harasis again went in search of him and returned at midday with news that they found his tracks and explained exactly what he had done. Said bin Salim, hopelessly lost, had circled round us during the night and seeing a grazing camel had caught and mounted it knowing that if given its head and forced to move it would make its way back to its water. He rode it bareback throughout the day until he reached the well at Haushi then, after drinking, had returned along his tracks to meet us knowing that by this time we would be following him up. No one seemed surprised by this action, it is standard procedure in such a situation – but first find your camel.

After Haushi we crossed gravel *sir* for another fifty miles to Baraima in the Wadi Halfayn, where several Wahibah families were encamped. From there we followed the Wadi Halfayn for two days to Adam. Throughout the journey our camels would average 3½ miles an hour and we would normally ride for seven hours in the day, halting at night and in the middle of the day in areas of good grazing. Of our three male camels two did not complete the journey and my own, a *dalul* of fine proportions, became very exhausted towards the end and I had expected to complete the journey on foot but fortunately it revived after being fed lucerne at Adam. It had been in much demand by the badu we had met to serve their female camels (*nagas*).

We arrived back at Izki on 30 March 1964, having travelled over seven hundred miles since setting out from there thirty-nine days earlier. I left for England a few days later and subsequently heard that my pessimistic predictions on the survival of the oryx were proved correct when the last were found shot in 1971.

Now a middle-aged Northumberland farmer, I was lucky enough to be included in the SAF Association party that visited Oman in September 1984 and was able to visit the Jiddat again, twenty years after I had journeyed to it by camel. Ibrahim bin Saqar, who had been told of my visit, came to Muscat to see me and drove me down to Yaluni to visit the White Oryx Project. Unfortunately Mark Stanley-Price, in charge of the project, was away but his wife Karen brought me up to date with developments. In 1980 five oryx were introduced from Arizona and placed in a compound one square kilometre in size, close to the spot where I had seen wild oryx in 1964. More were added in 1981 and in January 1982 the enclosure gates were opened and the herd, numbering eight adults and two calves, were released. In March 1982 the first wild oryx calf was born. By June 1983 the herd, now increased to eight adults and four calves, had wandered extensively over the Jiddat tracked by Harasis wardens. By the time of my visit a second herd of similar numbers had been released, which included four Omani-born calves. I was able to observe both herds which had returned conveniently to the vicinity of Yaluni.

Bore-holes provide permanent water at Al-Ajaiz and Haima and enable the Harasis to remain on the Jiddat with their camels and goats whereas before they could visit only for short periods after rain. I stayed the night at Ibrahim's camp which was close to Yaluni. He gave me news of my other three companions on the journey. The Harasis Salim bin Khamis, generally known as Bu Rashid, prefers to spend most of his time with his camels in the Sahma Sands; of the two Al Bu Shams, Khalud remains at Sunnaina and Said bin Salim has for years been one of Shaikh Zaid's falconers and travels with him on hunting trips to Pakistan and elsewhere.

That evening, as I watched Ibrahim milk his camels and could hear the goat and rice being prepared for us by the women, I reflected on how much had remained unchanged in twenty years. Gazelle are still plentiful on the Jiddat and the oryx are back.'

The latest publication on the subject is Hartmut Jungius' 'The Arabian Oryx: its distribution and former habitat in Oman and its reintroduction', in *The Journal of Oman Studies*, vol. 8 (1985).

Al-Qabil

Al-Qabil possesses a huge towered fort next to a mosque where prayers were in progress. The streets, consequently empty, felt like the symbolic centre of a de Chirico townscape, the world aslant, right-angles gently out of perspective, silence growing like moss underfoot because you walk on sand, and each private house introspectively gazing into its own courtyard, its family isolation ward.

In al-Qabil Motel, actually in the next village of Dariz (also called Duraizah), I spoke to the Goanese manager Santiago Deniz, who had previously managed a huge accommodation site at the newly-constructed industrial city of Jubail, on the Gulf coast of Saudi Arabia. No telephones were yet on line, but Ericsson were in the process of

Al-Qabil. Behind the fort

laying the cables, so it was only a matter of weeks before the long isolation of Dariz ended. We tuned in to the B.B.C. World Service: advances in leukaemia research, the Haitian dictatorship of Duvalier had been toppled, a computer system allied to underground sensors had an 89.7% success rate in predicting volcanic eruptions in Japan. . . Yet outside the unobtrusive enclosure of the motel, little had changed in the Sharqiyah.

Al-Mudairib

Early next morning I headed for the nearby village of al-Mudairib. Wilkinson's study cited above indicates that the present settlement (though of course there may have been others earlier) was founded by the Hirth tribe in the second half of the eighteenth century. I took with me as Baedeker the article called 'Architecture and social history at Mudayrib' by the Bonnenfants and Salim ibn Hamad ibn Sulayman al-Harthi in *The Journal of Oman Studies* (1977), to which readers are referred for greater detail on the oasis. Based on a single *falaj* (but not the original), Mudairib is roughly circular as regards the built-up area, and stands at the foot of a hill surmounted by a fort. Sections of the town wall and a number of watchtowers can still be seen, as well as eighteen mosques, ten *sabl*, five ruined towers, or *burj*, four fortified city gates or *darawiz*, and two fortresses, as well as twelve principal houses generally decorated with simplicity outside and with fine decorations within. Relations with East Africa – not only Zanzibar but Tanzania and Kenya too – have left their mark on some motifs in painted ceilings, and the benefaction of a mosque from a Zanzibari resident, and charitable shares in irrigation taxes' being made available to benefit a mosque on the island.

A typical mosque in Ibadhite Oman will be a simple rectangle, between one metre and two above ground level. You will see no pulpit, because Ibadhism requires the only pulpit to exist in the principal mosque of the city in which the head of the faithful resides. The roof of some mosques can be reached by holes cut out of a wall; in others by horizontal rungs in a corner. A covered, arched gallery in larger mosques serves as transition between the open courtyard and the quiet, darkened prayer-room into which shafts of light flood through tiny windows. Each district has its own mosque, often built by an affluent or especially pious man of the district, and worshippers of that district tend to congregate there, making a mosque in Mudairib the spiritual equivalent of a secular *sablah*.

The various *sabl* are called after major lineage groupings, such as the

Mudairib. Watchtowers overlooking the town; in the foreground the suq and steps leading down to the falaj

194

Maharma, Ghuyuth, Khanajira, and Daghasha. In other parts of the Arab World, such a meeting-place, where it exists, would be known as a *majlis ʿamm*, or general meeting-place, a term which also covers the activity itself. Socially, it represents a halfway house between the private home and the large *suq:* it is a place for discussion and consensus among men, and its lubricant is not alcohol as in the west, but coffee, with sweets such as *halwa*, and seasonal fruits. Men nowadays use the *sablah* for meetings in the afternoon, after lunch and before the *dhuhr* prayers; or after going to the *suq* in the afternoon; or after dinner and before the last prayers of the day. Men foregather to recite passages from the Qurʾan al-Karim during religious festivals; there they celebrate a wedding, or mourn a funeral. There they discuss irrigation, construction, general investment, and conciliate disputes.

You can wander around Mudairib all day, from the well-preserved fortified gate of Kisham, protecting the north of the town, to the Awlad Hamid fortress east of the *suq*, on the hill, created around 1800 by ʿAbdullah bin Sulaiman bin Muhammad. You can look at the great house built around 1869 by Muhammad bin Amir al-Maharmi and

Mudairib. Corner of suq

called Bait at-Tashash, to the Bait Hamud bin 'Ali bin Talib dating back to about 1900, which now belongs to Salim and Muhammad bin 'Isa bin Salim bin Hashil al-Khanjiri, who live in Zanzibar; the house is now managed by their relation Saif bin Muhammad. Fine arches, intricately-carved doorways, and elegant painted ceilings (I was permitted within) all exemplify the prosperity of an age now past. These forts and houses are at risk from cheap new building materials which threaten to smother the uniqueness of Mudairib with a kind of cosmopolitan gloss, or worse still, a ubiquitous shantyism with corrugated iron, oil-based paints and formica, in place of traditional wood and clay. Mudairib is worth saving, but its future is bleak unless the Government enforces planning restrictions and spends a great deal on restoration of its architectural heritage.

'Ibra

Such urgent protection of the built environment presents even greater problems at 'Ibra, a much larger town, because the expense would be correspondingly greater. Yet there are few districts that would lend themselves so neatly to restoration as an open-air museum as the Manzifah quarter of 'Ibra Sufala'. 'Ibra's kinship and social organization has been studied by Colette Le Cour Grandmaison in *The Journal of Oman Studies* (1977), to which the reader is referred. Briefly, she shows that an oasis community arose around a watersource and depended for its survival on the protection of this source, normally a *falaj* network, by repair of the channel itself, and by a system of alliances between tribal groups to ensure that the *falaj* is used equitably, and never destroyed or wilfully damaged. The Muscat-Sur road completed in 1977 runs south of the police station and north of the whole of 'Ibra: 'Ibra 'Alayah ('Upper 'Ibra') and the much larger 'Ibra Sufala' ('Lower 'Ibra') inhabited respectively by the Masakira tribe and the Hirth. The traditional rivalry between these two groups have had major historic consequences. For instance, each sector possesses its own market and town centre; the Wali, or Governor, has his office apart from both as a matter of judicial neutrality, like the hospital and the school. This bipartite existence occurs in many other Omani towns. You can sense the unease which dominated past encounters by the walls around the palm groves, and the fortifications, now crumbling through disuse.

In his *Rapport commercial sur l'Oman* Ottavi estimated the population of 'Ibra combined in 1900 at about 3,000; the *Handbook of Arabia* compiled in 1916 suggested a figure between 4,000 and 5,000. In 1977 Le Cour Grandmaison assessed the figure at 600 for 'Ibra 'Alayah with another 700 in the Masakira fringe villages, and about 1,000 for 'Ibra Sufala' with another 300 in the Hirth fringe villages, or a total of some 2,500.

This figure disguises the fact that each of the neighbourhoods has no more than 150–200 inhabitants, the constraint on density being the *falaj* water supply available. Each little community has its own

196

'*Ibra Sufala'. Al-Manzifah. Old suq*

defensive walls, *dirwaza* or strong gateway, and fortified houses of the richer members of the community which can act as refuges if the need should arise.

'Ibra 'Alayah comprises five distinct neighbourhoods, and is irrigated by two *aflaj:* the Masmum and the Lima'. 'Ibra Sufala' contains twelve different *aflaj*, which accounts for the fact that its built-up area is more than double the size of its rival's.

The main Wadi 'Ibra runs to the east of the built-up areas, but there is a subsidiary wadi which splits 'Ibra Sufala' in two and runs south of 'Ibra 'Alayah. The finest old houses can be found in the almost deserted walled village of al-Manzifah, citadel of the wealthy Brawna and Ma'amra factions. I saw a new volleyball net, a car, and heard a scatter of children's cries like rust-red leaves in a forlorn autumnal park, but found no evidence of piped water or electricity, and the rectangular covered *suq* lay silent and empty as if it had been newly excavated from centuries below the sands. Wellsted wrote of this very scene in 1835: 'A daily market for the sale of grain, fruit and vegetables is held here, to which the Bedowins and inhabitants of the neighbouring villages resort in considerable numbers. The stalls at which the vendors take their stand are only occupied during the hours of business. They are small, square buildings, surrounded by a low wall, roofed over, open in front, and have a floor raised about two feet from the level of the street.'

The jealous ownership of al-Manzifah is proved by the extent of the defences: an outside wall with a double gate surmounted by a mosque, itself easily defensible. Attackers would have to veer sharply before reaching another gate, then an area stoutly walled, with towers at strategic intervals, and thickly-walled houses of high quality built tall enough to allow armed men to reconnoitre beyond the walls to potential enemies outside. Nine great named houses date from the eighteenth or nineteenth centuries: five to the Brawna (singular Barwani), three to the Ma'amra (singular Ma'amiri), and another Bait al-Buma, jointly to both, as tangible evidence of their enduring alliance. Unlike many Omani oases, 'Ibra does not derive its major income from agriculture, as witnessed by the abandonment of cereals and the decline of vegetable gardens. But dates are still cultivated; the tenuous ties that linked 'Ibra of the present to past 'Ibra are dissolving naturally, and can be reinforced only by the artificial insemination of new employment, such as light industry or the creation of an open-air museum of the Sharqiyah at al-Manzifah which would restore a sense of pride in local history and traditions that is disappearing gradually with migration to the Capital Area.

We took the main Sur-Muscat road from 'Ibra Sufala' and in 4 km arrived at the Hotel Walid, which is a restaurant specialising in chicken biryani. Watchtowers dot hilltops like unsleeping eyes, then in 2 km we reached the right turn 27 km to Wadi Nam, with the low range of the

'Ibra Sufala'. Al-Manzifah. Gateway

'Ibra Sufala'. Al-Manzifah. Old patrician houses

Eastern Hajar bobbing to our right, accompanied by the insistent drum rhythms of Abu Fa'iz on the Suri cassette. In another 20 km a sign showed Wadi Dima 62 km off right, and in 21 km more (43 km from 'Ibra Sufala') we reached the graded road leading left to al-Mudaibi after 63 km.

'Ibra Sufala'. View into al-Manzifah from the gardens beyond

South to al-Musalla

Arid scrub on the track south attracts few grazing animals, and the plain lies panting with haze in the midday heat. A driver must choose between the corrugated track, damaging his suspension; or the unmarked gravel on both sides, with the occasional sharp stone that might puncture a tyre. And then the string of oases including Samad and al-Khadr is glimpsed in the distance, proclaiming a *falaj* in use. I stopped by the fort that dominates Samad and was visited by Wellsted in 1835, when he was invited to breakfast there by the Shaikh. 'It was a large fort,' he recalls, 'very strongly built with the same material as the houses. The rooms are spacious and loft, but destitute of any furniture. Suspended on pegs, protruding about two feet from the wall, are the saddles, cloths, and trappings of their horses and camels. The ceilings are painted in various devices, but the floors are of mud, and only partially covered with mats. The windows, in place of the usual ornamental wood-work, are crossed by transversed iron bars; and at night, in order to protect the inmates from the keenness of the winds,

Samad. Castle

they are wholly closed by wooden shutters. Lamps formed of shells, a species of murex, are suspended by lines from the ceiling, and the whole was essentially different from what I have seen in other parts of Arabia. Our meal, after the usual style, was sumptuous and plentiful; but so strictly do the Arabs regards the laws of hospitality, that it required much entreaty to induce our host, a man of high birth, to seat himself with us. This originates in a prevalent belief that if he partakes of the meal he will neither have leisure nor opportunity to look after his guests, and he therefore insisted upon waiting on us in the capacity of an attendant. It was not until I told him that we would not commence unless he did so, that he could be prevailed on to join in, and then we perceived he could play his part as well as the best of us. On returning to the tent I found, as usual, a great crowd collected there, but they were kept in tolerable order by a little urchin about twelve years of age, whose father, a man of great influence in these parts, had, a few years before, been killed by the Bedowins. He had taken complete possession of our tent, and allowed none of his countrymen to enter but with his permission. He carried a sword longer than himself, and also a stick, with which he occasionally laid about him. I was excessively amused at the gravity and self-importance of this youngster, who appeared perfectly well acquainted with the numbers, resources, and distribution of the native tribes, and his conversation on these and other subjects was free and unembarrassed, and, at the same time, highly entertaining. It may be observed, generally, of the Arabs, and particularly of the Bedowins, that their boys share the confidence and the councils of the men at a very early age; and on several other occasions I have seen their youths exert their influence in a manner that to us would appear preposterous. But it is a part of their system of

Al-Khadr. Weaving factory of 'Ali bin Khalfan. Ground looms

Al-Khadr. Weaving factory of 'Ali bin Khalfan. 'Ali with sample

education to cease treating them as children at a very early period and they acquire, therefore, the gravity and demeanour of men at an age when our youth are yet following frivolous pursuits, and being birched into propriety of conduct and manners.'

And this system of education, as Wellsted calls it, extends so far into the social fabric exemplified by the *sablah* that boys are drawn, even today, into discussions of communal importance which it took a revolution to create in China.

Samad is well-known for archaeological remains dating back to the 3rd millennium B.C. I explored the village of al-Khadr, which nestles in the wadi bed, where it receives *falaj* water from below and run-off from rain on the mountain slopes. I dipped my fingers into the cool *falaj* near the fort in the built-up area; cleaning fish nipped my skin. Families live in a part of the fort, which has been partly abandoned. Beyond the quiet *suq* I met 'Ali bin Khalfan, who invited me into his weaving 'factory', in fact no more like the factory of Chaplin's *Modern Times* than Omani government by consensus resembles Aldous Huxley's faceless, bureaucratised *Brave New World*.

'Ali's factory looks like any other courtyard house, with covered quarters to protect ground-loom weavers from the effects of the sun, and shady palms surrounding the open area like parasols. The ten weavers he normally employs had left to look after their date plantations, for this was the time of pollination. Alone, he showed me the traditional looms above the pits dug for the weaver's body and legs. Lengths of cotton were shown at RO 8, RO 20, and RO 50, which seemed very expensive, and accounts for the popularity of textile imports from Pakistan and India, and the decline of the indigenous textile industry. The Ministry of National Heritage and Culture does support a limited number of weavers, as at Suma'il, but weaving remains a dying craft.

Out of Samad and al-Khadr, the road widens, becomes faster and smoother, and shortly arrives at al-Mudaibi, with a very large rectangular *suq*, almost abandoned, next to an even greater main square which has taken over the function of the town's commercial centre for the Habus and Jawabir tribes. Three shuyukh, each with a white *dishdasha* and white turban, a walking-stick and a *khanjar*,

Al-Mudaibi. Three elders

Al-Mudaibi. Underground falaj and mosque

greeted me affably, shaking hands vigorously and asking question after question. 'How was my health?' 'Where had I come from?' 'Where was I going?' 'What was my name?' 'What was I doing in al-Mudaibi?' 'What could they show me, or did I prefer to go everywhere on my own?' 'To which tribe did I belong?' They insisted on taking me into

Al-Mudaibi. Fort

204

the Al-Hashimi Restaurant and buying me refreshing tea, showing me the intricate silver decoration on their *khanajir*. Some observers might have felt a forlorn quality about that silent square, with its sense of the old being overtaken by the new, then left to falter and slip into oblivion. But I, luxuriating in the charm and vivacity of the three wise men of al-Mudaibi, exulted in the sheen of gold covering this glittering oasis, recollecting the words of the great modern Egyptian poet Ahmad Zaki Abu Shadi:

'From light we began, to light we return;
The world consists of minute waves of light.'

Light illuminated a simple Ibadhi mosque and its nearby underground *falaj* outlet; it throbbed thickly on mud walls, at which an old broomstick propped up a new airconditioner outside the majestic fort; and it cast its brilliant spell over two little boys playing at a *falaj* in the old fortified quarter. It enveloped me, so that I slowed my steps in time with those of the three elders, attuning myself to their dignity for once before I escaped back into my western straitjacket of stress and

Al-Mudaibi. Two little boys playing in a falaj

Zahab. Fortified house

complexity, the grey sceptical no-man's-land between one certainty and its equally certain contradiction.

Continuing south, 9 km from al-Mudaibi we came upon *badu* huts like hayricks, paradisiac palm groves, then the village of Zahab. Its major dwelling is the fortified home of Amr Said Sulaiman al-Harthi,

Sanaw. Old settlement

Sanaw. New town

partly restored and partly ruinous. Shown around by the courtesy of
'Abdullah Salim Said al-Mandhari, I found an abandoned wasps' nest
in a squinch. Rubble-strewn stone steps led up to a second floor, which
seemed to have been bombed, but all this was simple decay in time.

Five km south from Zahab we came to ar-Raddah, with its little fort,

Sanaw. Badu suq

then 10 km west we came to the prosperous, expanding town of Sanaw, with a BP station. The old town, cited by the *Handbook of Arabia* in 1916 as being on a hill with date-palms below, has been virtually abandoned. Apart from the numerous *badu* huts on the Raddah track, there was a modern town with a new mosque and a great market recently constructed in a rectangular plan, with shops around the perimeter, and a covered fruit and vegetable market in the centre, the whole in a precinct that can be locked at night. *Badu* girls and ladies stood encased in black, with beaked masks to escape thievish glances. Baby camels waited a buyer, their legs roped to prevent a lolloping escape. The best time to visit Sanaw *suq* is in the early morning; by eleven in the morning activity was draining away in direct relation to the increase in the sun's power.

Southward from both Sanaw and al-Mudaibi stands the last oasis before the vast expanse of desert, where life is as precarious as a handful of water in your clenched fist. Here is al-Musalla.

I sauntered through the lovely oasis, its tall palms waving elegantly in the zephyr. Following the erratic course of the carefully-tended

Al-Musalla. How the falaj works

Al-Musalla. Village elder

falaj, I found dragonflies hovering before my face. Nobody came; nobody went. Then, turning beyond a mud wall, I encountered a patriarch, whose white beard curled resplendently down from his benevolent face to the pure white *dishdasha*, half hemmed by shadow. Our lives coalesced for a moment, we shook hands, and wishing each other the best of health and a long life, we moved on to our own hemisphere.

4 JABAL AKHDHAR AND DHAHIRAH

A full moon gleamed in the starry sky above the limestone outcrop opposite Ruwi Hotel one January morning. It was six a.m., and a Nissan Patrol four-wheel-drive car swung up the ramp, with Ahmad Khalifah, a twenty-four-year-old driver from Suwaiq at the wheel. We took the dual carriageway to Seeb, 30 km away, then at the Sahwa Tower roundabout forked left for the highway to Nizwa, marked 141 km.

Five km beyond, the Married Quarters Village of the Sultan's Armed Forces could be seen to the right, and 3 km farther on we came to the Rusail Industrial Estate, a name that would raise no eyebrows in the developed world, but has made headlines in Oman where, as recently as 1970, there had been neither industry nor even main roads.

The early explorers in Oman would have been dumbfounded by the fine modern housing, the excellent new roads, the evident prosperity of the people in their vigour, good clothes, and expensive cars, but perhaps above all by the signs of expansion into the industrial future without a transitional period of pollution and exploitation criticised by opponents of Europe's Industrial Revolution.

Rusail is the site of the Rusail Cement Company, a vital sector of domestic economy during a phase of major construction. In 1979, Oman imported 0.52 million tonnes of cement at a cost of RO 13.6 million, rising to 0.69 million tonnes in 1980 worth RO 18.5 million, 1.4 million in 1982 worth RO 30.1 million, and 1.8 million in 1984 worth RO 22.3 million. Clearly, the import of cement was becoming a massive drain on the Sultanate, and the Rusail factory began production, to cover domestic needs, in 1984, with a capacity of 0.62 million tonnes initially. 98% of its raw materials were available locally, gypsum being found in Dhufar.

Another major saving has been made on detergents. The National Detergent Company, manufacturing Bahar detergents, Surf and Autosurf, is protected by an excise duty of 20% placed on imported laundry detergent powders.

From Rusail the road climbs into the mountainous country once called 'Oman Proper' to distinguish it from Muscat. 'Jabal Akhdhar' means 'Green Mountain', but there is nothing green about the barren

210

scrubland through which you pass now, and indeed the narrow definition of the term covers just Shuraijah and Saiq, villages two kilometres apart on the landward side of the range. The wider definition employed here to describe Jabal Akhdhar and the Dhahirah, with Sirr, covers the uplands between Suma'il and Dhank and the seaward slopes between al-Hazm and Nakhl.

Fanja' and Bidbid

Near the 'Wilayat Bidbid Welcomes You' sign just beyond Rusail, Fanja' is marked at 20 km and the temptation to rush on to Izki and Nizwa must be resisted. For Fanja' is not merely one of the most eagerly-supported football clubs in the Sultanate, having won the league championship in 1984. It is strategically placed in the Wadi Suma'il: without Fanja' a military leader cannot control the passes between Muscat and the interior, and between the Western and Eastern Mountains. The new town, with its cinema, minaretted mosque (a novelty in Ibadhite Oman, where a tall minaret never formed part of religious architecture) and busy market, makes a sharp

Fanja'. New mosque

contrast to the brown ochre of the old town. Park in winter by the round watchtower commanding the road, which glides above the wide sweep of the wadi, making traffic safe from flash floods. In summer, by contrast, you can drive along the wadi bed, noting the old *falaj* network two metres above the dry wadi.

A car can be driven off right from the main road to the old fortified walls, and left there while you explore the old village, now almost derelict as the local people have left mud-walled houses for the advantages of stone houses, with running water and electricity. However, some of the old houses have been restored, and provided with running water; find the east gate, with a wood-roofed porch, and enjoy the panorama of the palm plantations and fields planted with vegetables and forage for animals.

I talked to a farmer who had stopped for a quick meal at the Haddabi restaurant on his way to the capital from Qarut: he compared his ease of travel now with the discomfort of his arduous journey before the new road was completed in the 1970s.

Nine km from Fanja' stands the great castle of Bidbid, overwhelming the little village beneath its walls. It once commanded the main

Fanja'. Cinema

Bidbid. Castle

crossroads to Sur and Nizwa, but the new asphalt road has curved round the oasis, bypassing Bidbid. The restoration of Bidbid Castle is a characteristic triumph of the Ministry of National Heritage and Culture, following a Unesco report recommending that solely traditional techniques and materials should be used in the repair of ancient castles. Cement render, once applied on old buildings in Muscat and elsewhere, has been found to stop the mud walls 'breathing', and the resultant rising damp will lead to cracking. Now the fort's walls are rendered with customary *juss*, a compound of fired mud, straw and gypsum. *Saruj*, a local cement of baked lime and mud, was used to repair the *falaj*, because it is water-resistant. The ceilings are of rush matting with a coat of mud and gypsum. As the first major task undertaken by the Ministry, Bidbid was subjected to a great deal of scrutiny and passed the test with flying colours, to the extent that similar methods have subsequently been used elsewhere. Unlike the forts at al-Hazm or Rustaq, Bidbid's was not homogeneous; its two parts were clearly built at different periods by different owners;

Bidbid. Oasis, from the castle walls

possibly one was the tribe of Siyabiyin, but conflicting tales exist.

One needs a letter from the Ministry in al-Khuwair to visit Bidbid; armed with this I greeted the solitary guard, Khusayib bin Huwaishil. He rose from his stone bench kissed the letter with feudal reverence, and waved me inside with his crutch. I bowed my head to step inside the Lilliputian wooden door within the great double door (all the better to behead you, my dear!) and sped up the forty-two steps to the highest point of the watchtower, beyond which the green expanse of oasis seemed to fill the land as far as the distant mountains. A separate courtyard leads to two lesser round towers, with a *falaj* and a *shawisha* tree.

The gatehouse has two storeys, the upper being living-quarters for the garrison. In one room you can see the hole in the floor through which enemies who had burst through the gateway could be bombarded or shot at.

214

Suma'il

Leaving Bidbid, you return to the main road by the bus stop and a cluster of shops near the Shell filling station: Bidbid Restaurant, Foodstuff Super Market, Bidbid Ready Made Garments. After 7 km we entered Wilayat Suma'il, then in 3 km the turning to Qaiqa' ('Gegah' is the eccentric transliteration on the sign), a village 1 km off left. Watchtowers like raised fingers appeared above palm groves on the right, then 14 km south of the Qaiqa' turning we headed left 4 km towards the village of Suma'il, bumping along a wadi bed. This was the first authentic shiver of danger; not waiting for horror, like a city under siege, but prudently anticipating a range of lesser anxieties: landslide from the mountainside, snapping of an axle, veering and overturning of the Nissan car, smashing and drowning in sudden flood. I do not seek fear, but the opportunity to overcome it; I do not welcome death, but I live so intensely that at any given moment I can accept death as part of the bargain. A man who flies two hundred hours a year must accept that he is in greater peril than a man who flies only twenty. A man who drives on the edge of precipices and through ravines will be at greater risk than a man who drives on tarmac roads from home to office every day. If we benefit from favourable summer winds, we must expect a winter storm that may lead to shipwreck. In Oman I met many

Suma'il. Up to the castle at the end of the track

greybeards who looked neither at me, nor at any other man, but through the mists of appearances to the true vision, a reality that transfixed them so that they could hardly bear to turn away. To such men our daily thoughts of comfort must be less than trivial, less than the dust. Their dialogue with the ultimate is undeniable, immediate, radically simple. We refresh our trivial, over-complicated minds by the merest contact with their prowess. Doughty could respond to their submission only with the most elaborate language, drawn from Elizabethan and Jacobean prose and old interpretations of the Scriptures, but felt inhibited in his empathy by Christian prejudice. We should bear no such sectarian rancour, for these People of the Book in Oman accept the British as expatriate workers, as they accept Indians, Pakistanis, Bangladeshis, French, Germans, Dutch, Americans. . .

The castle of Suma'il defends a rocky outcrop far above the wadi bed. I saw donkeys, yes, but they were outnumbered by cars. Little fish swam in the open *falaj*, cleaning the waters in memorable symbiosis between man and nature.

Watchtowers above both banks of the wadi remind you that now, as always, Omanis are spending roughly fifty per cent of their national income on defence, for there is a Marxist government to the southwest and the Iran-Iraq War continues to endanger the Gulf as well as the next generation in the Middle East. Suma'il has a weaving centre, but is best known for the many varieties of dates it grows. Oleander blooms softly pink beside limpid pools.

Izki and Birkat al-Mawz

Back on the main Nizwa road, we passed an almost endless chain of fertile oases benefitting from run-off from the high rugged mountains: Wibal, Biyaq, Rissah, and finally the important town of Izki, reputed home of the most reliable Omani authority on early Omani history, Sirhan ibn Sa'id ibn Sirhan whose *Kashf al-ghummah* ('Dispeller of Grief'), compiled about 1728, has recently been reprinted in E. C. Ross's translation. We have seen that 'Ibra is divided: so too is Izki. The walled quarter on the left bank of Wadi Halfain is occupied by the Ghafiri Bani Riyam, and is called Yaman; the other, on the right bank Nizar, is inhabited by Bani Ruwafah, Hinawi. Irrigation is from springs and by the Falaj al-Maliki, with many tributary channels, allowing a plantation of more than ten thousand palms to flourish year in, year out. The castle was once strategically important, when the Sultan was threatened; the 1916 *Handbook of Arabia* notes that he then maintained 'a Vali there with a garrison of 20 men', but the fort is now desolate, without a guard. An enormous cemetery spreads its rough, unmarked, eternal stones across mute gravel.

Izki's traditional *suq* consists of cubicles rather than shops. A sheet of corrugated iron offers ephemeral shade to the wanderer. In the narrow streets of the *suq* (inexplicably common to both antagonistic quarters) I found cooking oil, sweets, plastic toys such as 'space saucers', metal rakes without handles, shopping bags, spices, animal

Izki. Palm-groves and jabal, from castle

fodder. Roadside restaurants run by Indians peer hopefully out on to passing traffic , but Izki *suq* closes inward, like a turtle's head into its shell.

I now decided to make for Birkat al-Mawz, 'Banana Pool', remembering how Wellsted had relished the oasis in 1835. It has a fort, he wrote, 'with a very capacious and good-looking castle, belonging to the Sheikh of Suik [Suwaiq], within its walls, around these are several large groves and plantations; the plantain trees are numerous, and hence its name. A very large rivulet, or feletch, furnishes an ample supply of water, and from its situation, Birkat enjoys a delightfully cool atmosphere from the Jabal Akhdhar, and bears the reputation of being very salubrious.'

The new road bypasses the village, so you must take a graded road under the *falaj*-aqueduct to reach the oasis, which consists of a wide road between two facing rows of two-storey houses which eventually comes to the fort described by Wellsted, now ruined. The *falaj* is completely restored with cement.

Izki. Shadowed suq, with new mosque in background

Shuraijah and Saiq

To reach the Jabal Akhdhar from Birkat al-Mawz you must try to obtain a permit in advance from the Office of the Deputy Prime Minister for Security and Defence, and this is not often conceded. For the Wadi Muaidin road is military, winding steeply until, at a height of over two thousand metres, it arrives at the conglomeration of villages such as Hail Yaman, Saiq, and Manakhir. Plums, pomegranates and wheat are cultivated here, as are grapes, peaches, apricots, maize and garlic. These lie below Jabal Shams (Mountain of the Sun). Shuraijah (Wellsted's 'Shirazi') clusters on the cliffs with al-'Aqr and al-'Ain, divided between the Ghafiri and Ibadhite tribes of Bani Riyam and Bani 'Umr. Wellsted had reached Saiq from Tanuf on his journey in 1835. Since Tanuf he had met with nobody at all, but at Saiq the inhabitants 'crowded out in great numbers to welcome us as we passed along. Several entreated us to remain for the night at their village; but I was anxious to pass on to Shirazi, which is described as being the most extensive and plentiful of all the valleys.' Wellsted describes Saiq with

218

the ardour of an Englishman who knows he is the first of his nation to set eyes on it: 'a wilder, more romantic, or more singular spot than was now before us, can scarcely be imagined. By means of steps we descended the steep side of a narrow glen, about four hundred feet in depth, passing in our progress several houses perched on crags or other acclivities, their walls built up in some places so as to appear but a continuation of the precipice. These small, snug, compact-looking dwellings have been erected by the natives one above the other, so that their appearance from the bottom of the glen, hanging as it were in mid air, affords to the spectator a most novel and interesting picture. Here we found, amidst a great variety of fruits and trees, pomegranates, citrons, almonds, nutmegs, and walnuts, with coffee bushes and vines. In the summer, these together must yield a delicious fragrance, and produce a picturesque, verdant, and beautiful landscape. It was now, however, winter, and the whole were denuded of their leaves, and had a cheerless appearance. Water flows in many places from the upper part of the hills, and is received at the lower in small reservoirs, from whence it is distributed over all the face of the country.'

Fiqain. Ascending into the tower by rope

Manah

Sixteen km south of Birkat al-Mawz, the main road comes to the crossroads at Firq (which is what 'firq' or 'mafraq' means): the village of Firq lies 1 km to the right, Nizwa Motel 2 km to left, the new asphalt road 864 km south to Salalah on Route 31, but I took the dirt road left showing Manah 14 km distant. On this track you encounter the village of Fiqain 3 km before Manah itself, with about two hundred houses of the Al Bu Sa'id. I stopped at the Fiqain fort, to which access is now solely by means of a rope up to an opening in a massive wall. At 12.40 a muaddin was calling the faithful to prayer; a crowd of youngsters invited me to scale the roof of the mosque opposite the castle for a closer view of the ruined fort, then again at ground level, we climbed, one by one, up the rope into the dark interior, pigeons startled from their cooing rest into anxious flight. I stayed in that distraught haven too short a time, guilty of the one cardinal sin as defined by Kafka: impatience. 'Because of impatience we were driven out of Paradise; because of impatience we cannot return.' If the comparison with Manah seems far-fetched, it would not have done to Wellsted. As he

Fiqain. The tower from the mosque roof

Manah. Modern house, with generator

crossed the flat open fields of Manah (or 'Minna' as he spells it), he was entranced by lofty almond, citron, and orange-trees, yielding a delicious fragrance on either hand, and 'exclamations of astonishment and admiration' were drawn from him.

'"Is this Arabia," we said; "this is the country we have looked on heretofore as a desert?" Verdant fields of grain and sugar-cane stretching along for miles are before us; streams of water, flowing in all directions, intersect our path; and the happy and contented appearance of the [people] agreeably helps to fill up the smiling picture; the atmosphere was delightfully clear and pure; and, as we trotted joyously along, giving or returning the salutation of peace or welcome, I could almost fancy we had at last reached that "Araby the blessed" which I have been accustomed to regard as existing only in the fictions of our poets.'

Were it not for unobtrusive air-conditioners, schoolboys with satchels swinging at their side or flat on their heads, or elegant new homes with their own electric generators, and Japanese cars, Wellsted would recognise the Manah of today, with its dusty roads, sand-brown

Manah. Fort

fort, mudwalled houses, fertile fields of sugar-cane, whiterobed Zanzibaris, and camels grazing on bushes beside the track.

Wellsted writes that Manah 'is an old town, said to have been erected at the period of Nushirvan's invasion [the Sasanid King Khusrau Anushirwan, 531–578 A.D.]; but it bears, in common with the others, no indications of antiquity: its houses are lofty, but do not differ from those I have described at Semmed and Ibrah. There are two square towers, about one hundred and seventy feet in height [of which only one, called al-Manarah, 'the minaret', survives in 1986], nearly in the centre of the town; at their bases, the breadth of the wall is not more than two feet, and neither exceeds in length eight yards. It is therefore astonishing, considering the rudeness of the materials (they have nothing but unhewn stones and a coarse, but apparently strong, cement) that, with proportions so meagre, they should have been able to carry them to the elevation they have. The guards, who are constantly on the lookout, ascend by means of a rude ladder, formed by placing bars of wood in a diagonal direction in one of the side angles, within the interior of the building. The country in every

222

Manah. Sulaiman bin Saif al-Masruri in his sugar factory

direction around this town is flat and even; and the commanding view
they obtain from their summit enables them to perceive from a long
distance the approach of an enemy.' Cole stayed in the same house in
'Minnik' as Wellsted, and the town was later visited by Miles, who calls
it 'Minha'.

The mihrab in the simple square mosque must be one of the most
elegantly decorated in Oman; but the clock, stopped at twelve past
twelve to its right, detracts from its harmony. There are two ceiling
fans to cool the faithful, and rush mats on the floor.

But what of the sugar industry at which Wellsted marvelled? Could
that too be alive and well? It was, though I found only one family
processing sugar still. This was the family headed by the charming,
venerable Sulaiman bin Saif al-Masruri, who has produced sugar from
his canefield in Manah for more than sixty years. He is a member of the
Masarir branch of the Sulaimaniyin descended from Sulaiman bin
Abad bin 'Abd bin al-Julanda bin al-Mustakbir bin Mas'ud bin
al-Harar bin 'Abd 'Izz bin Mu'awilah bin Shams bin 'Amr bin Ghanam
bin Ghalib bin 'Uthman bin Nasr bin 'Azd al-Ghauth.

I was in luck, because the fields were ripening quickly for the processing season in February, though the lack of rains over the past two years had slowed the growth of the crop. Sulaiman's fields were within easy walking distance, and the octogenarian strode confidently ahead of us, on the tracks he had trodden decade after decade in the timeless content of the farmer. Sulaiman bin Saif waved his arms across the fields westward. 'Over there they make sugar too,' he told me, 'at Nizwa, Bahla, and Yanqul. But everyone knows that the best sugar comes from Manah.'

I agree with William James that scenery seems to wear in one's consciousness better than any other element in life, and the scenery of Manah remains as radiant in retrospect as it did that January afternoon, when Sulaiman bin Saif displayed his cane-fields as if they were some magnificent work of art. But neither painting nor sculpture will encroach on that part of a Muslim's heart which glories in the natural world where he himself feels so utterly at peace. The ability to surrender, 'Islam', is the key to a Muslim's triumph over adversity; for whatever happens is the will of Allah, and complaints are not merely idle: they are blasphemous.

Adam and Nizwa

Adam was described by the 1916 *Handbook of Arabia* as 'an isolated town and oasis 25 miles SSW of Manah, on the edge of the Rub' al-Khali', but that situation has been changed by the black-topped road connecting the Capital with Dhufar, and tracks leave westward for Natih and the oilfields; NW to Bisya and Jabrin; and eastward to Sinaw and the Sur road. Adam is the most southerly Omani town with an important *falaj*, and during the summer the *badu* come to tend date groves: the Duru to the west of the town and the Janabah to the east. The main sedentary tribes of Adam are the Mahariq, the Bani Hashim, and the Al Bu Sa'id from whom Sultan Qaboos is descended.

Nizwa, roughly equidistant between Izki and Bahla, is a dual unwalled town divided between Nizwa 'Alayah and Nizwa Sufala' by the Wadi Kalbu, which is met by Wadi al-Abyadh near the market-place. The city, which assumed particular strategic importance with the building of the great fort by Saif ibn Sultan in about 1660, commands the passes leading from Jabal Akhdhar by Wadi Ma'idin and Wadi Tanuf. Nizwa was the capital of Oman from the eighth century to the twelfth, then again under Barakat ibn Muhammad in the sixteenth and under Ya'rub ibn Bil'arub in the eighteenth.

The upper town is fed by the stream Daris, and the lower town by the stream Ghunduq, but the dominant position of the Bani Riyam in the upper town has made it possible for them, at times of intertribal crisis in the past, to diminish the water supply to the Bani Hina and others in the lower town. Along the wadi bed a number of walled quarters comprise houses interspersed with gardens, fruit orchards, and date-palms; indeed, the palm groves extend as far as Raddah, eight km down the wadi.

224

Jabal Nakhl. Wolf trap (Courtesy of Charles Hepworth)

Jabal Nakhl. Village of Nakhl (Courtesy of Charles Hepworth)

Jabal Shams (Courtesy of Charles Hepworth)

Wadi Sahtan. Chain route (Courtesy of Charles Hepworth)

Wellsted on Nizwa

Wellsted visited 'Neswah' in 1835. 'I proceeded at once,' he wrote, 'to the residence of the Sheikh, who possesses great influence in these parts. We found him seated before the castle gate, with an armed guard of about fifty men, who were standing on either side. The whole town had followed us thus far; but directly Sayyid Sa'id's letter was produced, and it was discovered who we were, they immediately dispersed, and the Sheikh expressed his regret that I had not sent him an intimation of my intended visit, in order that he might have met me on the road with a proper escort. We accompanied him to his audience-room, within the fort, which was lofty and well furnished. A house was soon procured and here, for the first time since leaving Maskat, I enjoyed the luxury of being alone, and remaining with our every motion unwatched for, at the different other towns on our route, the Sheikhs and principal men thought they honoured us in proportion to the time they passed in our society.'

Wellsted is the first European to describe the fort of Saif ibn Sultan, reputedly twelve years in the making, and Cole, visiting 'Nuswee' during the following decade, was not allowed to enter the fort. Wellsted and his party were admitted, he tells us in an atmosphere of evident excitement, 'after much ceremony, by an iron door of great strength and, ascending by a vaulted passage, passed through six others equally massive before we reached the summit. In order to render appearances more imposing, a janitor behind each inquired the purport of our visit; and, being told we were servants of the Sooltan, he removed several locked bars and chains, and we then passed on. The form of the fort is circular, its diameter being nearly one hundred yards, and up to the height of about ninety feet it has been filled up by a solid mass of earth and stones: seven or eight wells have been bored through this, from several of which they obtain a plentiful supply of water, and those which are dry serve as magazines for their shot and ammunition. We found a few old guns here, one bearing the name of Imam Saif, and another that of Kouli Khan, the Persian general who took Maskat. A wall forty feet high surrounds the summit, making the whole height of the tower one hundred and fifty feet.'

As regards local produce, Wellsted noted a great quantity of sugar-cane, and the popularity of *halwah*, one of the sweetest of all confections. You can see it being stirred in great vats even today: the Omani taste for *halwah* has not diminished with the passing years. He did not mention other sweets, which you can ask for in the *aswaq*, such as heavy sticky *harisah*, made of sugar, almonds and semolina; round *mabrumah*, made of *kunaifah* pastry, syrup and pistachio nuts; square *baluriyah*, also made of pistachio nuts and *kunaifah* pastry; round *barazik* biscuits with sesame seeds or pistachio nuts or both; and pyramidal *bukaj*, cashew nuts in a flaky pastry, topped with pistachio nuts.

The inhabitants of Nizwa in Wellsted's time made copper pots, and there were also goldsmiths and silversmiths. 'A considerable quantity of cloth and some good mats,' he added, 'are fabricated from the rushes which grow on the borders of the streams. Preparing cotton in the yarn is the principal occupation of the females. In the cool season they may be perceived coming out from beneath the groves with their spindles after breakfast, to enjoy the warmth of the sun's rays. The men alone attend the looms. Besides mats, the females manufacture some pretty baskets from the rushes. The former serve them to sleep on, and the latter they carry with them to market to deposit their purchases in. A great many camolines are fabricated here, and also brought from Nejd. The best, worn by the Sheikhs, are of a light-brown, or cream colour, and sell for forty or fifty dollars; the black camoline, and those striped in alternate vertical bars of brown and white, eight to ten dollars. The camoline forms the most important article of their dress, and its quality denotes the condition and rank of the wearer.'

Nizwa in the 1980s

Wellsted found the principal articles exposed for sale at prices which corresponded very closely to those charged at other towns of the interior: rice, wheat, barley, beans, camel meat, beef, mutton, kid, sweet oil and ghee (clarified butter). In 1836 sweet oil cost 50 pice per pound (*baiza* is the Omani spelling) compared with RO 1.400 for 3 kgs today; ghee then cost 56 pice per pound, compared with RO 1 per kg today. Women in the *suq* wear colourful clothes. Corrugated iron placed on wooden supports stretches from one side on the *suq* to the other, and the cubicles are locked with shutters. Clutters of showcases and baskets jut untidily out into the walkway, in permanent shade dappled with slanting sunlight. You could be anywhere in the Arab World, from Ta'iz to Baghdad, and at any time, as the bearded elders riffle through textiles, and shrill-voiced women out in convoy argue about children's shoes. If the goods on sale come from all over the world, that too is nothing new, for trade via Muscat has always dominated urban life throughout Oman. Chinese and Persian pottery has been excavated at Julfar (the modern Ras al-Khaimah) as Hansman has shown in his recent monograph, and Suhar, Muscat, Sur, and Qalhat have in turn provided goods from all over the trade routes to the Gulf for the interior towns.

Even the little Indian Restaurant opposite the castle is called nostalgically Mat'am al-Khalij (Gulf Restaurant) as though, shading their eyes from dust and sun, the men of Nizwa could make out the distant sea from these heights. For the equivalent of £2.40 (just over US $4.00) in total we each enjoyed a plate of salad, another of rice, another of chicken, and a Sulaimaniyah, an Omani word denoting tea without milk. An animated old man, hopping like a sparrow from one side of the table to the other to make a point to us, giggling and affirming 'Wallahi!' to strengthen his case, insisted on providing us with soft drinks, disappearing through the door like a jack-rabbit back into a hat before we could reciprocate.

On the opposite side of the road from the restaurant, in the lee of the new Sultan Qaboos Mosque, stands the tree under whose spreading shade goat auctions are held. I returned to the principal *suq* to look for coppersmiths, silversmiths, spice merchants and found gold at Sulaiman Hamad 'Abdullah as-Sulaimani's. Old coffee pots fetch RO 8–10 here, and the *khanajir* start at RO 85, much lower than in Matrah. Weaving from the Sharqiyah, tin trunks from India: I wandered entranced as if in timewarp, greeting each friendly shopkeeper as well as smiling passers-by. Crouching I watched silver being worked in the shop of Ahmad bin Salih al-Hashami.

Then it was time to present my visitor's permit to the great seventeenth-century castle, consisting of a rectangular *husn* and a roughly circular earth-based platform towering not only above the *husn* but above the whole verdant oasis, surmounted only by the Omani flag fluttering from a high flagpost.

I was fortunate enough to be shown round by the Nizwa-born Jum'a bin Said, twenty-two years a guard here, during the time of Sultan Sa'id bin Taimur as well as during the subsequent renaissance. One hand clutching his rifle, Jum'a waved the other around the large rectangular space, defended by walls, capable of protecting hundreds of people and their flocks as well as the garrison of men with their stores and ammunition. The round *qal'a* was intended to support heavy cannon, its height increasing the range of its fire, while diminishing the risk of attack from below. Its size made it virtually impregnable, given also that ample water supplies are available not only by means of an underground *falaj*, which could in theory be cut off or poisoned, but also by several independent wells.

Though filled with earth, to withstand the impact of cannonfire, the *qal'a* has been penetrated from within by a zigzag stairway cut off by a door at each turn, where holes are so placed that stones or musket fire can be aimed at invaders. The earth base is fourteen metres high, and the walls rise another ten metres, offering a complete crenellated walkway from which musketeers could reinforce the cannon at any point of the compass. You can still see the well-entrances at the top of the *qal'a*, the prison in the centre, and the old mosque abutting the wall. Even were the fort of no historic importance, it would still be worth climbing to the round platform, and the walkway above it, for the stupendous views on all sides.

Below, I strolled around the new *suq* whose wooden columns rise to the new mosque. Then I drove westward to the watchtower overlooking the main road and the great groves, the cemetery, and watched the day crumble its last light into sudden grey, promising myself that, after a night's sleep in the Nizwa Motel, I should return to the dawn chorus in Nizwa 'Alayah, its walled quarters haunted by the mute cries of tribal warfare long past.

Nizwa could be taken as an example of interior towns where the more prosperous possess a winter stone or mudbrick house in the built-up centre, and another in the date-groves for summer occupation.

Nizwa. View westward from the fort

Town houses in Nizwa are usually two storeys high, with a courtyard acting as a light and ventilation shaft to the upper floor. Neighbouring houses will be built up on two or three sides, leaving only a single entrance from the street. The door opens into a porch, leading to the *majlis*, the sitting-room for male guests with tiny barred windows on to the street. The only rooms on this level will be store-rooms, and a kitchen, perhaps. Niches in the walls often take the place of cupboards, though this is changing with the acquisition of furniture. The upper floor will be dominated by bedrooms, with a parapet ensuring privacy between houses.

Oasis houses of mudbrick tend to cover a larger area, to catch more breezes, and do not adjoin each other. An exterior wall will conceal a well with a bucket on a pulley, a basin for laundry and a bath, and a cooking area, as well as a one- or even two-storey house, with an internal lavatory above a sewage pit. Dates dry on the roof, above rooms kept almost dark, with numerous tiny apertures rather than windows to allow maximal wind circulation with minimal access to the sun.

Four km east of Nizwa *suq* you will find the motel, next to the Baladiyah (Town Hall) and near the Toyota agency, below toothy mountain crags on the other side of the road.

Jacaranda blooms filled the motel garden between the central pool and flanking double bedrooms. Like the restaurant, my room was air conditioned, but I preferred the fresh breeze of the evening to lull me to sleep through open windows.

Tanuf and Al-Hamra'

The main road from Nizwa to Bahla shows a right turn to Tanuf, where Wellsted arrived on Christmas Day 1835. From Nizwa, 'following the skirts of the hills to our left,' he wrote, 'we passed several sterile plains which present nothing worthy of observation, and at three hours arrived at Tanuf, where the Sheikh resides, whose authority is paramount on the mountains. After halting we were at first lodged in the mosque, which, strange as it may appear, is generally used in Oman as a caravansarai; but, fearing that they might not relish our walking about with our shoes on, and it being rather too cool to go without them, I procured another house. Here I was soon joined by the Sheikh, who came with several others to dissuade me from my intention of visiting the mountains.' Wellsted, on his way to Saiq, was not in fact so dissuaded, but it is interesting to note that objections are placed in the way of visitors wanting to travel in some parts of Oman even now. Not only must one overcome the 'no objection certificate barrier', but once inside the country one must obtain a multiple visa to be able to reach Musandam or Buraimi, for example. Even in the Jabal Akhdhar, one must leave one's vehicle at a military checkpoint at the bottom of the road up to Saiq, and rely on government vehicles for a lift the rest of the way. Many of these warnings and obstacles have a basis in historical conflicts, such as the rebel war in the Qara mountains in Dhufar, or the rising under Shaikh Sulaiman bin Himyar an-Nabhani operating from Birkat al-Mawz in the 1950s; others arise from suspicion of strangers' intentions, justified in part by the aggression of Persians, Portuguese, and more recently Marxist infiltrators from the People's Democratic Republic of Yemen, the last documented in Ranulph Twisleton-Wykeham Fiennes' *Where Soldiers Fear to Tread* (1975). But mainly the warnings derive from a desire to be absolved from responsibility for what happens to a guest when he leaves the immediate vicinity.

I personally was warned at Tanuf about ascending the mountain passes, because Wadi Mistal on the other side was the haunt of thieves and scoundrels. When, later on, I was to reach the point in Wadi Mistal where one may continue only on donkey-back, I was similarly advised by the affable people of Wakan not to proceed any farther (despite the money they would earn by guiding me over) because of the rascally people on the other side. Nothing has changed: human nature continues to be essentially amiable and mistrustful. As Wellsted reported at Tanuf: 'Most frightful pictures were drawn of the passes; and I believe they thought that we, like many of their worthy countrymen who pass their lives on the plains, had never visited a mountain district. The natives were also described as being little better than savages, and especially hostile to the visits of strangers; but, finding all their arguments ineffectual in changing our intention, they took their departure, evidently somewhat disappointed.'

The 1916 *Handbook of Arabia* characterised the 'Tenoof' of Cole as 'a walled village to the NW of Nizwa, at the point where Wadi Tanuf leaves Jebel Akhdar; elevation 1,950 feet. There are two gates in the wall, on the E. and W. sides, and within are about 40 houses of the Beni Riyam tribe. The Sheikh's house is behind the town, on the edge of the Wadi Tanuf, up which a track (the *Tariq ash-Shass*) leads over Jebel Akhdar to 'Awabi. The site of the village underneath the cliffs is cramped and, shut in as it is by the mountains, the heat during the summer months is intolerable. There are date plantations and considerable cultivation.'

Old Tanuf was demolished at the Sultan's orders following the unsuccessful rebellion of Sulaiman bin Himyar, *tamima* of the Bani Riyam who were the first tribe to settle Jabal Radwan, as Jabal Akhdar was originally known. Tanuf was Sulaiman's headquarters, and as such suffered virtual annihilation, but his castles in Saiq (that eyrie) and Birkat al-Mawz on the plain also suffered in reprisal, and Sulaiman himself took up exilic residence in Dammam, in the Eastern Province of Saudi Arabia. Old Tanuf today is barely a ghost town: mounds of rubble here and there cannot evoke more than a murmur of the past. Even the shrillest of memories can lose its echo, and old Tanuf has lost its echo, as it lost its *falaj*.

Tanuf may be intolerable in summer, though with the spread of generators and air conditioning plant many Omanis now have access to year-round comfort, but in winter the oasis is nothing short of idyllic. A motorable track leads through the palm-groves, citrus orchards and cultivated fields towards a state agricultural project and the factory bottling the excellent spring water that I drank everywhere in Oman in preference to the more expensive Evian or Perrier waters.

You return to the main Nizwa-Bahla road from Tanuf, and take a right-hand turn into the mountains signposted al-Hamra' 16 km, driving on a two-lane black-top road all the way. After 4 km a sign shows Qal'at al-Masalha 5 km off right, then after 5 km a little watchtower rises like a brown pepperpot left. After 1 km more a track leads right to the village of Ghamr, and after 2 km the palms of Ghamr can be seen in the distance, from the new Agricultural Extension Office. It lies meekly level with the plain. In 3 km more you arrive at the Town Hall and the bus stop, with a sign to Misfat al-'Abriyin 5 km, a village named for the tribe that lives there, populous too at Bahla, and here at al-Hamra' ('The Red One', an Arabic adjective that gives the hispanicized form 'Alhambra' familiar in Granada).

Al-Hamra' is a homogeneous town, the 'Abriyin having constructed it in the seventeenth century, around a single, crucial *falaj*, which runs parallel to the Congregational Mosque and the walled *suq*. The town itself is unwalled, being defended by three watchtowers. Women were washing clothes and dishes in the *falaj*, and chatting among themselves, quietly so that passing men could not overhear. At the old mosque, mudwalled, with rush mats on the floor, I noted an ablutions basin beside the *falaj*. The roof has new wooden beams, and the two wooden doors have been freshly painted green, the colour of Islam. The 'minaret' is no higher than the mosque itself. Hamdan al-Balushi told me that the shaikh of the 'Abriyin is elected, whereas other tribes customarily prefer hereditary *shuyukh*. The villagers keep bees and grow sugar-cane, raise goats, and tend fields.

It is worth a detour to visit Misfat al-'Abriyin, 7 km beyond al-Hamra' into the jabal, since it is the only true jabal village accessible by vehicle, up a succession of hairpin bends.

At the end of the road you can stroll through the village, and with the right preparations continue over the mountains into the Wadi Sahtan which includes the villages of 'Amq, Fashah, Mabu and Maqamma, all inhabited by 'Abriyin. Misfat grows mangoes, limes, and oranges on pleasant terraces that have come to evoke for many the true, rugged beauty of the Jabal Akhdhar.

Bahla and Bisya

Bahla ranks with Dubrovnik and Aigues Mortes as one of the finest walled cities in the world. Bahla's wall extends for a total length of more than 12 km, enclosing not only all the walled quarters, but many though not all of the plantations. The government restored the wall in the 1970s, but unfortunately permitted the new asphalt road to cut through it, instead of creating the bypass that would have preserved Bahla against unnecessary 'progress'.

Clearly ancient from its connection with the Bahila tribe, the town became the capital of the Imamate at intervals, for example under the Nabahinah of the fifteenth century, such as Makhzum ibn al-Falah, who ruled from 1406. The very fort is pre-Islamic in origin, and may have been built by Persian invaders.

You do not need a permit to enter the ruined fort of Bahla, but a guard is posted there anyway, presumably to ensure that no pilferage of stones occurs. The old man, with his trusty rifle and cartridge belt reassuringly crammed with live ammunition, prepared traditional coffee for me, with flavoured dates. A serving soldier ambled up and, after greetings, explained in excellent English that he had worked at the Scorpion factory in Coventry, spent a less happy stint with the Egyptian army, and then two periods at Bovington camp. I climbed to the forlorn heights of this great fort, clambering over mounds of rubble to obtain the best views of the valley far below. No other visitor disturbed a lizard, set like a chess piece by an invisible hand, which

Bahla. Inside the ruined pre-Islamic fort

moved imperceptibly to another dusty square, and another, until it
disappeared into a crack like a coin down a drain.

Nearby I explored the area of the mosque and the walled *suq*, then in
the heat of the day found the new *suq*, with heaps of dried fish, lucerne
as animal fodder, oranges from Hamadan, potatoes, garlic, Lebanese
apples, and bananas, all left unattended while the merchants prayed.
The new *suq* has been constructed in traditional style, but remains
uncovered, in this climate an odd decision and one significantly not
hallowed by custom.

All the Indians eating with me in the Ahya Restaurant hypnotically
watched an Egyptian film, relayed from Abu Dhabi television station.
Around the walls I noted pictures of Sultan Qaboos, crowds round the
Ka'aba in Makkah, a Rothman's calendar, the opening surah of the
Qur'an, and a poster for Red Label teabags.

Ahmad Khalifah shyly watched me while we ate chicken with rice,
onions and tomatoes, and asked:

'Do you like this food? Is it normal for you?'

Bahla. Dried fish in the suq

I nodded. 'Yes, I like it, but all food is normal for me: I eat Russian food in Russia, Chinese food in China and Mexican food in Mexico. . . '

'And Indian food in Oman,' he interjected.

'That's it. And Indian food in Bath, Glasgow, and Belfast. We have Indian restaurants in Britain as you do in Oman.'

The Egyptian film disappeared into the distance with the Arabic for 'The End' filling the television screen. Then came the news, the UAE flag, the national anthem of the UAE and a portrait of Shaikh Zaid.

My rendezvous with Jocelyn Orchard's archaeological team had been arranged for dawn the next day. So I had to find the Strabag Camp to arrange the hour of departure for Bisya. I hazarded a question in Hindi to a group of Indians still riveted to the screen. Yes they knew of the Strabag Camp; indeed one of them lived there and was just on his way back after the meal. There we found the Birmingham team, who kindly provided us with beds for the night in Bungalow 12. The Camp, now the property of the Directorate General of Roads, was originally erected by the West German road

Bisya village. Fort

construction company Strabag as its base in the interior during the laying of the road from the Capital Area to Buraimi, via Nizwa in the 1970s, and it is a base for road maintenance crews still. By kind permission of the Directorate of Roads, the Birmingham team under Jeffery and Jocelyn Orchard have been able to use the Camp as their headquarters during their surveys and excavations in the Wadi Bahla, work which, sponsored by the Ministry of National Heritage and Culture, has been much aided by Wimpey Alawi LLC and other companies in the Sultanate.

As may be seen from the map in J. H. Humphries' paper, 'Some Later Prehistoric Sites in the Sultanate of Oman' in the *Proceedings of the Seminar for Arabian Studies*, Vol. 4 (1974), p. 58, fig. 1, Wadi Bahla is one of a number of areas in Eastern Arabia and the Gulf where evidence of important human occupation has been identified between the third and first millennia B.C.

Some 30 km to the south of Bahla, in the neighbourhood of the lesser town of Bisya, the Wadi Bahla converges with a fluvial network dominated by the Wadi Sayfam. To judge from the many stone-built

Qarn Qantarat Nizwa, near Bisya. Circular stone masonry, seen from the north (Courtesy of the University of Birmingham Archaeological Expedition to the Sultanate of Oman)

tombs that may be seen on all sides, as well as the vestiges of less easily categorised ancient structures, this area would seem to have been a favourable one for settlement over a very long period. On Qarn Qantarat Nizwa, a long low rock outcrop near the village of ad-Dabi, immediately to the north of Bisya, Jocelyn Orchard showed us an impressive circular walled monument built of massive tabular limestone slabs which had clearly been obtained from the outcrop itself. This monument, which has been dated to the third millennium B.C. by potsherds found scattered over its surface, is interpreted by the Orchards not as some kind of enclosure, but as the partially surviving stone masonry revetment of a high earth building platform, whose crowning building, perhaps a tower in this case, was approached by a ramp, or ramp supplemented by stairs, still traceable on the south side. In view of the fact that three more monuments of a similar kind are in evidence on separate hills and outcrops in the area of ad-Dabi, they would further see this ruin on Qarn Qantarat Nizwa not as a major structure in isolation, but as one of a series of platforms crowned with buildings marking the perimeter of a substantial third millennium settlement. What this settlement was like, in so far as this may be retrievable, will undoubtedly engage the Birmingham team for a number of seasons to come, but at present the vision is of a settlement not unreminiscent of Oman's interior rural towns today with their massed cultivation and surrounding protective watch-towers. Certainly, the constructional concepts embodied in these third millennium building platforms have endured down to recent times, for, close beside the road in Bisya, Jocelyn Orchard took us to see a vacant, two hundred-year-old mudbrick house on a high stone-revetted earth platform, which likewise was founded on a low rock outcrop.

Around the middle of the third millennium, the settlement marked by buildings on outcrops was succeeded by a settlement of the Umm an-Nar culture, which flourished into the early second millennium. How this Umm an-Nar settlement came to an end is still uncertain, but one theory at least is that increasingly dry climatic conditions may have been responsible, obliging the people to move elsewhere or to become nomads. Eventually, in the first millennium, settlers again returned to this part of the Wadi Bahla, settling primarily in the area of Salut, immediately to the north-west of Bisya. According to current opinion, it was this millennium which saw the first construction of *aflaj* in Oman, a development which legend connects with the name of King Solomon/Sulaiman bin Daud, who in Persia has been partially identified with the fabled ruler, Jamshid. To quote the story: while Sulaiman was being conveyed by the winds on one of his daily journeys between Persepolis (Istakhr) and Jerusalem (Bait al-Muqaddas), he observed at Salut a castle in such excellent state that it seemed to be newly built. Discovering the sole occupant of the castle to be an eagle, whose forbears had inhabited the place for many generations, he decided to spend ten days in the creature's company, ordering his attendant retinue of spirits to dig during that time ten thousand *aflaj* to provide the land with water, a legend retold in J. C. Wilkinson's *Water and tribal settlement in South-East Arabia* (Oxford, 1977).

Bisya area. A building so far unidentified, during the 1986 excavations (Courtesy of the University of Birmingham Archaeological Expedition to the Sultanate of Oman)

Before leaving the Bisya area, we visited the Birmingham team's current excavations, which are being carried out in an area of revived agricultural activity on the northern approaches to the town. Here, in 1981, a smallholder was obliged prematurely to abandon a well that he was digging to provide water for his land by his sudden encounter with a mass of stone slabs and masonry only half a metre below the surface. When, a few months later, his unfinished well was found by the Birmingham team during the course of field survey, it was at once perceived that what the smallholder had accidentally hit upon was an ancient building of major proportions, and since then two full seasons have been devoted to bringing its details to light. From the 150 square metres of the building already exposed, it is possible to see that it is approximately circular in plan with a broad flight of stone steps leading down into it on the north side. Aligned with these steps and framing a central aisle are two rows of massive rectangular piers, some of which, remarkably, still stand to around 2 metres in height. These piers formerly supported the building's corbel-vaulted stone ceiling which, under the impact of time and flooding, was eventually to crash down between them in a chaos of tumbled stones. Most of this fallen stone has already been removed, but as we were looking at the building a tractor-mounted hoist was still engaged in lifting remaining slabs from portions of the interior to left and right of the central aisle. Whether the building, which is as yet without parallel in Oman, is of third millennium date, as the character of the masonry suggests, or later, has still to be determined, as also has its purpose, but the Birmingham team anticipate that firm answers to these questions will have been obtained by the time excavation is completed at the end of their next season.

In Bahla *suq* you can find great earthenware storage jars resembling those in classical Mediterranean shipwrecks, but try instead to locate in the town and oasis those potteries where the jars are still made, piled up as if in a Roman villa off the Via dell'Abbondanza in Pompeii. If you are merely passing through, you can see a pottery factory next to al-Ahya Restaurant, on the main road. Ornamental keys can be found in the marketplace, but possibly the most characteristic wares of Bahla are the cloaks: the *manasil* of black goat-hair, shaped like bat's wings, with an embroidered design at the back; and the woollen *bisht* with gold edging, worn over the white cotton *dishdasha*.

The aura of history pervades Bahla: turbulent tribes, destructive floods (such as that of March 1885 when many date-palms and houses were swept away), and the confrontation between the Imams of the interior, who held Bahla, and the Sultan in Muscat, which ended in truce as recently as 1970. Natural disasters are never far away: a flash flood carried a petrol tanker along a *wadi* bed as recently as July 1985. Yet the weather dissimulates: mild breezes merely touch the forelock of great palms in the oasis, where farmers raise sugar (though less than before), wheat, barley, beans, mangoes and plantains, as well as the ubiquitous date. The temptation is to stay in Bahla, encapsulated beyond time, between the city rush of Ruwi and the commercial life of Abu Dhabi and Dubai.

Jabrin

But I had an appointment with the Palace of Jabrin, and five km beyond the zebra crossing in Bahla I turned off left, reaching the little village of Jabrin after 4 km more.

The palace of Jabrin stands in the northern quarter of a fortified courtyard walled with unbaked brick, entered by a single gate near the eastern corner, opposite a tamarind tree beside which Miles camped. The tamarind is still there, affording shade to men and beasts alike, as well as the overheated vehicles which have just come on the five-hour journey from the capital.

We know tantalisingly little of the *madrasah*, or religious college, established by Bil'arab here in Jabrin, though some scholars' names have survived: the poet Rashid bin Khamis al-Habashi al-A'ma, and the theologians Sa'id bin Muhammad bin 'Ubaidan, and Khalf bin Sinan al-Ghafiri.

But we do know that Bil'arab was besieged at Jabrin by his brother Saif bin Sultan, and died there in 1692, his tomb silently attesting to the tragedy and pathos of a ruler murdered by his own brother. Saif's capital was Rustaq, where he died in 1708, but Jabrin enjoyed another brief period of glory as Omani capital when it resumed capital status under Muhammad bin Nasr until his death in 1728, when Saif ibn Sultan II again restored the supremacy of Rustaq. Jabrin then began its long sleep, from which it was reawakened so recently by restoration teams from the Ministry of National Heritage and Culture.

Most of the rooms are state-rooms with beamed and painted wooden ceilings. High windows are closed in their upper part by a plaster grating for light and ventilation, while their upper part may be

Jabrin. Fortified palace. Courtyard after restoration (Courtesy of Andrew and Dorothy Kirk)

240

Jabrin. Fortified palace. Painted ceiling (Courtesy of Andrew and Dorothy Kirk)

closed with shutters, the space between being shelved as storage space.

The painted ceilings resemble carpets in their intricate, delicate colours: geometric designs dominate as elsewhere in Islamic architecture, but scrollwork and floral patterns add their own individuality. Galdieri, an expert on the Safavid architecture of

Jabrin. Fortified palace. Painted ship (Courtesy of Andrew and Dorothy Kirk)

Isfahan (1500–1720), found Safavid influences at Jabrin, notably in the swastika-armed squares known in Iran as *chahar lange*, and in a complex composition of octagons and squares.

Jabrin's most significant building, set like a jewel in its oasis setting, had always been called a fort until, in his detailed analysis (*Journal of Oman Studies*, vol.1), Eugenio Galdieri suggested that this 'masterpiece of Oman 17th century architecture' was in fact a palace. Writing in the same journal (Vol. 6, part 2), however, Enrico d'Errico later demurred. 'The layout of Jabrin,' he stated, 'is the result of an original complex to which the Imam Bil'arab al-Ya'rubi added a wing for his palace. This addition has visibly altered the original plan which would otherwise be similar to that of other forts, with the result that there are two adjacent courtyards, and the whole complex is divided into two blocks with different floor levels, these joined on the upper level by a narrow corridor obtained by the opening of a passage within the thickness of the wall.'

The fort is a rectangle, vulnerably isolated in the plain, though its own *falaj* crosses it from south-west to north-east. The northerly tower is entered by a steep stairway leading to a chamber for ammunition, with a hole for the distribution of ammunition in the battery floor. The tower walls are two metres thick, sufficient to withstand the impact of cannon-fire. The higher, southerly tower contains two gun-battery levels. Between them, these jutting towers defend all four sides of the fortified palace from attack.

The constructor was the son of Imam Sultan bin Saif, that ruler whose capital was Rustaq and burial-place Nizwa. This was Bil'arab bin Sultan, a studious and generous man who soon moved his capital from crowded Nizwa to the solitary pleasures of Jabrin. Badger,

Jabrin. Fortified palace. Barrel-vaulting after restoration (Courtesy of Andrew and Dorothy Kirk)

translating Ruzaiq's *al-Fath al-mubin*, reckons the date of Imam Sultan's death at 1668, and one must prefer this to the 1679 suggested by E. C. Ross, translating the *Kashf al-ghummah*, if only because Bil'arab was presumably already in control by 1675, the date inscribed on the richly-decorated entrance arch over the staircase to the tomb in Jabrin.

Even the types of ceiling show a degree of variety rare elsewhere in Oman: in the corner towers and at the northern and southern ends you can find barrel-vaulting; three rooms on the northwestern and northeastern sides on the ground floor have vaulted ceilings; while the rest are flat ceilings, their finishing varying according to the room's purpose.

Miles on the Border of the Great Desert

S. B. Miles, travelling to Jabrin (his 'Yabreen') in December 1885, had referred to the fort there as 'dusty ruins' in the article 'On the Border of the Great Desert', which now follows in its entirety, with the accompanying map. Suffice it to say that, provided with a permit from the Ministry of National Heritage and Culture in al-Khuwair, you may see for yourself that these 'dusty ruins' have been restored to their former glory, as can be revealed in the photographs taken since by Andrew and Dorothy Kirk, reproduced here with their kind permission.

ON THE BORDER OF THE GREAT DESERT: A JOURNEY IN OMAN

By Lieut.-Colonel S. B. MILES.

TOWARDS the close of the year 1885 the tranquillity prevailing throughout the dominions of the Sultan of Maskat, who had wisely taken advantage of the prestige that had accrued to him after the discomfiture

---

Although the journey described here was made a quarter of a century ago, it is fifty years later than that of Wellsted, and is really an important addition to our knowledge of a little-known part of Arabia, including the mysterious Great Desert.—ED. *G.J.*

of his enemies in 1883, to consolidate his power in all quarters, gave me the opportunity I had long wished for to make a somewhat extended tour through the provinces of Oman and Al Dhahireh in order to visit certain localities and districts which circumstances had not previously allowed me the gratification of exploring, and with a view to extend my personal intercourse with shaikhs and other personages in those regions with whom I had not hitherto been able to become acquainted.

On my signifying the Sultan with my wishes, His Highness was good enough to assist me in sketching out the tour and in promoting my views in every way. He was kind enough to appoint Seyyid Hamood bin Nasir, son of the Wali of Muttrah, as my companion and introducer. He also directed the shaikh of the Beni Kelban tribe, Nasir bin Mohammed, to accompany me throughout as guide and as kefeer on behalf of the Ghafiris.

On December 11 I proceeded by boat to Muttrah, where my followers and the camels were assembled, and there distributed the baggage for the escort into six light loads for as many camels, in order that we might travel expeditiously whenever convenient and desirable to do so. The riding-camels for the party were selected with regard to their fitness to perform the whole journey, as it was not intended to change them *en route*, while for my own riding the Sultan kindly chose from his own stud a valuable female dromedary called Samha, which proved one of the swiftest, most easy-paced, and docile animals I ever bestrode.

It was late in the afternoon before we moved out of Muttrah, and as my party was not quite complete I spent the night at Wataia, intending to make a real start the following morning. On December 12, accordingly, we marched from Wataia, and, skirting the hill range over a bare and uninteresting country, halted at Fanjeh, where we camped for the night, and, starting early the next day, arrived at Semail in the forenoon. Here we stayed for a couple of hours to feed the camels during the heat of the day, and then trotted on to Saija, a village belonging to the Beni Jabir tribe, at which place we intended to pass the night.

In the month of March of this year, 1885, a cyclonic storm or hurricane of unprecedented violence had burst over central Oman, causing widespread destruction and misery. It had been followed by a deluge of rain, which had swept down the valleys and poured a devastating flood of water through the villages and settlements, and had done incalculable damage to houses and cultivation, while hundreds of thousands of date trees had perished. Dashed by the cyclone against the precipitous walls of Jebel Akhdar, the clouds had broken and fallen in torrents of rain down the steep gorges and ravines, and had concentrated a mighty wave down the Semail valley, which had carried everything before it. The river is usually but a shallow, babbling stream, rolling down its sandy bed, from which it is drawn to fertilize and irrigate the productions of nature on the banks, and though the Arabs are accustomed to see occasional floods after heavy rain, there was no record among them of a similar visitation, and I was assured that the oldest inhabitant could remember nothing so calamitous. The tumultuous rush of water had come upon the people so suddenly as to take them completely by surprise, and give them little time for precaution or for removal of portable property.

In the narrows the river had swollen to an extraordinary height, subsiding again as rapidly as it had risen, the marks of its highest point being still visible on the banks. The destructive effects of the terrible tornado that had visited the land nine months before were manifest everywhere, and as we rode up the valley from Fanja we witnessed the wrecked houses, the headless palms, many of which were broken or bent to the ground, the tempest-stripped trunks of trees and ruined gardens; while the poorer natives, whose rude matting and palm-leaf huts had in many cases been swept bodily down the vale with all their belongings, were still bemoaning the calamities and losses.

Saija is picturesquely situated on the left bank of a ravine or gorge, and has about eight hundred inhabitants and some nice houses and gardens. Dwelling in seclusion, in poverty, and in dread of the ruling family, though secure enough, I believe, under the friendly protection of the Beni Jabir, I found here the remnant of a once domineering and masterful race, namely, the Yaareba, whose princes held royal sway, not only in Oman, but also over the entire Indian ocean, which the pirate craft of the Yaareba rulers darkened with their sails, holding it in terrorism against Europeans and natives alike for a century and a half. It was under the rulers of this dynasty that the Portuguese, who had long held possession of Mascat and the coast of Oman, were eventually expelled from that country in 1650, and it was under the same vigorous régime that the "Mascat Arabs," as they were called by the East India Company's officials, became so renowned for their naval expeditions and predacious raids, and so formidable against the sea-borne traffic of the Arabian sea.

On arrival at Saija I had despatched a messenger with a note to Shaikh Nasir bin Mohammed of the Beni Rowaiha tribe, who had received me hospitably at Zikki when he was the Wali or Governor there nine years before, asking him to provide me with a kefeer through his territory to that town.

The shaikh's house at Wibal is only 3 miles from Saija, and soon after we had started the next morning we met the courteous old shaikh coming towards us, and our greetings were very cordial and friendly, for I had not forgotten his kindly reception of me, and I was glad to see him again. He pressed me to stay a day at Wibal with him, and if I could have afforded the delay would willingly have done so to please him; but it was not possible, and for the second time I was obliged to decline the invitation. The shaikh then insisted on becoming our guide himself and accompanying us to Zikki; it was a long ride, and I tried to dissuade him from undergoing the needless exertion and fatigue, but to no purpose, and he came all the way thither.

Wibal is in the proximity of a low white hill, which is very conspicuous from the road, being surmounted by a watch-tower and skirted by a dense grove of palms. The Wadi Beni Rowaiha is here very narrow, with high ranges on each side, while the gradient rapidly increases until we get about halfway to Zikki, when we finish our climb up to the top of the valley and arrive at the crest of the great central watershed of Oman, and here we quit the bed of the Wadi Beni Rowaiha, which can no longer afford us a highway, and begin to traverse the Nejd, which has an altitude of 2200 feet.

The Wadi Halfain, of which we see more anon, has its source to our right in the inner recesses of Jebel Akhdar. There is said to be a donkey track from Zikki to Seek up the mountain, but the path is precipitous and perilous. After Nejd Mejberia, where we rested at a small grove and felej, we came to Wadi Mateh and then to Karut. At one part the road is raised 20 feet above the plantations and date groves, through which the torrent dashes along until lost to view among the trees. We seemed to be moving, as it were, amidst the branches of trees, and the change was a pleasant one from the confined vista of arid rocks to the fertile and open country now unfolding in front of us. Passing a strip of level and grassy plain, we enter a shallow watercourse and begin to descend the gentle slope on the western face of the watershed. It was two hours after noon before we reached the fort at Zikki, where the Wali recently appointed by the Sultan, Shaikh Mohammed, received me very cordially, and treated my party hospitably. There had been a change in the political situation here, owing to events that had taken place in the previous year, 1884, when the fort had been evacuated and surrendered to the Sultan under an agreement with the Beni Rowaiha tribe. The new Wali had garrisoned the fort with Beluches.

It became necessary here to change our baggage camels, a procedure which somewhat retarded our departure the following day. It was consequently rather late in the morning when the camels were reported to be ready to start; but I had employed the interval in taking photographs of the Shaikhs and people whom I had assembled in the courtyard. At first the appearance of the camera appeared to create the impression that something uncanny was intended, and it was impossible to get them to stand in front of it. This difficulty, however, was overcome by my own people mixing with the others, and then exposing themselves to the same fancied danger; but I think the good folks at Zikki were much relieved when they saw the object of their dread shut up again in its box and packed away on a camel.

Having said farewell to our host, we mounted in front of the fort, and began our day's march by descending the steep bank of the Wadi Halfain, and wading across the stream to the left bank to pick up the track leading to Minha. For some little distance we continued to pass scattered habitations, and from this point looking back the view of the castle and hujrahs of Zikki was imposing and romantic. The road at first runs alongside the Halfain stream, here very shallow, and at 1½ miles we come to Zikkait, a small fortified village of the Beni Riyam, enclosed by a quadrangular wall strengthened by two bastions at opposite angles. We pass to the left of this, and now unrolls before us a broad, slightly undulating plain of light sandy soil, sprinkled with herbage and declining with unbroken horizon to the south and west. The Wadi Halfain, which intersects it, and which was still flowing when we last crossed and left it behind us, reaches the sea at Ghubbet Hashish in time of flood. The encampment to which that zealous and observant traveller, Wellsted, was taken by his Beni Boo Ali friends on December 7, 1835, was situated in this wady, and is known as Ghor. With its tributary the Wadi Kalbuh, the Wadi Halfain forms the highway from Adam to Mahot, and is much traversed, as it is well supplied with pasture and herbage, and is seldom without water. Caravans between Adam and the sea take eight or ten days, and I was given the names of thirty halting-places. The Wadi Halfain debouches at a small creek a short distance to the north-east of Mahot.

After maintaining a fair trot for three hours, we approached the environs of Minha, a large and straggling settlement standing in a rich and well-watered district, and where I was surprised to find there was no wali or shaikh of any consequence, the dispenser of justice and arbiter of disputes being the metowwa and kazi, Mohammed bin Mesood, who is reputed to be the most learned man in Oman. I had not intended to halt at this place for more than an hour or so, being desirous of pushing on at once to Adam, the frontier town of Oman, which had not been visited by Wellsted or any other European traveller. Some delay was, however, occasioned by the information received here that Adam, which like Zikki is occupied by two tribes constantly at feud and warring with each other, was in a disturbed state, a conflict with some loss of life having taken place a few days previously. Occurrences of this kind are but too common in Oman, where the people hold life very cheap, and there was nothing in the affair to hinder my plan of going thither, especially as it seemed probable that my presence might serve to distract the Adamites' attention for a time from their quarrel, and give time for their passions to cool.

As we meant to return to Minha, it was suggested by my Arab companions that a portion of the baggage with some of the escort should be left behind, to be picked up on our way back, and that we should take the remainder with a few of the Jenebah tribe as kefeers. This was accordingly arranged, and Seyyid Hamood engaged six Jenebah Bedouins to accompany us. The detention involved by this change kept us till nearly sunset, and we then started on again and rode 5 miles to a hamlet called Izz, where we camped for the night, crossing on the way the Wadi Metaiyin, which runs past Birket-el-Muz and joins the Wadi Halfain. Izz is encircled by a protective wall, and the population, which consists of members of various tribes, numbers about 300 souls.

The people were very friendly, but I preferred the open air to their close and stuffy houses, and with the shaikh's permission we took possession of some of the Arish or date-stick huts, of which there were many on the plain a little distance off. These huts are occupied by the Arabs and their families during the sultry summer months, when the heat in the village becomes stifling beyond endurance. They are small and decidedly draughty, but are clean and comfortable. I found it, however, piercingly cold during the night, owing, I suppose, to the extreme radiation from the sandy soil, and the dryness of the atmosphere, and I was glad of all the wraps I could muster. The morning of the 16th we mounted early and advanced in the direction of Adam, which lies almost due south of Minha.

Our road lay over a gradually declining sandy plain, sprinkled with dwarfish acacias, brushwood, and desert herbage, and as we had some distance to go we pushed our dromedaries along at a good pace. The cool breath of the morning was delicious, and as we sped along, the desert air, and the consciousness of being on the edge of the great wilderness, with a boundless expanse of open country unrolling gradually in front of us, caused an exhilaration of spirits which made the ride an extremely pleasant and enjoyable one. The fragrance in this wild region was very delightful, for in Oman, as in Yemen and in Hadhramaut, many of the plants are very aromatic. We met not a single traveller and saw not a single human habitation on our way; but the soil, though light, did not appear to be infertile, and wanted water only perhaps to make it suitable for man's industry. The vegetation of the plain supports a good deal of insect life, and feathered game was, in consequence, to be seen on the plain—houbara, partridge, plover, sand-grouse rise up in front of us, while timid gazelles and hares, startled from their beds, spring up occasionally from behind bush or sand-heap, or are espied by the keen eyes of the Bedouins trying to crouch and hide from observation.

After half an hour we crossed the Wadi Kalbuh, a fine stream, but not flowing so strongly as the Wadi Halfain, with which it unites further south. At 10 a.m. we came suddenly upon a well in the desert, in a locality where vegetation was more than usually thick, and here we rested a little while. It is known, after the name of the shaikh whose beneficence caused it to be sunk here, as Bir-al-Hamaid, and is said to be just halfway between Minha and Adam. There is no variety or change of scenery in this gently undulating plain until we begin to approach Adam, when a short range of hills rises up in front of us, cut in two by a deep vale through which lies our road. This singular dip was reached at 12.15, and here we came upon a cluster of hot springs bubbling up from holes and crevices in the hard rock, and sufficiently copious to supply the aqueducts that serve to irrigate the fields and groves of Adam. I observed that one of the felejes or subterranean streams had been newly excavated, and had large and deep shafts. Many work-people were in these pits, both men and women, and other men were engaged in drawing water, but they took no notice of us apparently. Looming above us on each hand to the height of perhaps 800 feet, and almost destitute of vegetation, are the hills I have mentioned—that to the right or westward being called Jebel Sulakh, and that on the left hand Jebel Mushmar.

From this point I obtained the first glimpse of the oasis I had come to satisfy my curiosity about, and I stood a moment to gaze at it. Away in front of us a long, dark, indistinct line appeared like a brown streak on the boundless plain, forming a pleasing and welcome contrast to the glare and the monotonous aspect around. Gradually, as we approach nearer, the mass becomes more definite, and resolves itself into trees, which conceal all the habitations except the huge castle that towers high and majestically over the palm groves, and overtops everything with its square shoulders. This border town is, in fact, a veritable oasis in the desert, and stands forward boldly, isolated and alone, as an advanced patrol of the Hadhr or settled Arabs on the margin of the great wilderness. I had despatched from Minha the day before two messengers to announce our approach, and as we advanced towards the settlement, which we reached about 7 p.m., we observed that a dense crowd had gathered to witness and to welcome our arrival; and it soon appeared that, in view of the hitherto unseen spectacle of a European as a guest among them, all parties had put a truce to hostilities, and were prepared to be neighbours and friends for the moment. I was received with every mark of friendship and cordiality which the hospitable instinct of the Arabs could suggest, my reception here and at other places being a proof of the friendly feeling and high estimation with which the name of England is regarded in this country.

Among the magnates of the Al Boo Saidis here was a cousin of our leader Seyyid Hamood, an agreeable and most courtly youth, who took it on himself to act the part of the host, and who, after ministering to our comforts, proceeded to show us round the place.

The quarter of the town occupied by the Al Boo Saidis, whither I was naturally first taken by my companions, is but a short distance from the house in which I was lodged, and we therefore set out on foot. I found it to be a walled hujra or enclosure of small area filled with low flat-roofed houses, and separated by narrow alleys, wherein dwelt about 150 families of this tribe. This quarter is called the Jami, and is, for defensive purposes, strengthened at one point by a tower or citadel. This hujra was not particularly clean or attractive, and the interest I took in it was centred on one point, viz. the house and birthplace of Seyyid Ahmed bin Said, the founder of the dynasty now reigning at Mascat.

A small house was shown to me, unpretentious enough, and here was born, about the time Queen Anne came to the throne of England, a man of humble origin (his father having been a camel-driver, and his tribe one of little renown or consideration at that time), who became the champion of his country, having taken the lead in the hour of danger and calamity, and saved it from a foreign invader. About 1742, owing in great measure to the traitorous conduct of Saif bin Sultan al Yaareby, the then ruler of Oman, who had called in the aid of the Persians against internal foes, Nadir Shah had been given the opportunity which he had long cherished and waited for, of conquering Oman. This conquest had been rapidly effected, for in their disunited and enfeebled condition the Omanis were no match for the Persians, who overran the country and occupied Mascat with a strong garrison.

At this time Ahmed bin Said held the position of Governor of Sohar, a town and fort on the Batineh coast, and being a man of uncommon boldness of spirit and force of character, eager and daring enough to take advantage of any opportunity, he soon wrenched the control of affairs from the ruling family, and eventually succeeded in compelling the Persians to evacuate the country. The enthusiasm and gratitude of the Arabs for their liberator, of whom to this day they never speak but with reverence and pride, caused him to be raised to the Imanate. The Yaarebi dynasty, part of the remnant of which I had met a few days before at Saija, was thrust aside for ever, and a new dynasty was founded, of which the present Sultan, Seyyid Faisal bin Toorki, is the fifth in generation and the tenth in succession.

It is a curious coincidence that the ruler of Persia at that time, Tamas Kuli Khan, better known as Nadir Shah, was, like Ahmed, an adventurer who, from being a soldier of fortune of the meanest birth, had raised himself by similar arts to the throne of a great empire. The Iman Ahmed died in 1783 A.D. at a good age, being succeeded by his son Said, a man of a different character, who after some years resigned power to his brother.

On leaving the Jansi quarter, I walked on to the other portions of the town belonging to the Mohariks and their allies, who greatly outnumber the Al Boo Saeedis, and occupy several hujras, the largest of which, containing the Shaikh's house, is called Till. Here I was met and welcomed by the Mohariki Shaikh, Hamaid bin Khamis, and also by the son of Saif bin Hamood, the Temeemeh of the Jeneba tribe, who was here as a bridegroom, having just espoused a daughter of the Moharik tribe. This latter was a very tall strapping young fellow, with a perpetual grin on his cruel face, displaying a set of large ivories and a crafty look in his eye, harmonizing well with his evil reputation as a slave-runner. I was somewhat surprised to find that I was received with even livelier demonstrations of cordiality than I had been greeted with at the Jami; one gets accustomed to being gazed at by a motley crowd in an Arab town in the interior, but I hardly expected to see the intense curiosity that was exhibited by the good people of Adam of both sexes and all ages, as I visited one quarter after another. In some places I was literally hemmed in by them as they stopped to stare at me, and not content with lining each side of the road as I walked through the narrow lanes and in the shady date groves, many of the children, after I had passed, would scamper round another way to get in front again, and thus obtain a second peep at the stranger. Determined as they were, however, to set no bounds to their curiosity, they were perfectly well behaved and good humoured, and their actions only caused us amusement.

Besides the six quarters of the town I have already referred to, there are two others occupied by the Sheibani and Beni Waeel tribes, making eight altogether, the population of the whole being, I reckon, about 4500. The large castle here, built originally by the Iman Ahmed about the year 1780 A.D., and repaired by Seyyid Azzan in 1869, is considered of little importance, and is weakly garrisoned by the Moharik tribe, to whom it now belongs. The little value placed by the Adamites on this fort is probably due to the remoteness and isolation of their position, which enable them to disregard the current of affairs in other parts of Oman, and to avoid complicity with tribal wars, reserving thus, with entire comfort and satisfaction to themselves, their powder and shot for each other.

Segregated as they are from others, and composed as the population is of such discordant elements as I have described, the social atmosphere of the inhabitants of Adam is by no means a tranquil or unchequered one, for the storms and calms of the political barometer recur with almost undeviating regularity. The Ghafiri faction, which is by far the stronger in this town, and is represented by the Moharik and their allies, is at constant and irreconcilable feud with the Al Boo Saeedis, who, when hard pressed by their enemies, as is often the case, are constrained to call in to their aid the Deru, a powerful Beduin tribe in the neighbourhood. On occasion of serious trouble arising, the Sultan of Mascat is ever ready to mediate and to act the part of grand pacificator; but beyond this His Highness does not interfere. He is, in truth, in a somewhat peculiar position towards the contending parties.

Being himself the head of the Al Boo Saeedi tribe, which is Hinawi, His Highness has, nevertheless, been supported mainly in recent years by the Ghafiri faction, and is thus not in a position to intervene actively in their quarrels.

With regard to the present juncture, the Sultan, who had, it seems, heard of the belligerent attitude of the parties soon after our departure from Mascat, had appointed Seyyid Nasir bin Mohammed to inquire into the causes of the present rupture, and, if possible, to arrange terms of peace, and I was glad to learn afterwards that the truce patched up in consequence of our visit had been preserved until Seyyid Nasir's amicable adjudication put an end to the war.

Adam is a town of some antiquity, and was probably founded by the Parthians before Islam. The word itself is probably old Persian, and there are indications of other Persian settlements in the neighbourhood. There are three felejes here, providing the people with a copious supply of water for all purposes. The springs are thermal, and I found the temperature of one of them to be 102° Fahr., while others ranged up to 112° Fahr., this heat being no impediment to animal life, as frogs and fish in abundance swim about in it. The elevation above the sea is 850, being 500 feet lower than Minha. Adam is the most advanced town of Oman towards the Great desert, and stretches out into the sandy ocean of Arabia. It forms a convenient market for the Beduin to resort to for provisions and other necessaries.

On the 17th I rode back across the strip of desert country to Minha, the nearest settlement to Adam, from which it lies distant about 25 miles. Minha has a population of some 2000 souls, but though thinly peopled, it covers a considerable area of ground. The soil is well watered and fertile, and produces abundance of corn and fruit of various kinds. It has been well described by Wellsted, who gives a glowing account of the extent and luxuriant beauty of the diversified cultivation around. He saw it in all its prodigality of production in the month of September, but at this time winter had thrown its dull and sombre garb over the scene, and though ripe fruit still clung to the trees in the orchards, there were no fields of golden grain.

My original intention had been to visit Nezwa before proceeding on to Bahila and Al Dhahireh; but though this plan would have saved us a day's journey, I resolved, after consulting with Seyyid Hamood, to march direct to Bahila the following day, my object being to inspect the intervening unexplored tract. I tried in vain to obtain an idea of the distance hence to Bahila, as no two agreed in their estimate. The Arabs have no means of measuring distance, and consequently time becomes the standard in reckoning a journey. The vagueness of this, however, may be conceived when it is remembered that these people have no means of measuring time any more than they have of distance, and thus every man had to judge by a standard of his own.

On the 18th, starting at 8 a.m., we reached Karsha, a hamlet of seventy houses, at 9, and half an hour later crossed the Wadi Kalbuh, here a broad but shallow stream. The country we were passing through is level and monotonous, bare and stony, sparsely inhabited and cultivated, though not devoid of desert vegetation. At 10.45 we stopped at Timsa, a small village situated under a hill, to drink coffee with the shaikh, who had, in a very friendly manner, intercepted us on the road with a noisy retinue to press his invitation, which indeed we were not at all disinclined to accept. Remounting our camels, we soon came to Kumaili, a dark pool of water lying under the projecting ledge of an isolated rock on the plain, which has been famed from time immemorial as a mystic well of peculiar power. This rock has a remarkable appearance, being cleft asunder, and disclosing a deep fissure or chasm, at the bottom of which lies the well; it is dreaded and revered by the superstitious Beduin as the habitation of magicians. The water is said to have become red from the blood of an unfortunate wretch who lies imprisoned by enchantment deep down under the spring, and it is affirmed that any one venturing to drink of it falls immediately under the spell and power of the magician. The water is, however, as I can attest, quite clear and potable, and all my party, except a few of the Beduins, drank of it unhesitatingly.

There are two kinds of magic in Oman, lawful and unlawful, both of which are articles of faith among the Arabs of all classes. Of the former one hears little, as few know anything about it, but the latter is Satanic and widely credited and dreaded. Though Babylon is referred to as the home of the black art, it happens that it is chiefly practised by negresses, who have brought over with them much of their African witchcraft, fetishism, and divination. In Muscat and other large towns the Arab ladies and negress sorcerers are said to hold midnight *séances* for the practice of "sehr," at which curious things take place, if one may believe the men. The husbands disapprove of this, and would check it if they could; but the women are more credulous, and are much afraid of negress witches. The inventors of magic are the two angels Haroot and Maroot, who are suspended head downwards in a pit under a huge mass of rock near Babylon. It is possible that the pit I saw near Bahila obtained its evil reputation from some fancied resemblance to this pit, the description of which is well known to Arabs, and which is referred to in the Koran. Hard by are to be seen the crumbling ruins of the tomb of a saint. I could obtain little or no traditional history about this spot, as the Bedouins were very reserved; but it is possible that it marks the site of an idolatrous and pagan temple of the pre-Islamite Arabs.

From this point we pursued our way over hard ground among low black hills without sign of habitation save one small hamlet named Foot, and soon after the outline of the distant town of Bahila, the name of which, I may observe, is spelt Barler in the maps of Wellsted and Badjer, begins gradually to steal upon the eye and form itself into a distinct picture. We sight first the high tower of a castle in the centre, with a long, low, white circumvallation of considerable extent studded with bastions at intervals and flanked by the bed of a stream, there being no gate visible on that side of the wall we were approaching. Presently within the irregular square space enclosed by the wall appeared a confused medley of palms and other trees, houses, mosques, green fields, and gardens, amidst which, and perched on an eminence in the centre, stands a massive and venerable pile surmounted by a lofty tower commanding and overlooking the entire surrounding country.

From the time they had sighted our party traversing the dark and barren plain the garrison had continued to fire guns from the walls at intervals as we approached, the roar of which reverberated in the recesses of the hills. We had to ride under the wall some distance, and then, turning an angle, we descended abruptly into the bed of the Wadi Bahila; and as we drew nearer the chief of the town, Shaikh Nasir bin Hamaid, with his mounted retinue, followed by a large concourse, issued from the massive archway which faces the wadi and forms the main entrance to the city, and rode slowly down the broad and sandy watercourse to meet us. Riding close up, the shaikh shook hands with me, and we exchanged formal greetings and salutations. I then expected we should ride together to the city, but the ceremonious part of the reception was not yet over. My party had, in the mean time, drawn into line on the edge of the bank under the wall, while the shaikh, now quitting my side, led his followers on to a smooth strip of ground and entered upon the more serious and important part of the programme, without which the istikbal or reception would not, in Beduin eyes, have been complete. The performance consisted of a display of horsemanship, twirling and firing their matchlocks at full speed, running races by twos with each other, and similar exercises, much to the delight of the populace and to our weariness, for we were tired and thirsty after our march, and would have been glad to dismount and rest. However, we had to look on and look pleased for the best part of an hour, by which time the shaikh's horse was fairly blown; and then we formed procession and marched slowly and solemnly on through the streets of the town, which were lined with gazing spectators, until we reached the spot where we were able to alight and unpack our baggage.

The little house allotted to me was convenient and in a pleasant garden, and here, an hour later, Shaikh Nasir paid me a complimentary visit; and when this interview and breakfast were over, I spent the rest of the day in riding about the town and obtaining a general notion as to the disposition of the area enclosed within the fortifications.

My return visit to the shaikh having been arranged for a certain hour on the following day, I proceeded to the castle accordingly, and was received by the shaikh with due Arab formality and courtesy. He was quite a young man, apparently not more than twenty-two, slight and agile figure and quiet manner; but what most arrested my attention during the interview, which was not a long one, was his heavy-featured, sallow, and forbidding countenance, which was entirely in keeping with the reputation he had recently earned throughout the country. His seizure of power by wading through the blood of his own brothers, though a sufficient evidence, in Arab eyes, of his determination of character and fitness for command, had undoubtedly rendered him an object of distrust and repugnance to his tribesmen; and as a glimpse of Arab life, I may as well relate here briefly the tragical events in which he played so conspicuous a part.

A few years before my visit, the old temeemeh of the Beni Ghafir tribe, Shaikh Hamaid bin Rashid, had died, and had been succeeded by the eldest of his three sons, Barghash. The old shaikh, Hamaid, had been esteemed and respected, but the sons did not share their father's popularity. Barghash had not long assumed power before troubles arose, chiefly in connection with the fort at Boo Einein, a much-coveted possession, and regarding which we shall have to speak later on. Barghash's relations with another powerful Ghafir tribe in that region, the Ibriyeen, had become complicated; his younger brother Rashid had married an Ibriyeen girl, and his own sister had married into the same tribe, but, notwithstanding this, he cruelly murdered two of the Ibriyeen. Partly in revenge for this and partly for political reasons, Rashid and the husband of Barghash's sister resolved to murder Barghash, and invited him to coffee at his sister's house near Bahila. Barghash unsuspectingly accepted, but his sister, having been apprised of the plot, met him on the road and warned him. Barghash returned home, and prepared at once to attack the would-be murderers, but they took refuge with the Ibriyeen. These events happened in 1882 or early in 1883, and Barghash, whose influence was rapidly declining, and whose conduct had alienated not only his own tribe, but all the Ghafiri faction, was soon to fall by the hand of his youngest brother Nasir. This youth, the present governor, was of a malevolent and intriguing disposition, and was so much feared and mistrusted by Barghash that the latter, suspecting his evil designs, took the precaution to provide him with a separate residence, and forbade his entering the castle. Barghash's fear proved to be only too well founded; but the measures he had taken were not sufficiently stringent to defeat the treacherous wiles of the young fratricide, for Nasir, having conspired with his weaker but equally vicious brother Rashid, contrived a scheme that was not likely to fail. In December, 1883, Nasir succeeded in bribing three of Barghash's slaves who formed his guard, and with

Bahla. Fort, 1885 (Miles)

Bahla. Fort wind-tower, 1885 (Miles)

the help of these men concealed himself one night in the castle. Creeping silently to his brother's apartment, he shot him dead while lying asleep, and then, leaving the women and domestics, who had been roused by the noise, petrified with terror, returned unmolested to his own house.

The next day Rashid, the second son of Hamaid, was proclaimed by Nasir and other conspirators Shaikh of the Beni Ghafir tribe. Rashid, an' indolent voluptuary, soon allowed the direction of affairs to fall entirely into the hands of Nasir, who four months later, in March, 1884, becoming weary and impatient of the situation, resolved to put an end to it. Watching his opportunity, and having prepared his own followers for the event, he treacherously shot the unsuspecting Rashid through the back one afternoon while they were riding out to Yabreen together. No disturbance ensued, and no hand was raised against Nasir, who took quiet possession of the castle. But though his ascendancy had remained undisputed, and he had been acting as Wali of Bahila on behalf of the Sultan, he had not so far been recognized as temeemeh of the Ghafiri faction in place of his brother Barghash, nor had his tribe condoned the murder of Shaikh Barghash. I was assured by some persons, however, that the Beni Ghafir were not sorry to lose their last chief, the incompetent Rashid.

The castle is a large, substantial, and handsome edifice, and is called Hisn Tamah, from the name of the Nebhani chief who is supposed to have built it, though there is reason to believe it is founded on a very ancient and perhaps Parthian structure. The Nebhani dynasty in the fourteenth and fifteenth centuries made Bahila the capital of Oman, and this castle was the residence of those rulers. It is ornamented with two towers, which command a splendid view of the whole valley, and one of which is, I think, the loftiest structure I have seen in Oman, and I was careful to photograph it. It is called Burj al Rih, or the Wind tower, and has apartments in it for use in the hot season, when the open and elevated windows let in the welcome breezes from all quarters. The town comprises twenty separate enclosures or divisions called haras, and contains a population of about 6000 souls. The larger or upper part of Bahila is occupied chiefly by the Beni Ghafir, the predominant and most numerous tribe, while the Sifala, or lower part, is inhabited by the Beni Shikail, Ibriyeen, and others.

The outer wall, embracing the town and cultivation, appeared to be an irregular parallelogram covering an area of about 2 square miles. It affords a useful protection against raids of their predatory neighbours in Al Dhahireh, who are ever making incursions into others' territories, but it is not defensible against a serious and organized attack by powerful tribes. It is traditioned to have been planned and built—as I was informed by our guide, Shaikh Nasir el Kelbani, who bears the title of Wazeer of Bahila—at a great cost by a woman named Gheitha about six hundred years ago. As I rode hither and thither about the place, I noticed fields of wheat, jowari, barley, lucern, sugar-cane, and cotton, besides the inevitable palm clumps and orchards of other fruit. The manufactures here are cotton loongies, or waistcloths, made of the khodrang or natural brown cotton, goat's-hair cloaks, mats, halwa, etc. The inhabitants have good houses, some of them three-storied, most of them built of sunburnt brick, and all of them flat-roofed. The people seemed to be generally well off and contented, but were still lamenting the damage done to houses, crops, and trees by the hurricane in the previous spring.

The Beni Bahila tribe, from which the town derived its name in ancient times, is probably the Bliulei of Ptolemy, and is now extinct in Oman, having become absorbed in other tribes. It is uncertain whether it was of Kahtanic or Maaddic extraction, as there appear to have been two tribes of the name. The ancestress of the Kahtanic Yemenite tribe was Bahila bint Kaab, who married first Malik bin Asor, and secondly Maan bin Malik. Maan's sons by other wives, however, also became known as Beni Bahila.

The Maaddic tribe of Bahila, which is probably the one that migrated to Oman, derived from Bahila bint Yasir bin Kais Ailan. The tribe had an unenviable reputation, and was one of the many tribes disliked by the Arabs. I found Bahila on the whole a very interesting place; the attitude of the people towards myself and party was especially friendly, and I should like to have remained some days here, but circumstances did not permit.

In the afternoon, accompanied by a cavalcade headed by Shaikh Nasir bin Hamaid, who insisted upon showing me the road, we started for Yabreen, which is about 4 miles down the Wadi in a northerly direction, and which I could not neglect the opportunity of visiting. The Beni Ghafir horsemen, following the example of their shaikh, amused themselves all the way by galloping about and running races with each other, as if they were out for a holiday, and I dare say my people would have gladly joined them in their exercises, and exhibited their own skill in the saddle, had they been on horseback instead of on camels, as Arabs dearly love and enjoy a display of horsemanship.

Yabreen fort is small, but in good repair, and is garrisoned by six Arabs. It is without a dependent township, and there is no settled population. Formerly it was of more importance, and in the palmy days of the Iman Seyyid Belarab bin Sultan al Yaareby, who built the castle about the year 1690 A.D., Yebreen was temporarily the capital of Oman, or at least the seat of government, and could then boast of an extensive madressa, or college, for the encouragement of learning, the renown of which is now smothered in its dusty ruins. Bahila and Yebreen stand in an angle of the great chain formed by the mighty mass of Jebel Akhdar to the north-east and its outlier Jebel al Koor to

258

Jabrin. Fortified palace. Exterior, 1885 (Miles)

Jabal Misht, 1885 (Miles)

the north-west, these mountains thus protecting and overshadowing them on two sides, while on the other flanks the vast wilderness, bare and open, unrolls itself interminably, until the sight fails, and sky and earth mingle imperceptibly together. As the sun sets here it reveals in all its glory and distinctness of outline the noble mountains I have mentioned, Jebel al Kur, which raises its peaks upwards to an altitude perhaps of 5000 or 6000 feet, and forms the geographical boundary, as well as natural barrier, between the provinces of Oman and Al Dhahireh. It stands isolated and almost entirely detached, as it appeared to me, from the Jebel Akhdar range.

Under a grand old tamarind tree near this castle of Yebreen we camped for the night very comfortably, our baggage having been brought up from Bahila during the evening, my plan being to turn back eastward and pay a visit to the large and ancient city of Nezwa, the ancient capital, before quitting the province of Oman, and crossing over to El Dhahireh. The elevation of Bahila by aneroid is 1550 feet, and of Yebreen 100 feet less. The main road connecting these two provinces passes to the south end of Jebel al Koor, and is called by some Nejd el Makharim, and by others the Nejd el Dhahireh, its low elevation—for it is said not to exceed 400 or 500 feet above the plain—rendering it an easy and convenient passage.

The pass at the north-east end of the mountain is of a very different character, being about eight times the altitude of the other, and infinitely more difficult and precipitous. It is known as the Nejd el Barak. After making due inquiries among the Arabs, and hearing their description of these passes, I resolved to cross over to El Dharireh by the northern pass subsequent to my visit to Nezwa. As will be seen presently, this decision, though adopted to harmonize with my own plans, and entirely fortuitous as regards local contingencies, turned out very fortunately for us.

Early on the 20th I despatched a Kossid from Yabreen with a note to Shaikh Hilal al Hinawi, to give him notice of our intended visit to Nezwa, and about 8 a.m. we mounted and began our march thither. Ascending the Wadi Bahila, we passed that city and proceeded thence in an easterly direction over a stony, bare, and lifeless plain, intersected by a few watercourses, sighting in our course but one hamlet with a patch of cultivation, wheat, sugar-cane, and indigo, situated in the Wadi Musabli, until we reached at 2 p.m. the village of Tanuf, which lies at the entrance of the wady and pass of that name. Here we were met by the Shaikh Suliman bin Saif al Riyami, whom I had seen some years before at Seek, and who had ridden out to intercept us, and to entreat me to stay a night with him, saying that in view of his long and staunch adherence and loyalty to the Sultan, His Highness Seyyid Toorky, it would be a discredit and reproach to him in the eyes of his tribe and of his neighbours if we passed his town without accepting

his shelter and hospitality.   I had not in truth foreseen this rather
embarrassing request, and though I could not but acknowledge the
force and reasonableness of it, I was unable to comply as the Governor
of Nezwa was already expecting us, and I could not in decency now
change the programme and stop on the way.   There was, indeed, only
one way of escape out of the dilemma, and I pacified the shaikh by
promising to halt one night at Tanuf on our way back from Nezwa.

It was 3.30 p.m. before we reached the outskirts of the city, for we
had been riding leisurely and had taken six hours to do the space which
lies between the old capitals, Bahila and Nezwa, the road having been
over rough ground nearly the whole way.   As we passed the outlying
watch-towers commanding the various approaches, the guards therein
fired their matchlocks to give warning, which was soon responded to by
the thunder of the guns of the great fort.   The appearance of Nezwa
from the direction in which we entered it was neither picturesque nor
imposing, but it was with feelings of pleasure that I rode then for the
first time into the ancient capital of the country in which I had so long
resided, and was thus able to see with my own eyes a city not only of
much interest in itself, but historically and traditionally associated with
so many of Oman's rulers, so many of her illustrious sons, and the
battle-ground of so many invading and devastating armies.   Nezwa is
one of the first sites occupied, and the one characterized by the most
salient incidents in the national life.   It is the theatre in which have
been enacted many of the most momentous events chronicled by the
historians of the country.

Descending into the bed of the Wady Kalbuh, and thence entering
the streets, which were thronged with people who had assembled to
witness our arrival, we were met and greeted very civilly by the
governor, Shaikh Hilal bin Zohair, who, after shaking hands, conducted
us to his public reception hall, which was situate within the gate of
the walled quarter where he resided.   Here we all sat down on cushions,
exchanging compliments and conversing, coffee in the meanwhile being
roasted, boiled, and distributed to the assembled guests according to
established Arab usage, but no pipes or cigarettes, as smoking is not
indulged in by the Ibadhis.   After a short interval I rose and was
conducted to my allotted abode.   An hour later I received the customary
visit of ceremony from the shaikh, who began the conversation by
informing me that the house I was lodged in belonged to a brother of
Seyyid Nasir bin Ali, the good old wali or governor of Mascat of twelve
years back, who was murdered in his house there in 1873.

Shaikh Hilal bin Zohair was a tall dignified man, very stoutly built
for an Arab, with pleasing manner and a fine, intelligent, and expressive
countenance, not devoid, however, of a certain craftiness of look.

He was considered at this time one of the most prominent and
influential shaikhs in Oman, not so much in virtue of his being the

chief shaikh of the noble and powerful Beni Hina, as on account of the adroit diplomacy he had shown in raising himself to his present high official position as Governor of Nezwa, which had hitherto been reserved for a member of the ruling family.

About five years previously the former wali, Seyyid Hamaid bin Saif, an amiable and sagacious, but rather eccentric old gentleman, who had held the post for many years, had succumbed to the intrigues and machinations of his enemies and had been succeeded for a short time by Seyyid Faisal. The change was followed by a series of injudicious interference, and disputes with the local tribes on the part of the Mascat government, and a struggle for Nezwa ensued, which eventuated in the great fort being taken possession of by Shaikh Hilal bin Zohair, who became the virtual governor of the locality.

The Nezwa policy of the government excited much irritation among the Sharkiyeh and other tribes, and was undoubtedly one of the chief causes of the great coalition which culminated in the attack on Mascat in September, 1883, by far the most formidable and determined effort on the part of his rivals that Seyyid Toorky had to encounter during the whole of his reign.

Shaikh Hilal bin Zohair had taken no part whatever in this rising, but had throughout maintained a loyal and friendly attitude towards the Sultan, in consequence of which he had been restored to favour and confidence by His Highness, and had been confirmed in the post of Governor at Nezwa. He died in 1894.

The shaikh paid me an early visit the next morning, and we had a long and interesting conversation, in which he communicated much information concerning the history, ancestry, and traditions of his tribe, about which he seemed never tired of talking, and about the topography of Nezwa and its neighbourhood, replying to my questions very courteously. He told me that the textile and embroidery industries here, once so famous and extensive, had entirely disappeared; that a little cotton, both of the white and brown species, was still grown; and that indigo was largely produced for dyeing the Bombay and American cottons, now so much imported.

After breakfast I took a walk through the town and environs to make myself acquainted with them, and found them so extensive that they required some time to realize the varied aspect of the whole.

The city is unwalled, and the space it covers is a medley of walled quarters, intermingled with groves of graceful palms, fruit, orchards, odorous gardens, and running streams, which, backed and sheltered by the grand mountains above them, present a remarkable picture of wild, natural scenery combined with luxuriant fertility and the evidence of human prosperity. The familiar features of the flat-roofed houses and palm-leaf huts presented nothing striking;

even the mosques are plain, unornamented, and unpretentious, as the magnificent Saracenic architecture of Mesopotamia, Syria, and Egypt is entirely absent from this simple land. I visited more than once the busy and thriving bazaars, which were well supplied with the commodities required by the people, and which, though of course a repetition of those at Mascat and Semail, Bahila, and Sohar, were nevertheless of singular attraction. Such glimpses of Arab life as watching the coppersmiths, braziers, dyers, and others working in their primitive Oriental manner, are always fascinating and calculated to arrest the gaze of a stranger. Among the artisans are makers of camel-saddles, potters, silversmiths, cobblers, cameleen weavers, carpenters, makers of hulwa—the national sweetmeat, for which Nezwa is famous, as it has a different flavour from that of Mascat, and is largely exported—blacksmiths, sugar or treacle makers, masons, mat weavers, and others. But the most noteworthy part of the bazaar, which is only shaded from the sun by strips of matting here and there, and is not particularly clean, is the copper market, which though inviting by the quaintness of its wares, is repellent from the incessant noise and deafening din of the hammering going on. The copper and brass work here made is distributed throughout Oman, the metal being imported in bars and sheets from Bombay, though formerly it was produced, it is said, in the country in considerable quantity.

The climate of Nezwa is very salubrious, with fresh exhilarating breezes by day and chilly cold nights in the cold season, and scorching heat, hot simoom winds in July, but dry and healthy always. The desert wind or simoon here blows from the south-west, and is very injurious to fruit trees, and sometimes carries blight to the corn. The elevation of Nezwa is 1450 feet.

Many of the tribes of the country are represented here dwelling together in amity, and compose the greater part of the population, but they are all in subordination to the two great rival clans which divide the power, dispute the supremacy, and keep their distance. These two tribes, the Beni Riyam and the Beni Hina, preserve an independent attitude, occupy separate parts of the town, and though rarely actually at war, maintain an ancient feud, and are mutually jealous and distrustful. The Beni Riyam, who are supposed to be the earliest inhabitants of the town, belong now to the Ghafiri faction, though originally from the Yemen, and are thus politically opposed as well as locally and tribally antagonistic to the Beni Hina. The Riyamis are by far the stronger of the two, hold the Alaya or upper part of the vale in which stands their hara, or quarter, known as Semed al Kindi, and their strong fortlet, the Bait al Silait. It tells in their favour that they are able to dam the stream and cut off the supply of water from the other part of the town.

Shaikh Hilal bin Zohair, since he seized the fort and became acknowledged as wali, or governor, has considerably strengthened his position, and reinforced his fighting material by sedulously encouraging in various ways members of his own tribe to bring their families in from other districts and settle here. The Beni Hina now occupy Akr, the chief of the six haras into which the Sifâla or lower town is subdivided, and as the great fort is in this quarter, and garrisoned by them, the position they occupy is not easily assailable.

The hara of Akr is enclosed by a high wall, and is entered through a well-built and massive gate called the Bâb al Sook; it contains altogether, I was told, 880 houses. Khorâsin, contiguous to Akr, is also surrounded by a wall; but Soal, which lies to the east of Wadi Abyad, is open and unprotected. There is also a hara called Khorâsitin, and another known as Harat al Wadi, from the fact of the Wadi Kalbuh running through it. The two wadis, Kalbuh and Abyad, divide the town and become confluent near the market under the fort. These two larger streams are fed by two minor ones, viz. the Wadi Musâbli and the Wadi Tanuf, the junction of Kalbuh and Tanuf being in Alaya, close to Bait al Silait. Nezwa is thus the meeting of four streams, and, as water is the great source of plant life in hot climates, Nezwa may be said to surpass all the towns of Oman in its supply of water, in natural wealth, and in the industry of its inhabitants.

As may be supposed, numerous feljes are drawn from these sources, which yield a most ample supply for irrigation and other purposes. It seemed to me that the most abundant flow was in the Wadi Abyad, or White river. I may here make the suggestion that the Wadi Kalbuh is the Alfulh of the Sicilian geographer Edrisi, who, however, gives it a wrong direction, as it flows not northwards but southwards, and joins the Wadi Halfain as mentioned above. Edrisi's Soal and Afra are no doubt Soal and Akr, the two largest quarters of Nezwa. His name for Jebel Akhdar, viz. Jebel Sharm, is merely a corruption of Jebel Shiraizi, the latter being the name of the chief town on the mountain. Nezwa is famous for its dates, which are nearly all of the Fard variety, and the vast multitude of palm-trees add immensely to the beauty of the settlement, and may be considered one of the chief sources of wealth to the people. Yet, notwithstanding the enormous production of these prolific trees, whose fruit almost takes the place of bread to the hardy Arab, the local consumption is so great that only a comparatively small quantity of the harvest is exported.

ON THE BORDER OF THE GREAT DESERT: A JOURNEY IN OMAN

By Lieut.-Colonel S. B. MILES

The Beni Hina tribe is one of the Al Azd stock, and emigrated from the Yemen at the same time as the Beni Riyam. Owing to the leading part taken by this tribe under its famous leader, the Dwarf Khalf, during the civil war which threatened the Yaarebeh dynasty at one time, it has given its name to one of the two great political factions, viz. the Hinawi and Ghafiri, into which Oman is at this day divided. Shaikh Hilal informed me that he had a thousand men-at-arms at his disposal at Nezwa alone, and he reckoned the whole population, including the Riyamis, at ten or twelve thousand, an estimate that seemed to me to be considerably under the mark. The great fortress, I was surprised to find, does not occupy an elevated position, but is situated in a thickly peopled quarter, and is hemmed in on all sides by dwelling-houses. It may be described as consisting of a large quadrangular enclosure, called the Hisn, at one angle of which stands the Kilaa, or citadel, a huge circular tower of solid stone masonry without window, loophole, or embrasure, but rising sheer, smooth, and unbroken to the roof, on which are mounted several parapet guns. The walls are of prodigious thickness and substantiality, and are well preserved; it is, therefore, easy to conceive that in a country where a prolonged investment and blockade are practically out of the question owing to the absence of commissariat organization, as well as to the non-existence of siege artillery, the castle might, if well provisioned, stand out for a long period. Its unquestionably massive structure is manifest at once, and its fame throughout Oman for its impregnability is intelligible enough; but the impression conveyed to my mind by its appearance after what I had heard of it from the Arabs, who invariably speak of it with conscious pride, was one of disappointment. I of course considered myself precluded from taking any measurements, but I may add that Wellsted's estimate of the height seemed to me to be too great by half. Commenced about four centuries ago on the site occupied by the ruins of an ancient tower of the Imams of the Nebhani dynasty, it was subsequently remodelled by some of the Yaarebeh Imams, and was finally brought to completion as it now stands by the Imam Sultan Belarab, who is alleged to have expended on it some eighty thousand dollars. The chief battle-ground lies in the open plain to the south-west of Jebel Akhdar.

The intertribal warfare of the Arab has generally been in the mountain passes when they could choose their own ground, but the plains have always been the arena when they fought in self-defence, and when battles have been forced on them by the invading armies of the Persians, Seljooks, Khalifs, or Wahabees. Resting at the base of the great mountain which overhangs and shields it on the north side, the town occupies a gentle slope intersected by many converging streams and nullahs, and screened from view towards the south by broken ground and low hills, the summits of which are often crowned with towers. This site, so extremely well adapted for defensive operations, gives it an unrivalled position, strategically and commercially, and it offers a haven of security and protection in this land of predatory and lawless marauders to the merchant, while it possesses an additional coign of vantage in commanding the mouths of the three passes leading from Jebel Akhdar, viz. the Wadi Tanuf, Wadi Miyadin, and Wadi Miyan, as well as in its proximity to Zikki, by which it draws to itself the traffic that is carried on up the Wadi Semail from the coast.

Here it was that the Azdite emigrants from the Yemen under Malik bin Fahm in the second century after Christ first established themselves, after having dispossessed and driven off the Parthian occupants of the land. Here it was that the messengers sent by Mohammed from El Medina delivered to the Julanda princes Abd and Jeifar the letters inviting them to accept the new faith of Islam, and abandon the idolatry of their fathers.

It was towards Nezwa as the capital that the many successive waves of invasion were directed, and here on the plains below were encamped at various times the imperial troops of Haroon al Rashid and other khalifs of Baghdad, the relentless and rapacious Seljooks of Kirman, and other Persian invaders. Here, too, the heretical Omanis fraternized with their schismatic neighbours the Carmathians of Al Bahrain, and it was in the Jami mosque of Nezwa, in the third century of the Hijra, that Abdulla bin Ibadh preached his peculiar doctrines and tenets so successfully that he gradually converted almost the whole country to his views. In this capital a long line of imams or sovereign pontiffs had ruled the destinies of the people and guided the main incidents and events of the chequered life of the nation during the period of a dozen centuries.

I was amused by the visits of the son of the owner, a bright little fellow of twelve, who, being naturally inquisitive, was full of questions about all he saw, and was delighted with the silver watch he received when I left.

I stayed a night at Tanuf, which is 8 miles from Nezwa, on our way to Al Dhahirah, as I had promised Shaikh Suliman when he intercepted us on the road, and had reason to be glad that I had accepted his invitation, as he received me very cordially, and gave me much information, though I could but ill spare the time.

Tanuf is a small, compact town of the Beni Riyam, fortified with a wall of fair height, pierced by two gates, the eastern and the western. The shaikh's house is behind the town, picturesquely built on the very edge of the deep ravine known as Wadi Tanuf, and which, cutting into the heart of the mountain, forms the natural road leading to Seek. Another deep gully or torrent-bed, also debouching at this point in the Wadi Mayin, which penetrates in a more northerly direction and forms a pass, albeit a very rough and precipitous one, across the chain to Rostak, is called the Tareek al Shas. The dark limestone cliffs, the cavernous glen, and the palm trees give a gloomy and sombre aspect to this spot, which has a romantic air of seclusion, and I was not surprised to learn that, shut in as it is, the heat here in the summer months is intolerable.

Shaikh Suliman informed me that Jebel Akhdar extended from the Wadi beni Rowaiha to Jebel Shoum, which was the highest peak in Oman, and he confirmed the tradition that the original Azdite name of Jebel Akhdar, as also that of Nezwa, was Radhwan, and was so called probably after a district in the Yemen. When it was first named Jebel Akhdar is unknown, but after the founding of the town of Shiraizi by the Dailemites about 975 A.D., the mountain seems to have been known for a long time as Jebel Shiraizi.

On the 22nd, having procured fresh baggage camels at Tanuf, and said farewell to the shaikh, who had treated us so urbanely and hospitably, we mounted just outside the gate, and turning our camels' heads to the north-north-west in the direction of the Akabat al Barak, by which pass we intended to cross over to the Dhahirah, rode off to pursue our journey. The path lay over a monotonous, drear, and stony plain, crossed by numerous nullahs. We passed several villages of the Beni Hina tribe, such as Homra, with a population of about four hundred, and other villages, whose names I omitted to record, and further on, between 10 and 12 miles from Tanuf, two hamlets of the goat-rearing shepherds, viz. Ghomriya and Belad Sait. The popular excitement among these rude and unsophisticated people caused us some amusement, and relieved the day's journey from the tediousness and fatiguing dulness it would otherwise have had.

As we approached Ghomriya, the sudden fire of the matchlocks and the whizzing of the bullets, a salute with which the sentries did not fail to greet us, caused a slight scatteration among our timid camels, and proved that our strange appearance had evidently given rise to as much apprehension as suspicion among the inhabitants, who meant their shots as a warning to keep a respectful distance. It was necessary for our kefeer, the Kelbani shaikh, to ride forward and represent who we were, before their fears were sufficiently allayed to allow us to pass by quietly. We soon discovered that the cause of the panic was due to the fact that a troop of Jenebeh Beduins had ridden rapidly by during the previous night from the east. The latter are at feud with the Beni Hina, who had mistaken our party for the Jenebeh returning to raid the district and lift the camels; hence the scare that had been created, and the vigilant distrust with which we had been regarded.

Some of my party at once examined the ground, and having detected fresh footprints of the dromedaries, could judge that the Jenebeh were in force. That they were actuated by no friendly intentions seemed probable, but whither they had gone no one knew. We also heard to-day that a band of seventy marauding brigands, collected from various predatory tribes in Al Dhahirah, had started on an excursion to harry and plunder the peaceful and prosperous peasants in the outskirts of Minha, and that they had, after crossing the Nejd el Makhârim, camped in the vicinity of Yabreen on the night of the very day we had left it. Shaikh Nasir al Ghafiri, the chief of Bahila, had despatched a kossid with a letter after us, giving warning about these robbers, who were said to be mounted on swift dromedaries, directly he had received intelligence of their proximity; but his kossid had missed us on the road, and had not had the sense or courage to follow us on to Nezwa. These marauding gentry are usually averse to hostilities, and prefer to swoop down on their quarry and carry off their plunder as expeditiously as possible before the villagers can assemble to oppose them. It was just as well that we did not encounter them, for, although my party was quite a match for them, it would have been regrettable had any bloodshed occurred even in self-defence.

After a ride of some hours over a rough country, very scantily dotted with vegetation, we reached the fort of Jebel Kur. Here we strike the Wadi Shama, and having crossed this we commence the ascent of the Wadi Ghol, which presents at this point a broad, sandy, and shallow bed not devoid of trees and other vegetation. These two wadis, the Ghol and the Shama, uniting lower down, are then known under the name of the Wadi Bahila, a torrent which, as I have above described flowing under the walls of that city, loses itself ultimately in the sands of the desert. The Wadi Ghol is not particularly serpentine, and I imagine it is so called because it abounds with snakes. Ghol is the common word used in Oman for this reptile, and in this wild hilly district they are tolerably numerous. Though poisonous species do exist in this land, the great majority of species are, I believe, innocuous, and cases of Arabs dying from snake-bite are extremely rare. Indeed, they appear to regard them but little, and seldom allude to them.

The bed of the Ghol, which is cut through high vertical banks, and contains a perennial supply of water, is occupied by several hamlets of the Yal Khameyyis, a Beduin branch of the Beni Hina, who lead a pastoral life and subsist mainly on their flocks of goats and sheep. It is to the vigilance of these hardy, active, and brave hillmen that the security of this pass is due, as, being ever on the alert, they are able to gather with marvellous quickness on the sign of danger, and thus hold in check the plundering bands of Al Dhahirah, who would otherwise be free to raid and devastate this district at their pleasure. After toiling for two hours up the bed of the wadi, which gradually becomes more steep and stony, we arrived at a shepherd hamlet called Mithar, where we bivouacked for the night at an elevation of 2600 feet. At this place was a very large zareeba or fold for the goats, from which we kept a very respectful distance. The thermometer fell heavily during the night, and I was glad of all the wraps I had with me.

The next morning we started at 7 a.m., and underwent a hard day's work of nine hours in crossing the pass; the severe gradient we had to climb, and the rugged nature of the rocky ground—for there was no pathway to speak of—causing great fatigue to the camels. Quitting Mithar, we recommenced the ascent of the Wadi Ghol, and soon after a haunted glen known as Dhul Jinn was pointed out to me as being the abode of efreets and evil spirits, and which is carefully avoided after dark by the natives, who firmly believe in hobgoblins.

For four hours we toiled slowly and painfully up the steep mountain-side, strewn with fragments and boulders of light surface-coloured limestone and mottled with scanty bushes of euphorbia, until, with a final spurt and the shouts of the camel-drivers, we arrived at 11 at the summit of the pass, the aneroid barometer indicating an altitude of 3700 feet; the highest peak of Jebel al Kur, just opposite, rearing its head apparently some 2000 feet above us. Here we halted for half an hour for rest and refreshment and to breathe the camels.

From this point we look down a small valley running south toward Adam, called by some Wadi Seifam, from a town in it of that name, by others Wadi Ali. The juncture of the spur at the head of this vale we crossed, and found ourselves on the north-west side of the Akabat al Barak; and now, for the second time, I gazed upon the little-known, wild, and unpromising plain of Al Dhahirah, the desert province of Oman. Bare and extensive, without any conspicuous feature to catch the eye or obstruct the view, it stretched away into the misty distance and offered a landscape as unattractive as it was unvaried. The descent did not prove so slow and difficult as the ascent, and we soon reached the shoulder of the mountain, a level stretch of sand and pebbles, about halfway down the slope, in which are located two shepherd hamlets, the one called Sint and the other Sunt.

Here the camels' loads were readjusted, and we made a fresh start, the road now falling into the Wadi Ain, at this elevation a rugged ravine, but further down the valley a large and seemingly important stream with running water. The direction it took was pointed out to the west-north-west, and under the name of Wadi Talaif it flows, studded with palms and villages, nearly to the sea at Abu Thabi. This river is, no doubt, the one mentioned by Edrisi noted above, as flowing into the Persian gulf at Julfar; but he was misled as to the source of it, as it rises, not at Nezwa, but in Jebel Kur. The low, sandy, maritime plain usually known as the Pirate coast, but formerly called the Julfar coast, took its name from the old Persian town of Julfar, which stood on the site of the present Ras al Khyma.

We were now able to move on at a fair pace, and at 2 p.m. came to a village called Hail, belonging to the Beni Hina tribe, situated almost at the foot of a very remarkable and, as it appeared to me, nearly perpendicular mass, perhaps 2000 feet in height, so evenly and delicately indented or notched along its ridge, that it had obtained the name of Jebel Misht, from the fancied resemblance its acicular peaks bear to the teeth of a lady's hair-comb. This comb-hill struck me as being so singular that I did not fail to take a photograph of it the following morning, though it was then at a greater distance and had lost some of its peculiarity.

Skirting the fields and cultivation of Hail, which with Jebel Misht we left on our right, we rode on for two hours more, and, just as it was getting dusk, arrived at our destination, Al Ein, a small hamlet of about 100 souls, so called from its copious spring. The sudden appearance of a mounted party, evidently well armed, threw the good people into a ferment, and in their excitement the guard in their watch-tower greeted us with a volley from their match-locks. We were so near at the time that one of the bullets grazed a

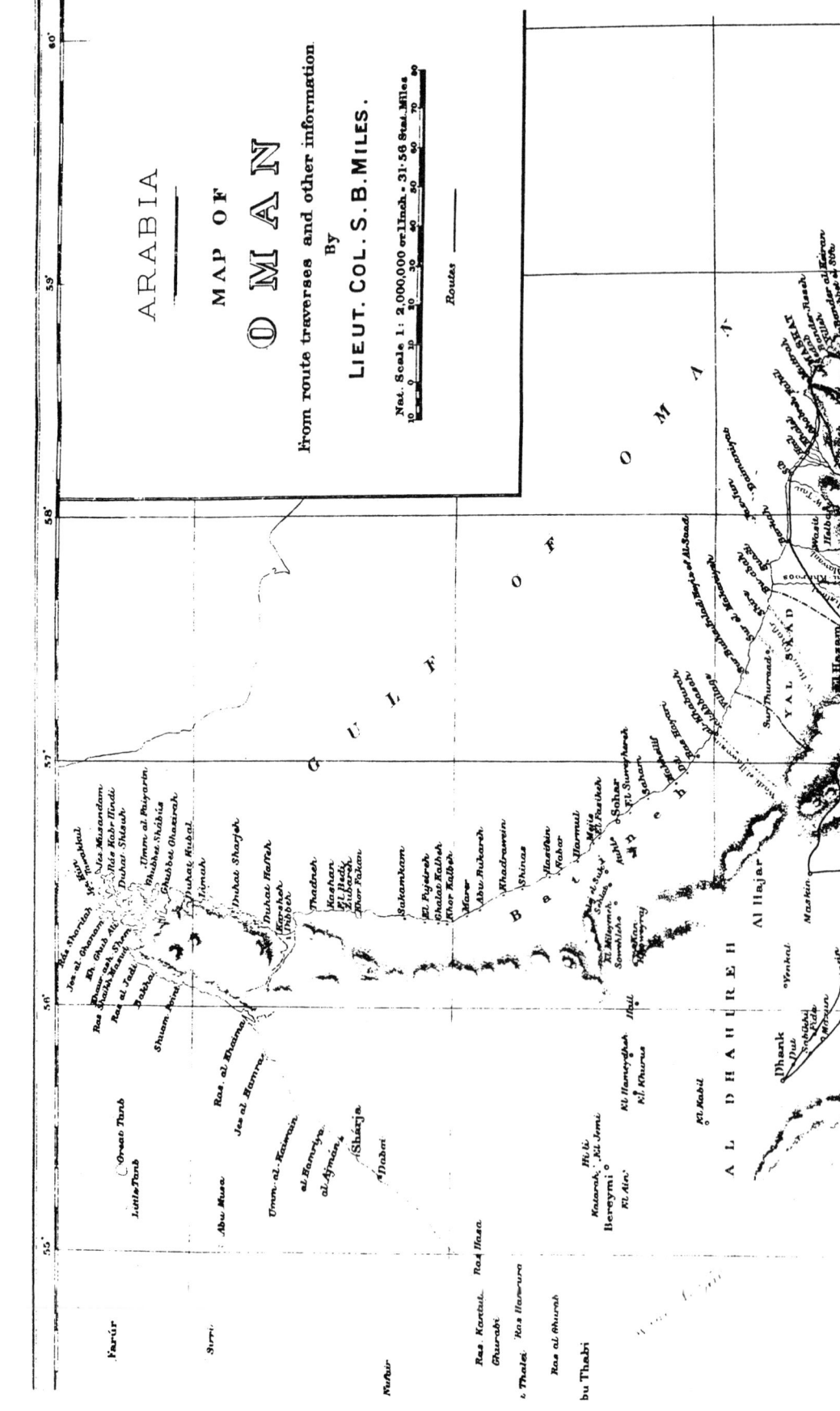

ARABIA

MAP OF

OMAN

from route traverses and other information

By

LIEUT. COL. S. B. MILES.

Nat. Scale 1 : 2,000,000 or 1 Inch = 31·58 Stat. Miles

Routes

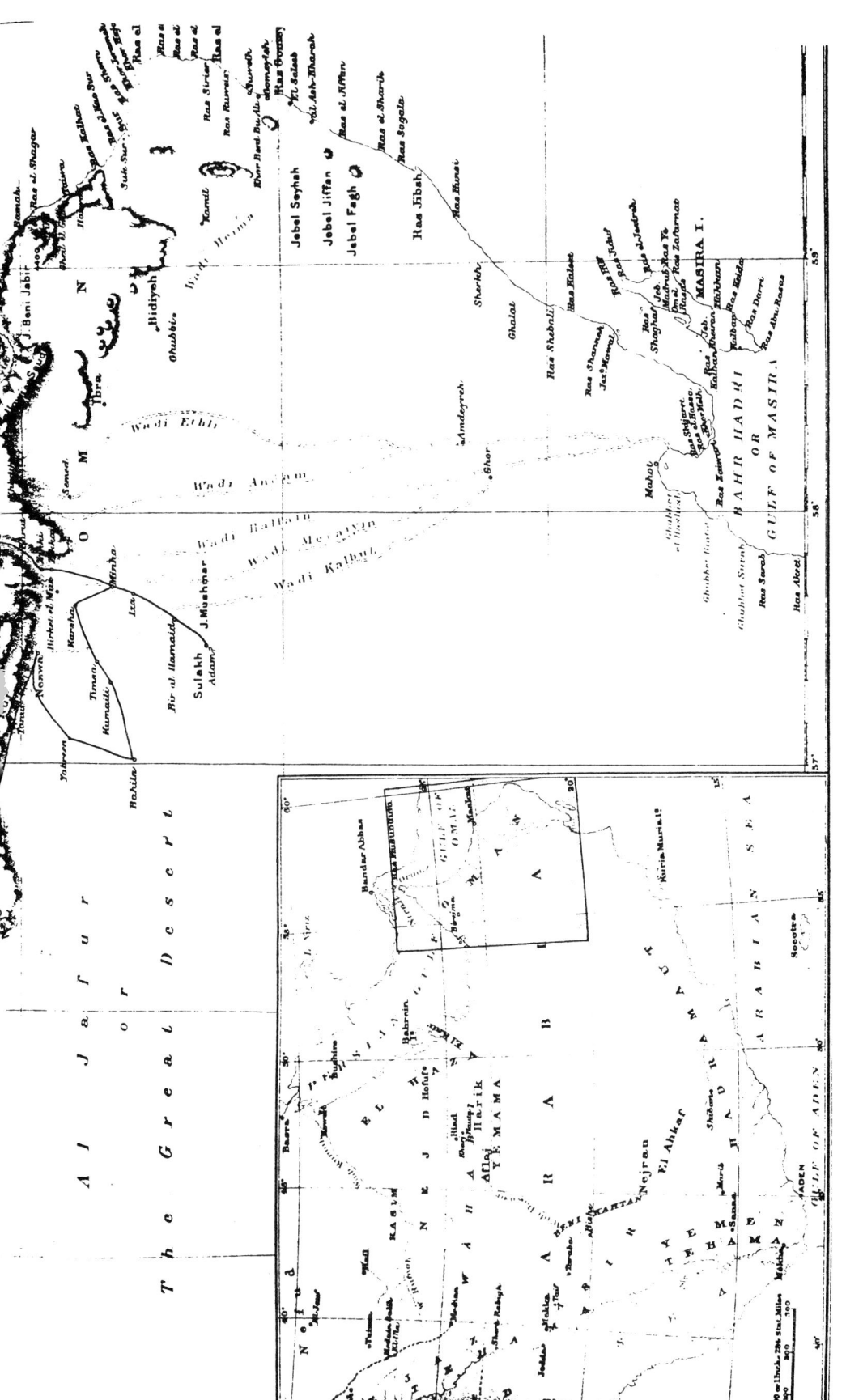

camel's leg, but no harm was done; and as such incidents were too common to be noticed, we were soon seated amicably on the bank of the nullah in the midst of the elders, who expressed regret for their precipitation. They gave me the names of the other hamlets in the ravines springing from Jebel al Kur, viz. Shedait and Nadan; but it is evidently a desolate region where the paucity of the population attests the poverty of the soil, and where the rarity of vegetation proves the scantiness of water. The only industry I could hear of at Al Ain was the manufacture of coarse pottery. The elevation of this place is 1400 feet.

After I had taken some photographs on the 24th, we mounted and resumed our journey down the Wadi Ain for some distance, and then turned off to the right over a sterile plain mottled with a few stunted acacias and euphorbias towards Bât, which we reached after 12 miles of travel. This town is inhabited by about 400 of the Beni Kelban, who cultivate wheat and indigo chiefly, and engage themselves in dyeing cloth, the dye stuff being made in large earthen jars called kabaia. We were not received here with any demonstrations of welcome, and the reason was soon apparent. It seemed that we had intruded upon a divided community, who had fallen out about a felej that irrigated their fields, and so bitter had the quarrel become that it had culminated in a permanent split, each party having constructed for itself a hujra, or walled enclosure, and having boycotted each other were now nurturing in their bosoms that implacable animosity so distinctive of local and parochial controversies. Here we halted for an hour to drink coffee, while our Kefeer Shaikh Nasir, who belonged to this tribe, interviewed the elders and busily studied the pros and cons of the dispute, in which he seemed much interested. We then proceeded on to Mohaira, where the watchmen on our approach saluted us from a distance with a desultory matchlock fire until Seyyid Hamood sent forward a man to expostulate, whereupon the recognition of friends having dispelled the cloud of anxiety, the sunshine of joy took its place, and the people, headed by their venerable shaikh, came out in procession to welcome us with noisy demonstrations, shouts, firing, and dancing. This being over, followed an invitation to pass the night there, which we agreed to do in view of the fatigue of our camels and other circumstances. The shaikh treated us cordially and lodged us comfortably. The village is small, perhaps seventy or eighty houses, but is refreshingly surrounded with plantations and orchards of dates, mangoes, limes, and plantains, etc., and the cultivation includes fields of wheat, jowari, lucerne, pulse, sugar-cane, and indigo, a bountiful stream of water sufficing to supply the whole needs of the population.

The following day we marched 8 miles on to Deriz, one of the chief towns of the Beni Ghafir tribe in this province, having a population of about 1200. On sighting our party, the shaikh and his people came out in a very friendly way to greet us. There is a small dilapidated fort here with two guns, and I was told there were hot springs in the vicinity. After a short halt at Deriz, I rode off with some of my party to visit the fort of Al Einein, or the Two Springs, a famous bone of contention in these parts for many years past. The castle is a large and lofty but rather decayed structure, with a walled courtyard flanked and strengthened by two towers, and though looked on by the natives with much respect, it did not appear to me a very strong place.

Within a short distance of the castle there is an extensive grove of date and other fruit trees, watered by three felejes or subterranean streams, half of which belong to the Ibriyin tribe, who are strong in this district and were the former owners of Al Einein, and half to the present possessors of the castle, the Beni Ali, who receive a yearly revenue of 200 bags of dates and 100 dollars in cash for the farm of their moiety. There are no inhabitants at Al Einein now; intestine dissensions and tribal wars have driven them away, and there is but one village left in the vicinity, Iraka, which lies just behind Al Einein and belongs to the Ibriyin.

About the year 1880 the Sultan, for political reasons, transferred his alliance from the Yemen or Hinawi faction to that of the Nizar or Ghafiri faction, a move which had a very disturbing effect in the country. In Dhahireh, where the general interest had been concentrated on Boo Einein, the change brought matters to a crisis, and a struggle for the possession of that fort ensued, which ended in its passing from one faction to the other. A large number of tribes were engaged in the contest, and the general commotion was out of all proportion to the value of the stake at issue, this being quite insignificant to any party. After much heavy fighting, the fort fell into the hands of the Beni Ali tribe, a result injurious to the prestige of the Sultan, whose power and authority continued to wane until they were restored by the decisive defeat of his brother Abdul Azeez towards the end of 1883, from which time they remained paramount until his death.

Turning again to join the party, we rode past the deserted ruins of Al Ghabbi, an ancient and historic town of some extent, and here I lingered for some little time, recalling to my mind the troublous times of 1875, when the place was sacked and destroyed by the Beni Ali tribe after a siege and a sanguinary conflict.

It was about 2 p.m. when we arrived at the town of Obra, where I intended to halt, and we had hardly dismounted and been received by the shaikh, when one of the baggage-camel drivers came up and reported that he had been waylaid by robbers and looted on the road. At first we were incredulous, as none of the rest of the party had seen anything likely to excite suspicion as we came along; but on inquiry it turned out that, trusting to the fact of his being a native of Obri, and imagining that he was safe from thieves in his own district, the driver had lagged behind with the intention of coming up slowly alone. He found, however, that he had overrated the fraternal feelings of his townsmen, and that robbers are no respecters of persons, for having been espied on his solitary ride, he had been dexterously cut off by two men, who had compelled him to unpack his load and exhibit his treasures. They relieved him of his matchlock and dagger, and took a bundle of clothes belonging to my servant, but did not touch the other articles in the load, which included among other things the photographic camera. After overhauling the pack the thieves decamped, as we heard afterwards, to a distant village, while our camel-driver, having repacked his load, had ridden post haste in to Obri.

The shaikh was greatly shamefaced at what had occurred, and, admitting himself responsible, promised full restitution; but, as I was well aware, he was quite helpless in the matter, and had no control practically over his people. Later in the day Seyyid Hamood learned that the two robbers were part of the band of seventy who had crossed into Oman on a marauding expedition, as mentioned above, and who, having just returned from there, were scattering homeward. It was refreshing to hear on the best authority that the raid had been a conspicuous failure, and that the would-be spoilers had not only gained little, but had been roughly handled into the bargain. They had, it seems, attempted a surprise on a village not far from Nezwa, and had managed to collect a few camels and women's silver ornaments before the alarm could be raised and the men summoned from their work; but the resistance had then been so stubborn that, though the villagers were in the minority, they had compelled the marauders to retire, leaving two of their number dead on the ground. Three of the villagers had been killed, and many had been wounded on both sides.

Obra is a large town subdivided into sixteen haras or sections, and is situated under a low white hill called Herbareh. The irrigation channels of the rich fruit orchards and cornfields which cover their luxuriant and well-watered oasis are supplied from two magnificent felejes leading from the adjacent hills. The date palm, whose graceful plumes are such a marked feature in Arab landscape, is here, as one may say, subordinate to the other fruits, of which Obra can boast an immense profusion and great variety. The oasis, however, will take long to recover from the effects of the great hurricane of March, 1885, which swept across it in the full tide of its fury, committing terrible havoc among the orchards, uprooting and hurling trees to the ground. Many of the best houses in the place were also demolished, including the wali's, whose residence was found to be little better than a heap of ruins. There is a small castle here with a high wind tower, and nearer to the white hill the site of a more ancient and ruined castle was pointed out to me, renowned in the land, and still fresh in men's memories as having sustained many years back one of the most furious sieges recorded in Oman history.

No little interest is lent to the bazaar of this town by the fact that it is not only by far the largest and most frequented in Al Dhahirah, but also possesses the characteristic feature of being the "Thieves' Auction Mart," where the booty and loot annexed and collected by the highwaymen of this Alsatia are disposed of to the highest bidder. I therefore paid it a long visit, and found that, in addition to the advantage of being a thieves' bazaar, it contained shops and booths of every description, and presented a very lively appearance, all the artisans being evidently in full employment, the town comprising considerable indigenous industries and productions. The market, dirty and malodorous though it be, is thronged with spare gaunt Beduins and townsmen, all well armed, jostling each other or haggling in a loud key with negro and other sellers about the price of commodities, and presenting an animated picture of Arab life. It consists of about three or four score shops, displaying all the articles usually required by Arabs, tradesmen working at their crafts, eager, bustling, and no lack of employment, negresses in rows squatting on the ground with baskets of fruit, kabobs, cakes, sweeties, and bowls of lubia soup. The wali, Mohammed bin Abdulla al Yakibi by name, was very civil and desirous of showing attention. He escorted me all over the town, the glories and beauties of which he was resolved should not be hidden from me, and was very careful to point out the lion of the town, viz. the Jami or Great Mosque, a plain, undecorated, unimposing structure, with nothing remarkable about it whatever, but which the inhabitants are very proud of, and consider to be the largest Ibadhi mosque in Oman.

As I have said above, Obra is a famous place for fruit. Limes, mangoes, sweet limes, dates, peaches, apricots, quinces, figs, bananas, oranges, pomegranates, pomeloes, almonds, plums, guavas, citrons, melons, and others, are grown and largely exported. Obra may, indeed, be looked on as the present capital of Al Dhahirah, and contains roughly about 5000 inhabitants. Al Dhahirah means the Ultramontane province, or that which lies behind the Batineh and the range known as Al Hajar. It enjoys the worst reputation for the lawless and predatory character of its dwellers, the tribes occupying it being one and all thievish, treacherous, and turbulent, subsisting to a large extent on the produce of their raids and incursions into the Batineh and other districts. In a country where every able-bodied man is habitually well armed and ready to fight and plunder, the quantity of available rascaldom is pretty considerable, and the quality of ruffianism is quite in keeping. It is a poor, wild, and thinly populated province, and is, though I have twice visited it, less known to me than any other.

In March, 1836, Obra was visited by Wellsted and Whitelock, who remained two days. The fact was perfectly well remembered by the people, and the spot on which their tent had been pitched was at once pointed out to me. It happened, unfortunately for them, that the time chosen for their visit was not an auspicious one, as the Wahabees were then marching through the town towards Oman, and the two officers narrowly escaped ill treatment at the hands of these fanatics, having been pelted with stones, driven out of the town, and eventually compelled to return to the coast. The elevation of Obra is 1180 feet.

As it was considered not improbable that we might meet with roving bands of Arabs on the road to Dhank, which was our next destination, Seyyid Hamood arranged with the Wali Shaikh Mohammed al Yakibi to accompany us thither, and he also took the precaution to engage one or two men of the Beni Kattab tribe as kefeers. The shaikh, however, was not ready to start when we were mounting our camels the next morning, so we set off without him, and had ridden more than halfway before he joined the party. Our course now lay due north, and soon after leaving Obra we crossed a low ridge of hills with an average height perhaps of 400 feet, the road for the rest of the way lying over a slightly undulating plain, sterile, waterless, and unadorned, save with the scantiest vegetation. Trending away at some distance on our right hand, and running parallel with our course towards the north, is the range known as Al Hajar, the intervening ground being rough and hillocky, intersected by shallow ravines. On our left, towards Bereymi, no vestige of a hill is seen to break the level line of the horizon, and the ground falls with a gentle declivity. It is a wild and little-known part, with a few villages, such as Sedaikee, Senanah, and others ; but I could not afford the time to visit it.

Here we may be said to quit the margin of the Great Desert, and to turn our faces towards the inhabited regions. This wilderness, on the eastern border of which we are now standing, stretches away to the westward for about 700 miles, forming the largest and most inhospitable expanse of sandy waste on the continent of Asia. Broadly speaking, it is devoid of rivers, trees, mountains, and human habitations, unexplored and unexplorable, foodless, waterless, roadless, and shadeless, wind-swept, and a land of quietude, lethargy, and monotony, perhaps unparalleled in the world. The extent may be best described by saying that it covers an area ten times that of England and Wales. The surface of the desert is said to be generally undulating, presenting here and there low ranges of hills, with groups of green and shallow wadies or oases, usually dry, but with an outline marked by verdure. The greater part would seem to be absolutely bare. In this desert the vegetation, it may be noted, chiefly consists of euphorbias, various acacias, such as camel-thorn, sayel, ghafe and samr. Tamerisk, dwarf palms, probably a species of chamerops, oleander, senna, colocynth, the rhamnus or sidr, rak, coarse grass in tufts, the useful calotropis or silk tree, which provides cordage and medicine, etc. The rainfall is scanty, but the Arabs make use of pools of water left after rain in the watercourses, and frequently obtain water in the ravines by digging. There are also springs of brackish water welling up here and there; the vegetation, however, obtains its chief supply of moisture from the dews, which are extremely copious in this country.

The desert is generally spoken of by the Omani Arabs as Al Jafur, or the unfrequented space, and as Al Ramool, or the sands; sometimes as Al Ahkâf, or the sandhills; also as the Sahar, or desert. In Arab geographical works it is called by the expressive name of Roba al Khali, or the empty habitation. The Bedouins, too, have of course their own names for the various districts and oases distinguished by special characteristics. The Beduin who roam the desert are mostly contingents from the tribes on the border surrounding it, and none of them, I believe, can be considered as truly nomadic. It is, however, difficult to believe that any part of the desert can be wholly inaccessible to the hardy Beduin and his friend the camel, with their wonderful power of endurance under privation of all kinds, as they can travel comfortably 60 or 70 miles a day, and carry food sufficient for several days, the scarcity of water being the one great obstacle they have to contend with.

In the eastern part of the desert the extensive tracts clothed with verdure are of great value to the people of Oman, as the Beduin who range it are able to rear vast numbers of camels, and thus perform a valuable service to their brethren dwelling in towns and villages, by providing them with an adequate supply of these useful animals. The Omani Beduins who are chiefly engaged in this pursuit are the Deru, the Aal Wahibeh, and Awamir tribes, and the sale of young camels at Adam, Obra, and other markets constitutes their chief support; they also rear large numbers of goats. The Affar Beduins, who are a section of the Awamir, are said to scour the desert more extensively than any other Omanis; they are very wild, truculent, and under no human control. They appear to live in an almost chronic state of semi-starvation. I have seen many of them and heard many stories about them. The large and powerful Jenebeh tribe, whose chief towns are Soor and Al Ashkharah, and who still possess the coast-line from Ras al Had to Dhofer, as they have done for two thousand years or more, are a seafaring race and do not rear camels. The speed and other high qualities of the Omani camel place it on a par with the famous Mahra breed, with which it is indeed identical, and which is deemed by the Arabs to be unsurpassed. That in former times attempts were made to establish caravan routes across the desert can hardly be doubted, but no efforts in this direction have been made for many years past.

In his Alte Geographie Arabiens Sprenger discusses the various routes through the desert, and mentions that in the year 616 H., 1219

A.D., some Beduins from the Neyd visited Dhofar in the time when Al Haboodhy ruled in that town, and started thence across the desert to Yemama, but never returned. The number of mammals to be found in these parts of the desert where vegetation exists does not, probably, exceed a dozen in species. The *Oryx beatrix*, known generally to the Arabs as the bakr al wahshi, or wild ox, but called by the Omanis the boosohla, is found in most parts, but is said not to approach within 50 miles of the sea-coast. There are two or three species of gazelle—one is the graceful reem, or white doe, which is rather scarce in the Omani desert; another species, the thabi, is of a rather darker colour and abound in the oasis. The wild ass, known as the himar al wahshi, inhabits the Antal and Yuba districts between Adam and Ras Madraka, and is not common elsewhere. The flesh of it is said to be more esteemed by the Arabs than that of any other game. The hydrax, or coney of scripture, has lately been discovered near Dhofar. The wolf, hyena, black hedgehog, hare, jerboa, and field rat make up the list of all the animals I could hear of.

It began to rain heavily soon after we had started, and the wind was so bitterly cold and piercing that by the time we had arrived at Mazun, a Belooch town 18 miles from Obra, we were all shivering, and I was so benumbed that I could hardly dismount. This place is inhabited exclusively by Belooches of the Hôt tribe, and contains no Arab. A large portion of the population came out to gaze at us, but their demeanour was unmistakenly unfriendly, and they did not invite us to enter the town, which I accordingly did not visit; but we lighted a big fire to warm ourselves, and those who could manage it changed their clothes, for we were drenched through. The Belooch stood around scowling at us as we remained, but were otherwise inoffensive. This was the only place I visited on this journey where I was not welcomed, for the little ebullitions of excitement that had occurred at some of the villages on the road as we passed were simply the result of fear caused by our sudden appearance, and were not due to unfriendly feeling, as was shown by the reaction that took place when they learned who we were. Mazun seemed to be a large town, with a high wall round it, but without fort, bastion, or tower that I could see. The elders told me that the Hôt had established themselves there seven generations back, perhaps about 200 years.

Turning our backs unregretfully upon the Belooch town, we pursued our journey for 8 miles further, still going north, and arrived at Dhank about 4 p.m. We found that a good many people had assembled in the open space where we had to alight, and the shaikh gave us a very friendly, though undemonstrative reception. Dhank was of much greater extent than I had expected to find it, and I was charmed with its quaint situation and beauty. The number of people is said to be about 4000, composed mainly of four tribes, viz. the Naim, Washâsha,

Azeez, and Al Boo Shamis, of which the first-named may be considered the predominant. The head shaikh and Wali of Dhank, and Shaikh Mohammed bin Ali al Naimy happened at this time to be absent at Al Beraimi, and the shaikh in charge was Khalfan bin Ali, whom I had not met before. The settlement is on the bank of a rapid stream flowing half round the base of a low transverse ridge or eminence of dark basaltic rock, which divides the town into two parts, the upper and lower, the houses being mostly interspersed among, and more than half concealed by, rich clumps of palms, orchards, and gardens. The bazaar is insignificant, and is hardly a third the size of that of Obra. The industries here are, however, most varied, though the trade appears to be less than at Obra. Besides indigo dyeing, there are several textile industries, the principal being woollen and cotton fabrics of different kinds, while bishts or camel-hair cloaks, with embroidered collars, and camel saddle-cloths are also worked here.

The castle in which Shaikh Mohammed resides is situated in Aliyya or upper Dhank, and is known as the Imam's house, built originally by the Imam Sultan bin Saif al Yaareby in the middle of the seventeenth century. Owing to the configuration of the ground, its plan is somewhat peculiar, being very long and narrow, and the curtain walls strengthened with six turrets. Some additions were made by Seyyid Azzan bin Kais in 1869, and it is now in fairly good condition.

A great deal of indigo is cultivated here, and, indeed, throughout Al Dhahireh. I was surprised at the extent of ground under it. The people showed me their dye vats, and told me that not only homespuns, but foreign imported cloths were dyed here. The house in which I was lodged was surrounded on all sides by the huts and sheds of the cloth-weavers, whose rudely constructed, creaking looms kept up an unmusical din well into the night, and this, accompanied as it was by the melancholy cry of the little booma or screech owl, in the palm grove, which latter is in Oman considered a bird of good omen, formed a concert which effectually prevented sleep. The elevation of this place is 950 feet above the sea. The Wadi Dhank is joined a little way off by the Wadi Boo Kerba, and these, flowing in a north-westerly direction, become confluent with the Wadi Safa, a large nullah in the ramool or desert pouring into the Persian gulf, nearly parallel with the Wadi Ain. These streams diminish their volume considerably in the hot season, but nature then compensates by the heavy fall of dew at night, which imperceptibly saturates the surface and quenches the thirst of the vegetation, thus reviving it and causing it to endure through a time of drought that would otherwise be fatal to it.

We had now reached the furthest point I had purposed visiting on this excursion, and as I was pressed for time, I reluctantly requested Seyyid Hamood the next morning to lead us in the direction of Mascat, and accordingly, mounting our dromedaries, we began to ascend the narrow stony bed of the Wadi Dhank, which is densely fringed with palms, until we emerged from the gap in the rock through which it flows, and at once a pleasant change in the scenery burst abruptly upon us. The romantic and sequestered glen in which Dhank lies is sombre, the air heavy with moisture and fragrance, and the dark foliage of the trees combines with the purple rock to enhance the gloom.

After passing the gap, the wadi, we find, spreads out to a width of about half a mile, forming an open, picturesque landscape, with a fine stream of water in the middle of the bed, of which the wall-like banks rise to about 100 feet. The little plain beyond is memorable in recent history as being the battlefield that witnessed the bloodiest conflict fought between the two great rivals for the throne of Muscat, viz. Seyyid Toorky bin Saeed and Seyyid Azzan bin Kais. It was in 1869. Seyyid Azzan was at that time Imam of Muscat, having ousted his cousin, Seyyid Salim bin Thowainy, from power about a year previously, and, being thus master of considerable resources, had been able to muster an army of four thousand men. Seyyid Toorky's force was numerically much inferior, but this deficiency in strength was more than compensated by the unusually bitter hatred and spirit of revenge that had been inspired by the persecuting zeal of Azzan and his metowwa Khaleyli, who had inflicted numerous cruelties and humiliations on their religious opponents. Seyyid Toorky, moreover, though by nature less energetic, was a more wise and prudent man, and handled his troops more skilfully. On the other hand again, Azzan was in possession of a serviceable brass field-piece, which had been dragged to the spot in charge of a Persian topchi, and which added greatly to the pride and confidence of his followers. Though strong and well posted, and aided by the thunders of the one-gun battery, Azzan's force was unable to resist the impetuous and fierce attack of its despised enemy, who, rushing on with cries and shouts and wielding their double-edged swords, mowed down their foe in heaps and gave no quarter. In dire dismay Azzan's army broke and fled in confusion, the rout was soon complete, and the active pursuit added greatly to the slaughter. The loss of the vanquished was about three hundred killed. Seyyid Azzan and his priestly advisers hardly drew bridle till they reached Mascat, and he never recovered the prestige he lost in this affair. Within a year he was besieged in his capital, and was slain in a conflict outside the walls of Muttrah by his exasperated enemies. One of my party had been present in the engagement, and his description of the fight as he pointed out to me the positions of the combatants and the salient features of the battle intensified the interest of the scene.

At 4 miles from Dhank we came to the village of Dut, and 7 miles further on is Fida, whose narrow belt of palms and other fruit trees fringes the wadi for a long distance. Both these places belong to the Beni Zeed. Between Fida and Al Einein lies a sterile, uninhabited tract, almost entirely destitute of vegetation and water. Leaving the wadi at Fida and turning to the right, we direct our course to the east, pausing at the top of the bank to catch a distant view of the picturesque town, castle, and palm grove of Yenkal, which my companions had described and were eager to point out to me. It is a populous and prosperous settlement, about 5 miles further up the same valley, and its fort is famed in Oman history as a venerable stronghold commanding one of the chief roads and passes through the mountains. I regretted not being able to visit it. At the town of Aridh, which we reached about 5 p.m. and where we halted for the night, there are numerous hot springs with curative virtues. It contains a mixed population of the Beni Shekail, Beni Omar, and Beni Kelban tribes, and is now in the possession of our guide Shaikh Nasir, like its neighbouring town Makiniyat. Both these places belonged formerly to the people of Deriz, but were captured and annexed by the Beni Kelban in 1867. The air assumed by Shaikh Nasir was one of increased dignity, and his face beamed with smiles as he welcomed us to his town, and he did the honours of hospitality with great *éclat*. The aneroid showed an elevation of 1250 feet.

On the 28th we left Aridh at 7.50 a.m., and after riding in a north-easterly direction for four hours and a half over barren, lifeless, and rather broken ground, gradually but perceptibly rising as we approached the range, we reached Maskin, a village of the Beni Kelban and a poverty-stricken place at the head of the Wadi al Kebeer, and in a notch of the watershed or crest. This chain, which divides the kingdom broadly into two parts, maintains throughout its whole length the same remarkable and characteristic features, presenting on the western or land side a more or less gentle declivity which merges into the desert, while on its eastern or seaward face it bears a wild and lofty aspect, falling abruptly in steep, precipitous crags to the mountain plain below. In the former case the valleys are shallow, verdant, and populous, in the latter their appearance is very different, being rugged, headlong torrent beds, and passing over a rough and stony passage difficult to climb and but thinly studded with hamlets, as cultivation is almost wholly precluded until the more level plain is reached. The view from this spot was really noble and extensive, and a short rest here gave me an opportunity of enjoying a landscape of a very enchanting and diversified character.

Our objective after leaving Maskin was the city of Rostak, which lies at the foot of the stupendous walls of Jebel Akhdar, and our best route would have been to follow down the winding but well-trodden bed of the Wadi Beni Ghafir, which is a natural highway, but to this our guide raised objections on account of the intertribal dissensions then raging in the valley. Under the guidance, therefore, of two men who rode in front, we struck diagonally across the bluff brow of the mountain, and from noon to dark continued zigzagging in a generally south-east direction down the uncommonly rough and broken face, twisting and turning now down a narrow torrent bed, now along a smooth and level ledge bristling with euphorbias, calotropis, and fragrant oleander, now among crags and boulders, making slow progress, but without any serious mishap, until at a sudden bend we arrive at a clump of palms in a glen, and as these always betoken Arab habitations—for the Arab and the date do not exist separately —I determined to halt, and we were soon bivouacked round a big fire. The wretched hamlet we had stumbled on is called Mahbeh, and we found it was supplied by a copious spring gushing from the rock. Slowly as we had worked our way down the mountain-side, we had outstripped the baggage camels, which toiled painfully after, and were not brought up by their drivers for some hours. About three in the afternoon we had passed, on our right hand at some distance, the lofty peak known as Jebel Shoum, which was pointed out to me by my companion as the highest summit of Jebel Akhdar, and which certainly seemed from this spot to overtop the rest, though not by much. I now looked on it for the first time, but it was soon lost to view behind nearer summits. Shaikh Nasir informed me that Jebel Shoum rose up very abruptly above the ravine called Wadi Hajar, opposite Wadi Ein, and in this case it must stand on the north-west side of the cluster.

On the following morning we entered and descended for some hours the Wadi Beni Ghafir, in which a fine stream of water flows continuously. Then striking out of this valley, we came upon a small walled town called Dahis, with a fortlet in the centre perched on a commanding eminence. Beyond this our path falls into the Wadi Sohtan, a pleasant and fertile glen which had remained for years a cause of strife between the Beni Ghafir and the Ibriyin tribes till a few months previous to my visit, when the quarrel culminated in an outbreak of hostilities, terminating after a long and bloody conflict in the rout and expulsion of its ancient and rightful occupants, the Ibriyin. The valley was still in a somewhat disturbed and unsettled state, as no man could be seen that was not fully armed and with lighted match, but we passed through the Beni Ghafir villages unchallenged and unopposed, though without the usual friendly demonstrations of welcome. I was told that the Wadi Sohtan joined the Wadi Rostak to the north-west of that town. After further winding among a labyrinth of low dark hills, we entered the outskirts of the city of Rostak, and, dismounting in the square space in front of the chief house, were cordially and frankly welcomed by Seyyid Hamood bin Azzan, and I was assigned a lodging in a handsome stone sablah just opposite Seyyid Hamood's house. But before taking possession of our quarters we all sat down, as in duty bound, in Seyyid Hamood's gateway to partake of coffee and exchange the news, I for one being very glad indeed to be off my camel and to get a rest after our tedious journey from Aridh; and while we were conversing up came Seyyid Saood, the second son of the late Imam Seyyid Azzan bin Kais, and of course every one present jumped up to receive him. He was soon seated amongst us, and his face plainly betokened the pleased surprise he felt at seeing so large a party of strangers in the gateway. Indeed, Seyyid Hamood and others had been equally astonished at seeing us, as they remarked that their latest news of our movements reported us to be still beyond the mountain range. Saood seemed to be a fair, bright, and handsome youth of about eighteen years, and of pleasing, dignified manners. He had, since his father's death in 1870, been with his elder brother Hamood, under the care of his paternal uncle Ibrahim, who usually resides at Rostak, but who happened to be at this moment absent at the fortress of El Hazam in the Batineh. I reckon Rostak to be the third largest town in Oman, Mascat being the first, and Semail the second. It holds fifteen villages in fief, and with these the population is estimated at 14,000. The bazaar is substantially and regularly built, and contains eighty shops with every description of merchandise for sale. In more than one of them I saw trays of fried locusts.

Rostak is a place of great salubrity, and is famous for its cool and temperate climate. Fanned as it is by the sea-breeze on one side, and sheltered from the scorching blasts of the great desert by the noble mountain which towers so majestically and precipitously on its south-west side, the air is dry and mild, and it would, like Nakhl, make an excellent sanatorium for Mascat. It owes much of its reputation, however, to its thermal springs, of which it possesses two very copious ones, ranking among the hottest in Oman. We passed them as we rode into the town, and saw that they were both enclosed in large circular tanks or reservoirs. They are much resorted to by the people, who have great faith in their medicinal virtues. The waters contain in solution a good deal of calcareous matter, and it is astonishing how many use them both internally and for ablution. Being close together, they probably derive their origin from the same source or fountain, and find their way through the fissures or dislocations of the rock.

I saw here vines, papayas, guavas, and mangoes, etc. Lucerne, wheat, millet, maize, and barley are grown, besides maseybili, lubia or Egyptian beans, bakili and other legumens, pulses, and vegetables of several different kinds.

Concerning the mango, I may here quote a curious observation of an old traveller, Mr. Parsons, who in his 'Travels in Asia,' p. 210, says, "This (August) is the season in which mangoes are ripe, which are so very excellent in their kind as to be preferred to any from India.

Rustaq. House of Sayyid Hamud bin 'Azzan, 1885 (Miles)

Rustaq. Castle, 1885 (Miles)

The stones of the Mascat mangoes are an acceptable present to those gentlemen in India who have gardens large enough to allow room for their growth. We bought a thousand mangoes for two rupees (five shillings), and endeavoured to preserve some of the largest to present to our friends in Bombay; but they would not keep sound during the few days we remained here. We picked out the largest of the stones, which Captain Farmer and myself divided between us."

How the Mascat mango came to be so much appreciated by the English at Bombay is a mystery, as it is rather coarse and stringy, and is far inferior to the Bombay variety, which was brought to perfection by the Portuguese, and is the best in existence. Mr. Parsons was a passenger in H.M.S. frigate *Seahorse*, which was lying in Mascat harbour in August, 1775, and one of whose officers, at that time a midshipman, was the great Horatio Nelson.

The general appearance of Rostak is superior to that of other Oman towns, and struck me as very fine and attractive, with its large white mansions, its venerable castle, its handsome though undecorated mosques, all embedded in magnificent plantations of shady mango, date, plantains, and orchards of fruit trees, backed by the frowning cliffs and crags of the grand cluster of Jebel Akhdar rising tier on tier above it. Traces of the great hurricane were everywhere visible in the town, and the upper story of Seyyid Hamood bin Azzan's house, a large and substantial stone structure, of which I took a photograph, bore evidences of the rough treatment it had experienced at the time.

The castle is a large and imposing structure said to be of great antiquity. It stands at some little distance from the town, but not, as is usually the case, on an elevated or commanding position. I gazed with considerable interest on this time-worn pile, which, though rather dilapidated, had the reputation of great strength among the Arabs, and is perhaps the oldest inhabited building in this part of Arabia. Within a large courtyard, the angles of which are strengthened with bastions, stands the keep or citadel, consisting of a body with four turrets. One of these, called the Burj al Rih or Wind tower, was built by one of the Yaarebeh princes, another turret was built by the Imam Ahmed bin Saeed when he established his residence here about 1750 A.D. The most ancient part of this stronghold, which bears a close resemblance to an old Anglo-Norman castle, is probably that part with its turret known as Burj Kesra ibn Shirwan, and the erection of which is ascribed to the period of the great Sassanian monarch Noushirwan, or, as some say, Khosru Parviz, in whose reign the famous expedition about 600 A.D. to expel the Abyssinians from the Yemen was undertaken.

The command of this expedition was entrusted to a Persian noble named Khuzrad Narsis, generally known by his title Wahraz, who, on his way down the Persian gulf from Obolla, detached a force of four thousand men, probably about one-fifth of his whole army, to invade and occupy Oman. The Persians landed at Sohar, and were so far successful that they were able to subdue the Batineh coast as far as Rostak, where they remained for about thirty years until the advent of the Moslem troops under Ikrima and Hodhaifa compelled them to evacuate the country.

The erection of the castle is due, according to Arab tradition, to the governor left in Oman by Wahraz. Its appearance is imposing and its structure substantial, and with successive repairs has remained for many centuries. Castles or fortlets similar to that at Rostak, of which there are about twenty in Oman, form the military and political centre of a province, and were built to overawe as well as protect the surrounding country. In domestic and intestine wars, whether tribal or dynastic, as well as against foreign aggression, they have played a very important part, and many a long and vigorous siege has become memorable in the annals of the country. The castle had generally a walled enclosure with or without mural towers, and the plan of the work was, of course, greatly influenced and affected by the disposition of the ground. Flanking towers or bastions are sometimes found, but the secondary defences are seldom strong. The concentric system is said to be of Arab origin, and was used in England after the Crusades.

The appellation of Rostak signified a market town, and we conjecture that the Persians were attracted to it, not only by the hot springs, but also by its central position as a meeting-place or market for the Arabs. The headquarters of the Persian Government were, according to Ross's 'Annals of Oman,' near the great port of Sohar, but it is locally traditioned that this castle of Rostak was ordered to be built by the king, who caused an iron chain to be suspended therein connected in some mysterious way with his palace at Ctesiphon. Any of his Arab subjects who had reason to complain of injustice or oppression were at liberty to come to the fort and shake the chain, the immediate consequence of which was investigation and severe punishment of the offender by Noushirwan.

Rostak has always been a place of political importance, and it was for a long time the capital of Oman, but it was in the seventeenth century, in the days of the Yaarebeh sultans, that it reached its climax of power. The founder of the Al Boo Saeedi dynasty, which succeeded, retained it as the capital until the end of the eighteenth century, when Mascat took its place. Since then it has been the appanage of the Kais branch of the family, and is now in the possession of Seyyid Ibrahim bin Kais, who has made more than one attempt to imitate the example of his brother Azzan and wrest the sovereignty of Oman from the present branch. At Adam the house in which the Imam Ahmed bin Said was born had been shown me, and here at Rostak, where he died in 1783 after a reign of over forty years, I saw his tomb. It bore a long inscription, of which I obtained a transcript, in which the date of the Imam's death is given, according to the Mohammedan calendar, as 19th of Muharram 1198 H.

Though the wadi in which the town stands is here called Wadi Rostak, it is known as the Wadi Auf higher up, from a small tribe of that name occupying the ravine and claiming to be originally from the Hejaz. Lower down, I was told, it is called the Wadi Fareh. Rostak is 800 feet above the sea. From this direction Jebel Akhdar appeared to be a tabular range without any salient peaks. The bluff point marked on the chart, and the altitude of which is given at 9900 feet, cannot be the same as Jebel Shoum, as the latter is said to be invisible from the sea. It has an angle of 270 from Rostak. The formation of the rock-masses here is very similar to that of Nakhl, and is no doubt metamorphic, the wavy appearance of the sedimentary rocks exhibiting the power of lateral pressure or of the oscillation of the Earth's crust during the time it was in a fluid condition. The hot springs, it may be added, are without taste or odour, but leave a slight deposit. The cold springs, which are also abundant and which help to supply the town and irrigation conduits, are of the temperature of the ground, and have a different origin from the thermal ones.

Rostak is distant about 90 miles from Mascat, and as the time I had allowed myself was getting short, I determined to ride there in two days, as the camels were still fresh and in good spirits. On the morning of departure, therefore, after taking some photos of the castle, we started off; but as none of us were acquainted with the road, we travelled for a considerable distance on the way to Sowaik, until we were put right by some villagers. At 11 a.m. we reached Mansoora, and soon after passed a small square town on a hill. At 1 p.m. we arrived at Jemma, a large village with a tower and castle, the latter standing on an eminence. Here we stopped an hour for rest and refreshment, while I seized the opportunity to look at the castle, which has an historical interest of its own. It was in 1807 that Seyyid Bedr bin Saif, who had usurped power with the aid of the Wahabees on the death of Seyyid Sultan three years before, was at Jemma with the two sons of Sultan—Salim and Saeed. Instigated and encouraged by others, Saeed took advantage of an unguarded moment to stab Seyyid Bedr, who, being alone and unsupported by his dependents, leaped from the window and tried to escape, but was pursued and killed. Seyyid Saeed rode at once to Mascat and seized the throne, which he retained till his death in 1856.

The ground from Rostak to Barkah, which we reached at sunset, has a gentle declivity, and our camels sped over the sandy soil at a rapid pace, being conscious, no doubt, that they were homeward bound. The distance is about 45 miles. Mascat, which is about the same distance, was reached the next evening, December 31.

Sulaif and 'Ibri

Once back on the 'Ibri road, seven km from the Jabrin turn the lunar landscapes rose like piles of gray-brown slag, though patches of scrub revealed the presence of at least some water, and scattered camels, grazing apart. Farther on the light played tricks on mountain colours: burnt sienna, white, oil-grease kaleidoscopes miraging once and then vanishing for good. Thirty km west of the Jabrin turning we entered Wilayat 'Ibri, and in another 10 km passed the sign to 'Amla 14 km away; in 15 km more at Kubarah the sign to al-'Amm 27 km off right, and then in 9 km the landscape flattened out on both sides, while in 27 km we hove in sight of Sulaif, just visible off left, on the other bank of Wadi al-'Ain near its confluence with Wadi Sunaisal, but merging in colour and height with the jabal outcrop. The village outside the walls has survived the abandonment of old Sulaif and its fort; the suq is shuttered and silent, but mercifully not yet demolished to make way for cement and plastic, neon and formica. Yes, there is Khairuddin Supermarket (oddly named, for nobody outside Oman would accept that designation) and as-Salam Restaurant, but the fish in the *falaj* nip and nibble as their ancestors did when the *falaj* was first dug, and a wiry, anxious tabby cat out of Thurber's angular cartoons tiptoed as if ready for instant flight: the primeval feline, barely domesticated even now. The *Handbook of Arabia* in 1916 found about 350 houses in Sulaif, but subsequent decades have seen a steady drift towards 'Ibri, parallel with the urbanward drift noted thoughout the whole world during the same period from Tokyo to New York, and Sydney to Mexico City.

The women of Sulaif, a village inhabited by Manadhirah and Suwawifah, went unveiled.

'Ibri lies three km off the main road, from a turning near the new office of the Ministry of Trade and Commerce. Past the Municipality (marked in Arabic 'Baladiyat 'Ibri'), we took a sign to 'Town Centre'. As you switch attention from one side of the road to another, marvelling at the swelling spate of new buildings, you might miss a sudden unmarked turn off right along a dirt road, and now beware those 'sleeping policemen', or humps in the road designed either to slow down your vehicle or bounce you up to hit your head on the car-roof, according to choice. The *suq* was vibrant and pulsating at nine a.m. and we strolled along the old walls, admiring a huge fish held up by a conceited youth, and passing 'Photostat', Hotel Rajadhani', 'Colour Studio, 'Hamad H. Balushi Textile Division,' a *vecchietto* arguing with a goat, an old lady lolling against a lorry, with a golden ornament through her nose. Goats were tethered to telegraph poles awaiting their lead-part in the auction. I mingled with the crowd, 'Fifty-two rial', 'Fifty-two rial', 'Forty-five rial', 'Forty-five rial', and the circuit wavered slightly as a shrewd owner bent his ear down to an offer from the crowd, crouching at the front, standing at the rear, and sauntering carelessly around the arena, alert as a gazelle. A van emptied sheep across the road, and willing hands secured them. I

'Ibri. Goat auction

peeled orange after orange, sucking up liquid in defence against the sun.

Old men wearing *khanajir* exchanged kindly greetings with me outside the Bank of Credit and Commerce International (Overseas) Ltd., and as I smiled back and offered the usual replies I recollected Wellsted's fears in 'Ibri (he spells it 'Obri' which approximates the local *'ain* sound more closely than does the usual transliteration). 'I had often before heard of the inhospitable character of the inhabitants of this place,' writes Wellsted, whose interpreter was a Shi'a Persian six feet tall and stout in proportion, an odd choice bearing in mind that the Wahhabi were then in control of 'Ibri, and that – as Wellsted by now realised – 'the Wahhabis have a more thorough detestation of the Persians, as sectaries of 'Ali, than any other class of Mussulmans.' Wellsted noted, 'The neighbouring Arabs observe that to enter Obri a man must either go armed to the teeth, or as a beggar with a cloth, and that not of decent quality, round his waist'.

Times have changed, of course, and pacific 'Ibri does not affect the modern visitor with such premeditated alarm, or sudden bouts of xenophobia. Traders will offer you in the best possible humour woollen camel rugs and camel bags, and you can see the old traditional craft of indigo-dyeing, or sample the usual great range of spices laid out so temptingly.

Dariz and Bat

I took the main road out of 'Ibri to Dariz, so sleepy after the animation of the goat-market, and then a track behind the village into the Wadi al-Hijr, in search of Bat. It was in 1973 that a Danish archaeological team under Dr Karen Frifelt and Professor P. V. Glob first carried out investigations and excavations, and Dr Frifelt's important survey, 'Evidence of a third millennium B.C. town in Oman', appeared in *The Journal of Oman Studies* (Vol. 2, 1976). A settlement and a large necropolis found at Bat have been dated to the third millennium B.C., and related to roughly contemporary sites at 'Amlah in Wadi al-'Ain, and Tawi Silaim, as well as to the Umm an-Nar culture on Umm an-Nar itself and at Hili, Buraimi. The stone-built tombs developed from the beehive type to the more and elaborate and spacious Umm an-Nar type, and about a hundred were explored in the first season, within the necropolis proper. The Danish team returned for further work in 1975–6, and concluded that the round towers 'can hardly have been the common ordinary dwelling places of the villagers or townspeople of that time'. Instead they suggest analogy with the Nizwa tower, until very recently the Wali's residence. 'Possibly', writes Dr Frifelt, 'something like that was the case with the prehistoric towers. But it is also conceivable that they were in some way connected with the water supply or irrigation.'

It is worth walking the extra kilometre to the latest settlement of Bat, if only to savour the wild wind of history, chaining mankind to those valleys and hills where the presence of water secures a livelihood.

Il Vento Sfrenato	*The Wild Wind Unreined*
Il vento sfrenato,	The wild wind unreined,
impalpabile,	impalpable,
sorgendo e svaniscendo,	soaring and vanishing,
che ci fa schiavi,	the enslaver,
come signor invisibile.	like an invisible lord.

But was my intutition about the essential continuity of Bat really true? Karen Frifelt thinks it is: 'Though the wadis run dry most of the time, certainly in summer and also during winters with scanty rainfall, there is a *falaj* system in working order besides older systems in disuse, and there are several wells in the area. Maurizio Tosi has pointed out that the protohistoric farm economy of these foothills included cereals and pulses, requiring more water than do dates.

On my return to Dariz, I was greeted by Shaikh Saif ibn 'Ali ibn Said ibn Muhammad al-Ghafiri. A funeral feast was in progress around the *sablah*, and I was offered halwa with almonds, oranges, bananas, and apples. A *falaj* ran through the old *husn*, now no longer needed, of course, for fortification, its round tower now only a home for pigeons. Shaikh Saif, waving me off when politeness permitted farewell, told me to visit Shaikh Muhammad ibn 'Ali in al-'Ainain, but with only five hours of daylight left for Dhank, Yanqul, and the mountain road down Wadi Hawasinah to al-Khaburah, I spent only enough time in al-'Ainain ('Two Springs') to tread the drifting sands around the main castle, and a smaller one nearby. Al-'Ainain may be a living village, but as in so many Omani oases, the pendulum of time swings slowly, and the heat of midday spreads lethargy before it, and after it deep into the afternoon, until the first wisp of breeze flutters a flag.

Dhank and Yanqul

At the crossroads I took the road to Dhank, by the new mosque, where the sign reads 310 km to Matrah and 135 to Nizwa. This main road towards Buraimi runs parallel to Jabal Dhank, except that it fades mistily northward as you drive. Southward all is flat, one wadi sloping almost imperceptibly into the next, scattered scrub on all sides affording mean pasture for the few animals wandering in search of nutriment. A castle rears left of the road at al-Subaikhi, 25 km from the crossroads, then in a further 10 km the sign shows al-Mazim (al-Mazan on the map), another 5 km to al-Ma'an to the right, another 3 km to al-Aflaj to the right, and another 2 km to al-Yafif to the right: a string of oases nestling for watery safety below the mountains to benefit from run-off in winter, ending at Dhank, 12 km off the main road, south of Jabal Abyadh ('White Mountain').

Cox's Excursions

The first European to travel, and describe, the way (such as it was) between Ras al-Khaimah and Dhank, was Major-General Sir Percy Cox, and his account of 'Some Excursions in Oman' is here reprinted by courtesy of the Editor of *The Geographical Journal*. Published in 1925, the article in fact refers to journeys made a quarter of a century earlier, Cox having been Political Agent and Honorary Consul at Muscat from 1899 to 1904. His portrait, reproduced by courtesy of H.B.M. Ambassador, Muscat, hangs on the staircase of the British Embassy there.

SOME EXCURSIONS IN OMAN
Major-Gen. Sir Percy Cox, G.C.M.G., G.C.I.E., K.C.S.I.

W HEN recently invited to read a paper describing some short
journeys made a good many years ago in the Hinterland of Oman
I had great hesitation in consenting, as I doubted whether my travels,
which hardly amounted to more than excursions, would be found to
provide sufficient material, from a geographical point of view, to justify the
reading of a paper before this Society.   I rather felt that what was new
would not be very interesting, and what would be interesting was not new.
However, after going through my notes and photographs taken at the
time, I have ventured to respond, and trust you will find that what the
result lacks in geographical importance is to some extent made good
in human interest.

I beg here to express my obligations to the Secretary of State for
India for the facilities kindly afforded me to see and draw upon the
reports and photographs which I submitted officially at the time.

The two short journeys with which I am going to deal took me,
first, from Abu Dhabi to Baraimi and thence onwards along the fringe
of Oman on the desert side to the Jabal Akhdhar or Green Mountains
and home to Muscat by the Wadi Samail, a total distance of about
400 miles ; secondly, from Ras al Khaima to Baraimi, and thence direct
to the coast at Shinas and on by the seashore to Sohar.

The section from Abu Dhabi to Baraimi had been previously traversed
in 1901 by the Rev. Samuel Zwemer, a missionary of the Dutch Reformed
Church then working at Bahrein, and that from Baraimi to Sohar also
by Mr. Zwemer on the same expedition, and by Colonel Miles travelling
in the opposite direction in 1875.   Both travellers related their experiences
in the *Journal* of this Society, and I shall have occasion to allude to
points in them later on.

As regards the country further south, Lieut. Wellsted, in 1835, having
rested for a few days at Burka on his return from the Jabal Akhdhar,
started afresh from the Batina coast intending to make for Baraimi,

then in Wahabi occupation. From thence he hoped, somewhat sanguinely, I think, to be passed on, under Wahabi auspices, to their capital in Nejd. On reaching Ibri, however, he found the town full of Wahabis and somewhat in a state of ferment, and the local Wali on behalf of the Muscat Government being unable either to further his onward plans or to protect him if he remained, he had no alternative but to retrace his steps, shaking the dust of Ibri off his feet in the face of a somewhat hostile demonstration from the inhabitants.

Just half a century later, in 1885, Col. Miles, coming round from Muscat by the Wadi Samail and Nizwa, passed through Ibri to Dhank, but the time at his disposal did not admit of his proceeding further north and he was obliged to bend his steps homewards from that point, making for the coast at Barka *via* Rostaq. Save that one of his followers who lagged behind was waylaid and robbed just outside the town, Col. Miles, travelling undisguisedly as British Representative at Muscat, and possessing a competent knowledge of Arabic, fared better at Ibri than his predecessor had done, but even he had reason to complain of the surly attitude of the people of Mazum a few miles further on.

From the foregoing review of previous travel it will be seen that the section between Ras al Khaima and Dhank was the only portion of my route not previously traversed by a European. That being so, I may in general spare you details incidental to all caravan travel in desert or semi-desert tracts, and merely touch upon such points of interest as my notes or photographs recall.

The journeys in question having been undertaken from my headquarters at Muscat, while I was serving there as Political Agent of the Government of India, I should like to transport you to the Sultan's capital for a few minutes in order to give you some local colour before we embark for the provinces.

The harbour and approaches to Muscat from the sea form an extraordinarily attractive picture, and for me possess a charm entirely their own. The first indication a traveller gets of the place as the steamer comes up the coast from the southward is a glimpse of the old Portuguese fort at the south-eastern corner of the harbour, now known as Fort Jalali. Then passing on towards the eastern arm of the harbour, or cove, as it really is, and rounding the island promontory you see on your port bow the barren rocks bearing the time-honoured record of ships which visited the port of old and painted their names on the barren rocks. Then you have before you the front and entrance of the same old fort, Jalali, with its battery of antiquated cannon directed from their embrasures at random across the harbour, while the closely packed town lying at the toe of the horseshoe opens out to view as the ship passes in between the points.

The picture now before us includes the Sultan's custom house and palace on one side with the British Agency at the eastern end of the foreshore, its foundations forming the sea wall at that point. It will be realized that when, as often happens in winter, a strong breeze from the north drives breakers into the harbour, their continual thud against the walls of the building is somewhat trying, but as in the case of most chronic noises one gets used to it after a bit. At the other or western extremity of the foreshore lies Fort Merani, the other of the two old Portuguese forts built to command the sea approaches. Though the older of the two, it is neither so solid nor so picturesque as Fort Jalali, and the old cannon located on an emplacement halfway down the rock are now relegated to saluting purposes. The fort however contains an interesting stone tablet let into the masonry recording the building of the structure in the eighth year of the reign of Philip I., King of Portugal, by the command of the Viceroy of the Portuguese Indies, and its completion by its founder and first captain, Belchior Calaça, in 1588.

A full translation of the inscription, from which the above summary is taken, is given by Captain Arthur Stiffe, R.I.M., in his "Ancient Trading Centres of the Persian Gulf," in the *Journal* of this Society.

Any one who has the energy to climb to the top of the rugged hills at the back of the town is rewarded with a fine view of the harbour and the town within the walls, and will notice how every foot of available space seems to have been taken advantage of.

Apart from the coming and going of men-of-war and merchant vessels, which were perhaps the main factor in the life of the place in my day, the harbour was a never-failing source of interest in one way or another. In the hot weather it was always visited by small parties of fishermen from the island of Socotra, who camped about in the neighbouring coves and made a living by fishing from their catamarans, and diving for jetsam in the harbour. Civilized fishing-rods were among the articles they periodically recovered from the bottom—carried away, I always supposed, by big fish, when the owners, fishing from the decks of ships in harbour, had fallen asleep or left their rods hitched to a stanchion. These Socotrans used a large rectangular fishing-basket which they anchored to the bottom with heavy stones, and employed for trapping red and grey mullet, apparently the only fish confiding enough to patronize them.

Up to quite recent years the dwindling survivors of a fine old type of clipper sailing ship used to ply from Muscat and other parts of the Persian Gulf to Calcutta and even to China. There were about a dozen of them left when I first went to the Gulf, but one by one they have all now disappeared. Within my own ken, one of the last foundered on the voyage to Calcutta, while another sank in the fairway in Muscat harbour and had to be cleared with dynamite. The Arab captain of one of them claimed once to have made the voyage from Muscat to Calcutta in seventeen days, with a strong wind behind him. This would give an average of about 7 knots.

Let us now walk through the town past the Sultan's palace and the shops of the Indian traders—Hindu Bunnias who do most of the business in foodstuffs; then out beyond the town walls by the Tuyan gate and into the valley beyond wherein lie the sweet waters of Muscat. Just outside the gate we pass a little establishment of the greatest importance to the housewives of Muscat—the retail oil and fuel shop. In such barren surroundings as exist, firewood is inevitably a very scarce and expensive necessary of life, and is sold almost by the stick, and it is surprising to see how few sticks suffice for the modest daily needs of a Muscat kitchen. A couple of hundred yards further on we come to the small oasis created around the sweet-water wells, which even in times of drought never seem to fail. As one regards the town from the harbour it appears set so close against its background of burnt-up rock that one can hardly believe that it boasts such a verdant spot as this tucked away behind it. On the Muhammadan Eeds or festival days most of the inhabitants disport themselves outside the town, in this and other small oases. From the wells in the wadi bed, over which a small watch tower keeps guard, the populace draw their water by hand and bucket, but in the tiny vegetable gardens clustered round the irrigation is done by bullocks toiling up and down a steep slope and filling a large leather skin at each lap. There are many such wells at Muscat, and the bucket wheel of each develops a special wheeze of its own which the owner cultivates and recognizes, so that when not in his garden he may know by the familiar sound that his employé at work on the well is not slacking. When the wells are in operation the chorus of wheezes is phenomenal, but one almost gets to like it—in the distance.

It will at once strike the visitor to Muscat that the population contains a large negro element. The reason is that in bygone days Oman was probably the most extensive market existing for the slave traffic by dhow from the African coast, and one of the most creditable achievements of Great Britain in those waters has been its persevering suppression. A few escaped slaves are still freed annually by H.M. ships or at British Consulates, but the numbers are gradually decreasing.

According to the International Slave Trade Convention, slaves taking refuge under the flags of the Signatory Powers, whether ashore or afloat, become automatically free; and though nowadays other flags are no longer in evidence, there were at one time a good many slaves in Muscat who had been released by French as well as by British men-of-war in the Indian Ocean. The result was that during the period of somewhat acute rivalry which prevailed between Great Britain and France in the few years preceding the *Entente*, these communities of freed slaves were most staunch supporters of the flag to which they owed their freedom, and the attitude of cordiality or the reverse prevailing at the time between

Muscat. Fort Jalali, 1900 (Cox)

Bani Habib village, 1900 (Cox)

the British and French representatives at Muscat was speedily reflected in the slave communities, and on special occasions free fights among them were not unheard of. Altogether the ex-slave communities were a merry crowd, and a frequent source of interest. For instance, the British freed had their own band, which on state occasions, such as the King's birthday, would come and discourse sweet music at the British Agency ; while, on the occasion of a memorable tour to Muscat and the Persian Gulf by the Viceroy of India, the late Lord Curzon, a number of them appeared on the sea front as a sort of boat carnival, dressed up to represent boats of various patterns, including one displaying a picture of Queen Victoria on its side.

I have explained to you how rock-bound the town of Muscat is. In consequence all caravan traffic with the interior has to be conducted through Matrah, the commercial port of Muscat, lying at the head of a more extensive but less sheltered bay 3 miles to the north.

So much for the capital of Oman. Its picturesque surroundings naturally lend themselves generously to the water-colour artist and the photographer. Unfortunately, the camera takes no account of the temperature, which for some months of the year is almost insupportable. But Muscat had its compensations in other directions, and my wife and I spent nearly five interesting years there.

When the Sultan heard of my lively ambition to make an expedition from Abu Dhabi to Muscat, he entered heartily into the proposal, and said he would be glad to seize the opportunity of coming with me as far as Abu Dhabi, as he had several matters connected with tribal politics which needed discussion with Shaikh Zaid bin Khalifa, at that time the grand old man of the Pirate Coast, on whose friendly offices I depended for safe passage for nearly half the journey. I was of course only too willing, and we proceeded there in company : he and his retainers in his steam yacht the *Nur al Bahr*, and I in H.M.S. *Redbreast*, one of the small gunboats of the East India Squadron.

As we steam up the coast and round the promontory of Oman towards the Trucial Coast I will touch on one or two points of interest which we pass *en route*.

The hinterland between Muscat and Khor Fakkan consists of a fertile maritime plain lying below the foothills of the main range, averaging about 20 miles in width. Along the shores of it lie a goodly number of fishing villages, and a few small townships of considerable local importance, situated as a rule near the mouth of a wadi descending from the watershed above, and so providing water for irrigation. At several of these ports, generally where a wadi coincides with an important route from the interior, are to be found small communities of British Indians : partly Hindu Bunnias, and partly Muhammadans of the Khoja persuasion, descendants of immigrants from Sind and disciples of H.H. the Agha Khan or recent seceders from his faith. These communities it was the

Entrance to Elphinstone Inlet, Musandam, 1900 (Cox)

Idaiyah, NW of Jabal Samaini, 1900 (Cox)

Masaikin Plain, N of Buraimi, 1900 (Cox)

duty of the British representative to look after and occasionally visit. Unfortunately, most of this coast is liable to heavy surf when the wind is off the sea, and landing from boats is then very difficult and often impossible. It was consequently always my habit, on my visits to these ports, if flat calm did not prevail, to wear a bathing suit under my clothes ; and I have several times, when my boat was unable to approach the shore, had to wade and swim in with my clothes on my head. On one or two of these visits I have reached the beach in this undignified garb under a ceremonial salute of two or three guns from the local Arab Governor. These old rusted iron guns, generally lying on the ground without carriages, are to be seen in twos and threes in most of the small ports of the Persian Gulf—relics, I suppose, of Portuguese days. After setting a match to the touch-hole, the gunners have to be pretty nimble to get clear of the recoil, and I have known of several accidents resulting from a loader inserting a new charge in his excess of zeal while the remains of the previous one were still smouldering.

Whenever possible I would give a few days' warning of my coming, so that neighbouring Arabs having cases to be settled in their dealings with the Indians might have time to make rendezvous with me.

The most important of the Batina coast ports is Sohar, the residence of the Sultan's provincial Governor, and the headquarters of an administrative Division. It has a distinguished history in the annals of Oman, and it was there that the present Al Bu Sa'id dynasty of sultans originated.

An important seaside industry at many of these ports is the preparation of shark flesh for transport to the interior. It is dried in the sun into a sort of biltong, and is largely imported inland, where it is a staple article of food. I have sampled it once or twice when my men were eating it and found it very tasteless, and exceedingly tough and stringy ; at the same time, as the natives seem to like it, it is a very useful and portable form of provender for one's following.

As we near Khor Fakkan the maritime plain comes to an end, and we approach the point where the hills come down to the sea and begin to form the rugged and ironbound promontory of Ruus al Jibal, or "mountain headlands," ending in Ras Musandam, or "Anvil Head."

Near the end of this promontory are two deep bays, one on either side of it. Locally, of course, they have their own Arabic names, but on our charts, and among Englishmen, they are known as the "Malcolm" and the "Elphinstone" inlet respectively. They are very much alike the one to the other, but the "Elphinstone" Inlet, which wanders inland for nearly 10 miles, is of the more interest to us, as it contains the ruin of a British cable station which was located there in the first half of the nineteenth century. The scenery in the inlet reminds one greatly of a Norwegian fjord, and we may explore it in passing. Far inside is to be seen the long-deserted telegraph building. The Indo-European

Telegraph Department to which it belonged found they could not keep their employés alive, much less in health, owing to the intolerable heat of the summer aggravated by the solitary conditions of life, and the station accordingly had to be transferred to the island of Hanjam, where we now have an important cable and wireless station. This part of the promontory is occupied by small hamlets and fishing villages belonging to a strange tribe called the " Shihuh " (some of the Ichthyophagoi of Ptolemy), whose origin has always been a subject of doubt and surmise. Individuals of the tribe whom I tackled on the subject stoutly repudiated, as they naturally would, the suggestion that they were a mixed race, but such evidence as there is, whether of type or dialect, certainly points to an admixture of the Arab with the Persian and Baluch. Be that as it may, they are a very simple and unsophisticated community eking out a precarious livelihood and subsisting on fish, with a little rice, and in the more isolated hamlets, particularly on a shellfish of the genus Strombus, which is found in great numbers in all their coves. As you tour the inlet an occasional boat ventures to put out from some little hamlet on the shore, to offer wood or fish for sale. Their habitations in these places are like tiny crofters' huts formed of loose blocks of stone, and not high enough to stand upright in.

Leaving the inlet and passing on our course south-west along the Trucial Coast, we touch the fringe of the Great Pearl Bank. It is in this deep arm of the Gulf that most of the great pearling industry is carried on, sending to civilization annually pearls to the market value of nearly 3 millions sterling. The pearl fishery is practically the whole life of the inhabitants of the western shores of the Gulf, of whom upwards of 20,000 are employed in it ; and one of the most important tasks which we have shouldered, as being inseparable from our position of predominance and control in these waters, has been the resolute preservation for them of this industry ; firstly, against piracy, and secondly against its invasion by instruments of modern science, such as diving dresses and steam dredgers. The fishery is carried on in two categories : the deep-sea fishing (speaking relatively) and the inshore fishing ; the one worked by moderate-sized sailing vessels on the deeper banks, often many miles from the mainland, and the other in rowing boats, a few miles from the shore. The former carry a fortnight or three weeks' provisions and remain out till it is time to replenish, unless driven in before by stress of weather. For the inshore fishing the boats come out from their home port for the day only and return with their catch at dusk. On the course from Ras Musandam to Abu Dhabi, if passing at the right season, one falls in with both categories.

Their methods and equipment are extremely primitive ; the latter consisting merely of a horn nose-clip, leather finger-stalls, a basket, a knife, and a communication cord. The divers help themselves to sink with a heavy stone, which is hauled up as soon as they reach bottom.

We are now passing the Trucial Coast ports, of which Ras al Khaima, Sharja, Dibai, and Abu Dhabi are the chief. We are concerned at the moment with Abu Dhabi only. On arrival there, the Sultan and I landed from the vessels which had respectively conveyed us, the Sultan staying with the Shaikh, while the Arab guest-house was placed at my disposal. The simultaneous descent upon them of the Sultan of Muscat and the British representative was such an unusual event that the equilibrium of the community was considerably disturbed, and anxious as I was to get under way, I realized that there was no business to be done in the way of getting a caravan together until the Sultan had departed and things had simmered down. Meanwhile His Highness promised to explain to the Shaikh what I had come for, and arrange with him at what point on my route the Khafirs which the Shaikh at Abu Dhabi would send with me to the confines of his sphere of influence, would hand me over to the Sultan's Khafirs coming from the Muscat direction on His Highness's behalf. The latter only remained one night ; the following morning the Shaikh of Abu Dhabi gave H.H. and myself a display of some of his blood stock at a sort of gymkhana on the sands behind the town, at which I managed to get one or two pictures.

In the evening the Sultan departed, and I then began to get busy with my plans.

The relative position of the Principalities of the Trucial Coast, *inter se,* depends almost entirely on the personalities of the ruling Shaikhs for the time being. Unfortunately, owing to the chronic prevalence both of family and tribal feuds, they have for the last few generations been a short-lived class : nevertheless, at the period of my journey the Shaikh of Abu Dhabi, then over eighty years of age, had successfully retained his seat for more than half a century, and consequently enjoyed a very commanding position among his fellows. His practical influence extended not only to Baraimi, where he owned large date gardens, but also far beyond. I had met him before and knew that I might depend upon his friendliness in general, but it seemed inevitable that in the first instance, in the natural wish to save himself any inconvenient responsibility in regard to me, he should endeavour to deter me from my journey on various grounds, *e.g.* excessive heat, dearth of camels, absence of water and grazing, etc., and it required two or three long interviews and much persuasion to induce him to further my plans. However, at last he not only consented but insisted that I should travel as his guest as far as Ibri, after which I was to become the responsibility of the Ruler of Muscat. Before we parted for the night he had promised to do his utmost to obtain camels and arrange for me to start the following day.

The next morning, while the camels were being collected, I walked about the town with the old Shaikh and his sons, and was conducted through the bazaar, which seemed a very good one for such a benighted spot, containing almost everything needful in a small way from a note-

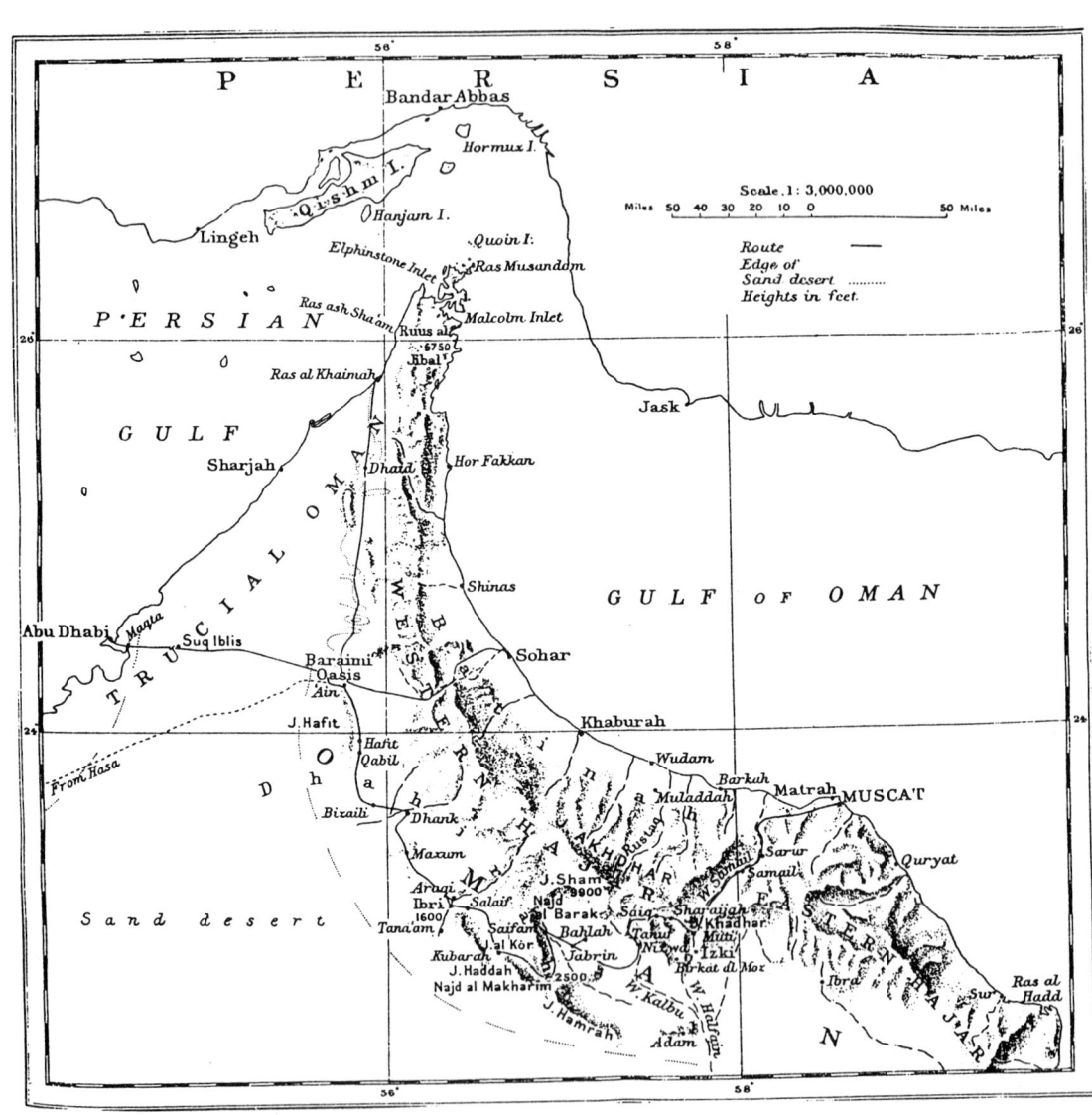

Map of Northern Oman, 1900 (Cox)

book to a rat-trap. The daily life of the Shaikh provided a very interesting example of the patriarchal system. It will be understood that a man in his position is frequently obliged by the exigencies of tribal politics to make sudden matrimonial alliances which must sometimes place considerable strain on the strict interpretation of the Koran ; and here in Abu Dhabi this venerable Benedict had several different establishments dotted about the town between which he divided his attentions. It was his habit, in the hot season at all events, to conduct the administrative business of his principality and dispense justice sitting out in the open in the early mornings outside the particular *ménage* forming his domicile for the time being.

I found the coffee habit on these occasions very trying. The coffee-pot and sweetmeat-tray were produced on every possible excuse, and small cups of the strongest black coffee spiced with a few cardamum seeds administered continually. It was just the same when travelling : my caravan never seemed able to go a couple of hours together without halting for a few minutes to make coffee, yet they would readily go without solid food from morning till night, when it suited them.

About 2 p.m. my camels began to arrive, but it was 5 p.m. before we got away. Shaikh Zaid accompanied me for a short distance and sent his son Sagar with me as far as Maqta, the ford across the tidal creek 10 miles from the town. Sagar was a handsome and intelligent young man, thirsting for knowledge of the great world, and, apart from some small grievances which his father had instructed him to ventilate, the chief topics of our conversation were the development of wireless and the idea, recently mooted by the late Rev. J. H. Bacon, the aeronaut, of crossing the Great Arabian Desert by balloon. I had seen mention of it a few days before. Sagar left me about 8.30 p.m. on his own side of the creek, and I just managed to get across with the water up to my girths, guided by the old caretaker of the tower which stands in mid-stream to guard the ford.

I see that Dr. Zwemer, referring to this ford, says that "ordinary camels with their riders have a close escape from drowning every time they cross " (I suppose from the current), but I was blissfully ignorant of the danger when I crossed. I lay down as I was on the further side of the creek and slept till 1.30 a.m., finding it very cold owing to the heavy dew, though the season was about the hottest in the year. Then we started over the sand-dunes, passing the landmark called by Zwemer "Suq Iblis," or " The Devil's Market," about 7 a.m. My companions did not know it by that name: they called it "Narsáila." The question of topographical names among Arabs is always a difficult one ; it often happens that different communities know the same features by different names. As the caravan's track passed 200 or 300 yards from the sandstone outcrop, I turned aside to examine it closer and see if there were any birds or other live things on it, and found lying at the base of it at a spot where I suppose he had found a little shade, the fresh skeleton of a negro, with some tufts of woolly hair still adhering to the skull. I inquired from my companions if there was any particular story attached to these poor bones, and they informed me that it was the skeleton of a young slave who had recently fled from Abu Dhabi and must have died of thirst. I thought to myself, " A slave in Abu Dhabi is verily between the Devil and the deep sea." The Shaikh acting as my Khafir or pilot and responsible to the Shaikh of Abu Dhabi for my safe passage was one Muhammad Bin Hilal adh-Dhahiri, the Dhawahir being a tribe whose habitat is in and around the Baraimi Oasis. I seemed to know his face, and on comparing notes I found I had met him some years before at Sohar, when I was there on a visit with the object of recovering some ingots of silver which had been pirated from a wrecked steamer and had been traced to Sohar.

To-day for the first time I heard " the music of the sands " as the caravan toiled up and down the sand-dunes : it was like the booming of a great organ in the distance, and, to one hearing it for the first time, an extraordinary phenomenon. At nightfall we bivouacked in the sand-dunes, and the following day reached some wells known as Huwail, close to which was a single date-mat shelter occupied by an aged Badoo of the Bani Yas tribe, who made his livelihood by helping to water camels. He possessed a fine she-camel and foal, and on our arrival apparently announced his intention of killing the foal, as he said, " in honour of Shaikh Zaid's friend." I was having my tub when I got the message, and shouted out immediately that I did not want the meat and that the foal was not to be killed, but it was too late. The old man had stabbed it in the chest and cut its throat before I could stop him, and the wretched mother moaned restlessly for her offspring for the rest of the day. My followers were no doubt inwardly delighted, as they got a sumptuous meal while I had to make generous amends to the old man.

During these two days I saw frequent tracks of the Reem Gazelle (*Gazella Marica*), and also a small herd of Muscat Gazelle (*Muscatensis*) which passed a few hundred yards ahead of us. On the evening of May 4 we reached the outskirts of the Baraimi Oasis, the last two hours of the march being through very heavy sand-dunes. All the afternoon Jabal Hafit had been in evidence on our right front—apparently a perfectly barren mountain. On descending from the sand-dunes and just before reaching my pilot's summer camp, we crossed a Batha or scour bed about 300 yards broad, which my companions simply called " the Batha," and stated that it had its origin in the Bani Kaab hills and reached the coast just west of Abu Dhabi. It is probably a confluent of the Wadi Telaif referred to by Col. Miles in that connection. Bin Hilal stated that it periodically became a running stream.

We were making for the latter's summer camp, and as we approached his family tents it was a very pretty sight to see the affectionate delight with which the two children of his younger brother ran out to greet him and their father. Until they saw our party approaching in the distance they had had no warning of our coming, and now, racing over the sands towards us, they fell rapturously on their father's neck and hugged him—two pretty and well-bred little Arab children. On arrival I was at once deposited in the family guest tent, of very coarse black-and-white blanketing, and there our party sat awaiting coffee while one person after another dropped in to welcome us. It was interesting here to note the various forms of greeting in vogue. The slaves and menials kissed our hands; both youths and adults kissed Bin Hilal on the lips, just touching them; others rubbed noses, a practice I had never seen on the Muscat side of Oman—probably imported, like the slaves, from Africa.

The camels which had brought us had only contracted to come as far as Baraimi, and another lot was supposed to have been ordered to meet us there, but Bin Hilal gave me no hope of getting any before the 7th, this being the 4th.

My journey from Abu Dhabi had taken fully forty hours of actual travelling, whereas Dr. Zwemer records that he did it in thirty-three, or that thirty-three is the ordinary time for a caravan. Although the position of Baraimi, afterwards determined accurately by Lieut. Scott and myself, makes it only 95 statute miles in a direct line from Abu Dhabi, I could not make it less than 100 miles' actual marching by the course which we had followed, calculating the rate of the caravan, which I frequently timed, at $2\frac{1}{2}$ miles per hour. Travelling very light, it would no doubt have been possible to ride through in thirty-three hours, but I cannot imagine it being done in that time with a string of laden camels over this heavy sand. I found it exceedingly hot at Baraimi. The temperature at breakfast was 110° F., and I noted that all the natives in and round the camp were wearing curious thick socks of many colours, made locally, to protect their feet from the intense heat of the sand. Old men of the community sat about knitting these socks at odd moments, to the manner born, and they seemed to wear well in the loose sand.

This is all I have to tell you regarding the section from Abu Dhabi to Baraimi, and leaving me there waiting for a fresh lot of camels for the onward journey through the Dhahira Province, to the Jabal Akhdhar and Muscat, I will ask you to imagine yourselves back at the coast; this time not at Abu Dhabi but at Ras al Khaima, from which point I made a second journey to Baraimi in 1905. The circumstances in which it befell me to visit Baraimi again after such a comparatively short interval were these. At the time in question my lamented comrade, Mr. J. G Lorimer of the Indian Civil Service, was engaged at the headquarters of the Government of India in compiling, for official purposes, a book of reference dealing with the external spheres of interest of that Government, and it was the duty of local officers concerned to render him all possible co-operation. I received a letter from him one morning explaining

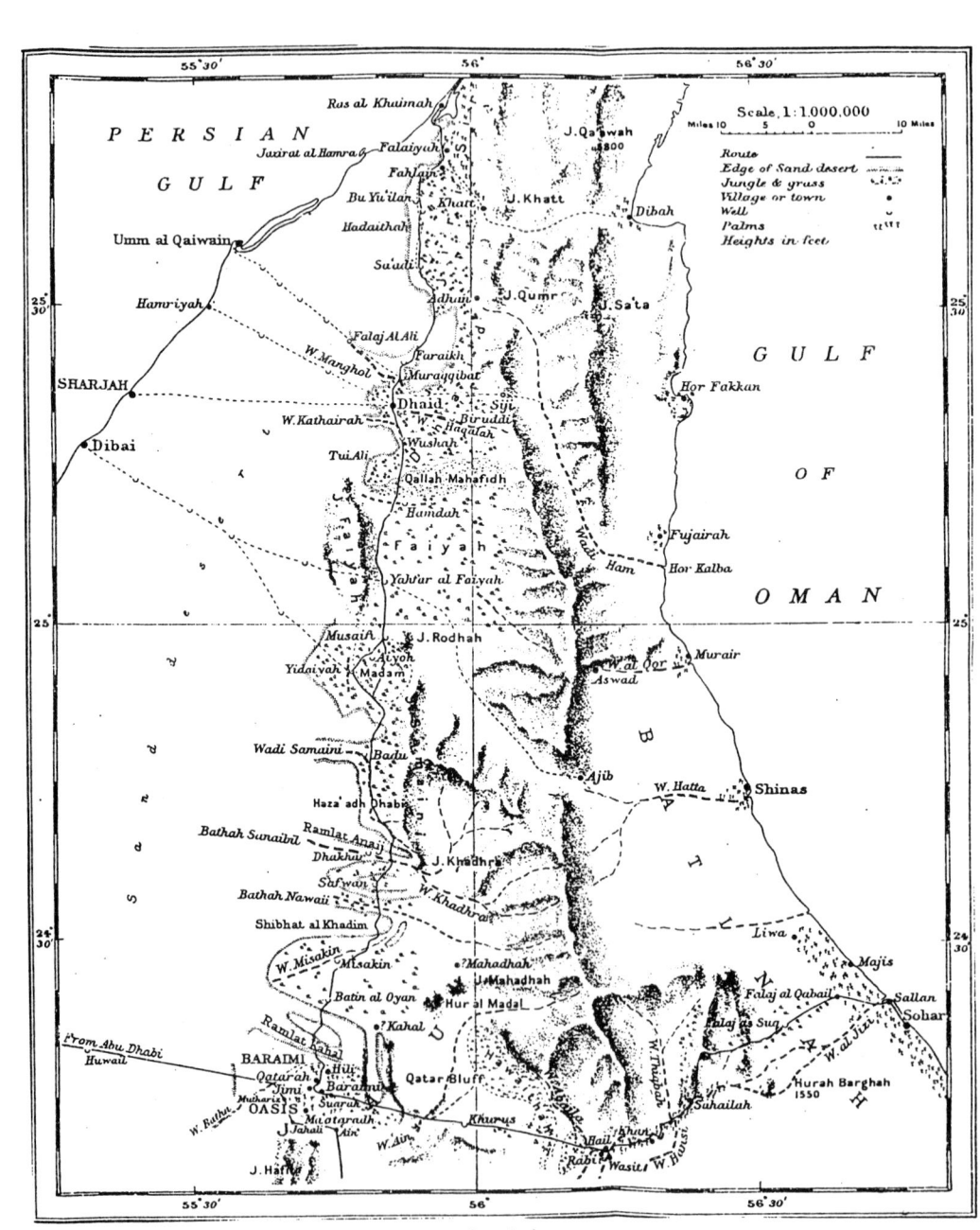

Map to show Cox's route from Ras al-Khaimah to Buraimi

that the officer of the Survey of India engaged at Simla in constructing an up-to-date map of the region to accompany his compilation,
Lieut. Fraser Hunter, was very anxious that the latitude and longitude
of Baraimi should be accurately determined, as the precise position of
many other points hinged on it. Mr. Lorimer asked if I could possibly
arrange to get this done. Being in a position at the moment to absent
myself from headquarters for two or three weeks, I thought the simplest
course would be for me to undertake the task myself, with the co-operation
of the Commander of R.F.M.S. *Lawrence*, the Indian Marine Despatch
Vessel on duty with the Resident in the Persian Gulf.

Having once traversed the wilderness between Abu Dhabi and
Baraimi, I decided to try a new route and start inland from Ras al
Khaima ; and it was accordingly arranged that the *Lawrence* should
convey me there, and that one of her navigating officers should be placed
at my disposal, together with one of the ship's chronometers and a
sextant for taking the necessary observations. Lieut. C. A. Scott was
the officer deputed, and in due course he and I were landed at Ras al
Khaima with our frugal baggage and precious instruments. As soon as
we were well away the ship was to steam round to Sohar, on the Batina
coast, and await our arrival there.

In order to give the chronometer the best possible chances, we had
devised a sling arrangement consisting of two stout rubber rings about
1 foot in diameter, connected like two links of a chain ; the chronometer
was suspended from the lower ring, while a bamboo pole was passed
through the upper ring and the ends of it carried on the shoulders of a
specially selected level pair of Arab coolies. Lieut. Scott or I marched
behind these coolies the whole way, one of us always being in vigilant
attendance on the instrument, to ensure that it was not jolted or bumped
on the ground when the caravan halted. It was a very tedious task, as
you may imagine, tramping, often through heavy sand, on the heels of
the coolies for 100 miles or more ; but the trouble expended was fully
justified by the result, as we found on rejoining the *Lawrence* that the
instrument had maintained an excellent rate, and Lieut. Scott had taken
several very satisfactory sets of sights, including of course the position
of Baraimi. Nowadays the possession of a small portable wireless set
would have made such a laborious process unnecessary, but at that
time wireless was not developed to that extent. The position of Baraimi,
as subsequently determined by Lieut. Scott, makes the distance from
Ras al Khaima, as the crow flies, approximately 110 statute miles, that
is, 10 miles longer than from Abu Dhabi. According, however, to the
readings of the survey wheel, or " perambulator," which Lieut. Scott
and I took with us, the distance actually covered over the ground was
125 miles, and this receives corroboration from the time test of 2½ miles
per hour, as it took our string of camels forty-nine marching hours, as
compared with forty hours for the 100 miles from Abu Dhabi.

I will briefly describe Ras al Khaima before we start inland.

The principality begins where Muscat territory ends, on the coast below Ras ash Sha'am, while its southern boundary runs inland with that of the adjoining principality of Jazirat al Hamra, and then crosses the peninsula to Diba, which is an appanage of Sharja.

I have referred earlier this evening to the Jowasmi pirates, who were such a menace to the peace of the Gulf during the early part of the nineteenth century, and incidentally under strong Wahabi influence. Ras al Khaima was their stronghold, and it was against that port that a British expedition under Sir W. Grant Keir was despatched in the winter of 1819-20. The Sultan of Muscat, welcoming any step likely to check the progress of the hated Wahabis, co-operated heartily with us both by land and sea, and the result of the expedition was the speedy submission, not only of Ras al Khaima but of all the other Pirate Coast chiefs. It was this successful operation that brought about the Maritime Truce to which I have previously referred. The principalities of Ras al Khaima and Sharja are still under the rule of members of the Jowasmi clan, but the numbers of this stock are now so diminished that the Jowasmis can no longer be regarded as a tribe.

The town of Ras al Khaima is built on a long narrow spit of sand running parallel to the coast and enclosing between it and the mainland a wide lagoon which provides a very convenient and sheltered anchorage. It was this quality, no doubt, which made it such a popular resort for pirate craft in days gone by.

We experienced no difficulty in obtaining camels at Ras al Khaima, and only spent one night there, leaving our quarters in the Shaikh's fort the next forenoon in the presence of an interested crowd.

The terrain encountered when caravaning from Ras al Khaima to Baraimi differs considerably from that between Abu Dhabi and Baraimi. On that side it was practically an unmitigated waste of sand the whole way. Here you have almost level tracts of sandy or stony plain alternating with re-entrants of sand-dunes, the former greatly preponderating and for the most part plentifully clad with jungle trees and vegetation, and dotted with wells at comparatively frequent intervals.

Moreover, according to my inquiries, it would appear possible, by making *détours* of no great magnitude, to avoid most of the sand-dunes altogether, except at those infrequent intervals when a heavy rainfall has made the surface of the plains too greasy to be safe for camel traffic.

But with no previous experience or information regarding the route, and my main object on this occasion being to nurse the chronometer and get the position of Baraimi fixed as quickly as possible, it was not worth my while to take chances by making tentative *détours*, so I never went far afield from the ordinary caravan route.

Though sometimes possessing no particular landmark or apparent limit, all these separate tracts to which I have referred have their individual names and accepted boundaries in the minds of the local inhabitants, names of places being chiefly taken from the most important wells in the area.

The first tract you enter on leaving Ras al Khaima is the cultivated plain of Sir, the commencement of which is seen across the lagoon. It runs for about 7 miles south of Ras al Khaima, and contains ten villages boasting some 2500 inhabitants and gardens containing about 10,000 date trees. Irrigation is from wells.

This Sir district, of which Falaiya is the most southerly village, is brought to an end by the convergence of the sand-dunes towards the hills. The plain of Jiri is then entered. From this point to Baraimi no settled habitations or communities are encountered except in the Dhaid oasis. The rest, as I have already stated, consists of alternations of jungle-clad plain and undulating sand-dunes—the latter, however, only representing about 20 per cent. of the whole distance.

Dhaid is an appanage of Sharja, the chief Jowasmi principality on the Trucial Coast, and the Shaikh of Sharja is represented by a Wali who resides in a walled enclosure or fort of the usual type over which flies the " White-pierced-red " flag of the Trucial chiefs. The principal functions of the Wali are to enforce order generally, to maintain peace between the different tribal communities in the colony, and to control the distribution of water from the fine Falaj or underground aqueduct which waters the oasis. This Falaj takes off from the Wadi Haqala in the foothills of the spinal mountain chain 18 miles away, and passes through the precincts of the fort as it enters the settlement. The village consists of about 150 families practically all dwelling in date-mat huts, and composed of contingents from three tribes, the Tanaij, the Naim, and the Bani Qitab. The area under date cultivation is about 3 miles in circumference. The Jowasmi Shaikhs and their families used formerly to spend some of the summer months in Dhaid, but nowadays it is said to be so continually swept by hot winds that it has ceased to be used as a summer resort.

Except the tract known as Qalla-Mahafidh, which is an elevated stony plain commencing (from the north) about 20 feet above the general level, and gradually descending again in its depth of 3 miles down to the general level, the country between Dhaid and Baraimi differs little from that north of Dhaid, and only calls for passing comment, but the last lap as it were, i.e. from Misakin into Baraimi, is perhaps worth a brief description. This Misakin plain lies in a depression between sandhills, plentifully studded with " samr " acacias, tussocks of " arta " grass, and other desert vegetation. From this point Jabal Hafit, the isolated mountain beyond Baraimi, comes into view for the first time some 20 miles away ; also two noticeable sugar-loaf hills thrown off from the spinal range, named respectively Hur al Madal to the south-east, and Jabal Mahadha further away to the east. After traversing the Misakin valley for 4 miles you rise on to the sand-dunes of Batin al Oyan, a succession of mild undulations 4 miles deep. At 8 miles the gentle character of the dunes develops into a most formidable belt of steep sand-hills and valleys which form a perfect maze. Tracks here are obliterated as soon as made, and caravans must keep close touch as it is very easy to lose direction. After 7 miles of meandering through this tedious wilderness a commanding point is reached from which one looks down on the commencement of the Baraimi oasis 2 miles away, partly hidden by the jungle of acacias. We entered Baraimi on this occasion from the northwest, through the recently formed Bani Yas colony called Mas'udi, started by the eldest son of the Shaikh of Abu Dhabi, and made our camp at the village of Jimi, that being the settlement nearest the centre of the oasis and convenient as a point for fixing its position. When one speaks of "Baraimi," one means not the village of that name but the whole oasis, composed of ten separate villages situated roughly in a circle about 6 miles in diameter. Baraimi village, being the original settlement, has given its name to the whole. It is the headquarters of the Naim tribe, the original owners of the oasis, and possesses the usual square fort of sun-dried brick with a tower at each angle. This fort cannot compare in strength with those in other important centres of Oman proper; but still, such as it is, it is considered strategically important and offers a serious obstacle to a tribal invader approaching Oman from the west. At the time of my visits the population of the oasis was estimated at about 5000. The habitations, which are for the most part dotted about in clusters among the gardens, are built of mud and date screens combined. The water supply is from numerous fuluj or underground aqueducts coming in from the hills to the east, and one or two also from Jabal Hafit. Light though it is the soil is evidently most prolific, and it was calculated that the oasis supported not less than 60,000 date palms besides all the fruits and vegetables to be found in that region, e.g. grapes, melons, limes, figs, pomegranates, a few mangoes, and in the way of crops, wheat, barley, and jowari and quantities of lucerne. Trade is partly with Sohar, on the coast to the eastward, but mainly nowadays with Sharja and Dibai, which latter is now a port of call for the British India Company's steamers.

Baraimi boasts no bazaar properly so called: local trade is done almost entirely by barter at an open-air market which is carried on in the Khidáma quarter of Baraimi village. Here the Shaikhs of the Naim, though not now the most powerful tribe in the oasis, still enjoy a special position by virtue of their past history; but the real power in the neighbourhood is the Shaikh of Abu Dhabi, whose material possessions and consequent influence in the oasis are yearly increasing.

As I have mentioned, on the occasion of this my second journey to Baraimi we left for the coast, to rejoin the ship at Sohar with our precious chronometer directly we had fixed the position of the oasis to our satisfaction. The route to Sohar has been fully described by both Col. Miles and Dr. Zwemer, and in any case time does not admit of my touching on it now ; but I will ask you to accompany me onwards through the Dhahira province in continuation of my earlier journey from Abu Dhabi. You had left me waiting at Baraimi for a new string of camels.

We got away in due course on May 7, steering due east *via* Muthariz and the Shaikh of Abu Dhabi's new settlement at Jahali, and thence on to the camp at the village of Ain (3½ miles), which lies at the north end of a spur of Jabal Hafit. Jahali had only been in existence for about six years ; it possessed a nice new fort and a walled date and fruit orchard, at that time in its infancy, but very promising. It extends to the foot-hills of Jabal Hafit. Ain is also one of Shaikh Zaid's villages, and he keeps a Jemadar or caretaker there. Across the plain to the north are to be seen the Naim villages of Jimi, Qatara, Baraimi, and Su'ara. On arrival at Ain I was as usual deposited in the guest tent, and found it extraordinarily hot. People crowded in till I felt almost suffocated. While we were sitting there, waiting for the inevitable coffee, one among some Awamir Badawi, who were sitting amongst the company talking to me or about me, pulled off his chefiyeh, or headcloth, and out of a loosely tied knot produced a live scorpion, which he and his companions proceeded to toss to one another and allowed to crawl about all over them. I thought it had probably had a piece of thread tied round the base of its sting, as I had often seen done in India, but they declared not, and certainly none was detectable. Then another Amiri, at the request of the audience, mumbled a long poem, the recital of which was said to prevent any harm ensuing from the sting of the scorpion ! I was a good deal worried here by the behaviour of my followers. I had brought plenty of food of my own and had no intention of sponging on any one, and the hire I paid for the camels was specifically calculated and expected to include food for the camels and men, but these camel-men thrust themselves upon the hospitality of any habitations they passed, delaying the caravan to eat hot meals and expecting me to make the necessary return to the Shaikh. There was no end to it, but it seemed the only way to get along. Next day we got off at 6.30 a.m. and trekked for 2¾ hours, about 8 miles, in a direction parallel to Jabal Hafit, bivouacking among the tents of some of Naeim and Awamir Badawi. The former go to the gardens at Hafit during the date season, but live in tents during the rest of the year.

Before quitting Jabal Hafit I should like to try to clear up a misconception involved in Dr. Zwemer's reference to it. In his paper of January 1902 he alludes to Jabal Hafit as " the first spur of the Okdat Range," and goes on to speak (from hearsay, of course, as he did not pass that way) of the " Okdat Range " as if there were one long range stretching as far as Ibri, 95 miles away. The fact is that Jabal Hafit is a solitary hogback hill about 2000 feet high at its summit and only about 20 miles in length, extending approximately from Ain to Qabil. Its highest point, or rather points, as there are two twin summits, are latitudinally level with the cluster of villages comprising Hafit and consisting of Hafit proper, Gharbi, and Bu Gharah. The use of the expressions " Jebel Okdat " or " Jebl Akabat " is unintelligible to me. I found the word " Okdat " to be one of several synonyms in common use for " sandhills " in those parts ; while " Aqabat " is a word used throughout Oman for a mountain pass, or defile between two ranges.

I think the explanation of Zwemer's misunderstanding is that south-west of Hafit, but quite separate, and not in prolongation of it, there is a line of sandhills (Okdah) which on a clear day might have been pointed out to Dr. Zwemer from the Baraimi road and appeared to him from there as if they were in prolongation of Jabal Hafit.

While here a messenger reached me from my *locum tenens* at Muscat, *via* Sohar, informing me that all was well, but announcing that H.M. first-class cruiser *Amphitrite* was expected (on her way out to China), and urging my early return to meet her. This was not very welcome news in the circumstances, nor did I see how I was going to get back in time without changing all my plans. However I replied that I would do my best, and having dismissed the messenger we started off again, making almost due south to Hafit village, which we reached after one hour's trot. We then struck across to the left through bush country to Qabil, where we spent the night, going on to Bizaili next morning. On the way we were fired on from a Badawi encampment, the occupants taking us for a raiding party, but no harm was done. The country we were now passing through reminded me greatly of the Haud of Somali-land, but without the charm of always being in sight of game, and it continued almost till we reached Dhank, three and a half hours' journey from Bizaili. At Dhank we camped outside the town in the open. The usual simoom blew all the afternoon, and the heat was excessive. About 4 o'clock, however, I went to the top of the hill behind the town and scanned the surrounding country towards Ibri and Jabal Hafit. The latter was just visible and nearer due north than I had expected. Dhank possesses some fine date gardens and a running stream. There is no proper bazaar, but as at Baraimi there is a public place where goods are put up for auction or exposed for sale. There was no apparent reason why we should not have camped in the date gardens, but I had the greatest difficulty all along in getting my companions to sleep in the vicinity of the towns ; they always wanted to get away into the open, and I came to realise that they had reason on their side, bred of experience.

My next halt was Mazum, a small settlement of Baluchis allied and inter-married with the Bani Qitab. It consists of about three hundred houses. It was at this village that Miles experienced some incivility, but I had better luck. *En route* however I had a little misadventure. My pilot had previously been expressing his views as to the pace of our respective camels, and I had agreed that as soon as we came to a stretch of good going, I would match my mount against his. I may mention that I had a very fine and well-bred riding-camel lent me by the Shaikh of Abu Dhabi. The morning after leaving Dhank we came on a suitable bit of going, and accordingly we decided to have our match and started our camels off at full gallop. My camel was fresh, and in trying to pull her up the headstring came away in my hand, and she went away with me into the blue. Sitting as I was behind the hump, according to the approved method of riding in Oman, I could do nothing to stop her, and I was fully occupied in keeping my seat and holding on to my sketching case, compass, etc., and trying to ward off mimosa thorns. The direction she was taking was at right angles to our line of route, and if in my wild career I had happened upon a strange Badawi encampment it might have gone hard with me. I could only just reach the camel's cheek or neck with the end of my riding-cane, and the best I could do was to keep her more or less on a wide circular course by poking at her neck and endeavouring to guide her gradually back towards our starting-point. At last in the course of half an hour I had gradually got her head in the right direction, and my pilot Hamud and another were able to cut in in front of me and head her off. I was somewhat exhausted, my clothes torn, and face and hands badly scratched with thorns, but I was otherwise none the worse. I did not offer to race any more, however !

The method of riding in Oman on a mere skin on the lightest of trees behind the hump is quite different to our method in India, and elsewhere, and though certainly less hard on the camel, needs more efficiency in the rider, and you will realize my plight when the headstring came away.

Once again on the march to Ibri we were fired on from a distant encampment, and bullets whizzed across our bows, but I was now used to this little amenity, which appeared to be the ordinary practice of a Badawi community—based on the assumption that any strange mounted party seen approaching must be enemies on the raid. Apparently the etiquette is to stop and return the fire if you really are an enemy, otherwise ride on and take no notice.

As we neared the town, which had evidently received warning of our coming, a number of horsemen galloped out and gave us a demonstrative welcome, continuing to perform weird evolutions in front of us until we reached the broad shingly wadi just outside the town, where some two hundred men were drawn up in line and fired a sort of *feu de joie*—and not at all badly. We were then escorted into the town by a very friendly if very dirty crowd. The whole populace seemed to have turned out, and women and children lined the path. Burkas (yashmaks) were not much in evidence, and the ladies seemed comely as a whole, but the greasy dirtiness of their apparel was indescribable. Every one, man, woman and child, insisted on shaking hands, in spite of the heat, and would take no refusal.

As I proposed to remain here a day or two, I insisted on camping in a date garden instead of out in the open in the sun, as my companions would have preferred. In the course of time a suitable place was chosen. The inhabitants were very friendly and good tempered—in fact, too much so, for they crowded round my tent all day and allowed me no privacy whatever. Some of them never left me even for food or prayers, and when I suggested that their dinner would be getting cold, they received the hint with an amused smile, but pleaded to be allowed to remain as they had never seen anything like me before. (And the present generation probably had not.) Wherever I went I was accompanied by a small crowd. In the afternoon, when it got cool enough, I walked up to the top of the hill overlooking Ibri from the south-west and took a number of angles to various points. At the top of this hill one gets a good view over the plain embracing the villages of Ghabbi, Araqi, Bait al 'Ainain, Dariz, and the Jabal Misht and the Jabal al Kor, pronounced " Keör." At dusk the ladies of the place were again present in force, and sat round and watched me attentively while I had my dinner. As soon as I had finished the local Shaikh came in, accompanied by four or five other men who were to act as night guards on my camp. Before turning in we discussed Oman affairs and the general insecurity of life which prevailed. Here, in a large town like Ibri, the townspeople are fairly immune from the operation of blood feuds so long as they remain within the town, and they do not allow parties of outsiders to sleep inside the walls. They are afraid of visitors and visitors of them. In fact, in many places where tribal feuds exist men never sleep in their own houses, but always somewhere outside. The reason for this is that if a man is known to sleep at home his enemies know where to find him, and it is a common ruse for treacherous enemies to come to a house at night and call up the occupant, who jumps up without thinking, and the intruders are thus sure of their man. On the other hand, sleeping outside one man cannot be distinguished from another, and thus all are collectively safer.

Ibri is the most important settlement in the Dhahira district, and boasts a tremendous expanse of date groves, a good bazaar, and a fine mosque. It used to have the reputation in Miles' time of being a thieves' market, but seems to have since turned over a new leaf. I was told, however, that in spite of their time-honoured Wahabi associations a good deal of date wine was manufactured here at Ibri; while indigo is extensively grown for home use, half the community having their fingers and faces stained with it. As before explained, from here onwards I was to be the responsibility of the Sultan of Muscat, and it had been arranged that his representative, Shaikh Rashid Bin Uzaiz, should meet me at Ibri, while my camels for the onward journey were to be supplied by the Daru tribe from Tana'am. Shaikh Rashid joined me in due course, and as soon as we had secured our new camels, on May 15 I resumed my journey. Our first stage was Salaif, a picturesque village having a situation very much like that of Ibri, in a narrow gorge-like wadi with a tower perched above the village on a precipitous overhanging rock. As usual we slept outside away from the habitations, but next morning, the 16th, after taking early tea I had my paraphernalia shifted to a shady date grove, and then went up to the top of the hill with two local guides to take some bearings with my compass. On the crest of this hill there were the remains of some stone fortifications and embrasures, said to have been erected by the Wahabis when they were here early in the nineteenth century. They apparently captured Ibri, but were unable to reduce Salaif. My informants spoke of Al Mutairi as the Wahabi general, and declared that there were persons still alive who remembered the episode.

From Salaif we marched to Kubara, following the Wadi Salaif for 10 miles and then turning south. Kubara is under the lee of Jabal al Kor and about 10 miles distant from it. We found the inhabitants of this village in a very bad way owing to drought. They had had no rain for three years, and their water was failing and date trees not bearing fruit. Poor people ! we found them very civil and friendly, and for the first time during our trip we were spontaneously offered chickens and eggs for sale. It was pitiful to see the effects of the last three years' widespread drought : Fuluj failing, date trees drooping, and much land gone out of cultivation. In the morning, while the tents were being struck, I beat a retreat to the nearest date garden for shade, and there I was soon surrounded by a party of housewives from the village. One middle-aged dame did most of the talking, and asked me many questions about my country and discussed the burning question of the drought. They were a very nice, polite lot, and for a wonder did not ask me for medicines or regale me with their ailments. After about half an hour's edifying conversation one of my men came to say that the caravan was ready, so with a " God speed you " from the ladies I went my way.

Both at Baraimi and onward through Dhahira I noticed that the natives, when addressing me or referring to me in conversation, either when not aware of my identity or not sufficiently sure of themselves to address me as " Consul " (or " Balyooz," which is the term used for Consul in Oman and the Gulf ports), always applied to me the word " Nasarani " = Christian : for instance, " Good morning, O Christian," or " Hie ! Christian, there is no road that way ! " The term did not appear to be said or meant disdainfully, as the speakers were quite friendly, and I took it to be merely a relic of Wahabi fanaticism which was of course predominant in this district in the last century.

From Ibri onwards, or rather from Dhank onwards, the route which I followed differed very little from that of Col. Miles, so I do not propose to deal with it in great detail. There are one or two points, however, wherein the field-book and compass sketch which I made of my route give the lie of certain features somewhat differently to that given by Col. Miles in the map accompanying his paper in the Society's *Journal* in August and October 1910. Unfortunately, his paper had not appeared at the time of my journey, or the submission of my report, and it was only when it appeared afterwards that I noticed the differences. I can only say that I made a careful field-book and compass sketch and recorded full notes, and the sketch was plotted out immediately on my return, and these are still available, so that, other circumstances being equal, I think my version is likely to be the more accurate. I refer particularly to the lie of Jabal al Kor with reference to Jabal Ahkdhar and Jabal Hamra. The march from Kubara to Saifam proved a long trek of about 26 miles. At 6½ miles we were abreast of the Hadda peak, and reached the pass in 7 miles more. The Hadda peak is not a solitary mountain as it appears when viewed end on from Kubara, but is the northern extremity of a long range forming part of the Jabal Hamra. There is a gradual rise up to the najd at the north end of the mountain, height about 200 feet, and then a steep drop of about 250 feet, during the next 1½ miles into the basin or heart (Jauf) of Oman Proper down below. The route up to the najd was quite good going, and camels could trot nearly all the way. As it nears the apex of the triangle culminating in the najd the path passes quite close to Jabal al Kor, and from here one sees an enormous cleft in the side of Jabal al Kor just opposite the Hadda peak forming the source of the Wadi Ain. In the basin which we have now reached two villages are to be seen, Dan and Nadan. The terrain is hard plain covered with sharp broken talus shingle and dotted with the usual mimosa trees.

On this part of the march one of the escort riding in front did a smart bit of business : he dropped out of the line ahead of us, shot at and killed a hare with his rifle, retrieved it, and mounted again without losing more than a dozen places in the marching line.

The descent into the Jauf of Oman was down a torrent bed, and took us half an hour. On reaching the bottom the path turned right round to the left, that is, to the north-east, along the south side of Jabal al Kor. It was 8 o'clock before we got to Saifam, which lies in a wadi bed of the same name right under the peak at the highest point of Jabal al Kor and not far from the southern mouth of the Najd al Barak, or " Pass of the Lightning," which can be discerned if one moves a little out into the plain. This was the alternative route taken by Miles.

On our arrival at Saifam Shaikh Rashid learnt that the Shaikh Nasir bin Hamad, the 'Utubi Shaikh of Bahla, had come down to Jabrin to meet us, and was awaiting my arrival there, so Shaikh Rashid thought we ought to push on first thing in the morning. We accordingly started off at 6.30, and reached Jabrin, 7½ miles away, at 8 o'clock, the caravan coming in at leisure.

Jabrin is State property, and is now merely the summer quarters of

Shaikh Nasir and a few other Bahla people. The fort looks imposing from the outside. On arrival we were met by a relative of the Shaikh and taken through the village to an open space outside under a fine acacia tree, where they had some camel and horse races for our edification. I noticed that the population here seemed to be mostly negroes. As I was only staying until the afternoon, it was not worth while pitching a camp, and they had a room carpeted in an outhouse in one of the date gardens for us to rest in. People dropped in and out all day. About 2 o'clock I started off with one of my escort and a petty Bani Riyam Shaikh from Tanuf to see Bahla. It was about a 4-mile ride, partly through a thickly wooded glen. It is a most imposing place with a very fine old castle, of which Miles gives a description and a photograph in his paper. I only rode through the town and went over the fort to enjoy the view from the top, and then returned to Jabrin to pick up my caravan. From Jabrin I marched to Nizwa, with a halt halfway at Radda. Radda and Farq to the north are both in the Wadi Nizwa, while facing southwards you are looking towards Manah and Adam, the former being visible in the distance. The whole plain in which Bahla, Jabrin, Kubara, and Nizwa lie is like a large crater studded with low volcanic hills, bounded on the north by the Jabal al Kor, on the east by the Jabal Akhdhar, and on the west to south by the Jabal Hamra, the habitat of the Daru tribe. This latter small range looks geologically of a completely different character to the surrounding systems, and appeared of a bright terra cotta colour.

At Nizwa we had the usual preliminary reception. The Wali was one of the Muscat Royal Family named Saiyid Saif bin Hamad. He it was who got the Nizwa back into the possession of the family by murdering Hilal bin Zahir (for whom see Col. Miles' paper). He received us at the fort gate, and after giving us coffee took me to a date garden where a comfortable shed had been prepared for me. The caravan came in in due course, and after the hard travelling of the last few days I was glad to get into comfortable quarters.

To-day I bade good-bye to some of the men who formed my escort from Ibri. During the journey I had talked a great deal with them of the nature of the Great Desert and the possibility of crossing it, and two of my companions, Shaikh Muhammad bin Mansur of the Al Wahibah and Mazar bin Naif of the Awamir, had repeatedly protested their readiness to take me across, declaring that in the winter we should have no difficulty with good camels and a good bundobust. They would propose to take 15 men, 5 Al Wahibah, 5 Awamir, and 5 Daru, and of course only picked Badawi camels able to go for many days without water. They declared that as far as getting me across was concerned they could guarantee success; all the misgiving they had was as regards our reception by the Badawi when we reached the other side: that, they said, must be my responsibility and would need careful previous arrangement in order to ensure friendly treatment. These conversations imbued me with serious thoughts of contriving an expedition to attempt the crossing from Adam or the furthest possible base on the desert fringe in that direction, but my transfer to Bushire soon afterwards, and the publication of an article by Mr. Bacon suggesting a crossing by balloon, turned my thoughts in other directions. Probably the riddle will be solved by air in the course of time, but it does not appear to be a practicable proposition for the existing aeroplane.

After a quiet night I went out early with the Akid or commander of the fort to look round Nizwa and take a few photos, including one of the old Persian Bazaar known as the "Suq el Faris." I also took a picture of a cultivator churning indigo vats in the date gardens, and a view of two of the main street and fort. I then entered the fort with the Wali and was shown all over it, taking a view from the top and a photo of one of the Hazrami garrison occupying his leisure, in these piping times of peace, at the loom, in one of the gun embrasures.

Wellsted has given a description of Nizwa fort. It is a very strong one and renowned in the annals of Oman. It contains a capacious water reservoir, but at present no battery to speak of; there is, however, an interesting old gun in the basement bearing the Portuguese arms, but with a Persian inscription round the touch-hole mentioning the name of Padshah Shah Abbas. The Wali says the gun is said to have been brought from the island of Hormuz, and that would seem probable.

While at Nizwa I received another letter from Muscat telling me that H.M.S. *Amphitrite* had not left England till May 6, so she could not reach Muscat for some days after the date first announced: this was a great relief to me. Shaikh Rashid also received a letter from the Sultan by the same messenger, in which H.H. gave him the news that 200 Omanis had been killed by the Portuguese in East Africa. On this occasion a fleet of Omanis dhows from Sur, while engaged in loading up a very large consignment of slaves, was cornered by a Portuguese man-of-war and all captured or killed and upwards of 700 slaves released. 150 of the Omanis who were captured were sent to the Portuguese penal settlement in Angola on a sentence of twenty-five years, and I imagine that those who survived are still there.

During the forenoon of this day, May 20, it clouded over and in the afternoon some nice rain fell. Every one was transported with delight at this welcome visitation, as there had been a long-continued drought. Birds and frogs held high revel, and it was the greatest pleasure to feel and smell the rain.

I noticed that the population in general were in a highly nervous state and easily alarmed. One only had to hear a rifle-shot or two and every one got the wind up. It was significant of the unsettled state of the country and unsatisfactory relations prevailing between one tribe and another. For example, while out for a walk on the outskirts of the town, I said I should like to go up to the top of a nice-looking tower about 500 yards off to get a view of the oasis, but my companion said it would not be safe to go as it was a Bani Riyam tower, and we should certainly be fired at—and this was within a stone's throw of his own house.

I was sorry to leave Nizwa where everybody had been very friendly and civil to us. Between Nizwa and Samad al Kindi, my next stop, only 2 miles away, is an unbroken stretch of date gardens. It was considered politically desirable that I should halt at Samad al Kindi for the following reasons : The Shaikh of the place, Hamdan bin Saif, who proved to be a fair aristocratic young man of about thirty, was at feud with his cousin at Tanuf. In the days when their grandfather Shaikh Sulaiman ruled both Tanuf and Samal al Kindi, his two sons, Saif and another, had a bitter quarrel. It developed into a resort to firearms, and in the fray Saif shot his brother and his brother's daughter and lost one or two of his own adherents. He thereupon fled to Samad al Kindi and remained there, while the dead brother's family stayed on at Tanuf. The feud of course smouldered on, and one day Himyar, son of the dead brother and brother of the dead girl, succeeded in shooting his uncle Saif dead while at prayers, and then returned to his stronghold at Tanuf. The result is that both families now go in fear of their lives, and venture out very little. It was feared that if I paid attention only to one branch the others might lay themselves out to defeat my plans in order to do a bad turn to their rival. I had to spend the night at Samad al Kiudi, and as the gardens were full of mosquitoes, Shaikh Rashid and I went and slept out in the open plain. In the morning I got off betimes and arrived at Tanuf, a distance of about 7 miles, in a couple of hours, our direction seeming to be due north the whole way. Tanuf is situated at the base of the mountains blocking and commanding the entrance to the ascent up the Wadi Tanuf to the plateau of the Jabal Akhdhar.

I found Shaikh Himyar a fair-complexioned young man of much the same age and type as his cousin in Samad al Kindi, but smaller in stature. He gave me a salute from his tower with an old cannon, and there were the usual evolutions outside the town, followed by a gathering in the guest house, a long building (50 feet by 15 feet), while coffee was being served. Small talk for public consumption on these occasions soon gets exhausted and the long waits are very boring. Arabs themselves do not seem to mind sitting speechless for an indefinite time, tapping the ground with their canes, but it is extremely boring for Europeans.

As soon as I could I adjourned to my own quarters. The Shaikh came in to see me after breakfast. I had difficulty in getting him down to anything definite as regards my journey, and we parted without having arrived at an understanding. I came to the conclusion that greed was at the bottom of his intractability, and that he was out for blackmail, and this was soon confirmed. On leaving me he went to my pilot, Shaikh Rashid, and said that the Sultan some time ago had

Nizwa. Fort, 1900 (Cox)

Tanuf village, 1900 (Cox)

Jabal Khadr, 1900 (Cox)

promised him a Lee-Metford rifle but had never sent it, and he wanted
to know why. Shaikh Rashid offered him his own, saying he knew
nothing about the Sultan's promise, but in any case would send him a
new one from himself and me when we got back to Muscat. The Shaikh
did not come in to me again during the day, but was busy making every
sort of ridiculous difficulty, including the old chestnut which I had so
often heard, that as the result of Col. Miles' visit five and twenty years
before, a blight had fallen on the fruit. Also that the people were
alarmed by my camera and prismatic compass—all bunkum. Finally,
he made specific pecuniary demands on this and that score, which
amounted in all to 500 Maria Theresa dollars, about £50, to be paid
in advance. I replied that I was the Sultan's friend, travelling for my
own pleasure and enjoyment in H.M. country, accompanied by his
most well-known and respected retainer, and bearing all necessary letters
of introduction, and I absolutely refused to pay any blackmail, either
500 dollars or any sum at all. He must either let me through without
any offensive conditions or I would go straight back to Muscat and lay
the matter before the Sultan. I heard no more that night, and turned
in somewhat worried but hoping for the best. In the morning with the
dawn donkeys were produced and our frugal baggage loaded up, and
about 6 a.m. in came the Shaikh in a most amiable mood, smiling as if
there had never been any unpleasant question between us, and announced
that he was going to accompany me himself. Accordingly, at 6.30 we
started on our way, Shaikh Rashid, Shaikh Himyar, my servant, and I,
all on foot, with four donkeys with our food and bedding following on
behind. The ascent was very steep to begin with, and distressing for the
first two hours owing to the heat and the sun blazing straight into our
eyes. At 8.15 we had an easy while the Arabs had a cold breakfast of
dates and dried shark flesh. They cut or hammered off great chunks of
the shark meat, and after beating it into a fibrous state picked it to pieces
and ate it. After about four hours' walking at 2½ miles an hour we
came to a rain-water cistern in the rocks, formed by blocking up the
side of a deep cranny in the nullah. Here under the shadow of an overhanging rock we halted for the hot noon-day hours, and in the afternoon
intended to do the second half of our journey to Saiq ; but about 1 o'clock,
just as I was ready to start, some villagers brought in a goat, and it
appeared absolutely necessary to my party that the goat should be eaten
immediately, so we were considerably delayed in getting off, and after
walking for two and a half hours at about 3 miles an hour, we found it
was too late for us to reach Saiq that night. (I may mention that my
companions generally used the diminutive form of Suwaiq for this
village.) Meanwhile we reached the Wadi Habib, which gives its name to
a small village of about thirty houses of Bani Riyam. It was built up the
side of a deep nullah and reminded me very much of a Kashmir village.
As we descended the series of steps down the side of the nullah in single

file, the villagers all passed up one by one, each clasping our hand in one or both of theirs and uttering their rustic greeting, " Marhaba-Bakum " (" welcome to you "). On getting down to the bottom of the nullah we were shown to an open carpeted platform (for the entertainment of guests) in the bed of the wadi, surrounded by a mass of pomegranate and walnut trees. One fine walnut tree formed a canopy to the guest house, as the acacia tree did at Jabrin. All this time it was thundering and lightning and the sky looked very threatening. Our hosts inquired if I would like to sleep out in the open or under the trees. I replied, as it looked like rain I would prefer being under the tree or even in the shelter of a house. They assured me it was not going to rain, but our donkeys had hardly arrived and shot my things down under the tree when it came on to rain in torrents, and our baggage got very wet before we could get it into shelter. The villagers were very kind and friendly, however, and helped us to carry our things into shelter and fetched water and wood for us, and with a change of clothes and a fire we were all soon comfortable and ready to turn in. I did not sleep very well, as I was tormented by some biting insect which attacked my hands and wrists ; I could not hear any buzzing, so concluded that my visitors were those of the silent order.

In the morning I got my tea at daybreak, and then strolled out with my camera and gun and collected one or two birds while the donkeys were loading up.

Saiq proved to be only 4 miles distant, and I found it a disappointing place, like the ordinary Baltistan village : stone houses plastered with reddish earth and with flat mud roofs, and with a succession of orchards in front. There was a certain amount of wheat and barley cultivation in terraces, and a good show of mulberries, peaches, and apricots, but the main cultivation was in the form of a fine show of pomegranates. After halting for a short time to look round we continued our march to the village of Sharaija—so pronounced by the inhabitants.

Miles and Wellsted spell the name Shiraizi, connecting it with Shiraz, but I am inclined to doubt whether it has any such connection. I believe it to be the same root as Sharga or Sharja, and to be a diminutive of that form. It is a larger and more picturesque village than Saiq, situated at the head of a fine canyon facing south-west and looking down towards Birkat al Moz. The sides of the wadi are very massive and steep, and it is altogether a grand bit of scenery. According to Miles, who went from here to Birkat al Moz, the terraces of cultivation go right down the Wadi Mi'aidin, for 1000 feet or more. My poor old pilot Shaikh Rashid was beginning to look very tired and sorry for himself ; probably he had never had to walk so much before in his life. After partaking of the usual coffee we were conducted through the busy portion of the town to the so-called guest house, two compartments on the ground floor which had been used as stables being set apart for our

Shuraijah, looking towards Birkat al-Mawz, 1900 (Cox)

Muti, at the foot of Jabal Akhdhar, 1900 (Cox)

use, one for our donkeys and the other for ourselves. The floor was covered with litter, which we proceeded to clear away as far as possible. I expect this is the place that Wellsted mentions, and in which he would not stay. The villagers told him this was where travellers were always accommodated. Since Miles' visit in 1885, only one European had visited Sharaija, one whom they called " Butros." They referred no doubt to the missionary Peter Zwemer, the brother of the Rev. Samuel Zwemer whom I mentioned at the beginning of this paper as having travelled from Abu Dhabi to Baraimi. Peter Zwemer died of fever in America very soon afterwards.

Just below the guest house was a fine circular reservoir, fed by a spring, in which at the time several lads were bathing. The season, it will be remembered, was the end of May and excessively hot in the sun. I sat and looked on longingly until our donkeys came, when I got out my bathing suit and had a dip, but the spring water was too ice-cold for me to stay in long. While sitting at the pool I talked to some villagers, and one of them of course repeated the old story about Col. Miles and the blighted fruit trees. I tried to demonstrate the foolishness of it, and went on to say that anyhow on this occasion please God my coming was propitious, for wherever I had stopped I had brought rain after prolonged drought, as for instance at Nizwa and at Wadi Habib last night. The words were hardly out of my mouth when a storm began to work up overhead, and before long we had heavy rain, and it continued all the afternoon; pleasant for them but not convenient for me, as it obscured the view and greatly hampered my photographic activities. My Tanuf friend Shaikh Himyar here took his leave, as he feared for his life if he went further afield then Sharaija, saying that if his hostile relations heard that he had got down to Muti they would be able to cut him off. He apologized humbly for his uncouth conduct at Tanuf, and pleaded that it was due to his ignorance of our usages ! He certainly had been very helpful since we started, and his presence had ensured me a friendly reception on the Jabal, and so on Shaikh Rashid's advice, which I felt bound to follow, he was promised a Lee-Metford rifle and 100 dollars, to be handed over to his representative on our arrival at Samail. He sent a local Shaikh, Yunas by name, to represent him as far as Muti, which was the limit of the Bani Riyam.

That evening we went on to the Sharaija suburb called Ain, where we slept in a dirty little stone hovel ; it was not so bad however as it looked, and no insects kept us awake. We were up and away with the lark, and I was just congratulating myself on having a long day in front of me when we were stopped, by arrangement apparently, at the next hamlet of Aqar by a deputation bringing coffee and food for myself and followers. Shaikh Rashid swore that the repast should not take more than half an hour, but it was 7.30 before we got off again. From Aqar our way lay at an angle of 150° across the Wadi Miakal, some way down which lies the village of Hail, to which however I did not digress. Our path went up and down one ravine after another, with a general rise to the crest of the range just below the peak called Khadhar. We must have gone up 800 feet since leaving Sharaija, and at our highest point the summit of the peak appeared another 500 feet above us, and our course took us almost round it. Then standing practically on the summit of the range between 7000 and 8000 feet high you have a grand panorama spread out before you—Muti, Qarut, Birkat al Moz (the town just hidden), and then more to your left the Wadi Halfain and other scours from the watershed tailing out like ribbons into the plain and losing themselves in the distant haze of the desert. This is the point at which I would like to start a survey if I had to make one, and then go across to Sharaija and Saiq and thence along the crest northwards to the end of the range. The northern portion of the plateau has yet to be explored, and it was a great disappointment to me that the advent of H.M.S. *Amphitrite* made it impossible for me to remain longer on the Jabal. It had taken us two hours from Sharaija to reach the crest ; our descent then commenced. For the first 300 feet there was no made track, but the slope was easy and gradual ; then we reached a col with a little misjid perched on it. From that point the drop of some 3000 feet into the Wadi Halfain is down a road of steep (and too steep) steps said to have been made by the Persians. It took us two hours to get down, practically without a stop, and I found it particularly trying on one's thigh muscles. Moreover we were descending from a nice climate into a furnace. It was 2.30 by the time we reached the bottom, and my poor old pilot Shaikh Rashid was quite played out. On the way down the only persons we met were two young women going up the hill to collect " boot," a species of coriander.

On reaching Muti we found that the rest of our party who were to have met us here with our camels and caravan had not arrived from Izki (Ziki), so we sent a messenger to tell them to hurry up as they ought to have been awaiting us.

After taking some tea I took a walk round about and looked at the view up and down the nullah, and took some compass angles and a photograph. Meanwhile the caravan and escort turned up, together with a lot of Bani Ruwaha tribesmen. Shaikh Rashid was convinced that they had come to squeeze us, and was therefore anxious to get off at once, tired though he was. Meanwhile two messengers arrived with letters from Muscat saying " all well " and ship not yet arrived.

Starting forthwith, we reached the najd of Qarn ad Daru (two small hillocks) at midnight, Wabal at 2 a.m., and going all night arrived at Samail at daybreak, and from thence we rode into Muscat as quickly as we could during the day, arriving just in time to greet H.M.S. *Amphitrite*, in which I proceeded on a " show-the-flag " tour round the Gulf ports—a very hot tour, I am afraid, at that season of the year for a ship's company straight out from home.

If I may trespass on your patience for a few moments longer I should like to touch briefly on the natural history of Oman before I conclude.

Only two of my predecessors appear to have taken any lively interest in the subject, namely my friend Col. Jayakar, of the Indian Medical Service, and Col. Miles. The former, a Mahratta by origin, was Agency Surgeon at Muscat for some five-and-twenty years, including my time there, and was a man of great industry and scientific bent. He made a very fine collection of the fishes of those waters, which he sent to the British Museum together with other smaller collections which he made from time to time.

I do not think he ever left the precincts of Muscat itself, but he always kept a vigilant eye for any natural-history object likely to be of scientific interest, whether bird, beast, or fish.

Col. Miles, situated like myself, used what opportunities he could snatch during his tours inland to collect specimens, but it will be realized, I think, that travelling under the conditions that he and I did, satisfactory collecting was no easy matter.

As regards mammals, the following seem of particular interest in the Oman fauna :

Oryx Beatrix. The Beatrice Oryx named after Her Royal Highness Princess Beatrice.

Hemitragus Jayakari. Jayakar's Thar, named by Mr. Oldfield Thomas from two skins sent home by Dr. Jayakar. It has never been obtained or seen in its native haunts by any European sportsman.

Gazella Muscatensis. The Muscat Gazelle.

Gazella Marica. The Reem Gazelle.

A small hare (*Lepus Omanensis*).

A Hyrax and a Civet.

The latter is apparently the same form as that found in India and beyond and on the island of Socotra, and though now indigenous was probably at some period imported.

Among the birds only about 120 species have been collected by Col. Jayakar, Col. Miles, and myself, so that a great deal remains to be done.

Perhaps the most interesting items are :

The Muscat Bee Eater ;

The Arabian See-see partridge ; and

The black-headed Arabian Chukor partridge.

When it becomes a little more accessible to travellers, Oman will offer to the naturalist a field which has been very little worked.

Before the paper the PRESIDENT said : Our lecturer this evening is a gentleman who requires little introduction to an audience of this Society—Sir Percy Cox, whose acquaintance with the Persian Gulf and with the countries which hem it in on north and south has indeed been a prolonged one, starting from the time when he first proceeded as Consul and Political Officer to

Muscat in the year 1899. During that period, as he has risen from post to post—from Consul to Consul-General ; from Consul-General to Political Resident in the Gulf ; and then again after a period of service as Secretary to the Government of India in the Foreign Department, and as Chief Political Officer with one of the Indian Expeditionary Forces to the post of Acting Minister in Persia ; and finally, to that of High Commissioner for Mesopo- tamia—Sir Percy Cox has rendered service to Great Britain of which his fellow-countrymen may well be proud. But it is not in the guise of a states- man or diplomat that he comes before us to-night. He is with us as a traveller who has made excellent use of the opportunities which his official position has provided from time to time of exploring some of the lesser-known parts of the Middle East. The particular journeys which Sir Percy Cox is to describe to us were undertaken a good many years ago into the thirsty hinter- land of Oman. Few European travellers had penetrated to that little-known part before, and those who had had given somewhat meagre and perhaps not too accurate accounts of its geographical features. Sir Percy Cox to-night will be able to fill in the lacunæ left by the very few travellers who preceded him, and possibly also to make some corrections in the geographical data which they have given us. I have now much pleasure in calling upon him to give us the account of his journeys.

Sir Percy Cox then read the paper printed above, and a discussion followed.

The PRESIDENT : There are not very many who can speak with first-hand knowledge of Arabia, but prominent amongst those who can is Mr. Philby. I will ask him if he will say a word or two.

Mr. H. ST. J. B. PHILBY : I am sure you have all listened, as I have, with the very deepest interest to the account that Sir Percy Cox has given us of his travels in the Oman hinterland. There are a great many points which suggest comment on my part, but I must be brief. First, as to the question of climate. Sir Percy Cox has told us something about the horrors of " Elphinstone " Inlet, where the Indo-European Telegraph Department were scarcely able to keep their employees alive ; he referred also to the heat at Baraimi, but said nothing very discreditable about the climate of Muscat, where he and Lady Cox are said to have spent five years. I suppose he refrained from telling us of the climate for a particular reason ; for fear, for instance, that nobody would believe him. I think you will agree with me that neither he nor, if I may take the liberty of saying so, Lady Cox are particularly convincing advertisements of the unhealthy nature of the Persian Gulf climate in which they have passed not only five years, as one might have thought from to-night's lecture, but rather more than a quarter of a century. I am sure that both would willingly return and spend another quarter of a century in the same parts if they had a reasonable chance. I am sure that Lady Cox, at any rate, would prefer to do so, rather than stay in our own wretched climate. I should like to suggest that the secret of success under such conditions is obvious. You must like a place and your work in it so much that you have absolutely no time to think of its drawbacks and disadvantages.

Sir Percy Cox has referred to his travels as mere excursions. Well, they may have been ; but mere excursions in countries which are unknown, or very little known, are apt to have a very considerable geographical importance, and this is no exception to the general rule. The fixing, for instance, of a place like Baraimi is a very definite geographical achievement, and so far as I know, though I may be wrong, there is no other recorded instance of the successful manipulation of a ship's chronometer at any distance within the interior of Arabia. That is a very great achievement. Then the journey of more than 100 miles or so from Ras al Khaima to Dhank over territory which, except for Baraimi, had never been travelled by a European is something of a geographical achievement. And when I saw the map on the screen this evening it seemed to me that there were a good many additions as compared with the maps one is accustomed to see of that part of the world, and every addition represents a definite advance on the total sum of human knowledge. It seems to me a great pity that all the interesting information we have received this evening has been allowed to lie hidden in the dusty archives of the India Office for a quarter of a century. At any rate, better late than never, and after to-night they will be on public record in the pages of the *Geographical Journal*.

There is another point I should like to make. Although other preoccupations of perhaps greater importance have militated against Sir Percy Cox extending his Arabian excursions further into the interior, he has really made ample amends to humanity for this very serious shortcoming on his part, for he has never missed a chance of placing in the way of others opportunities to travel that circumstances denied to him. Only a short time ago Major Cheesman, to whom the Society has to-day awarded one of its great honours, was able to tell you that it was to Sir Percy Cox that he owed the chance he had of making that great expedition to the Jabrin Oasis, and I would like to take this opportunity of adding that it was also to Sir Percy Cox that I owed my year of very delightful exile in Central Arabia. I would go further and say that if, as I venture to predict, the twentieth century does for Arabia what the nineteenth century did for the exploration of Darkest Africa, and if that very large and ugly yellow horrid blank which we saw to the south of the central part of the Arabian map ceases some day to be large, yellow, ugly, or blank, the credit will be largely due, although indirectly, to Sir Percy Cox.

There is one point to which Sir Percy has not, I am sorry to say, had time to refer, though it is raised in his paper, which will in due course appear in the *Journal*. Sir Percy Cox quite calmly suggests that the riddle of the Great Desert, one of the greatest riddles remaining for geographical solution at the present day, could be solved by aircraft. I think that is a perfectly horrible suggestion. I hope we shall hear no more of it. But again referring to his paper and not to what he said to-night—I am sorry he had not time to say it—it is very interesting to me to note that the information he got on the Oman side of the desert exactly confirms and corroborates in detail the information which I was able to collect on the other side at Wadi Dawasir; that is to say, on the north-west side. It is of very ominous significance that the very three tribes that he mentioned in his paper in connection with crossing the desert are the very three tribes that my friends from the other side told me they were in the habit of raiding. That is very interesting, but in the circumstances it would, of course, be very important to make quite sure of the kind of reception one is likely to get on the other side before starting. But there is—and Sir Percy Cox is able to support this—no really insuperable difficulty in the way of such a venture, though any attempt ought to be made, not by aircraft, but in the ordinary way of riding there on the back of a camel, and not behind the hump! And I would say that I think one has every right to ask His Majesty's Government not to add unnecessarily to the natural difficulties of such a venture by putting official obstacles in the way of those who are foolish enough to want to do anything so silly. After all, there are some people in these days of overcrowded populations who could easily be spared from this world, and if the worst came to the worst—and the person taking the risk would take care that it should be as small as possible—one would not really expect His Majesty's Government to send battalions of infantry into the waterless wastes to avenge one's death. I speak with a certain amount of feeling and reason on that particular point.

Sir Percy Cox, you will remember, had to hurry over the last part both of his lecture and his journey ; the reason for the latter was that he had to catch H.M.S. *Amphitrite* in order to join her for the purpose of a cruise in the Persian Gulf, the object of which was to show the British flag. I venture to think he did not justify the precipitancy which he showed in the matter. He, it is true, went down to join the *Amphitrite* to show the British flag, which after all was well enough known in those seas, but the *Amphitrite* could, I am sure, have shown the flag under less expert guidance than that of Sir Percy Cox. I would suggest that he was engaged in a very much more important task in showing the flag in an area which on his own showing was an area which did not know the flag. It is interesting to remember that a succession of British visitors to those parts has resulted in a progressive improvement in the manner of the inhabitants thereof. For instance, as Sir Percy told us, Wellsted had a poor reception at Ibri, whilst fifty years later the next visitor, Colonel Miles, had quite a good reception at Ibri, but rather an uproarious one at Mazum. Sir Percy went through both with flying colours ; in fact, he got to Tanuf before he had trouble, and that only of a mild order. If only every British political agent in Muscat, once in the course of his incumbency, would make an excursion into the interior until he was stopped by the rudeness of the natives, the interior of Oman would in the course of a few centuries come to be as well known as it ought to be considering the length of time that it has been under the benevolent protection of Great Britain ; and the same applies to an almost greater extent in the case of the Aden hinterland.

There is another point of considerable importance, and one on which I feel somewhat strongly. You will realize from what Sir Percy Cox has said that all his predecessors in the exploration of that part of the country, the hinterland of Oman, number no more than three, and those travelled there at very wide intervals—1835, 1885, and 1901. In the circumstances I think it would be permissible and excusable to call Sir Percy Cox's attention to a somewhat serious omission. I seem to remember some time ago—and I refreshed my memory on the subject to-day—that Palgrave, in speaking of his travels in Arabia, mentions having made certain little excursions of the same description that Sir Percy told us about to-night, into the interior of Oman from Muscat. They were nice little excursions. I was trying to puzzle out the geography of them this afternoon, but the misfortune is this : that all the places at which Palgrave stayed, and where he was entertained, appear to have disappeared from the map. I should like to have heard what Sir Percy Cox thought of those excursions, and I regard his omission as a little strange in view of what he appears, according to the official report of the Society, to have said at the meeting at which Major Cheesman read a paper. He there made a definite attempt to support the reputation of Palgrave. It is a long time since I talked about Palgrave, and I do not propose to talk very much about him this evening. As, however, I was not present at the meeting at which Major Cheesman took advantage of my absence to resuscitate the Palgrave bogey and was supported by Sir Percy Cox and Dr. Hogarth, I would like to take this opportunity of exploding a little theory in regard to which Dr. Hogarth is very persistent whenever he rises to defend the memory of a man who, after all, cannot defend himself. I have no objection to his doing that. But Dr. Hogarth explains Palgrave's admitted digressions from the path of strict truth on the ground that he had the great misfortune to be shipwrecked, and that during that shipwreck he lost all his notes. A statement of that kind made before a Society of this description by a person of the kind that Dr. Hogarth is ought to be considered absolutely convincing, for Dr. Hogarth is the greatest living authority on Eastern questions, and on Arabia in particular. I am very sorry indeed that Dr. Hogarth is not here this evening, because if he had been I should like to have asked him on what authority he makes the statement that Palgrave lost his notes on the occasion of his alleged shipwreck on the coast of Oman. He certainly did not get it from Palgrave himself, who goes out of his way to explain that he was rather proud at not having lost his notes on that occasion. As a matter of fact, on 23 January 1863, after his reputed wanderings in Arabia, he parted from his friend Barakat, who afterwards became an Archbishop, and to that man he consigned all the notes he had taken during his wanderings, and he goes out of his way to tell us that he took that precaution because he had a presentiment that he was going to be shipwrecked, and he adds that if he had not taken that precaution his book would never have appeared. The notes were perfectly safe and sound when Palgrave, after his shipwreck, rejoined his friend. He was duly shipwrecked according to his presentiment, but he had taken the precaution to preserve his notes; the only ones lost were those between 25 January 1863 and the date of the shipwreck, March 10. But even that line of defence does not explain Palgrave's vagaries in describing Oman, which he visited after the shipwreck. There is one more point suggested by the mention of Palgrave. Sir Percy Cox had considerable difficulty in regard to the name of Jabal Hafit, and says that Dr. Samuel Zwemer calls it " Jebel Okdat " or " Jebel Akabat." But why does he single out Dr. Samuel Zwemer? After all, the authority for the name of Okdat was Palgrave, and one of Palgrave's companions was actually a resident of the village of Okda. That seems to have escaped Sir Percy Cox. As a matter of fact, he is no doubt right in saying the name of the hill is Jabal Hafit after one of the villages at its foot; but if there is a village Sir Percy Cox has not seen on the other side of the foot called Okda, the hill might quite well be called Okdat also. The Arabs moreover have a peculiar method of nomenclature. They may call a place by what it looks like. Sir Percy told us Jabal Hafit was a ridge about 20 miles long, with a pass or col in the middle. That is what the Arabs would call " Aquabat," meaning a pass. I am not sure about his interpretation of the word Okdat in Arabic. It may mean " sandhill," but I should expect it rather to refer to a difficult pass between sandhills ; the word " Okda " ('Uqda) of course means a knot.

The PRESIDENT: If there are no other speakers I will convey to Sir Percy Cox the gratitude of this meeting for the time and trouble which he has taken in resuscitating the notes of his journey and giving us so delightful an account this evening. We now regard him as a most admirable photographer, for he showed us this evening one of the finest series of views of desert country which it has ever been my fortune to see.

Note added by Sir Percy Cox :

Owing to the lateness of the hour it was not possible for me to make any rejoinder to Mr. Philby's observations ; but as it is of some little geographical importance to clear up as far as possible existing doubt as to the name position, and extent of the " Jebel Okdat," it may be convenient if a summary of the record accompanies this paper. The italics are mine.

1835. Wellsted, in the map accompanying his ' Travels in Arabia,' shows a Jebel Okdat range, but in quite a different position from that discussed above. Wellsted's range begins 40 miles south of Baraimi and continues for 20 miles' in a south-south-westerly direction. He shows no hill feature near Baraimi, which he places about 15 miles west of its true position. He did not himself reach Baraimi, so what he recorded was necessarily from hearsay.

1863. Palgrave, in his 'Eastern and Central Arabia,' writes: (a) (vol. 2, p. 301):

"On the morning of 16 February 1863 we sighted the Oman coast between Aboo-Debee and Dobey—long, low, and sandy, but well lined with palm groves and villages arranged along the glistering shore. *Far in the distance like a cloud rose the heights of Bereymah or Djebel Okdah.* . . ."

(*Note.*—From Palgrave's position at the time, in a sailing vessel some miles from the coast ("in sight of the coast," he says), the Bereymah Oasis could not have been much less than 100 miles distant.)

(b) Vol. 2, p. 339. Referring to his companions, on his voyage by dhow from Sohar to Muscat, during which he was ultimately wrecked, he writes:

"Besides these we had on board ten other fellow-passengers: two from Djebel Okdah . . . I now learnt from the men of Okdah many particulars regarding Bereymah and Dahireh, which may here find place. Of Bereymah itself they said that the town was strongly nestled amid the passes of the Djebel Okdah ; . . . that Djebel Okdah itself was a lofty mountain equalling in height the range of Roos-el-Djebel : that the soil was light and the vegetation not less luxuriant than the Batineh ; that beyond the Djebel Okdah extended, east and south, a long series of hills parallel to Djebel Akhdar."

(*Note.*—In the foregoing Palgrave does not mention or necessarily imply that there is a village of Okdah, and neither Miles nor I heard of any ; but as regards the mountain range, even if the meaning—" sandhill "—which I suggest for Okdah, with good authority, is accepted, it is no doubt conceivable that Okdah—" Sandhills village "—and Djebel Okdah—" Mount Sandhills "—may be used synonymously for Hafit. Palgrave places Bereymah in approximately its correct position, but he shows a mountain feature, representing or including his Djebel Okdah, stretching from a point 30 miles north of the oasis to Bahla, 120 miles south, *i.e.* 150 miles in all, whereas the feature, whatever its name, is actually 20 miles in length.)

1871. In the map issued by the Hakluyt Society in 1871, in connection with the Rev. G. P. Badger's 'Imams and Seyyids of Oman,' Jebel Okdat is evidently inserted on the basis of Palgrave's erroneous location of the feature.

1885. Now let us hear Colonel Miles on the subject. In his Muscat Administration Report for 1885-1886, he writes :

"The Wadi Dthank and the Wadi Boo Karba both join the Wadi Safa, a large watercourse in the Ramool desert, running to Abu Thabi, almost parallel to the Wadi Ain: *the Jebel Okdat of Wellsted is purely mythical; there is no such range in that direction whatever except Hafit, near Beraimi.*"

When Miles passed up to Baraimi from Sohar in 1875, he came in full view of Hafit or whatever the mountain feature was in that neighbourhood, and when making the above remark he makes it from first-hand knowledge ;

furthermore, he shows Jabal Hafit as an isolated mountain and in its correct proportions in the paper on his journey contributed to the *R.A.S. Journal of Bengal*, vol. **46**, p. 41, 1875-76.

1902. Dr. Zwemer, in the *Geographical Journal* for January 1902, p. 62, writes :

> "On Tuesday we reach the oasis of Beraimi, a 4-mile stretch of fertile palm country under the shadow of Jebel Hafid, *the first spur of the Okdat Range.* It seems, after careful inquiry from several Arabs, that the true name or at least a second name for this mountain range is Jebel Okabat. The first name signifies ' knots,' the latter 'deep defiles': both names are appropriate to the rugged outline of the range as seen from the desert. . . . *Beyond Beraimi, the road along the Jebel Okdat range passes the following villages :* Hafid, Senanah, El Felai, Dank (or Danj) Jabil, Subaihi, Mamur, *Abri.* . . ."

I think it may be fairly deduced from the above, and from my personal observation and notes, on my way from Baraimi to Ibri, that while the expression Jabal Okdat may well be used by local Arabs synonymously for Jabal Hafit, there is without doubt, no continuous mountain range in the locality in question other than that referred to by Colonel Miles and myself as Jabal Hafit, a feature starting 5 or 6 miles outside Baraimi and stretching for 20 miles southwards. Wellsted went off the line in the first instance, and apparently Palgrave, Badger's cartographer, and Zwemer followed him.

P. Z. C.

Dhank consists of two parts: 'Alayah ('Upper') and Sufala' ('Lower'), the former consisting of about 400 houses in five walled quarters and the latter of about 300 houses in seven. There is a long, narrow fortress in Upper Dhank, and the people belong mostly to the Na'im tribe, cultivating rich date plantations. In the mid-1980s the visitor is struck not so much by the picturesque beauty of old Dhank as by the great new schools, spreading over a wide area, and the magnificent new wadi-retaining wall along the wadi bed, protecting the villagers and their crops from flash floods. We drove along the bed of Wadi Dhank, the blacktop road having given way now to seasonal tracks, past the village of Dut, inhabited by Sa'idah and Bidah; Fida', belonging to the Bani Zid, with their rough soccer pitch and impressive castle isolated features in a seemingly endless green ribbon of date-palms along the left side of the wadi and finally Yanqul, after we had waded through pools in the wadi to test whether the four-wheel-drive vehicle could pass through. That's a lie, surely: we simply felt the need for cool waters to heal hot, sweaty limbs, aching from the bounce-bump-bounce of graded roads. Yanqul, a stronghold of the Nabahinah who dominated Oman in the early seventeenth century, allows the lungs to breathe freely. In a contaminated and polluted world, its pure mountain air provides invigorating solace. The *suq* was closed for prayers at 2.40, and the Restaurant Sulaiman bin Salih al-Farsi presented closed shutters, so we ate at Hamdan al-'Alawi's,

Dhank. Wadi-retaining wall

enjoying the *lavash* (thin unleavened bread) one can eat wherever Persian or Urdu are spoken, yoghourt, goat's meat, onions, and sliced tomatoes. To allow in a merciful breeze, the Indian customers had drawn back the mosquito nets on the windows, enabling the mosquitoes to enter without going round by the open door.

Wadi Hawasinah

We filled the petrol-tank at the Fida' filling station on the good main road, and followed the track ever gently downward towards Miskin (2,750 feet above sea-level) and the Batinah, via Khudal, al-Aridh with its big new school, small fort on the right, and swept down towards Miskin, through barren rock, mirages rising like *djinns* to meet us on the road ahead. Fifty-five km. beyond Fida' the asphalt road ends, and one can understand why: there is so little traffic. Even so, the people whose business is with the Batinah as opposed to 'Ibri and the Dhahirah would be well served by a continuation of the blacktop to the

334

coast and it may happen one day. At least as far as Miskin, a word meaning dwelling or home, and inhabited mainly by the Bani Kalban. I observed new houses, fresh-painted white and brown, among the flourishing date-groves, wheatfields, lucerne fields, and orchards where mangoes and figs were growing fat and juicy. Old watchtowers reminded us of past bloodshed, bitter rivalries, when no man felt secure from other tribes, and loyalties shifted by the season. In a few kilometres the track divided, unmarked; we chose the right then, unsure, called out towards a *badu* encampment (not entering uninvited) for instructions. No: we had taken the track towards Rustaq by mistake, and had to head back the way we had come, and take the right-hand fork, passing a *badu* girl collecting firewood. The track winds above and beside the Wadi Hawasinah, with optional wanderings down into the wadi bed during the dry season. If you choose the upper route, the rocks are more sharp and the bends more tortuous, the whole process slower but safer; the lower route will be unexpected even to those familiar with the same way some months earlier, for stones and even boulders may have shifted, rendering a familiar way now impassable, now suddenly easier. It looks to the driver as though Wadi Hawasinah has been set as a kind of ultimate examination of patience, ingenuity, reflexes, and doggedness; as if the Grand Master had laid out a chessboard with almost infinite squares and endless choices, each of which might wreck your vehicle or your body, and even end your life in hysterical vengeance for your effrontery in attempting the task. Boulder-strewn, overhung with crags, the track defeats all but the most determined adventurer. How hard must we brake to avoid an oncoming vehicle around this hairpin bend? Should we risk descending into the wadi bed to avoid puncturing the tyres on these granite claws, or will the car sink there into soft sand? We are perpetually certain that we shall reach Khaburah at the end of the day, but our progress is so necessarily and painfully slow, hitting rock after rock, skidding on surfaces polished clean by wind and weather, waiting while a goat skips across the track, that we almost consider getting out to walk, but then we should have to return to fetch the vehicle, so we scrape on, cracking our skulls on the roof, grazing our elbows on the door, tick-tocking from side to side like a pendulum in shipwreck.

At last we arrive at Miqzi, a village not on the map, in a kind of dazed triumph: after all, how many travellers have managed to get this far, from Dhank and Yanqul to unmapped Miqzi? Even if I were to pass out now, my bruises indicating some rare type of mediaeval torture, I should have been well pleased. But no, the car is still capable of going ever onward, and we descend again in a welter of grey, darkling rocks, abandoned stone dwellings, and eyed watchtowers resting in the twilight: Ghaizain, Shakhbut, a thick village wall through which the track passes on the right of the wadi, al-Qusf off to the left, its name like all the others a kind of barking defence against intruders, and as the darkness sank on the mountains and valleys like a tent, the sheer slopes subsided gently and we found ourselves speeding along the plain to lights, to lights on the coast. Here we were at Khaburah, on the Batinah road leading to the capital. Somewhere between Suwaiq and Wudam al-Ghaf we pulled off the road, dragged sleeping bags on the beach where waves lapped far off, and the moon winked at me as I fell asleep. Mountains overlooking the great Wadi Hawasinah rose and shone with roselit dawn in my dreams, and then at last came the first silence of the day.

I was awoken by the voice of Abu Fa'iz, singer of Sur, booming out from the cassette player within the Nissan Patrol. Gulls squawked out of unison, one winging low across the sun in sudden silhouette. Lucidity refreshed every particle of the morning, as though darkness could no longer revive after such clarity. I washed in the lazy surf, watching as fishermen dragged in *huri* after *huri*, each with its gleaming Yamaha motor. Fish glistened with the sheen that only fades after hours on dry land. The men, barefoot, with eyes crinkled up in wrinkled faces tanned dark brown, hailed me cheerfully, offering their catch more in hope than expectation of a sale.

Al-Hazm

It was time to explore the foothills and wadis of the Western Hajar. Instead of taking the tracks south from the Batinah highway along Wadi Mabrah, Wadi Bani Ghafir, or Wadi al-Far', I chose the main road that leads to Rustaq via al-Hazm. The reason is pure history. Rustaq was the capital of Ahmad ibn Said, founding Imam of the Al Bu Sa'id dynasty, who died in 1783 at Rustaq, and was buried in its fort. Rustaq has been the market town for the whole of the Western Hajar ever since, with an influence disproportionate to its modest size. From the Batinah the plain remains completely flat and barren, with few animals or *badu*, until you come to a turning left 9 km to Jamma, dominating Wadi al-Far', and right 1 km to al-Hazm, along a rough road. We reached the fort, still in use for police purposes, at 8.30 in the morning, and presented the necessary letter of permission from the Ministry of National Heritage and Culture. The tiny village has no government electricity, so the fort has its own generator. I waited while a new power cable was being erected. One kindly policeman let me turn the knobs on his remote control radio, which emitted those faint burbling, howling noises that one associates with haunted manors in horror movies. Another showed me his antique rifle, explaining its mechanism while pointing it accidentally near my stomach. All wore *dishdashas*, and presented me with ceremonial coffee before the tour could begin. It had begun in earnest at the massive wooden door, sufficient to repel all those early invaders in the decades after the great castle was constructed by Sultan ibn Saif in 1708. Elected Imam in 1711, he chose to remove his capital to al-Hazm in 1711, and was buried here in 1718, in a walled-up tomb. Entering on the right, you pass a well, then a round tower with a water-channel (which still runs with fresh water as it was always intended, during sieges), then a tin bath full of cannonballs, and a round store-room for dates.

The gentle gradient of the wide stone staircase was clearly planned so that horsemen could ascend while still mounted, then even more powerful wooden doors. Three sets of prisons can be visited, each more awe-inspiring than the last, and a missing step in a stairway (intentional, surely?) reminded me of that heart-stopping missing stair in the *Kidnapped* of Robert Louis Stevenson. Secret tunnels are said to lead outside the fort to safety, but drifting sand must have silted up such exits, if they ever existed outside the realm of rumour. A mosque, with a Quranic school above it, is reached by stone steps. Stone carvings, plaster mouldings: everywhere details of this castle reveal hands at the behest of a royal master. Then I was showed into a room where boiling honey, oil or water could be poured down on to invaders who had managed to arrive thus far. A passage led to a round tower with three cannon, one roughly mounted and pointed to the plain, one pointing thoughtfully to the massive inner walls, and the third mounted too low, so it would hit the stone out below the window. I picked up a stray locust that had ventured into al-Hazm stronghold and flicked it carefully out of the window, into fresh air. Below I saw lemon trees, though their scent was too faint to detect. Beyond a narrow prison, the Imam's room, and soldiers' dormitories I found the second round tower, again with seven windows and three cannon, one showing the name of Don Felipe II, Rey de España, with the date 1591, seven years before the King's death.

Out of the window lies the plain, panting in the haze like a tawny lioness. History can behave like gossamer tracery, as at the Palazzo Ducale and Bridge of Sighs in Venice; or it can suffocate in darkness like the Palace of the Popes at Avignon. At al-Hazm I felt mislaid, as though in a grotesque vision I had been born here, 'untimely ripp'd' as Macduff had been, and kept incarcerated for no fault of my own. A lizard eyed me from the window-sill, moving its body as its heart beat. 'Palpita la lucertola', I kept repeating as a talisman. 'The lizard palpitates' and I too feigned unknowing, treading warily behind the armed guard, careful to say nothing that might result in my apprehension on a charge of trespass, rebellion, or treason. The solitude of al-Hazm was to become a kind of delightful dread, yet another sense in my nervous system, that affects me even now, ivory-towered in an English university city set with hedges, lamp-posts and semi-detached houses. I recall as if I am there at this instant walking out on to the burning-hot roof of the second round tower (again that duality, as in Hinawi-Ghafiri, Alayah-Sufala'), finding two more cannon, and yet another beside which I gazed out on the village, fed by underground *aflaj*. Near the entrance I found 1126 (corresponding to 1714 A.D.) carved in Arabic numerals on a wooden door.

The sun rushed up to meet us as we bade farewell to our kindly hosts at al-Hazm, idiotically, because they were going nowhere: what we actually said was 'Ma'a as-salama, wa shukran jazilan' ('With peace, and many thanks'). We saw ash-Shubaikah on our right, back on the main road, then Mizfah was signed off left, and al-Ma'idah 12 km off left, before we came into sight of the cluster of villages known as Rustaq.

Rustaq

Rustaq has a bus stop at the suq near the fort with departures daily to the Capital at 6.30, 12.30 and 16.30. The dawn bus reaches Khaburah at 7.05, Suwaiq at 7.40, Musan‘a at 8.15, Barka road at 8.40, Seeb at 9.15, Ruwi bus station at 10.10 and Matrah fish market at 10.20. Near the bus station I was welcomed into Nakhl Sweet Factory by Jum‘a bin Muhammad bin Humaid: a great vat of sweet, sticky *halwa* was being stirred, and I bought a small jar to taste. A van from Unikai Dream Ice Cream of Ruwi tootled past the open doorway.

Rustaq, an oasis certainly inhabited from pre-Islamic times, is dominated like most Omani towns by its castle, Qala't Kisra or Kisva or Kisfa Sharwan, depending on the correct name of the Persian military governor, or *marzuban*, who built it early in the 7th century. We went to see the hot springs first. Called 'Ain Kisfa presumably after the same *marzuban*, passing below a great rock outcrop, and finding it (not as hot as those near Muscat) newly tiled, beside a small cemetery. To go there, take the turning off the main road at Salah Cold Stores corner. I stopped to admire a beautiful traditional house with carved wooden windows, again respecting the Omani Government's sound policy towards new building: it can be ultra-modern, with air-conditioning and a garage and garden, but it must still look as at home in Oman as do the forts of the Jabal or the *huris* of the Batinah. Rustaq has never failed for water, making it a sensible, centrally-located oasis from which to rule the interior. So thought Nasr ibn Murshid, who moved the capital from Bahla to Rustaq when he took power as the first of the Ya‘rubid rulers in 1034 A.H. (1624 A.D.) and that is where the capital remained, with a few interludes (Jabrin twice, al-Hazm, Nizwa) until the death of the second Al bu Sa‘idi, Imam Said ibn Ahmad, who died in 1779. The central fortified complex, so easy to defend, incorporated not only the castle, but store rooms and rooms for people and space for animals to withstand a siege as long as necessary, as well as the Friday Mosque. The greatest ruler of Rustaq was Saif ibn Sultan, appointed Imam in 1692 after a series of bloody quarrels with his brother Bil‘arab. In his *al-Fath al-mubin*, the Omani historian Humaid ibn Ruzaiq al-Bukhit records that Saif owned 28 ships, of which the largest had eighty guns and may have been captured from the Portuguese. He reputedly owned seven hundred male slaves, and a third of all date-groves in Oman. He ordered 30,000 date palms to be planted, over 6,000 coconut palms, and repaired many of the *aflaj* which had fallen into disuse. Humaid's figures roughly coincide with those of Lockyer (*An account of the trade in India*, 1711) and Hamilton (*A new account of the East Indies,* 1730) whose notes related to 1715.

Since this book is concerned with travels in Oman, we should also

338

note travels of the Omanis, mainly within the Gulf area, towards India, and specifically to East Africa. Imam Saif, in 1696, despatched a naval force to rescue the inhabitants of Mombasa from subjugation by the Portuguese. His seven warships and ten dhows carried 3,000 Omani fighters, Negro slaves, and Baluchi mercenaries, who finally broke the terrible siege in December 1698, after thirty-three months, and stormed the walls. The only defenders who had survived the great plague of Fort Jesus were eight Portuguese soldiers, three Indians and two African women.

Remember while visiting Omani villages and oases that ancestors of the people you meet may have lived and died far away. They may have known Zanzibar, Kenya, Coromandel, not to mention Hurmuz, Gwadur, and Basra. They will have made the blessed Hajj to Makkah al-Mukarrimah, and travelled overland to Dubai and Abu Dhabi, Fujairah and Khor Fakkan.

Rustaq is not only strategically important: it possesses no fewer than five principal *aflaj*: Abu Talib, Ma'azah, Hammam, Sikhrah, and Tahiri. You can see an inverted siphon, and you should not miss the hives, which have flourished from at least the seventeenth century, and probably much earlier. A governmental beekeeping project helps apiculturists to teach farmers new techniques and foster the best of the old ways. Green alfalfa fields and large date plantations refreshed eyes weary of sandy brown. We drove out to Burj al-Mazra'a, 'Tower of the Farms', the western limit of Rustaq, and saw colourful parachutes, red and white, of the Oman Parachute Squadron, dotting the sky like dark, heavy, slow raindrops far off.

Wadi Bani Kharus

A permit from the Ministry of National Heritage and Culture is necessary to visit Rustaq castle, which has recently undergone restoration; but even from the outside, the sun-dried mud-brick walls and turrets are attractive, romantic, and essentially Omani.

Every village around Rustaq, and every wadi has its own distinctive attributes if you care to discover them. Take 'Ain 'Amq, for instance, ('Deep Spring'), a terraced village high up Wadi Sahtan, on a graded road some way off the main road. Here, the Khaburah Textile Project of 1981 had expanded in 1983 to cover the villages of 'Ain 'Amq and Mabu. A weaving expert from Khaburah came twice a week to teach the women weaving, and the number of weavers soon rose to seventeen, first on ground looms, then on tapestry frames, and finally on floor looms. The looms had been made in Rustaq, and the wool for weaving was spun there too: goat-fleeces for rougher products and sheep-fleeces for finer quality. The village women, once limited to tending goats and the fields while their husbands were earning a living principally in Rustaq, now have a chance to earn extra money with extra skills, and are offered a 50% subsidy if they wish to buy their own floor-looms.

'Awabi and Wadi Bani Kharus

From Rustaq you can take Route 13 to 'Awabi, a distance of 17 km spent hugging the foothills of Jabal Akhdhar, a sign 'Beware of the Camel' offering one of the few signs of activity: but we saw no camels, even so.

One splendid journey for the experienced driver with access to a four-wheel-drive vehicle is the route from the Rustaq-'Awabi road into the fastnesses of the Jabal, sometimes becoming steep to dangerous, though improvements have recently been made to widen it at the most difficult passing-places. You can take other tracks off the main graded road to see individual villages. It is possible to avoid returning the way you went by continuing up the track to the very end of the wadi, where it joins Wadi Sahtan, but I advise you to do this only during good daylight.

'Awabi is situated on the left bank of the Wadi Bani Kharus, and belongs chiefly to the 'Abriyin and Bani Kharus tribes. At an altitude of 600 metres above sea level, the oasis is comparatively cool in summer, like Rustaq, and it too is noted for its bees and its fort, Bait al-'Awabi, which controls the road from Nakhl. I found twenty new houses, mosque and shops being erected by the Ministry of Social Housing. The drive along the wide track in the bed of Wadi Bani Kharus is pleasant, if dusty; I did not see any of the rock drawings carved into the limestone at the wadi mouth, but then I had not been impressed with other schematic rock carvings in Oman which, like those in neighbouring areas of the Gulf, show little artistic aptitude. Instead we sped swiftly up the fast sandy track to the village at the head of the wadi, of three- and four-storey houses: a walking track leads from here to Tanuf over the Jabal, or across to Wakan and Muti at the head of Wadi Mistal.

Wadi Bani Kharus

Wadi Mistal

Miles on the Jabal Akhdhar

Back in 'Awabi, past gardens of maize, millet, lucerne and barley, we took the sign for Nakhl. S. B. Miles visited Nakhl first in 1876. 'Across the Green Mountains of Oman', here reprinted by kind permission of the Editor, *The Geographical Journal*, describes that visit, with accounts of 'Awabi, Istal, Saiq, Shuraijah (perpetuating the myth that it was named as a diminutive of Shiraz), Izki (his Zikki), and Sum'ail.

ACROSS THE GREEN MOUNTAINS OF OMAN

By Colonel S. B. MILES.

JEBEL AKHDAR, or the Green Mountain, as the Arabs call it, lies about 60 miles south-west of Muskat, and is the central and culminating point of the great chain forming the backbone of Oman. Viewed from the nearest part of the coast, some 40 miles away, this gigantic mass looms, a most conspicuous and majestic feature, in the distance, its dark precipitous front rising abruptly, cliff above cliff, in wild and desolate nakedness to a height of nearly 10,000 feet. The sight of the mountain is so impressive that from the moment I first saw it I made the determination to visit it, and the accounts I subsequently heard of its inhabitants and pensile gardens increased my desire to make its acquaintance. But the country at that period was very unsettled, tribal and dynastic wars being of perpetual occurrence, and it was not for several years that a fitting opportunity for carrying out my design presented itself.

In the middle of the year 1876 I found myself able to undertake the trip, and the Sultan having caused instructions to be despatched to the Governor of Nakhl to make arrangements with the Beni Riyam who possess Jebel Akhdar, and other tribes on the way, to provide an escort, I sailed from Muskat on June 27 in a native boat for Sib, where I landed and camped for the night in a garden near the shore. The next morning, having procured camels, I started in company with two sheikhs, who had come from Nakhl to meet me, for Burka, where I arrived about mid-day. After an interview with the governor and a short rest, we resumed our journey and struck inland in a direction almost due south, the path leading along the Wady Hammam.

We were now traversing the eastern part of the maritime plain known as the Batina, which, except in the oases, is an uncultivated desert, the surface being in some parts sandy, in others stony. At 10 miles we came suddenly upon a small but rich oasis named El Wasit, near the junction of the Wadis Hammam and Maawal, where we experienced, in the cool and humid atmosphere of the date groves, a most refreshing change from the heat and glare. We now ascended the Wady Maawal, and, skirting the base of the Nakhl range, arrived, after another hour or so, at the village of Hibra, where we stopped for the night at the house of Sezzid Ali, a grand-nephew of the Sultan. The Wady Maawal is known in the upper part as Wady Mayin, and has several villages, viz. Jenab, Hibra, Afy, Musalmat, Wasit, etc. The Maawal tribe to which it belongs is Kahtanite, and numbers about twelve thousand souls. In the fifth or sixth century of our era this tribe began to assume a predominant position in the country, as the Julanda princes who then ruled over Oman belonged to a Maawal family, a circumstance which gave the tribe a status analogous to that held by the Âl Bu Saidis at the present day. The Julanda dynasty is supposed to be referred to in the Koran (Sura 18). It acquired great reputation in the Moslem world from the action of the brothers Abd and Jaifar, who reigned at Nezwa in the Time of Ignorance, and who accepted the new faith immediately on receipt of the prophet's letter of invitation, and championed it against the idolaters in Oman. The persecutions of Mowiya and Hejaj brought the dynasty to an end, and compelled the last of the line to seek safety in exile. The Maawal tribe was ever unfriendly towards H.H. Sezzid Turky, and was severely chastised in 1883 for their share in the attack on Muskat.

The following day we entered the ravine leading to the secluded glen where lay embosomed the town of Nakhl, or, as one may call it, for the three words have the same signification, Palmyra or Tadmor. The approach was one of striking and impressive singularity. We were now close under the lofty mass of Jebel Nakhl, which, rising abruptly, towered over us to a height of 5240 feet, and as we rode up the winding torrent bed, it seemed as if we were about to penetrate the very bowels of the mountain. No sign of human habitation, no cultivation, no gardens were visible, nothing but dark and desolate rocks met the eye; the silence was profound, and I was wondering where the town could possibly be, when from above, in front of us, several matchlocks were suddenly discharged in our direction, and I perceived a watch tower perched on a steep pinnacle 200 feet high, standing guard, as it were, over the entrance, and from which the sentries had fired to give notice of our approach. Rounding an angle, we were now confronted with the massive ramparts of the fortress, which, warned by the watch tower, immediately began to fire a salute from a battery of twelve-pounder iron guns, the sound of which reverberated sharply from the rocky walls of the glen. The town itself, however, still remained invisible until, skirting the pinnacle, we passed under a two-arched viaduct, when the whole settlement, houses, palms, gardens, orchards, and cultivation, burst upon the view, presenting a scene of a very picturesque character. It was getting hot when we arrived, and though the sun poured its scorching beams upon the black rocks around us, and heated the air to an insufferable degree, I could not help stopping for some minutes to gaze upon the scene, and to admire the remarkable strength of the position.

I was very cordially received by the Waly, or governor, Sezzid Salim bin Khalfan, who came down to meet me and conducted me to a small house in a pretty garden that had been prepared for me, and for the use of which I was indebted to the courtesy of one of the Sultan's officials. Here I was detained three days by the dilatory habits of the Arabs, and then, having received letters requiring my presence at Muskat, I returned thither, by the shortest route. A week later, July 9, I started again, accompanied this time by Mr. Maguire, the B.I.S.N. Company's agent, and, taking a direct course, we made our first stage at the village of Halban, not far from Jebel Tau, the abutment of the Nakhl range, where we camped for the night. We found Halban a place of unusual industrial activity, the people being busily engaged in the manufacture of indigo dye from plants extensively cultivated in the neighbourhood. It is prepared in large earthen jars, and the dye is used to give the imported textile fabrics of Manchester that dark blue colour so highly favoured and almost exclusively worn by the women of Oman of all ranks and conditions. In the evening we inspected the village school, which was near our camp, and was held in the open air like an Indian pâtshâla, the sacred peepul being here represented by a large mango tree. The pedagogue was an old moolla, rod in hand, and among the twenty-one children sitting at his feet we counted five girls, one of whom was learning to write. Among the crowd of visitors that came to see us was the Sheikh of Al Tau, who pointed to his town in a west-south-westerly direction, and gave us a pressing invitation to visit it, saying it was three times as large as Halban. The next day, skirting the base of the mountain, we rode on to Nakhl, crossing numerous nullahs or ravines, and passing two villages, Farah and Lajaali, on the way.

The houses at Nakhl are built of sun-dried bricks or stones plastered over; many are high and spacious, and, though with but slender pretensions to architectural beauty, not destitute of exterior decoration. The lintels are often carved, and the doors ornamented and strengthened with pointed iron knobs or bosses. The windows are never glazed, but are closed at night with strong wooden shutters, and are sometimes furnished with mashrabiyehs, sometimes protected by strong iron cross-bars. The interiors are badly planned; the stairs are narrow and steep, and the upper apartments long and narrow. Plaster or cloth ceilings are not in vogue, but the teak beams and rafters are often handsomely carved or painted in various devices. The windows are usually placed very low, so that the occupants reclining on the floor may look out of them, and at the top of the room circular holes in the wall serve to assist ventilation. On the floor are carpets or mats and cushions, but other furniture is scarce, and tables and chairs, of course, not to be seen. Strong wooden brass-bound boxes are the receptacles for apparel and valuables; and round the room are ranged broad shelves, on which is displayed a quaint and wondrous assortment of cuckoo clocks and other timepieces, coffee-pots, china figures and ornaments, English and Indian toys, and a variety of other curios, highly valued by their owner. In this ardour for collecting, as well as in the style of house decoration, Persian taste is very perceptible. Most of the houses have a small garden attached.

A noticeable fact here, it may be observed, is the mixed character of the population. This comprises, besides Arabs, Persians, Negroes, and Zattut, a very large proportion of Bayâsir, a race supposed to have emigrated originally from Hadhramaut. The Bayâsir are an industrious and peaceable folk, and many of them are wealthy, but are held as aliens by the tribal Arabs, and are never entrusted, I believe, with positions of authority and command. When a Bayâsir happens to meet a sheikh on the road he will not go up to kiss hands and give salutation without first dropping his sandals by the side of the path, after the manner of servants and inferiors.

Nakhl is abundantly supplied with water, and, indeed, one of the most characteristic features in its physical aspect is the existence of copious thermal springs, the medicinal virtues and efficacy of which are famous in the land. The chief fountain, or rather group of fountains, is called Hammam Thowâreh, and lies at the head of the ravine, a little distance from the town, among fragrant gardens and shady palms and mangoes. The sight is very curious. In the midst of a numerous family of smaller springs, the father of fountains gushes up out of a cavity in the ground, apparently about 6 feet in depth and 2 in diameter, and pours forth a volume of water, roughly calculated by me at 200 gallons per minute. This is the hottest, as well as the largest, spring at Nakhl, and has a temperature, I found, of 106° Fahr. There are at least twenty other springs in its vicinity, yielding altogether a very bountiful supply, but the temperature of those I tested did not exceed 104° Fahr. They are all tasteless and inodorous. On the other side of the town is a similar assemblage of springs called Hammam Odaisee, the most prolific of which issues from a hole in the rock and is led into a tank, from whence it flows to irrigate the fields. This spring was 105° Fahr., and another near it was 102° Fahr. Notwithstanding the high repute and universal belief in the curative properties of these waters, I did not observe, either here or at Thowâreh,

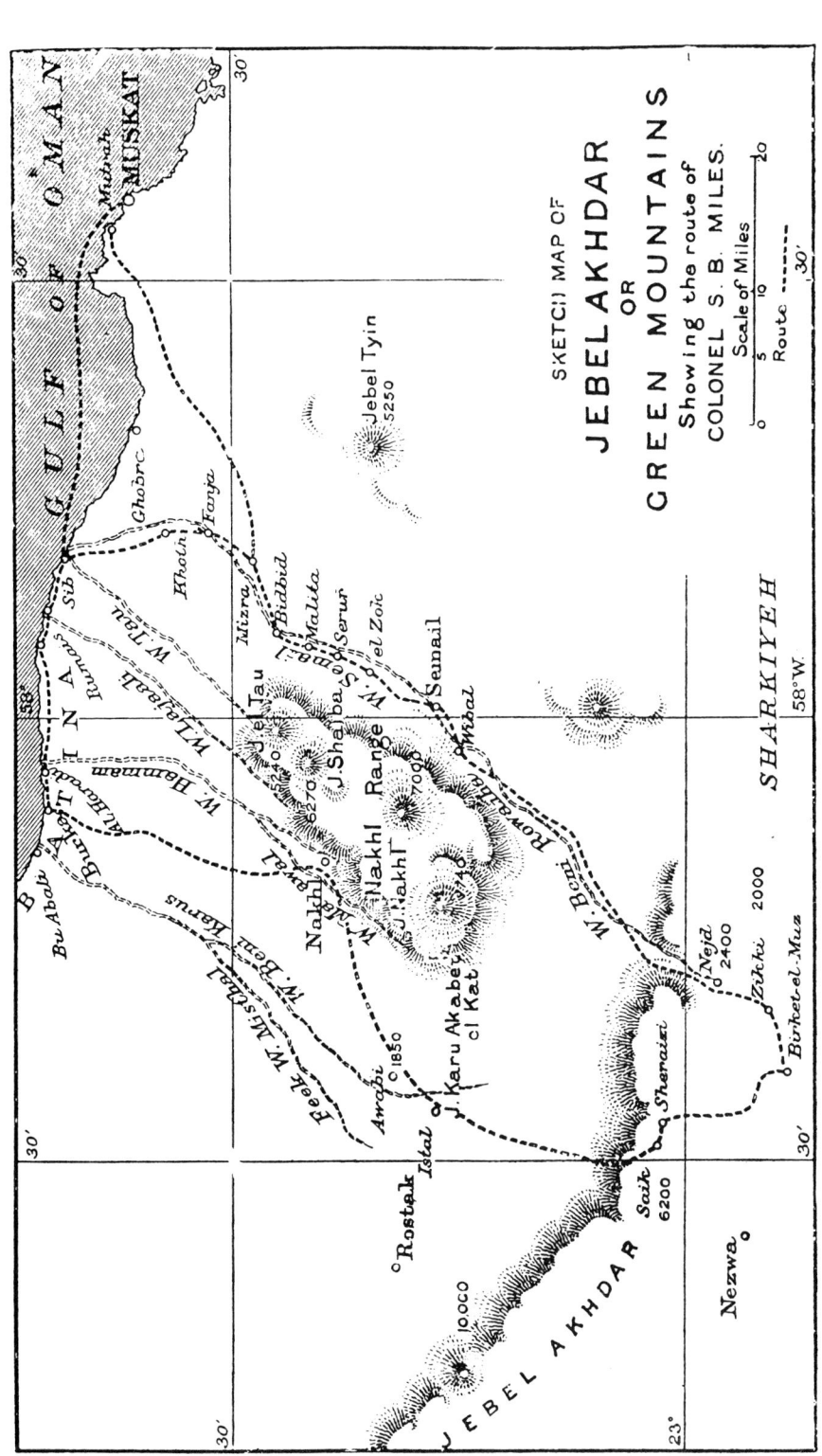

SKETCH MAP OF

JEBEL AKHDAR

OR

GREEN MOUNTAINS

Showing the route of
COLONEL S. B. MILES.

Scale of Miles

0   5   10   20

Route -------

any facilities for bathing, such as are to be seen at Bosher, although the local faith is that hygienic effects can be more rapidly and permanently obtained by ablution than by taking the water internally. The supply of water is sufficient not only for the domestic requirements of the people without the need of wells, but enough and to spare for the irrigation of the gardens, fields, and date-groves of the settlement. The natives assert that the heat of the water decreases in summer weather and increases in winter, and they have a firm belief that the supply is unvarying and inexhaustible. The rivulets issuing from these hot springs abound in little fish, even where the water is still warm ; they are about the size of minnows, and are of two species.

Mechanical ingenuity is not the forte of the Arab, and I was somewhat surprised to find, in one of the streams near the path, a flour-mill turned by water-power. It was rather a primitive and diminutive affair, but it was the first thing of the kind I had then seen in Arabia, though I have since noticed similar ones near Rostak. The mill consisted of a circular upper stone, bevelled up to a thin edge, revolving upon a stone floor, and attached below to a vortex wheel, which was set vertically with oblique floats or blades. The grain was put in unhusked, and appeared to be ground very slowly.

The castle is built on an eminence overlooking the town, between Jebel Laban and the pinnacle rock, the ascent up the ramp to the main gateway being very steep. This gate leads through a strong outer rampart pierced with embrasures and armed with a battery of iron guns mounted on rickety carriages, and, passing this, you find yourself in the courtyard, in which stands the keep, consisting of a high curtain flanked by two towers, from whose lofty battlements a superb view is presented towards the coast. It has three gates, and is in rather a dilapidated condition. Standing within the gorge and protected by the heights around, the castle is well placed, and is considered invulnerable by the Arabs, who have a deep sense of its military importance. Owing to its position, Nakhl has indeed played a by no means insignificant part in the history of Oman during the past three centuries, particularly in the closing years of the Yarebeh dynasty, and has sustained many sieges. The ruins of two other forts are to be seen here, one on the north side called Jeneb, dismantled by Sezzid Saed bin Sultan, and, after being repaired, was finally demolished by Sezzid Azzan in 1868; and another which stood near Thowâreh, and was razed to the ground by Sezzid Toorkee in 1874. Enfolded in the iron embrace of the deep gorge in which it lies, the town is shut in on all sides, and, occupied as the confined space mainly is with palms and houses, there is little room left for cultivation. The town has a regular and well-supplied market, and is divided into five quarters, viz. Atik, Hadhain, Safrat el Ijal, Khoryeh, and Hujret el Kurein. The date trees are extremely prolific, and the fruit is esteemed of superior sweetness and lusciousness—virtues which are ascribed by the people to the copious and constant irrigation of the plantations.

Out of a population of about 6000 the pure Arabs do not exceed 1500, representing the following tribes: Yaarebeh, Harrâs, Beni Kharus, Sereeriyin, Al Khozair, and Hadhârim. The number of fighting men is 800, mostly Bayâsir. In the lower classes there is much admixture of Persian blood. Each hâra or quarter of this town had a "sablah" of its own. This institution is very popular, and consists of a small shed, or covered platform, raised above the ground and open on all sides; it may be 15 or 20 feet square, with a light roof of mats and palm leaves resting on wooden posts. I have seen some, however, more substantially built, and bearing some resemblance to a mosque. Situated centrally in the village, it forms the council hall where the sheikhs and leading Arabs assemble daily to discuss local politics and chat over the events of the day. The tobacco-pipe of Turkey and Persia being almost unknown in Oman, the inevitable coffee-pot is in full requisition, and the sheikhs' slaves may be seen close by busily engaged in roasting, pounding and cooking the berry for the company.

There are numerous hand-looms at Nakhl, in which coloured lungies and puggrees are woven, as well as cotton cloth of the natural brown variety called "khodrung." The blue yarn required is dyed here, but the red and yellow yarns are imported from Bombay. Embroidered silk belts for ladies are also made here. Another industry is the production of porous earthen vessels for cooling water, the quantity made in the Nakhl factories being almost sufficient for the home demand. The clay used is a bluish marl, brought from the neighbouring village of Musalmât, mixed with sand. The kalib, or potter's wheel, has two discs, the lower one, which is called "raha," being turned by a treadle. The clay to be moulded is placed on the upper wheel, and is fashioned by an iron instrument called "moshal," the finishing touches being done with a sort of comb called a barit.

During my stay I visited some of the schools, of which there are five here, and saw the children imbibing instruction in the usual Moslem style, repeating aloud sentences of the Koran or rules of grammar read out by the mollah. They attend in the morning, and may be seen at an early hour hurrying to school, boys and girls together, some with a "minfa," or wooden Koran-stand, on their heads, some with a painted board or camel shoulder-blade, on which they learn to write, under their arm. The instruction given is of a very elementary character—reading, writing Arabic grammar, the Koran, and a little arithmetic being the only subjects. But the boys of the learned and wealthy are often educated at home by a mollah, and advanced further. The lack of method is partly compensated for by the precocity and tenacity of memory shown by the boys. Nakhl deserves attention for the comparatively advanced state of education among the people, there being a larger proportion of persons in this town able to read and write than in any other in Oman. There is a good number, also, of professional scribes. Books, consequently, are not so scarce here as elsewhere. The higher position of learning here is attributed to the influence of the Persians, who occupied this part of the country during the time of Nadir Shah.

Nakhl has, for the greater part of the year, a temperate and pleasant climate, being preserved from the scorching winds that sweep over the great desert, and partly shaded from the sun's rays by the impending mass of mountains to the east and south. It is, moreover, cooled and refreshed by the sea-breeze from the coast; but we found from personal experience that the heat, though dry, could be sometimes stifling and oppressive in the extreme. The elevation, taking the mean of two aneroids, is 1100 feet.

On July 11 our kafeer, Sheikh Selim, having intimated that he had procured an escort, and that he was ready to conduct us to the foot of the pass, we mounted horses and left Nakhl at 3 a.m. the next day. Our route lay along the Rostak road, which, following the sweep of the range, led us west by south over a barren country with an intricate system of low hills, ridges, and ravines. Some of these ravines are inhabited by warlike tribes, and it may be convenient here to give the names of these ravines as they occur in succession from Sib westward. The first of them is the Wady Tau; the second is the Wady Lajaali, also called Wady Halban, and having its exit at Romais; and the third is the Wady Hammam, or Nakhl, which joins the Wady Maawal, and reaches the sea at Al Harâdi. The fourth river, the Wady Beni Kharus, becomes a confluent in its lower course with the Wady Misthal, and flows into the sea at Bu Abâli. The Wady Misthal belongs chiefly to the Beni Riyam, who occupy Fik and three other villages.

Soon after leaving Nakhl we pass to our left Towye, a hamlet lying at the foot of the Akabet el Kat, a rugged way, little better than a goat track, but the only one over the range to the Semail valley. At 8 a.m. we reached Felej el Khosair, in the Wady Beni Kharus, where we dismount for breakfast, while the sheikh trots on to Awabi to make various arrangements. This little dell is highly cultivated, and produces an abundance of fruit and vegetables. The fragrance pervading the air from the shrubs and sweet-scented herbs was very pleasant and enjoyable.

Leaving Felej el Khosair at 4 p.m., we rode on for two hours till we came to Towye Saih, a hamlet of the Dahaul Arabs, where the ravine suddenly expands into a small plain three-fourths of a mile in extent, and a mile beyond this we reach El Awabi, where we halt. The wady here turns sharp to the left through a deep and narrow defile, and at this point on the left bank, completely commanding the passage, stands the castle Bait Awabi, in a position of great natural strength. Commanding the most accessible pass up Jebel Akhdar on the north, this fort has ever been a bone of contention between rival chiefs and factions. In particular it forms a constant source of trouble and hostility between the Ibriyin and Beni Riyam tribes, the former holding possession of it, while the latter would like to destroy it. Shortly before my visit the fort had been attacked by the Beni Riyam, who had mauled it very considerably, without, however, effecting a capture. One of the towers, I noticed, had been almost levelled to the ground. At one time the Ibriyin, fearing a strong coalition against themselves, offered Bait Awabi to the sultan, Sezzid Turky, who declined it. Subsequently His Highness changed his mind and asked for it, but in the mean time the tribe had changed their minds, and refused to give it up. It was eventually purchased by the sultan for a large sum, and the transit dues, which had been previously levied by the Ibriyin on goods passing through the defile, were then abolished.

Awabi lies about halfway between Nakhl and Rostak. The settlement covers a fairly large area, and every available spot has been reclaimed for tillage. The wheat and jowari had just been reaped, but other crops were still standing, and the fields, kept neat and regular, bore witness to industry and good husbandry. Stall-fed cattle of the small humpbacked kind are numerous, and almost every house appeared to have a cow or two. They are fed on barley, dates, and lucerne; and, though there is plenty of coarse grass, they are not allowed to roam about the hills. Awabi has a population of about 2000, with a fighting strength of nearly 400. It is occupied by three tribes, viz. the Beni Kharus, Ibriyin, and El Harras, who appear to dwell together amicably. The headman of the town, Sheikh Jabir, was very attentive and obliging, but was not very communicative. He was much depressed by the chastisement recently inflicted on his people by the Beni Riyam, and pointed out to me, sadly, the havoc they had committed among the date palms, the prostrate trunks of which were lying about in hundreds.

The shoran or bastard saffron plant grows plentifully in these hills, and the dried flowers are used by the women, who generally go about unveiled, to daub their own and children's faces with. It stains the skin yellow, and does not improve their appearance, though perhaps they think otherwise. The ladies who can afford it employ true saffron for the same purpose.

The preparation of dried dates, known in Oman as "bisr," and in India as "kharak," is carried on largely at Awabi, and as the season had now commenced, I took the opportunity to observe the process, and was taken round the factories by the sheikh. The dates selected are almost exclusively of the "Mubsili" and "Khanaizi" varieties, and are picked before they are quite ripe. The factory had a chimney about 15 feet high, and contained several open, circular, copper boilers, capable of holding five gallons each, and nearly full of water. Into these vessels the dates are put, and allowed to simmer over a slow fire. As the water in the copper decreased from evaporation, it was filled up again, but it gradually became inspissated by the extraction of the juice of the date. The fruit is left in the water about half an hour, and is then taken out and spread on mats or cloths in the sun to dry, after which it becomes hard and of a pale red colour. It is exported in large quantities from Muskat to India. We were here transferred to a new escort, part of our old one returning hence to Nakhl, and camels were here substituted for horses, as being better suited for rough hill work, and the gradient between this and the next stage being very severe.

After leaving Awabi, the elevation of which is 1850 feet, the next morning we were led past the fort in a direction generally tending south-west, the tortuous ravine gradually narrowing to a cleft with perpendicular sides 600 to 800 feet high, and though evidently swept occasionally by impetuous torrents, pleasantly fringed, and adorned in places with rhamnus, tamarisk, oleander, asclepias, and other trees and shrubs that have clung tenaciously to the ground and withstood the rushing waters. At two and a half hours, 6 miles from Awabi, we come to Istâl, a village of the Beni Kharus; and a little further on a heap of ruins on the right bank above us, indicating the site of an ancient castle, named Hisn Salut, is pointed out. This fortlet, which was finally destroyed less than half a century ago, in one of the many tribal wars for the possession of this defile, had a chequered history, and as we passed it the Arabs of my party had a lively discussion on bygone events associated with it. The formation, as disclosed by the Wady Beni Kharus, appeared to consist, at first, of an argillaceous slate, giving place, as we ascended, to a dark sandstone variegated with reddish or brown streaks, the stratification being sometimes confused and dipped at great angles, sometimes crumpled by pressure, and apparently metamorphic.

Opposite the village of Istâl is a curious high ridge with a serrated crest, called Ikhbal. Above this the structure changes, limestone predominates, and a pleasing transformation takes place in the scenery, the hills assuming a verdant appearance that was denied to the lower slopes. In this part of Oman many of the wadies are cut through conglomerate, which usually forms the bed, and the Wady Beni Kharus is no exception to this rule.

Between Istâl and Aleya we passed six small hamlets occupied by shepherds and mixed tribes, owning large flocks of goats. At Aleya, where the aneroid barometers showed an altitude of 2400 feet, we dismounted and halted in order to begin the real ascent of the mountain about nightfall, the sun's rays at this season of the year forbidding any attempt at alpine climbing in the heat of the day. It was destined, however, that I should traverse the mountain alone, as during the afternoon a kosid arrived with the post-bag, and my companion received letters recalling him to Muskat on business. He was accordingly obliged to relinquish the design of coming with me any further, and it was arranged he should return to Muskat the next day. The rugged conformation of the mountain sides here is picturesque, but the area available for cultivation is very limited, for the space is confined by the intersecting ravines that furrow the slope.

As may be naturally supposed, the ground has not been reclaimed without considerable difficulty and labour, the inequalities of the surface in some places necessitating the fields and orchards being raised and banked up. The soil is very fertile, and the earth is merely scratched by the plough, which is small, light, and simple. I saw a man turning up the stubble with a plough drawn by a single bull. The ears of corn are cut off close, and are threshed with flails made of date-stalks. Fruit grows here in profusion, and we noticed the citron, vine, lime, orange, and other kinds. A few coffee plants may be seen here, and it is the only place in Oman where it still lingers, the flourishing plantations that formerly existed having now all disappeared.

The sheikh informed me in the evening that he had made an arrangement with the owner of six asses, just arrived from above, to take us to Nezwa for fifty dollars, and to this I willingly agreed; but as six animals were not nearly sufficient for our requirements, and as no others were procurable, we had to hire porters to carry up the rest of the baggage. The sturdy Arabs who are accustomed to do this work, carry the burden on their backs in a net supported by a band or rope across the forehead, and seem to think little of their performances.

It was a little after midnight on the 13th that, after taking leave of my companion and of the sheikhs who had so obligingly conducted us thus far on the journey from Nakhl, I commenced the ascent of the Akabat el Hajar or Lhojar, as it is popularly termed. It was quite dark as we began the ascent of the precipitous mountain in front of us, up a rugged watercourse, and I wondered how the Arabs could see their way along the narrow path. At times we had a stiff climb, and zigzagged up the acclivity at a perilous angle; at other times the track was less steep and easier. As we leave the torrent bed and continue to progress upward, the gradient becomes much more formidable, and the mountain now presents sheer perpendicular cliffs and bold buttresses. The path would be here quite impassable for beasts of burden had it not been artificially improved by the construction of successive series of rough steps formed of huge slabs of stone, and by the curbing and revetting up of the road in parts where it overhangs a precipice. The stupendous nature and difficulty of the work and the skill and enormous labour bestowed on it claimed my wonder and admiration, but it was in vain that I endeavoured to gather any local tradition respecting its origin. The absence of tradition and the character of the work led me to regard it as of Persian conception and execution. A scattered Arab tribe like the Beni Riyam, always at war with its neighbours, could not have done it, and I think the most probable period to which it can be assigned is that of the Dailemite conquest of Oman in the tenth century of our era.

We found the asses we had hired of the hillmen to be surprisingly strong and sturdy animals; they were as surefooted as mules, and so active that they performed the ascent, a toilsome climb of 5000 feet, in five hours without exhibiting much fatigue. Unlike any others I had seen, these creatures were more stoutly built and much more spirited than their humble kindred of the plains. Their owners seem to treat them well, and do not, I believe, take them beyond the foot of the mountain. It is said that these asses are descendants of domesticated animals that have run loose in certain localities, three of which were mentioned to me; the asses in these colonies are probably in much the same half-wild state as the ponies in the New Forest and on Exmoor. Where the nature of the path permitted, the one I rode was accustomed to make short and rapid spurts, and then stop to regain breath. It was an impetuous, vicious little brute, always trying to bite somebody, and had acquired in consequence the name of Dheyaib, or the "Little Wolf." The saddles used are heavy and clumsy, and the load put on them appeared to be almost as much as a mule would carry. Owing to the sharp hard rim and deep cavity of its hoof, the ass is well fitted to climb rocky hills, and its value to the Arab in this mountainous region is great.

After about five hours of incessant toil, we reached the summit of the pass as the sun rose, and my weary party camped by the side of a little rill for two hours for rest and refreshment. I could have enjoyed a longer stay, as at some points it was possible to obtain a glorious panoramic view of wild and majestic scenery. We were here standing on one of the highest ridges of the mountain, and the escarpment, as it appeared to us, dropped almost perpendicularly from crag to crag, sheer to its base; the sea, though far distant, appearing from the dizzy height to be close under us. I found that one of my aneroids had given way during the ascent, but the other, graduated to 10,000 feet, was all right, and showed we had attained an altitude of nearly 8000 feet.

On resuming our journey, we descended for some distance a gradually shelving and undulating grassy plain with many large trees, which, however, were too far off to be recognized, and then came to a broad and verdant vale, which intersected our path and led away to the south-west. We saw a few shepherds tending their flocks in the distance, but the only wayfarers we met on the road were a party of women carrying bundles of grass on their heads for their cattle. After travelling in an east-south-east direction for 10 or 11 miles, we dismounted at a small mosque and grove of trees by the wayside, and while resting here were joined by some of the sheikhs of Saik, Mohammed bin Saed and others, who had come out to welcome us and escort us to their town. The temperature at this spot at noon was 85° Fahr. in the shade, and there was an exhilarating and bracing freshness in the air truly delightful. From this point several high peaks were visible, but there was no village nor sign of human habitation near.

In the company of our new guides we now moved on again, and, passing on the way the spring and felej that supply Saik with water, we found ourselves, on turning a corner, suddenly brought to a stand on the very brink of a yawning chasm, dropping vertically to a depth of 400 feet below us, and effectually barring our progress. I looked about wonderingly, when the sheikh, taking me by the arm and pointing to a white village with a pretty green setting, lying in a sequestered nook at the foot of the cliff, said, " There is Saik ; I will show you the way down to it." It was certainly the most singular situation for a settlement I ever beheld, and the mode of access to it was not less remarkable. Steps cut in the rock led to the bottom of the cliff, and down this long and slippery staircase my little steed tripped nimbly and steadily, but I was not sorry when we reached the ground. The whole community was there to receive us, and quarters were assigned to me in a small house that was vacant, while my party camped in an open space outside the village. This curious cluster of houses has a population of about four hundred, who subsist by growing corn and fruit, and exchanging their surplus produce for dates, cloth, hardware, etc., for they have no manufactures. They have many wells, and have also a small felej to irrigate their vineyards, fields, and orchards.

The cereals are wheat and jowaree, and two crops are gathered in the year. The rose, myrtle, and jasmin luxuriate in the gardens. Strong but rude trellises support the vines, which were still very abundant, though they were said to have much decreased of late years from blight or phylloxera. This misfortune is attributed by the natives to the machinations of an Afghan, who, about twenty years previously, had endeavoured to preserve grapes by adopting the Kabul method of packing the fruit in cotton-wool. A consignment was sent to the Zanzibar market, but the venture did not prove a success, and the attempt to start a trade in boxed grapes was abandoned. A year or two later the vines happened to be attacked by disease, and the people sagely concluded that the Afghan had cursed their vineyards after the failure of his speculation.

After receiving and dismissing a crowd of visitors, I went in the afternoon to pay a return visit to Sheikhs Nasir and Suliman, sons of the old temeemeh or chieftain of the Beni Riyam, Saif bin Suliman, by whom I was cordially received and regaled with coffee and conversation. They took me over their house, which, though the largest in

Saik, is an unpretentious structure, two stories high, built, like the rest of the houses at Saik, of stone cemented with yellow clay, and surrounded by a pleasant orchard and garden. I had hardly returned to my cottage, after visiting the sheikhs and taking a walk through the town, when a violent thunderstorm with hail and heavy rain burst upon us and lasted for some time. The rain on the rocky plateau above soon concentrated in the watercourses, and the tumultuous cascades that began to tumble down the steep walls of the chasm afforded a fine spectacle. The people informed me they had already had several smart showers, and that there was generally a fair amount of rain during the monsoon.

At 4 p.m. to-day the thermometer stood at 82° Fahr., at 6 p.m. it was 80°, and at 10 p.m. 74°. I found the elevation of Saik to be 6200 feet. The word Saik in Arabic signifies "a cleft or chasm."

This was the first point in the interior at which I had touched the route of Lieuts. Wellsted and Whitelock, and, having made inquiry as to whether there was any remembrance of their visit, I was gratified to meet with some who had personally witnessed their arrival forty-two years before. At my request, Sheikh Mohammed sent for two old men of the village and brought them to my cottage. I found that they retained a clear impression of the event, and their statements amusingly indicated how narrowly the strangers had been watched. The Arabs remembered that the officers had tents with them, used brass instruments to gaze at the sun, and spent much of their time in writing. Among other objects of wonder and curiosity, it had been noted that they possessed a bottle full of snakes.

The sheikhs of the Oman tribes have, in general, but little power over the other members in time of peace, but the temimeh of the Beni Riyam may be regarded as an exception, and he is indeed, from this and other causes, one of the most prominent sheikhs in the country. He is seldom on good terms with the Sultan of Muskat, and defiantly appropriates to his own use the produce of the Bait el Mal or crown lands, which belong of right to the ruler. In connection with this family, I am sorry to have to record one of those domestic tragedies so common in Arabia. Sheikh Nasir bin Saif died a few years after my visit, leaving three sons, the eldest of whom, Mohammed, inherited his father's position as temimeh. In 1886 Mohammed and his second brother were murdered by their uncle Suliman, who usurped authority and held it until 1899, when Nasir's third son retaliated and put his uncle Suliman to death.

The Beni Riyam occupy the towns of Sheraizi, Saik, Nezwa, Zikki, Birket el Muz, and Tanuf, and number about 17,000, of whom 3000 are fighting men. The tribe is Himyaritic and of Kamar origin, its eponymus being Riyam bin Nahfan bin Tobba bin Zaid bin Amr bin Hamdan. Their ancient domicile was in the province of Hamdan, in the Yemen near Jebel Atwa, on the summit of which, where fire issued from the ground, was the temple of Riyam, a fane of great sanctity with idols of the sun and moon, and a great resort of pilgrims. According to the Arab geographer Hamdani, the ruins of this temple were still to be seen about the year 950 A.D. Tradition relates that the Beni Riyam emigrated in remote times from the Yemen, in company with their cousins the Mahra, and marched on to Oman, leaving the Mahra on the way, in the land they now occupy. After crossing the peninsula, the Beni Riyam were fortunate enough to secure as their new habitation a vantage-ground in the most inaccessible and central part of the country, and called the mountain, after the name of their old capital Radha, Jebel Radhwan, since changed, when or why I know not, to Jebel Akhdar. The fire temple seems to have passed out of the memory of the tribe, who now regard Radhwan as the name of an ancient prophet that arose among them. A portion of the tribe is said to derive its descent from Malik bin Fahm, who invaded Oman in the second century after Christ.

On the 14th I moved across to Sheraizi, which lies a mile to the eastward, and was received by the headman, Sheikh Salim bin Abdulla, whose present was a basket of peaches. The town is much larger than Saik, and offers a most striking contrast to it in the position it occupies. Though less quaint and romantic in appearance, Sheraizi is more favourably and finely situated, being perched near the head of a ravine, like an eagle's nest, on the brow of a lofty cliff which falls rapidly to the valley beneath, commanding an extensive prospect. The town is built on so steep a declivity that the houses appear to overhang one another, and the only communication is by means of narrow, dirty, irregular steps leading up and down from one row to another. The houses are small and mean-looking; they are constructed of stone, and are sometimes, but rarely, plastered with clay on the outside, as at Saik. Just under the town there is a copious spring of pure water, which gushes from the rock to fill a circular cistern, sufficing for the requirements of the inhabitants, and serving the conduits that fertilize the terraces below.

I spent most of the day in wandering about the place and examining the extensive hanging gardens which are spread along the precipitous valley walls and form the most interesting and beautiful characteristic feature at Sheraizi. To the left of the town, beneath it, and on the opposite slope of the valley, the whole face of the hillside, to the depth of 1000 feet or more, is cut into a parallel series of ledges or terraces, most symmetrically arranged and highly cultivated as vineyards, orchards, and cornfields. These curious pensile gardens, the like of which I had never seen before, with their varied foliage and ripening fruit, apricots, grapes, figs, and pomegranates, formed a very attractive and pleasing sight, and had evidently been most carefully constructed, the terraces being stepped up with revetments wherever the natural features of the ground had not availed, to maintain the earth in position. Owing to the sharp angle of the slope, the ledges are in general very narrow, perhaps 10 or 12 feet in width, and considerable ingenuity has been displayed in their disposition and in overcoming the difficulties of the ground. The labour bestowed on them, however, would have been futile without an abundant supply of water, and in this respect nature has been prodigal, the cultivation being easily and perpetually irrigated by the numerous mountain streamlets, which are taught to meander from one ledge to another in turn, being confined by little embankments along the margin. Extensive as these terraces appear, the space they cover is not very large, and the inhabitants would be glad of more soil. The limited scope for tillage necessitates economy, and the corn may often be seen intergrown with leguminus, and sometimes even with melons and other cucurbitaceous plants.

Opposite the town, and on the other side of the intervening valley, which is called the Wady Miyadin, and flows to the south-south-east, is a conical hill with a ruined tower and mosque on the summit; and to the right, lying south-east by south, is El Jabul, a curious peak, which, I was told, was looked on as a potential stronghold, having served as a refuge for the tribe in old times. It is now untenanted, and no vestiges are left of ancient buildings. It is capable of easy defence, as the path leading up it is so steep, narrow, and rugged. The scenery of this valley, with its regular and richly varied terraces, like giant steps on the mountain-side, is as beautiful as it is extraordinary, and of a character of which it is impossible to give an adequate idea.

In the afternoon clouds again gathered round the peaks, and we had a pelting thunderstorm, which was very grand; the lightning was very vivid, and the cataracts of water seemed to gratify the Dailemites immensely.

The women here, and throughout the valleys of Jebel Akhdar, fetch water for domestic purposes in large copper vessels instead of in earthen jars, as in other parts of Oman. They also use copper cooking-pots almost exclusively, the employment of this metal being due, no doubt, to the difficulty of bringing earthenware up from the plains, and to the absence of clay in these mountains suitable for good pottery. The hillmen procure these vessels from Nezwa, where I afterwards found the manufacture of copper ware to be one of the principal industries.

During our ascent of the Akabet el Hajar to the plateau, the change in the character of the vegetation had been complete, marking the difference of elevation by the substitution of that of the temperate zone for the tropical and subtropical verdure of the plains. The plantain, the mango, and the stately date palm, which forms such a conspicuous feature in the scenery of Arabia north of the eighteenth parallel, and which we last saw in the Wady Beni Kharus, have all disappeared, and in their place we have the walnut, pine, apple, and pomegranate. The fig, peach, apricot, vine, almond, and lime also flourish. M. Aucher Eloy mentions the cherry, but I did not observe it. The pomegranate is a handsome shrub, with dark green foliage and crimson flowers. It receives more attention at Sheraizi than any other product except the vine, and the fruit, though small in size, is of superior excellence. It forms the chief item of export from Jebel Akhdar, being shipped from Muskat to the value of 10,000 dollars annually for the Bombay market. The Arabs are very fond of the pomegranate, and employ it for making sherbet. The rind is very astringent, and is used medicinally as a febrifuge and anthelmintic. The terraces appeared to offer a peculiarly suitable site for growing coffee, and it was once extensively grown here, but from some cause the plantations failed, and the plant has now entirely disappeared.

I need hardly say that the assumption as to the nutmeg growing in these hills is erroneous, Wellsted's mistake having probably arisen from the name this spice bears in some European languages. Mace, the arillus or covering of the nutmeg, is another word indicating the belief prevalent in old times that Oman was the habitat of the spice, it being derived from Maceta, the appellation given to Cape Mussendom by the Greek geographers. Like so many other articles still bearing Arabic names, *e.g.* sugar, pepper, olibanum, rice, etc., the nutmeg was introduced into the Mediterranean by the Arabs.

The vineyards at Sheraizi are very extensive, and are regularly irrigated and manured. The vines are trained on rough trellises at certain distances, and produce both varieties of fruit, white and black. The grapes, which ripen here in August, were hanging in goodly bunches, and it did not seem to me that the people had any reason to complain of the crop, but they grumbled a good deal. Although much of the fruit may be eaten, and a portion made into raisins, there can be no doubt that the bulk of the crop is intended for the manufacture of wine of an inferior kind, which is entirely reserved, I believe, for home consumption. I did not, of course, witness the process, as it was too early in the season, and I had no opportunity of tasting the vintage, but I should imagine it, from the description I heard of the method employed, to be far from satisfactory. After crushing the grapes and mixing water with the whole mass of pulp, juice, skins, etc., they leave the liquid for about three weeks to ferment. The wine thus made is considered fit to drink in about three months after the fruit has been plucked. The wine thus made is consumed in the long winter evenings by the Sheraizi men, whose wine-bibbing propensities are notorious and reprobated throughout Oman. The Arabs of the interior, being a temperate and abstemious race, regard the constant manufacture of wine in their very midst by these Persians as a scandal to their religion, and do not permit the production of spirituous liquor in any other part of the country, so far as I am aware.

Jebel Akhdar was explored botanically to some extent in 1836 by M. Aucher Eloy, the intrepid and indefatigable French naturalist, who ascended the mountain by the Akabet el Hajar, and returned to Muskat by the Wady Semail. In his zealous and enthusiastic search for plants, he traversed the greater part of his journey on foot with infinite toil, reaching the coast again at last enriched with many specimens of new species, but weary, fever-stricken, and footsore. In his journal, which was edited by M. Jaubert in 1838, he gives a graphic account of his sufferings and adventures. M. Aucher Eloy tells us he found many pretty flowers well suited for introduction into Europe; he collected about 250 species, and reckons there may be 500 altogether in Oman, an estimate which will, I suspect, be found much below the mark. In the vicinity of Nakhl the botanist found a tree of the genus Niehburia, and a very pretty flowering shrub, which he named *Vogelia leprosa*, also a beautiful new Dichyptera. The vegetation on Jebel Shaiba, the altitude of which is 6270 feet, was new to him. At El Hajar was a new Bœhravia, and at Aukaud a new violet, a new genus of primula, and a new species of Lunius, which he named *Aurea*. At the dispersion of M. Aucher Eloy's collection, the Kew Herbarium took 2600 specimens, and the British Museum 1907.

This mountain is estimated by Lieut. Wellsted to have a length of 30 miles from east to west, and an extreme breadth of 14 miles from north to south. The highest point visible from the sea was reckoned by the nautical surveyor at 9940 feet, but I have not been able to ascertain the local name of this peak with precision, some Arabs giving it as Jebel Hauz, some as Jebel Shum. The abruptness of the north and south sides render this grand rocky mass inaccessible from the plains, except by means of the torrent beds that ages of surface drainage and the waters of living springs have scored in its rugged flanks. Three of these natural highways leading to the summit can be travelled by animals with burdens, and form, consequently, the main passes. On the north side the Akabet el Hajar has been already mentioned as the one by which we ascended; the southern side offers two, the Wady Tanuf and the Wady Miyadin, the former being longer, but of the same character as the Wady Miyadin, which will be described further on. There is also, on the north side, the Akabet Fik, and on the south side the Akabet Shash, and perhaps others, but they are little better than goat-tracks. Eclipsed in height by a few peaks only, the northern flank of the mountain presents the most elevated ridge, and this feature is characteristic of the whole chain, which is much more lofty, abrupt, and precipitous throughout its length on its northern face than on its south flank towards the desert. The plateau declines gradually from north to south, and then falls rapidly to the plain in cliffs which, though wild and striking, are less imposing in their grandeur than those on the sea-ward front. Owing to this conformation, the ravines thrown to the south by the watershed are the longest and most numerous, and drain off by the two great arteries, the Wady Miyadin and the Wady Tanuf, the bulk of the surface or rain-water that falls on the mountain. Among the tributaries received by the Wady Miyadin are the Wady Sarut and Wady Saik. The chief tributary of the Wady Tanuf is the precipitous Wady Beni Habib, in which are two villages, Ain and Akr.

To two or three only of the peaks visible from Sheraizi could the Arabs give names; they pointed out Jebel Hauz, a few miles to the south-west, and Jebel el Ham, a tall peak to the northward. Jebel Akhdar must have been very different at the distant period when it received the appellation, then, no doubt, an appropriate one, of "The Verdant Hill," from the drear and arid aspect it presents at the present day. Though immense masses of exposed rock, destitute of vegetation, give the mountain generally a savage and unattractive appearance, there are some parts that are well wooded, and the plateau we crossed had much long grass and herbage. Many of the deep ravines, more-over, are said to possess dense thickets of thorny undergrowth and euphorbia, and the extent of cultivation at Saik and Sheraizi strengthen the conjecture that the range in former days was better clothed with arboreal vegetation. The destruction, if it ever took place, of the forests that once covered the surface of the plateau, would have given full scope to the denuding power of the rain, and the long-continued effect of this would be to wash the fertile soil into the valleys below. This would prevent the renovation of the forests, and thus we have a bare landscape instead of a tract shaded by extensive woods. Again, the denudation of trees must have caused reciprocal action in reducing the rainfall.

If any useful minerals exist in Jebel Akhdar—and the only one I noticed was iron—they are little explored and utilized by the inhabitants, who devote themselves either to agriculture, in the case of townsmen, or to rearing animals. The bold and hardy shepherds, who are by their calling sprinkled about everywhere with their flocks, number several thousands, and form the chief fighting material of the tribe. Though so widely scattered, they assemble with great celerity and promptitude when summoned by the sheikhs for war.

There is a paucity of animal-life in these hills. Wolves, hyenas, wild goats, ibex, wild cat, and leopard are said to be found; but the last named, if existing at all, is very rare. Kites and vultures may be seen circling round in the sky, but other birds appeared to be scarce, both as regards species and individuals.

Almost from the commencement of the rise of the Arab empire in the seventh century, the possession of Omen was coveted by the khalifs, who regarded it as an integral portion of their dominions; but the people of Oman never freely accepted this view, and preferred independence, holding that subjection to the central government merely meant the payment of a heavy tribute without any corresponding advantage. The result was that for three centuries Oman became the ever-recurring scene of sanguinary conflict and devastation, and was reduced to the position, except during some fitful intervals of repose, of a tributary province of the empire. But persistent as the efforts were to effect a complete subjugation of the land, there was one part, Jebel Akhdar, that long remained impregnable and defied successively the Khalifs Mowiya, Abbas, Harun, Motadhid and Mutti, whose troops overran all Oman, except this mountain stronghold, at the foot of which they surged and struggled, like angry waves against a rocky islet, in vain. The remarkable achievement of its capture was reserved for the Buwayid Malik of Fars, Adhad ul dowla. The uncle of this prince, Muiz ul dowla, had previously, in 354 H. (965 A.D.), invaded Oman, but hostilities had then been averted by the prompt submission of Nâfi the black, who, having murdered his master Eusof twelve years before, then held power with a Turkish guard. Nâfi was left in charge as governor, but was expelled soon after by a combined Carmathian and Omani force. This revolt was followed, in 355 H., by a second Buwayid invasion under Abul Faraj and Adhad ul dowla, who ravaged the country and brought it once more to obedience. In 362 H. (972 A.D.) the Buwayid army of occupation in Oman, which consisted partly of Persian and partly of negro troops, the latter numbering several thousands, broke out into open mutiny, and threw the country into anarchy. The news quickly reached Bagdad, but the Amir ul Oomra, Izz ul dowla, was powerless to take action in so distant a province, being himself at that time in a critical position, and his cousin, Adhad ul dowla, therefore, who had long wished to annex Oman to Fars, seized the opportunity to despatch a force from Siraf across the Persian Gulf to restore order. His general, Abul Harab, defeated the mutineers in three successive battles, and took possession of the country for his master. His treatment of the inhabitants, however, was so oppressive that they were soon in revolt against him. The national gathering was so strong that the newly elected Imam Sheikh Ward bin Ziyad and his deputy Sheikh Hafs bin Rashid were able to drive the intruders back to their ships.

For the moment Oman was again free, but the dark cloud of humiliation that followed this transient gleam of liberty was more calamitous than any previous one. Adhad ul Dowla met the disaster by sending an army under his wazeer Abul Kasim al Mathhad, powerful enough to crush all opposition. The fleet sailed first to Sohar, and then moved on to Kuryat, where the Imam Ward had concentrated the Arab tribes. On the plain between the sea and the Devil's Gap a great battle ensued, and the Arabs, worsted, but not subdued, retired up the Tyin valley, pursued by the enemy to Nezwa, where a second stand was made.

This conflict, more desperate and sanguinary than the first, resulted in the destruction of the Omani force. The Imam Ward was slain on the field, and the country fell prostrate at the feet of the victors. Jebel Akhdar alone remained intact, and in this mountain fastness the survivors now took refuge. No previous conqueror had ever ventured to attack these menacing and almost inaccessible heights, but the wazir Abul Kasim felt so elated and confident that he resolved to crown his work by storming and reducing this last citadel of the Arabs.

In two divisions, up the precipitous and rugged Wady Miyadin and Wady Tanuf, the Persians fought their way in face of the Arabs, who defended the mountain by hurling down rocks, slinging stones, and shooting arrows against their eager and relentless enemies. Step by step the Persians pushed on, and step by step the Arabs retreated, fighting desperately for many days. The summit was gained at last, and the final stake had then to be fought out in the open field. In this battle, which took place on a small plain called Sherif, above the Wady Beni Kharus, the Arabs are said to have numbered 10,000, but the Persian strength is not given. The struggle was long and bitterly contested, but the despairing valour of the Omanis could not prevail against the superior arms and training of their adversaries, who, after a terrific carnage, utterly vanquished them. The Arabs' cup of humiliation was now full, and Abul Kasim's conquest of the land was absolute and complete. The women and children of the Arabs became, of course, the spoils of the victors, and many of these Dailemites or Persians, attracted by the salubrity and fertility of the mountain, resolved to settle there, selecting for their new abode a village on the site of the present town of Sheraizi, which they re-named the "Little Shiraz," after the capital of Fars. As the Persian power waned before the Seljukian Turks, and as the Arabs recovered strength and freedom, the people of Sheraizi gradually became absorbed in the Beni Riyam tribe, of which they now form a distinct and dependent section. Though they have assimilated themselves to the Arabs, during the long period of their occupation, in language, dress, and habits, and are only to be distinguished by a somewhat fairer complexion and different physiognomy, it is evident they maintain themselves as a separate community and keep aloof as much as possible from the Arabs, seldom mingling, rarely intermarrying with them, and never descending into the plains. Though they are said to be a dissipated and depraved race, they are a peaceable and quiet folk. Their industry has been concentrated on agriculture, and the elaborate work of terraces, if not originally designed, has been at least vastly improved by them. It must not be forgotten, also, that they have conferred a benefit on the country by the introduction of many valuable fruits, as the pomegranate and the vine, the walnut and the peach, and the almond and the mulberry, most of which were brought over from Persia after the Buwayid conquest in the tenth century of our era.

Not being prepossessed with the sour-visaged people of Sheraizi, and being pressed for time, I remained only one day here, and bid adieu early on the morning of the 15th. The descent of the precipice commenced immediately after leaving the town, and I rode down the declivity, passing on the way a few hardy plants which struggled for existence at the edges of the ravine, where they derived a scanty nourishment, until we reached, at about 1000 feet, the Wady Sarut, just under El Jebûl, which rises perpendicularly from the torrent bed to a considerable height. Another steep descent of some 2000 feet down the rough and stony bed, half choked with great boulders and fragments of rock, led us into the Wady Miyadin, and, following this, we came to a village called Musaira with a small plantation of date palms, the sight of which, at such a high elevation, was a surprise. Although this pass is less precipitous than, and does not bear comparison with, the Akabet el Hajar, it has required an almost equal amount of rough engineering work, and the steps have been most laboriously and ingeniously constructed. Riding down it was no easy matter, but my steed made his way along the slippery path with great steadiness. A small but perennial stream flows along the Wady Miyadin, fed by the springs issuing from crevices in the rock. These springs appeared to be more abundant in the higher parts of the mountain. On reaching the bottom of the pass, we experienced a hot simoon wind blowing strongly up the valley from the direction of the desert. The air was most oppressive and stifling, but the thermometer only indicated 110° Fahr.

The banks of the wady at this part exhibited chiefly a dark bluish, veined limestone and a very brittle ferruginous shale. Miyadin, where we stopped for coffee, is a pretty village under high cliffs, with a felej giving a bountiful supply of water, and many date, lime, and other fruit trees. The hot breath of the simoon was here suddenly changed into a cool and refreshing current of humid air, and to this succeeded a thunderstorm with heavy rain. At 1 p.m. we were again winding down the ravine, which now presented a more gentle declination towards the south. Three miles further we came to Misfa, a small hamlet, and here the banks of the wady begin to recede and to decrease in height, vegetation at the same time becoming more abundant, tamarisk, rhamnus, palm, and acacias fringing the bed. For two hours more we rode along the gradually opening valley until, emerging from the hills, we found ourselves upon a spacious plain, now parched and dried by the burning rays of a summer sun, and with the horizon unbroken save by distant clumps of palms. Here we suddenly changed our direction and turned to the east, arriving at length at the gate of a stone-built castle, where we knocked for entrance. This castle was

Bait Rudaida, the residence of Sezzid Hamad bin Hilal, a second cousin of H.H. Sezzid Turky, and an amiable and intelligent youth of sixteen, who welcomed me in the most friendly and hospitable manner.

I was accommodated at first, after the usual complimentary interview and coffee, in a little mosque within the castle precincts, until a room had been prepared for me upstairs, our arrival here having been entirely unexpected. In the mean time I had leisure to look round and see the castle, which is of similar size and style to Bait el Felej, near Muskat. Surrounded by an exterior wall which forms the courtyard, the Bait consists of a long rectangular structure, two stories high, protected by defensive towers at the angles, and enclosing an inner quadrangle open to the air. Inside the courtyard are quarters for the garrison, and in one corner is a small mosque. If well guarded, it could well resist an Arab force, unprovided with artillery; but it was at this time in sad want of repair. The cause of this was not difficult to discover, the young prince being a ward of the Muskat government, and little care being exercised to guard his interests and protect his revenues and property. At this time, however, the boy was in high favour with the Sultan, and had lately been presented with a small iron gun, of which he was immensely proud. Bait Rudaida is close to the town of Birket el Muz, or "Pool of Plantains," from which it is separated by a low conical hill, capped by a watch tower, placed there to command the water-supply. Around the town is an extensive date grove, intermingled with orchards and cultivation. The population is about 3000, and the settlement is divided into three hujrahs, or sections, one belonging to Sezzid Hamad, and the other two to the El Amair and Beni Riyam tribes respectively.

In the afternoon I walked over the Sezzid's estate with him, and then through the town, which owes its prosperity and extent to the fostering care and liberality of Sezzid Hilal, on whom it had been bestowed when a mere hamlet, as an appanage in addition to Sowaik, by Sezzid Sultan bin Ahmed. The fields produce the usual kinds of corn and vegetables grown in Oman, but were not so regular or well tilled as in most other parts. The product in which Sezzid Hamad seemed to take the keenest interest was sugar-cane, and he did not neglect to show me over his luxuriant plantations. He possessed a rude mill for extracting the juice, and made many inquiries as to the best method of refining sugar. The cane is propagated here, as in India, from cuttings, not from seed, and it is believed to have existed in Oman from time immemorial. Frequent and copious irrigation is necessary for the successful production of sugar-cane, and the Sezzid's fields were watered by a felej or subterranean stream drawn from the hills. The felej is a kind of artificial river, and is one of the most ingenious institutions for bringing water to stimulate the prodigal hand of nature that could possibly be conceived.

The hills abound in fountains, but the soil is so porous and thirsty, and the evaporation from the intense heat so rapid, that irrigation of the valleys and lowlands by any other method would be exceedingly difficult. The ground, though apparently arid and bare, is often fertile enough, and only requires water. Vivify it by irrigation, and it will yield an abundant harvest. The system is wonderfully well adapted to the country and to the economic condition and habits of the Arab. After the initial labour and cost, it requires but little trouble to keep it in repair, and the continual expense and toil of raising water to the surface from wells is avoided. The construction of these underground watercourses is generally undertaken by the tribal communities in each town or village on a sort of joint-stock basis, each individual contributing his quota in money or in personal labour. They are universal throughout Oman, and there are few villages without at least one of these felejes. From a spring at the base of a hill, which may be many miles distant, the villagers conduct the water to their fields through a tunnel or conduit below the surface in the following manner. A line of vertical cylindrical shafts, 4 or 5 feet in diameter and 100 to 150 feet apart, is first sunk between the spring and the village, and these shafts or pits are then connected together by a channel underground in such a way that the body of water, flowing by gravitation, reaches the surface as it approaches the cultivation. The felej is always commenced near the spring, where the shafts are deepest, and the work carried on to the point where the water is required for distribution.

The plan is, of course, more troublesome and expensive than an open irrigation canal, but it has the advantage—one of great importance in this parched and desiccated land—of avoiding loss by absorption and evaporation. In long felejes the upper shafts are often 30 or 40 feet deep, and in some cases the connecting channels are lined with masonry or brickwork. This, however, depends very much on the nature of the soil, a circumstance which also dictates the distance between the shafts. Near the villages the pits are frequently made accessible by steps or a sloped path, to enable the women to procure pure and cool water for domestic purposes. The rows of mounds formed by excavating these pits are a conspicuous and peculiar, though a rather unsightly, feature in an Oman landscape.

The upper apartments of the castle were high and spacious, but very modestly furnished, and not scrupulously clean. The ground floor was devoted to kitchens and storerooms. The place looked dismally bare and empty, comfortless and neglected, and bore the appearance of a house whose glory had departed. It had once been rich in articles of luxury, collected by its founder in the days of his prosperity, but these had all vanished. The work of confiscation and spoliation of Sezzid Hilal's estate was completed by the Metowa under the Azzan-Khalaili *régime* about 1869, but the Birket el Muz property was restored to the rightful owner, Sezzid Hamad, as an act of justice by Sezzid Turky.

Round the room I occupied ran a broad wooden shelf, on which was ranged a miscellaneous gathering of lamps, clocks, china, medicine-bottles, etc., and among these curiosities was a dusty heap of Arabic manuscripts, of which I made a careful list. They were mostly religious works, and belonged, I found, to a learned Ibadhi Mulla, known as the Kazi, to whom had been entrusted the guardianship and education of Sezzid Hamad. This prince was devoid of political ambition. He never interfered, when he grew up, in the jealous intrigues and factious quarrels so rife among the chieftains of Oman, but led a quiet homely life, absorbed in books and country pleasures, until his retirement to East Africa, where he died in early manhood. His father, Sezzid Hilal bin Mohammed, of whose noble disposition and generous spirit some account was given by Lieut. Wellsted, was a notable personage in his day, and a warm friend of the English. He held a foremost place in general estimation as a member of the ruling family, and was regarded as a man fully worthy to guide the destinies of the nation had he been called to the throne. His memory was long cherished in the country, and seldom, I believe, has a man's death been more sincerely lamented in Oman than when Sezzid Hilal was treacherously murdered by his cousin Kais in 1864. The story of this tragedy was related to me as follows:

About two years before the close of the reign of H.H. Sezzid Thowaini, who was murdered by his eldest son Salim in 1866, Sezzid Kais of Rostak formed a plot to destroy him and seize the government of Muskat himself. He communicated his plan to Sezzid Hilal, who was too prominent and influential a personage to be ignored, but the latter, having always been loyal to Sezzid Thowaini, indignantly refused to join, and denounced the plot. Sezzid Kais then determined to be revenged, but, failing to find an assassin, had to undertake the task himself. Approaching his cousin Hilal, who, unlike most men of his rank, seldom wore a sword, in an apparently friendly manner, Kais suddenly drew his sword and struck him savagely on the head. Staggered by the unexpected blow, Hilal recovered sufficiently to plunge his dagger into the bowels of his assailant, who fell dead. Hilal was removed to Sowaik, where he expired shortly after. The fort at Sowaik was then assaulted by the adherents of Kais, and, though gallantly defended by Hilal's sister, was captured and annexed to Rostak.

We managed to get the camels ready and make a start at six the next morning, though our courteous young host, Sezzid Hamad, was loth to let us depart, and insisted on accompanying us a good part of the way on horseback before he would take leave. Our road lay over a level plain dotted with acacia and scant herbage, and intersected by shallow watercourses, the most considerable of which is the Wady Hajar, with a village of the same name. Two hours at a slow trot took us to Zikki, a large and important town on the banks of the Wady Halfain. The wali, or governor, Sheikh Mohammed bin Sinan, who at this time held the castle on behalf of the Sultan, came out to meet us, and received me very cordially. After coffee in a large subla inside the courtyard, the sheikh showed me all over the castle, and then, taking me to an upper apartment, informed me that he wished to give a general entertainment to the Arab escort. I was anxious to push on and cover another stage in the afternoon before halting for the night, but the sheikh was so earnest and persistent in offering hospitality, declaring that it would lower him in the eyes of his people to allow my party to pass his gate unfeasted, that I felt obliged to submit and acquiesce in his wish, though I knew from experience that on such occasions the culinary preparations demand deliberation, and that our day's march was at an end.

The town of Zikki has some beautiful plantations of palms and extensive cultivation, and stands in a very picturesque locality. It is divided by the Wady Halfain, and is supplied by one of the most copious springs of water I have seen in Oman. On the left bank the ground is low, fertile, and well tilled by its occupiers, the Beni Riyam. The right bank is much higher, and on this elevated site are built the castle and the walled quarter of the Beni Rowaiha. Outside the settlement are several watch towers and hamlets of the Âl Amair and other tribes. The population may be 8000, and is mainly composed of the two tribes mentioned, the Beni Riyam being by far the strongest. These tribes, locally known as " Yemen " and " Nizâr," live in a chronic state of antagonism and warfare with each other, ever vigilant against surprise, and ever ready for a skirmish, except in the short intervals of " salfa," or truce. The political relations of both tribes are somewhat confused, as the Yemenite Beni Riyam are now ranked as Ghaffirees, and the Maddic Beni Rowaiha as Hinawis.

The castle is a lofty structure, compact and massive, with walls about 5 feet thick. It was at this time scarcely a century old, having been built on the site of an ancient fort by Mohammed Jabri, maternal uncle of Sezzid Said bin Sultan, but it had an antiquated appearance from its battered and dilapidated condition, one tower having tumbled down bodily. Outside and in front of the castle were two mounted iron guns, old, honeycombed, and unsafe, but capable, as I can affirm, of making a prodigious noise, for they were fired as a salute on our arrival, and this is a quality of immeasurable value in a country where the use of artillery is so little known. The castle not only overawes the whole settlement, but, standing as it does in a commanding position at the head of the Semail valley, dominates and controls one of the chief highways and arteries of traffic. Its possession, therefore, has always been regarded as a matter of military importance by the central government.

The Wady Halfain is a perennial stream for some distance from its source, and flows to the south-east, reaching the sea at Ghubbet Hashish. About halfway down its course it meets the Wady Kalbuh and Wady Andam, and its grassy bed, which is much frequented by the Bedouins for grazing their flocks, forms the natural road from Mahot to Adam. The elevation of Zikki is about 2000 feet.

The Waly Sheikh Mohammed had duly performed his promise of collecting fresh camels for us, and we were on our way again at daylight next morning. I was about to take leave of him, when he announced his intention of accompanying us part of the way, and from this resolve I could not dissuade him. He rode a handsome and fiery black Arab, and kindly offered me one of his stud, but I preferred a dromedary, as we had a stage of nearly fifty miles before us, and the camel is superior to the horse in speed and comfort on a protracted journey. With our faces towards the north, we found ourselves climbing a gentle acclivity to the crest of the mountain chain, along a rough and stony watercourse, the stupendous crags and precipitous cliffs of Jebel Akhdar rising in dark masses on our left hand, deeply furrowed by ravines and clefts, while to our right the ridge trended away east by north. An hour's ride brought us to the Nejd or summit of the chain, which is 400 feet higher than Zikki. Various names were assigned to this pass by different persons I spoke to, viz. Nejd Beni Rowaiha, Nejd Mujberriya, and Nejd Soharma, but they were all of one accord as to the Nejd being the lowest point of depression along the range. The watershed in this vicinity provides the source of two of the longest rivers in Oman, flowing in opposite directions, viz. the Wady Halfain, already described, and the Wady Beni Rowaiha, or Wady Semail. Crossing the ridge, we descended a stony ravine leading into the main bed of the Wady Beni Rowaiha, and we now saw, stretching out in front of us to the coast, the largest, most populous, and, politically, the most important valley in the land. This long, rich, and splendid valley, lying between two mountain ranges, forms one of the main channels of communication between the coast and the interior, and, though here and there barren, is studded along the banks of its ever-flowing river with a succession of towns and villages, bordered by palms and cornfields, orchards, gardens, and cultivation. Its aspect is much diversified, exhibiting at some places vegetation in great exuberance, while at times we rode over desolate tracts of sand and pebbles without a sign of house or tree. The ranges that form the flanks are somewhat irregular masses of varying height, soaring at some points to lofty peaks several thousand feet in altitude.

At one part the hills approach and contract the valley to a narrow passage, at another they retire to let it expand. Flowing in a generally north-eastern direction, the Wady Semail is fed by innumerable springs and rivulets, and is swelled after rain by the tributary, but transient streams poured into it by the ravines and torrents. The volume of water that reaches the sea would, of course, be much more considerable were it not for the enormous quantity drained off by the inhabitants to irrigate their fields. The source of the river is in Jebel Akhdar, and the flow is tolerably fast, but without any sudden falls to Semail, from whence the descent is more slow and gradual. Computing the curves and windings, the entire length can hardly be less than 100 miles.

Passing from one fertile strip to another, and wading across the narrow bed of the stream many times in our winding path, we continued to sink more into the heart of the valley, which became more populous, cultivated, and attractive, as we advanced. After five hours' hard travelling we arrived at Wibal, where Sheikh Mohammed left us. He pressed me to stay the night, and I was half inclined to accept the invitation, as I felt rather fatigued, and would have liked the opportunity to see the place, but I decided to move on, and after making the usual presents to the sheikh, I took leave. Wibal stands on the left bank, under a conspicuous white hill surmounted by a watch tower. Behind it to the west rises the imposing Nakhl range, one of the peaks of which, called Jebel Karu Akabet el Kat, with a rugged pass close by leading to Rostak, has been estimated at 7000 feet.

For two hours more we trotted on in the fast-fading light, and it was nigh sunset as we approached the outskirts of the Semail Aliya, or Upper Semail, and began to pass through patches of cultivation and gardens, intermingled with the indistinct forms of houses and date-leaf huts. It was not easy to pick our way among the intricate network of irrigation channels and the labyrinth of palms, and thinking it better not to penetrate further, lest we should disturb and alarm the people by the sudden appearance of a mounted party in their midst at nightfall, we turned off to camp by the stream.

The Beni Rowaiha are the remnant of a renowned and noble tribe, the Beni Abs. Of the race of Adnan and the stock of Ghatafan, they claim descent from the Beni Hâshim, and on their first arrival in Oman called themselves the Beni Hâsham. The patronymic of the tribe seems to be Rawaha bin Rabia; they are, however, often still spoken of by other Arabs as the Wilad Abs. Though one of the most recent immigrants, having probably entered Oman after Mohammed, the Beni Rowaiha have gained possession of one of the best-watered valleys in the country, and now hold an influential position, being a powerful community of about eighteen thousand souls, peopling thirty villages. Surviving to some extent to this day—for the breach has never been completely healed—the feud between the Abs and Dhobyan had its origin in a quarrel about a horse-race in Nejd, in the sixth century of our era, and the war that then took place is known to fame as the

"War of Dâhis," the two tribes being at this time enemies and neighbours in Oman, as they were in Nejd 1400 years ago. The story of the War of Dâhis, which lasted for forty years, is a typical illustration of the internecine strife that occurs at intervals in Arab nomadic life, even at the present day. The details of it, as collected from many Arab poets and narrated by Fresnel and Caussin de Perceval, are very curious, but are too long for insertion here, and I can only give a bold outline.

In the year 562 A.D., the Sheikh of the Abs, Kais ibn Zohair, made a successful foray on the Thalaba tribe, and as a ransom for the booty and captives taken, demanded and received a famous horse named Dâhis. The extraordinary fleetness of Dâhis became the theme and boast of the tribe, and the envy of their neighbours, and it was not long before a match was made to race him with another horse. This was done by a cousin of Kais, who agreed with Sheikh Hamad bin Bedr of the Dhobyan tribe, to run Dâhis against a Fezara mare named Ghuba for a wager of ten camels over a course of fifty bow-shots. The match had been made entirely unknown to Kais, who, when informed, highly disapproved of it, and wished to withdraw, but eventually the wager was increased to a hundred camels, and the length of the course to a hundred bowshots. As the day fixed for the race approached, the horses were kept without water, the plan being that the horse which first plunged its nose into the water-trough 10 miles from the starting-point should be declared the winner. The racers were to run riderless, and to make them gallop their best, maddening thirst was to take the place of whip and spur. Over the yielding sandy plain the superior strength of Dâhis told, and he was soon well ahead of his rival the mare, which, though fleet, had less staying power. He would undoubtedly have won the race had it not been for a ruse of the Dhobyan Sheikh, who had concealed a man in a hollow in the course, with orders to check Dâhis and throw him off his stride. The trick succeeded, and Ghubra was first at the goal. Kais was informed of the stratagem by onlookers, and was beside himself with rage and vexation. As the race, however, had taken place in the country of the Dhobyan, he was powerless to do more than protest, and after vainly endeavouring to induce Sheikh Hamad to repair the injustice by restoring the wager, he returned home. The Abs were so hot and eager for revenge that the first blow was soon struck, and the first victim was a brother of Sheikh Hamad, who had acted so perfidiously. On this event hostilities would have ensued, of course, if the Abs had not immediately paid the Dhobyan the bloodwit of one hundred camels. Sheikh Hamad accepted this payment for his brother's death, but, after doing so, avenged himself by treacherously slaying a brother of Sheikh Kais.

In the war that now followed, the first battle, known as Dhul Marâkib, was a triumph for the Abs, but in the second the Abs were disastrously beaten, and obliged to give hostages. A long truce was then concluded, at the termination of which, in the year 576 A.D., Sheikh Hamad, instead of restoring the hostages given by the Abs, foully murdered them.

The attack made by Sheikh Kais, directly he heard the news, was so sudden and furious that the Dhobyan were taken by surprise, and suffered a loss of twelve killed. After this the position of the Abs became so critical, for they were much inferior in strength to the Dhobyan, that they resolved to migrate. They were pursued by the Dhobyan, who were plundering the baggage, when the Abs, making a sudden onslaught, routed them with great slaughter, killing Sheikh Hamad bin Bedr, whose treachery had caused the war. After many wanderings and adventures, the Abs arrived at last in the territory of their old enemies, the Amir bin Saasaa, by whom they were kindly and hospitably received.

The Amir were at this time at feud with the Beni Temim, who had meanwhile allied themselves with the Dhobyan, and as war was now inevitable, the two coalitions collected their strength for the final struggle, the force put into the field by the Dhobyan and Temim tribes being, it is said, the largest ever assembled in Arabia. The Abs and Amir tribes retired to a precipitous defile called Shoab Jabala, where they awaited attack. The enemy, confident in numbers, attempted to storm the ravine, but were thrown into confusion by a device, and the Abs, rushing down, utterly routed and dispersed them. This famous fight took place in 579 A.D., and practically terminated the campaign, though the tribes continued a desultory war for about thirty years longer, when they were finally reconciled by mediation, and the Abs returned to their former abode. Sheikh Kais ibn Zohair, however, scorned to make peace with his enemies, the Dhobyan, and retired to Oman, where he turned Christian, and became a monk or recluse.

Such is the brief story as preserved in ancient poetry, but the oral tradition of the tribes has somewhat varied it in course of ages; for instance, many Arabs believe now that Dâhis was ridden by Sheikh Kais, and the mare Ghubra by Sheikh Hamad.

Early the next morning, July 17, I crossed the boundary and entered Semail Sifâla, or Lower Semail, where I was met and greeted by the Wali Sezzid Nasir, whose acquaintance I had made two years before at Soor. After coffee I informed him of my intention to proceed on to Muskat at once, and asked him to procure fresh camels without delay. To this he demurred, begging me to stay with him till the following day, and saying it was his duty to entertain my Arab following in proper style before he could let them depart. I did not relish the delay, but it was impossible to refuse the Sezzid's hospitable courtesy, and I therefore consented to the feast on condition that the camels were forthcoming by 1 p.m. Sezzid Nasir then led me to a summer abode in a small garden at the edge of the stream, where I was furnished with the customary carpets and cushions, and where he left me to enjoy the luxury of a bath before breakfast.

The "Dayara," or circle of Semail, may be called the capital of the valley, as it contains the residence of the wali, a strong castle, and is the chief centre of population in it, being occupied by various tribes. It is, however, not a single compact town, but rather an aggregation of twelve adjoining, unwalled villages, with their plantations, fields, watch towers, and homesteads, forming, as a whole, a picturesque and luxuriant settlement, extending along the valley for 10 miles, with an average breadth of 1 mile. The names of the different villages composing the Dayara are—Semail Al Hajir, Ghobra, Harras, Zok, Mizra, Sital, Jemmar, Bistan, Sifeh, Jebeliya, Dubk, and Sil el Saheileh. These are divided into two nearly equal parts, known as Upper and Lower Semail, the former being held by the Beni Rowaiha and other Hinawi tribes, and the latter by the Ghaffiris. The castle stands in Upper Semail. The boundary-line separating these two hostile camps is sharply defined by a small transverse ditch called Sherkat el Haida, and across this ditch many a fight has taken place, for the tribes are constantly quarrelling and skirmishing. In these little affairs the combatants usually commence operations by firing at each other across the Sherkat from behind cover, and then, heated with the fray and stung to fury by the taunts of their adversaries, engage at close quarters, using their long double-edged Omani swords with great effect. Sometimes, when the river is very low, the Hinawis above try to dam up the stream, and thus cut off the supply of water from their enemies, the Ghaffiris, and this plan, when successful, which, however, is rarely the case, quickly brings about a suspension of hostilities and a truce.

There is a daily bazaar or market held close under the north side of the fort, and consisting of the usual food-supplies required, fruit, vegetables, meat, and salt fish, but no cloth or hardware shops. Semail, indeed, can boast of but little trade. Authority is too weak, and the general feeling of insecurity too prevalent, to allow of much traffic being carried on.

The settlement is rich in "Fard" dates, one of the finest varieties of this fruit produced anywhere. It is the kind most appreciated and esteemed by the Americans, who are good judges, and a very large quantity of boxed Fard dates is annually shipped to the New York and Boston markets. At the period of which I am writing (1876), Muskat was regularly visited by sailing ships from Boston for cargoes of pressed dates in bags, but of late years a change has taken place, and the Fard dates are now packed in boxes, and exported to the United States by steamer. Fruits grow here in great abundance and variety, and of excellent quality. Muskat and other markets are largely supplied from Semail.

The only manufacture in the place worth mention is cloth-weaving, and the creaking of the loom may be heard in every hamlet. Lungies, puggries, and khodrungs are the chief articles produced, the cotton of which they are made, both white and brown varieties, being extensively grown in the valley. The loom is somewhat heavy and clumsy in construction, and is horizontal (not vertical like the Jewish looms we read of in Scripture), and the weaver sits and works at it in a shallow pit, with half his body below the surface.

The annual revenue derived by the Sultan of Muskat from Semail for local expenditure is said to be 6000 dollars; two-thirds of this amount being yielded by the zakât, or tax, and the remaining one-third representing the produce of the Bait al Mâl, or crown lands, which are usually farmed out.

The population of Semail is probably from 20,000 to 25,000 souls, but in such a large and scattered place is difficult to estimate.

The eastern range has fewer peaks, and presents a more broken contour, than the western. It is known by many different appellations as we pass along it, its highest peak (5250 feet) being called Jebel Tyin. Just above the village of El Zok the road to Ak branches off, and it takes three hours of very rough travelling to reach the village of that name. The Wady Ak is regarded as the key of Muskat from the direction of the Sharkiyah, or eastern province, as it offers the most direct route to the capital. When on the war-path to attack Muskat, the Sharkiyah tribes, if permitted, pour down this steep and rugged defile into the wild entanglement of hills and ravines beneath, and thence into the Wady Semail; but the Nedâbiyin tribe, to whom the Akabet el Ak belongs, is usually subsidized by the Sultan to hold the pass. The valley is here very much shut in, but I obtained a fine view of the gigantic mass of Jebel Akhdar from the village of Ghobra.

On an isolated rocky eminence rising sheer above the floor of the valley, to which it presents on the western side a precipitous cliff, 300 feet in height, stands the ancient castle, whose imposing aspect adds much to the local scenery. The position is a commanding one, and well suited to enable the castle to serve the threefold purpose for which it was no doubt intended, viz. to overawe the turbulent part of the community, to command the passage of the valley, and to protect the whole settlement.

The hill has been scarped at the base on all sides, and the plan of the castle has been entirely influenced by the nature of the ground. The massive barbican, or gateway, in which are apartments forming the residence of the akeed, or captain of the garrison, is on the eastern and lowest side, and is joined to the keep at the south-western corner by low curtain walls, embracing an area of irregular shape and considerable extent. The keep is a large circular tower of solid stone masonry, built on the highest point of the rock, affording a superb view over the valley. In the barbican are two wells cut through the rock, giving an unfailing supply of water, and in the keep is a capacious reservoir, always kept filled. I counted eight iron guns in the fort altogether, three mounted on field carriages, the others dismounted. The castle was in a battered and shattered condition at this time, having sustained a bombardment during the recent operations undertaken by the Sultan to recover possession of it from a rebellious relative, but the wali informed me he had received orders from Muskat to repair the breaches in the walls. The wali is the castellan as well as the governor of the district, but does not reside in the castle, the guardianship of which is entrusted to the akeed, who at this time was a Belooch with sixty men under him. Sezzid Nasir told me he took care to interfere as little as possible with the tribesmen, and confined himself to maintaining peace and order and settling disputes. I gathered that the position he held was one demanding much tact, patience, and discretion, and was not a very enviable one, but he was evidently treated with great deference and respect by all.

Our host's hospitalities occupied all the forenoon, and I found my party extremely reluctant to make a move; but, though the day's march before us was no shorter than that of the previous day, I resolved to push on, and having taken leave of my courteous friend Sezzid Nasir, I mounted my camel and started.

From Semail castle the road winds down the valley in a north-north-easterly direction for 12 miles, as far as Serur, from whence it runs nearly due north. Serur is a rich and pleasant oasis of some extent in the possession of the Beni Hina tribe. At this point the stream disappears from sight, and sinks in the porous soil to flow underground for some distance, when it again reappears. A mile beyond Serur is the hamlet of Malita, and at another mile we came to Bidbid, a charming little oasis with a multitude of dates, rising like a green islet out of the broad barren sandy bed. In the centre is an old fortlet, untenanted, and fast crumbling to dust. Below Serur the hills begin to retreat from the river-bed, especially on the western or left bank, and the valley now gradually merges into a broad and open plain.

At Mizra, a village with an isolated rock and watch tower, the road to Muskat branches off to the right; but I may as well continue, from notes made on a subsequent trip, the description of the valley down to Sib, before going on with the narrative of the present journey. After leaving Bidbid, the path runs along the left bank over stony ground for about 5 miles, and then leads into a populous and thriving oasis called Fanja, the wady at this part being known as Batha Fanja. This town belongs to the Beni Hina and Hedâdebeh tribes, and is surrounded by a luxuriant belt of palms and well-cultivated fields, extending perhaps 3 miles in length. It lies 20 miles from Semail, and the population exceeds three thousand. The town owes its prosperity to the existence of excellent potter's clay in the neighbourhood, suitable for cooking utensils and glazed earthenware. The large jars used by the indigo-dyers and similar vessels are also produced at these potteries. Just beyond Fanja a masonry aqueduct, 3 miles in length, called the Felej el Dhowaikar, runs at some height along the elevated right bank, and leads to a village and fortlet of that name, now in ruins. It was constructed, I was told, in the middle of the eighteenth century, by the Imam Sezzid Ahmed bin Said, to whom Dhowaikar belonged, and appeared to be an unusually costly piece of engineering for Arabs to undertake, but when, long years ago, Dhowaikar was destroyed and deserted, the aqueduct fell into desuetude and decay, and has since been breached in many places by mountain torrents. At three hours from Fanja, and about 35 miles from Semail, we arrived at the castle of Khoth, which is picturesquely perched on a solitary hill, rising out of the river-bed, with a hamlet and patch of fruit orchards below. It belongs to the Hedâdebeh tribe, and is considered a position of some importance, as the castle overlooks, and to some extent commands the lines of communication between Muskat and the Batina. On the left or western bank, the Wady Sakhnan, known in its upper course as the Wady Beni Jabir, unites here with the Wady Semail. It can be reached in the vicinity of Semail by a low pass, and affords, thence, an alternative and short road to the Batina plain. The country between Khoth and the sea, a distance of about 12 miles, is bare, uninhabited and uninteresting.

Resuming our journey from Mizra, we took an easterly direction, and skirted on our right the hill range, which here trends along the coast, the path leading over a tract of broken and stony ground, very irksome to the camels, and necessitating a slower pace. Passing a low ridge known as the Nejd el Shubba, we rode for about 16 miles over a narrow maritime plain, much furrowed and intersected by ravines, and having several hamlets belonging to petty tribes, to Wataya, the site of a palace built by Sezzid Thowaini, now in ruins. It was now getting late, but in another hour or so, passing Rui and Bait el Felej, we arrived at Mutrah.

Nakhl and Wadi Mistal

Nakhl boasts springs as hot as those of Rustaq and in its Wadi Mistal surely the most beautiful climb in all Oman, to the villages of Wakan and Muti, their views down into the plain breathtaking at any time of day. If I had to live for the rest of my life in one Omani oasis it would be Wakan of the apricots, sweet grapes, and luscious bananas. Donkey tracks spread over the crystal-air, scented jabal to Wadi Halfain, and to Saiq, and I was offered a tempting donkey-ride across to Tanuf. Colours – what colours? Lilac, mauve, shades of brown from yellowish to deep tan, infinities of green, and a compound of grey and cream that I have never seen before or since, as though Wadi Mistal possessed a palette marginally but crucially different from all others.

To find Wakan, take a sign marked 'Wadi Mistal 37 km', just outside Nakhl. You will end up in a depression named Ghubrah for the main village in the wadi bed, dominated by a picturesque castle. Then the valley widens, and a new six-room primary school and new hospital come into view, and the long and breathtakingly-beautiful climb starts to Wakan; with 270 inhabitants of the Bani Riyam. The track is well-maintained, and well-used. In Wakan, I asked to see the main mosque, and was shown into it through the courtesy of 'Abdullah Khusaif. Simplicity is the keynote of Ibadhism; here the roof is made of palm-fronds, two windows let in the brilliant light low down in the right-hand wall (as you face the mihrab) and one low in the left-hand wall, while two tiny holes in each wall much higher up let in narrow shafts of sunshine. Wooden steps form a rudimentary 'minaret' in the right-hand corner. The floor is covered with rushes; the entrance is by two wooden doors. The centre of the mosque is adorned with two arches, the whole conducive to prayer, to meditation, to peace, to submission, a tacit repudiation of all ostentatious religious architecture from San Pietro in Vaticano or S. Giovanni in Laterano to the cathedrals of Coventry or Liverpool.

Then suddenly the upward surge to Muti and Wakan is over, dusk draws in its skirts over the distant mountains, and we set off downhill for Ghubrah and the main road, passing Afi and before long entering Wilayat Barka, on our way back to the Capital.

5 BATINAH

Just as the Dhahirah means 'outside, outer, or the rear', with reference to the mountain ranges paralleling the northern shore of Oman, so the Batinah is 'inside, inner, or the front.'

Neatly echoing a symbol of Islam, the Batinah is crescent in form, and roughly three hundred kilometres long, between the Capital Area and the border at Khatmat Malahah. The most fertile zone of Oman, it possesses almost continuous plantations of date-palms and market gardens providing comfortable subsistence for its inhabitants, who form the densest population of the Sultanate outside the Capital Area. Constant supplies of food throughout the year have always attracted seagoing vessels of all types wishing to stock for voyages, and there are substantial towns from Seeb to Barka, al-Musan'a, Suwaiq, Khaburah, Saham, Suhar, and Shinas, at regular intervals. Fishing has not only supplied valued protein to the local people, but a reliable income from sales to the interior. Rain has brought alluvial soil down from Wadi Hatta to the Shinas area, Wadi Jizzi to Suhar, Wadi Hawasinah to Khaburah, Wadi Far' to al-Musan'a, Wadi Ma'awil to Barka and Wadi Suma'il to Seeb. The water table is close to the earth's surface but was supplemented in ancient and mediaeval times by a system of *aflaj* bringing rainwater from the Jabal Akhdhar and Jabal Hajar al-Gharbi. Animal-powered wells called *zajarah* have largely been replaced by governmnent-sponsored improvements, though Wilkinson has suggested that even now the level of prosperity in the zone has not equalled that in mediaeval times. Electrification, and an excellent fast new dual carriageway along the coastline from Muscat to the west (though not to the east as yet) has ensured that employees choosing to work in the Capital may still keep their homes and part-time jobs in fishing, farming, and animal husbandry in the Batinah. Moreover, this is the main road to the United Arab Emirates, and thence to Saudi Arabia and the other Gulf States, now as ever linked by trade, religion, and common politico-social ties. Nomads and semi-nomads found it easy to obtain seasonal work in the date-gardens, or casual jobs whenever drought forced them from their traditional grazing grounds. Most semi-nomads have taken to settling along the coast road since Independence, working mainly as truck-drivers, builders and labourers, or shopkeepers, according to Hartmut Asche's 'Moderne Wandel und nomadische Bevölkerungsgruppen in der nordomanischen Kustenebene Al-Batinah', in Fred Scholz's invaluable *Beduinen im Zeichen des Erdöls* (1981)

As regards seagoing trade, only Suhar played a really significant rôle, especially in the 9th and 10th centuries. Barka, Seeb and Suwaiq were also garrisoned by the Portuguese, but chiefly to control unruly elements along the coast. The population is mainly Sunni or Ibadhi, but there are communities of Luwatiyah at al-Khaburah (mainly), and al-Musan'a and Barka, as well as at Matrah; some Bahrainis and Iranian traders in larger towns; and Baluchis from coastal Makran (in modern Pakistan), an area closely connected with Oman throughout the centuries by trade and immigration, especially taking into account the Omani ownership of Gwadur enclave from 1784 to 1958.

From the Capital Area, the first major sight along the coast is Seeb International Airport, on the right-hand side, followed by the Royal Flight Airport, and then the village of Sib – like every settlement except al-Muladdah on the seaward side of Route 1 to the frontier at Khatmat Malahah.

Barka and Bait Na'man

Wellsted visited Barka in February 1836, noting: 'At present Burka is principally remarkable for its fort which, owing to its great height and size, is very conspicuous from seaward. It mounts thirty pieces of artillery; but so little attention is paid to them or their carriages, that not one half could be fired; yet is it deemed by the Arabs impregnable; nor, probably, so far as they themselves are concerned, are they mistaken . . . Seyyid Hilal's harem was confined here at the period of my visit and we were not, in consequence, admitted beyond one of the towers near the entrance.'

I also found it impossible to gain admittance, for the simple reason that the fort was being restored by the Ministry of National Heritage and Culture to its former glory. A friendly guard consoled me by showing how he was weaving rushes, his ascetic triangular face cast down in concentration, his whole body swathed in a *dishdasha* that must have made him extremely hot even at that early morning hour.

Wellsted estimated Barka's population in 1836 at four thousand. 'A considerable portion of these, as on other parts of the coast, are employed in fishing, and the remainder attend the date trees. The bazar is very extensive and the Bedowins flock in from the surrounding country to make purchases of grain, cloth, &c., and almost every article procurable in Maskat may be obtained here. The anchorage at Burka is an open roadstead, affording no protection against the prevailing breezes. The same remark applies to nearly every town on the coast and they have, in consequence, few bagalas of any burthen trading along it. Merchandise is brought from or conveyed to Maskat in small boats, of from thirty to fifty tons

378

Barka. Fort under restoration, 1986

burthen. Vessels of this size upon the approach of bad weather are
hauled up on shore beyond the action of the sea with little difficulty.
A revenue of from three hundred to four hundred dollars is annually
drawn from Burka. It arises principally from dates on which, as well
as on all other exports or imports, a duty of ten per cent is levied. The
Imam [then Said ibn Sultan. *Ph.W.*] maintains a small force of about
two hundred men here: their wages are partly paid out of this
impost . . . It rained hard all day, and we were unable to leave, but
derived considerable amusement from watching the busy scene
passing in the bazar. The seller disposes his several articles in a heap
before him, and seats himself quietly on his haunches beside it. A
buyer approaches, and the affair is settled probably after not more
than half a dozen words have been spoken. But mark the contrast: at
the distance of a few yards is also seated a female, who sells grain;
one of her own sex approaches for the purpose of buying: that war of
words has lasted nearly an hour, and yet appears no nearer a
conclusion than at first. Here there is a boy with a basket of dates on
his head, bawling forth, as he totters under the weight of his load, the

superiority of his commodity and its price. There, a man parading to
and fro with a turban and a pair of sandals. At a distance are some
butchers' stalls, and beef, mutton, &c., are doled out to those who
crowd round it, by means of a pair of wooden scales, having stones as
a substitute for weights.'

A Barka fisherman with a fine haul would have looked familiar to
Wellsted a century and half before. Against a backdrop of smooth
sand, gliding waves, smooth blue sea and cloudless sky, he held up
for me his trophies: 'One rial! One rial!' A *suq* leading down to the
sea dates from after Wellsted's time, and here I found merchandise
from all over the world, from India, Pakistan, Taiwan, Spain, Italy,
Hong Kong, and the People's Republic of China. Between the *suq*
and the beach I looked at the old *barasti* shelters and houses, now
regrettably in decline, with increased expectations. Regrettably, for
nothing conveys the atmosphere of al-Batinah more unequivocally
than a *barasti* hut. Paolo Costa, writing in the *Journal of Oman
Studies* (Vol. 8, 1985), has justly compared the sophistication of the
Batinah *barasti* to the architecture of the Marsh Arabs in Iraq, where
techniques handed down from successive generations have made
possible highly elaborate and comfortable homes at very low cost.
Some palm-fronds are stripped of their leaves, and others not, but in
either case the *zur* or stems are tied together to form a *da'am* or panel
between two and four metres long. This may be rolled for storage, or
strengthened with rigid braces and erected as fencing, roofing, or
winter quarters. External support might be provided by buttressing
logs.

Palm-frond wind-catching towers, called *bad-gir* in Persian and
possibly derived from South Iranian tradition, used to be common

Barka. Fisherman

along the littoral, but the widespread introduction of air conditioning has decimated these picturesque survivals. The two dominant types of *barasti* dwelling were the winter house (*'arishah*) and the summer quarters (*karijin*), often both found within the same walled compound for an extended family, which would also incorporate toilets, washing areas, cooking facilities, and a storehouse. Some temporary homes are constructed for the period of date-harvesting in the summer. One can only emphasise and echo Dr Costa's urgent plea for restoration of existing *barasti* houses, and the creation of new ones, so ideally adapted to the climate of the coastal belt. The eradication of subterranean and dry-wood termites alike must remain high priority if *barasti* shelters are to survive in any numbers, as these represent a hazard more potent than wind or weather.

I found a whole school of *barasti* shelters in the village of Bait Na'man, about 4 km from Barka by an interior asphalt road, then a sandy track, signed to Na'man. The term 'bait' or house does insufficient justice to this splendid ruin of a square seventeenth-century fortified country house, extended in the eighteenth century by the addition of two towers and an extra outer wall of masonry. It thus became a castle to class with the outstanding defensive fortresses of Jabal Akhdhar such as Rustaq, and was used a staging-post by Imam Ahmad ibn Sa'id on his journeys between Muscat and Rustaq. We are indebted to Dr Paolo Costa for a full history and description of Bait Na'man in the *Journal of Oman Studies* (Vol. 8, 1985).

Of three storeys, the castle has round towers at opposite corners, in the mode already familiar from Jabrin and al-Hazm. It resembles Jabrin, too, in having been built at the corner of a large square walled enclosure, here on a low tell. No permit is required to visit the monument, which is unguarded and unprotecting by fencing.

The builder was Imam Saif ibn Sultan, who died in 1711, a man whose dynasty, the Ya'ariba, was responsible for the creation of al-Hamra' in Jabal Akhdhar and Birkat al-Mawz in the foothills, and for the revival of 'Ibra, Quriyat, and Barka itself, not to mention Jabrin and al-Hazm. Bait Na'man, a rustic retreat close to the sea and two easy stages from either Muscat or Rustaq, is known to have attracted Imam Ahmad ibn Sa'id (1749–83) to stay there for at least two days en route between his residences; and it became the permanent residence for some time of his sons Saif and Sultan. Of all the episodes in the eventful history of Bait Na'man, none is more crucial than the uprising of Badr bin Saif, who claimed the right to rule instead of Sayyid Sa'id bin Sultan. In 1807 the two men attacked each other with daggers inside the building, then the wounded Badr escaped by jumping out of a window and escaping on horseback until he was overtaken by men loyal to Sayyid Sa'id. In recent years Bait Na'man has been despoiled of doors, windows and fittings, to become a splendid, isolated ruin, redolent of the past.

Exhilarated by the proximity of sea and sand, by the brilliance of the sun and the warmth of zephyrs, we did not return to the main road, but sped along the shoreline towards the headland called Ras Suwadi. It was low tide, so the village of Suwadi was connected to the

Bait Na'man. The fortified mansion

low limestone island by a sand spit; at high tide it is cut off. You can hire a boat to Kharabah and Jus, two of the Daymaniyat Islands, but the rest may not be visited because they are protected as a nature reserve for birds. Suwadi is a shell-collector's paradise, like almost the whole Batinah.

Ras Suwadi. Low tide

382

Al-Musan'a and Tharmaid

The next village is al-Musan'a, one of a handful of towns (including Birka, Seeb, and al-Khaburah) where you can see bullfighting. This is not the variety connected with the Spanish *corrida*, but more akin to the bullrunning at Pamplona, in the sense that bulls do not face armed picadores and toreadores, and almost certain death, but they are let loose with nothing between them and their human audience. The idea is not, however, to chase people through the streets, but to engage each other's horns for up to five minutes at most. The 'meet' is on occasional Friday afternoons, and on such holidays two to three thousand spectators might gather in a great open square. Referees pair up evenly-matched bulls, and two at a time, they meet in momentary combat, before the referees decide which bull is the stronger of the two: no bull ever dies, but the stronger may increase rapidly in price from 500 rials to five times that amount if, that is, the 'catchers' manage to bring him back after the contest is over.

The fish market at Musan'a is covered: I greeted each vendor in turn, and they solemnly returned my salutation. Four rusting cannon show the futility of war outside the castle of Musan'a, so ruinous that you can climb into the upper sides to obtain a fine view over the old town and towards the sea. I ws now beginning to identify at first glance the *aswar* (the plural or 'sur'), those fortified enclosures built to protect women, children, and livestock in time of strife. Paolo Costa, to whom students of Oman will be forever indebted for unobtrusive aid to scholars, travellers, and archaeologists, has contributed a pioneering study of 'The *Sur* of the Batinah' to the *Journal of Oman Studies* (Vol. 8, 1985). More than ninety such have been identified in the Batinah alone, and Dr Costa reminds us that *aswar* are 'to be found near almost every minor settlement in most parts of Oman, wherever an unwalled group of dwellings needed a defended compound for the inhabitants to retreat to in the course of attack by a foreign army or during inter-tribal strife.' The *sur* is absent, by contrast in major towns with walled quarters or indeed (as at Bahla) a wall surrounding the whole town. Examples of the latter kind would include 'Ibra and Mudhairib, Izki, Nizwa and Rustaq. From researches in the Batinah, it appears that a village with its own *sur* generally numbers no more than twenty or thirty houses, partly because a *sur* cannot hold any more people and their livestock.

There is a particularly interesting example of a *sur* at the next village along the coast, Tharmaid, where virtually the whole wall and defensive towers remain intact, and are indeed being restored. Tharmaid is incidentally the site of a new naval base, so the juxtaposition of ancient and modern defence seems particularly piquant. As usual, the Tharmaid *sur* has a single entrance, though in this case (Costa 34) a complex three-storey gatehouse is recessed from the external alignment of the wall, posing special difficulties for attackers. Three corners of the enclosure have a round tower. Remains of rooms can be seen, built against the exterior wall which

Tharmaid. Sur

had a crenellated parapet and a sentry walk. The whole represents a shrewd, farsighted defensive policy impossible to fault, when one considers that, like all the others, this *sur* was never intended as a permanent residence or for long-term occupation. And do not miss the restored fort at Tharmaid.

Tharmaid. Fort

Suwaiq

Ahmad Khalifah, my driver, now paid me the compliment of inviting me into his own home, at Suwaiq. A concrete-block house with electricity, inside the usual walled compound, this was one of the latest generation of dwellings. I was ushered into the *majlis*, or sitting-room, next to the front door, a fan whirring to cool the air in the heat of noon. The *majlis* was bare except for a carpet on the floor, and pillows against the walls on which to rest back and elbows. Three tiny cups of coffee were followed by three cups of tea, bananas, and French apples; the voices of women excitedly whispering in the background reminded me of Wellsted's experience in Suwaiq a hundred and fifty years before. 'We arrived at Suik at 3.30,' noted Wellsted, 'and I found the Sheikh absent, looking for the Wahhabis who, it had been reported, were then in the neighbourhood; but we were most hospitably received by the Sheikh's wife, who had a house and every other accommodation very soon prepared for us. "You will please those gentlemen," said she to her slaves who were sent to attend us, "and let them want nothing, or look to your heads." We accordingly received every luxury which the Sheikh's kitchen could afford.' The shaikh in question was Sayyid Hilal, a thirty-five year old cousin of Sayyid Sa'id. 'In point of character, he stands of all the chiefs of Oman next to that prince; his figure is tall and commanding; he excels in all warlike exercises; is passionately attached to hunting and other field sports; and though somewhat spare in figure, is considered, in addition to his extraordinary agility, the strongest man of his nation: he is generous to profusion. I have have heard the Arabs in Maskat relate that, when upon a visit to the Imam he has received from him a present of eight hundred or a thousand dollars, in the course of a couple of hours afterwards he has bestowed the whole of it in presents to his followers.'

Ahmad Khalifah took me to Suwaiq Castle, an imposing structure. I shook hands with every elder in the *majlis*, and was fortunate enough to meet the Wali, Said bin Salim al-Badi, a man of great courtesy and warmth clearly as much loved as was Sayyid Hilal in Wellsted's time. Hilal's wife, whom we have met already, was renowned for her determined valour. While Hilal himself was under safe conduct in Muscat, a force from the Imam came to seize Suwaiq Castle, failing which her husband would be put to death. Calling this bluff, she is said to have scorned the attackers, saying 'Go back to those who sent you, and tell them that I shall defend the fort to the utmost of my power; and if they choose to cut him into pieces before me, they will find no alteration in my resolution.' Hilal was later exonerated from suspicion and reinstated at Suwaiq, but he was again absent when the Shaikh of Suhar, on his way to Rustaq, suddenly decided to invade Suwaiq during its lord's absence. Hilal's wife closed the gates, collected the garrison, and made such good use of the guns mounted on the towers that the two hundred Suharis were repelled, and Suwaiq was saved. Wellsted reported that 'Hilal feels the greatest respect for, and stands in some awe of this dame, without whose advice and concurrence he undertakes nothing of moment.'

I don't begin to understand why, but in Suwaiq I felt that I had *never been away*, and wrote a poem for this magical oasis, with its beach and its *suq*, its castle and *aswar*, its fields of lucerne and barley, its date-palms and *barasti* huts.

In the old *suq*, I was stopped in my tracks by the nearest human equivalent of a bouncing talking leatherbound book, named 'Abdullah bin Khamis bin 'Abdullah of the Mandhari tribe. Selling *halwah* at 2 rials for the special variety, 1½ rials for the 'original' variety in his words, and 1 rial for the 'weak' variety, he turned up his nose at the last, as if challenging me to offer anything so paltry. I remembered Artur Schnabel's wry reply when being told that twenty dollars for a piano lesson was rather a lot. 'I also do lessons at ten dollars an hour,' he answered, 'but I don't recommend them.' The *suq* tempted me with okra, onions, lemons, oranges, coconuts, apples, bananas, green peppers, lettuce: is there nothing that will not grow on the Batinah? It is often said that coconuts do not grow in northern Oman, but we have firm evidence in Humaid ibn Ruzaiq's *Al-Fath al-mubin* that Imam Saif ibn Sultan planted no fewer than 6,000 coconut trees in the Batinah at the beginning of the eighteenth century.

We passed an old man on a donkey arguing energetically with another. A youth sat beside a dozen thin Chinese mattresses, waving a cut wand at a tethered camel.

Whitelock on the Batinah

Francis Whitelock, a young lieutenant in the Indian Navy, accompanied. J. R. Wellsted on his visit to Suwaiq ('Suick'), though almost nothing is said of his presence. Wellsted's *Travels in Arabia* (1838; reprinted in 1978) have long been prized as a major source for travellers in Oman, but details of Whitelock's journey have never been reprinted since they first appeared in the *Transactions* of the Bombay Geographical Society (1836–8). This then is the first time Whitelock's voice has been heard again for a century and a half.

Notes taken during a Journey in Oman, and along the east Coast of Arabia. By Lieutenant F. Whitelock, I. N.

In November 1835, Lieutenant Whitelock obtained the sanction of Government to travel in the interior of Arabia and Persia, in order to make himself proficient in the languages of these countries, and arrived at Muscat in the beginning of the following month (December 1835). Having laid down no fixed plan, he determined in the first place to proceed to Semed, a town situated inland south of Muscat, where he was informed Lieutenant Wellsted was residing. He left Muscat on the 13th, and on the 16th of December arrived at Semed.

On the 19th December, in company with Lieutenant Wellsted, he started for Nezwah, and arrived there on the 22d. Here they remained a few days, and then proceeded to the Green Mountains, a description of which is given in this communication. They spent a few days among the mountains, and met with great kindness and hospitality from the people; they then returned to Nezwah by a different route. At the latter place they received information from Muscat, that H. H. the Imaum had engaged with a Shaik to conduct them from Braimee to Deryeh for 500 dollars, for which sum he agreed to supply camels, guards, water, &c. In consequence of this information, they considered it necessary that one of them should proceed to Muscat to close the bargain with the Shaik, and to procure the money required. Lieutenant Whitelock volunteered to undertake the journey, on which he started on the 1st of January, and arrived at Muscat on the 4th. He found great difficulty in obtaining the money from the agent, and was in consequence obliged to remain longer in Muscat than he should have done, for the place was very unhealthy at the time. Here the Author was attacked with fever, and finding that he was not likely to recover at Muscat, he left it, though in a very weakly state, and reached Nezwah, by easy stages, on 20th January. He found Lieutenant Wellsted seriously ill, and all the servants laid up with fever, and unable to move or render them the slightest assistance. Under these circumstances, they were, much to their annoyance, forced to abandon for a time their projected journey, for which every arrangement had been made with the Shaik of Lasha.

In order to re-establish their health, they determined to proceed to the sea coast, and accordingly started for Sib, situated a few miles from Muscat, and considered a very healthy place.

Having remained here a month, and finding their health greatly improved they determined to attempt the journey to Braimee, and thence to proceed on to Deryeh, if possible. They procured the necessary letters from the Imaum, who provided a Shaik to escort them. The Author and his companion started on the 25th February, and proceeded along the coast as far as Suick, then changed their direction towards the interior, and arrived at Abree on the 12th of March. At Abree they were treated most unceremoniously by the Shaik, who requested them to leave the town as expeditiously as possible, for he could not protect them, nor give them a guide or guards to escort them to Braimee, as he was afraid of the Wahabis. They afterwards learned that the Wahabis were actually in the town at the very time.

Finding it impossible to get to Braimee by this route, the travellers returned to Suick on the coast, and were most hospitably entertained by the Shaik. After remaining several days at this place, they determined to try if they could get to Braimee from Schinas, where they arrived on the 25th March in a boat from Suick. Here they waited several days for permission from the Wahibi Chiefs, but finding that a longer stay was necessary before they could expect to obtain it, they resolved to make the most of their time, and to proceed in a boat to Cape Mussendom, in order to visit Coomza. As they were on the point of starting on this expedition, they received intelligence that the Imaum was very anxious to see them, and had despatched messengers in quest of them. They therefore returned to Muscat, where, on their arrival, they found that the report which had induced them to visit the place was false.

Mr. Wellsted, finding his health still bad, and that there was little prospect of being able to proceed to Deryeh, determined to return to Bombay, for which place he started in April 1836.

The Author again set out from Muscat for Rostack, a town which he had not visited before. Here he was very civilly treated, but finding that he could not prevail on the Shaik to allow him to proceed through his territories to Braimee, he left Rostack for the coast; and on the 18th April reached a place called Messna, where he remained only one day, and then proceeded to Sohar, and thence to Schinas, at which place he arrived on the following day.

At Schinas the Author had an interview with the son of the Wahabi Chief, who was passing on his way to Muscat; he attempted to obtain a pass from him, but without success. On the following day, the Author hired camels, and set out from Sharga across the desert. From Sharga he visited Cossáb, then Bahrein, and Grane or Quoit. At the latter place he attempted to procure a boat to take him to the Táb or Endian river, which it was his object to visit, but not finding a boat here, he proceeded to Bussorah, where he was more successful.

The Author gives the following sketch of the general features of the part of Oman which he traversed in his various journies through that province.

' The range, of which the part we visited is styled the Green Mountains, appears to continue as far as Cape Mussendom, gradually decreasing its distance from the coast as it approaches the Cape, so that from Burka the range is distant 40 miles, from Schinas 15, from Khorefa Khan 10, and about a quarter of a mile from the beach at Cape Mussendom. The chain at the Green Mountains is composed of three parallel ranges, of which the central one is the highest. Here the hills are not detached, but connected together by gorges of considerable elevation, whereas in the part of the chain towards Cape Mussendom the hills are detached, though they follow the same general direction. The country between the sea and the mountains presents, immediately adjoining the coast, a belt of date groves as far as Khore Kulba, occupying a breadth of about four miles, beyond which the plain, at every point at which I crossed, is barren up to the foot of the principal range, but has detached hills scattered irregularly over it. Beyond the mountains to the south, we have sandy desert, which separates Oman from the province of Lasha and Nejid, the boundary of Oman in this direction being a line which runs directly from Ras-el-Had to Abothubi. The coast between Abothubi and Cape Mussendom is denominated by the natives generally the " Oman Coast " that which extends from the Cape just mentioned to Ras-el-Had, is called in the same manner the 'Batnah Coast."

Pengelley on the Batinah

The next first-hand report on the Batinah was produced by W. M. Pengelley, British Political Agent at Muscat from May 1861 to January 1862, and published in the *Transactions* of the Bombay Geographical Society (1860–62).

Remarks on a portion of the Eastern Coast of Arabia between Muscat and Sohar. By Lieutenant W. M. Pengelley, H.M.I.N., British Agent at Muscat. [*Presented by Government.*]

[Read before the Society, October 17th, 1861.]

The petty jurisdiction over the northern part of Oman having recently undergone alteration, it will be necessary to note the present dimensions of the districts supervised by the "Walys" (deputy, or subordinate governors) residing in the chief towns on the Battnah coast.

Four of the abovenamed officials are located between Core, Culbah and Muscat (namely at Sohar, Burka, Seeb, and Muttrah), whose duties are to decide in all minor offences; those of a more serious kind being referred to the capital to be adjudged by the "Wazeer" or "Kady" according to the nature of the case.

The above towns are supposed to be the centre of districts; the jurisdiction of the "Waly" extending north and south, namely, the "Waly" of Sohar (Seyf bin Suliman, late "Waly" of Sahm and Khaboorah) from Core Culbah to Kadeirah, a distance of nearly one hundred miles. The "Waly" of Burka (Seyd Saeed bin Mahommed) from Kadeirah southward midway to Seeb (about forty miles), and the "Waly" of Seeb (Seyd Saeed bin Hamad), from the latter extremity south to Muttrah about the same distance.

The authority of a "Waly" diminishes in proportion as it may be called into exercise at a distance from the coast, where culprits are sometimes arraigned before Arab chiefs, by whom disputes or misdemeanors are settled with but little regard either to justice or equity. Instances of the above too frequently occur, kindling ill-feeling and engendering warfare between various Arab tribes and the legitimate ruler of the country.

"Walys" are commonly chosen from amongst the older inhabitants, and elected without regard to affluence or social position, but simply from character as to probity and intelligence, coupled with a due amount of attention to the forms of the Mahommedan faith. They appear to be esteemed by the people, and afford general satisfaction, as not a murmur reached my ears, though I discreetly yet sedulously endeavoured to ascertain from the "felaheen," or poorer class, if any causes of discontent at the form of government, or complaints against the "Walys" individually, existed.

The general aspect of the country between Muscat, Burka, and Sohar is level, as the mountains in no part approach the coast within a journey of from one to two days.

Practical cultivation is adopted only in exceptional cases; the husbandman on sowing being usually content with irrigating the soil and leaving the result to nature.

Fresh water, everywhere plentiful on this coast, is obtained from wells, many of which have been sunk a few yards only from the beach.

To a casual observer there is but little variety in foliage of either trees or shrubs. The former, with the exception of the date, are rare. A stunted thorny mimosa abounds in the desert, and thinly skirts the edge of the cultivated belt of land parallel to the line of coast. The mango is to be met with at Seeb and Sohar; but the fruit, for lack of cultivation and grafting, is of an inferior order. A few cocoa-nut, lime, and almond-trees complete the arboreous catalogue. Some specimens here and there of "adonsonia," and other trees of a grotesque appearance, the nomenclature of which unfortunately I am ignorant of. Fruit, in its season, is abundant, and deemed by the natives to be very wholesome. Dates, pomegranates, peaches, and grapes are plentiful; they are imported to the market of Muscat from Rastak, Smael, Nakhl, and other towns in the interior.

Vegetables are scarce at all times; good onions, pumpkins, and bhendys are procurable, but only in small quantities; which may be accounted for by the fact of the mass of people subsisting solely on fish and dates, the former taken diurnally in the vicinity of the beach in such abundance as occasionally to exceed even the magnitude of the demand.

There are eight descriptions of animals only, both wild and domestic; in this part of Oman, horses, camels, asses, bullocks, goats, gazelle, hares, and a few pariah dogs; and though the country by nature in most part is admirably adapted for wheeled conveyances, none of the above (save in the artillery), are used for draft—goods being invariably transported on the backs of asses and camels. Horses are scarce; the only good ones are imported from Quies, Bussorah, and Baghdad, and ridden by the irregular soldiery, and a few of the more wealthy inhabitants.

An extremely singular-looking boat is in use by the fishermen on this coast, in which they trust themselves in all weathers, and sometimes perform voyages of from fifty to a hundred miles. It is constructed entirely from the stripped branches of the date, which (subsequent to a month's immersion in sea-water) are sewn together in the form of a boat. They are sharp at either end, and have a double flat bottom : the upper is used as a kind of deck on which the fishermen seat themselves. Necessarily the sea percolates freely through all parts, but the extreme buoyancy of the branches readily supports a crew of two men, together with, at times, a considerable cargo of fish. These boats are called by the natives (شَشَه) " shashah," and are propelled by two paddles, assisted occasionally by a small sail.

Burka is noted throughout Oman for the skill with which some of its inhabitants can trace footsteps. An individual was brought to me who, I was informed, supported himself and family in the above manner. He was able readily to distinguish the difference between the foot-prints of a man or woman, boy or girl ; and I was informed he seldom or never failed either in tracing a robber, or restoring to the owner any animal which might have strayed for miles into the desert. This kind of business is termed " athr."

The town contains about 3,000 reed and mat-houses, to which may be added one-fifth of the above number, constructed of rubble masonry, one or two only being chunammed.

The fort, which is built close to the beach, is the only building worthy the name of such. It has a time-worn appearance, and is evidently in a tottering condition ; probably it would be seriously damaged by the fire of its own guns, if shotted. There are seventeen pieces of ordnance mounted (12 and 24-pounders.), but none can be said to be effective.

Thoughout the summer Burka is full of inhabitants, at which time it is garrisoned by about 400 fighting men. During the date season, from June to August, all who are able quit Muscat. The weaver puts aside his loom, the boatman hauls up his boat, and high and low lend their aid in gathering in the crop of dates, in which all are interested ; for with scarce an exception those who have a dollar to spare invest the same in the purchase of a date-tree.

The "kharaj" or tribute to the state, being one-tenth the product of the tree should, if efficiently levied, yield a good revenue. Formerly such was the case, but of late years it has not been rigidly enforced, and the people tender almost what they please. For example, in the year 1855 the tax on the date crop between Shinas and Khaboorah, a distance of about 55 miles, yielded 35,000 dollars, and at that time from 30 to 35,000 dollars was the annual average. Now, although the fruit-bearing trees have considerably increased in number, only 14,000 dollars are tendered to and accepted by the state for the same portion of country.

The amount of business transacted daily in the Bazar of Burka is insignificant, being essentially retail. Here, as elsewhere in Omân, the ready cash is all in the hands of the Banyans, who ostensibly keep small shops for the sale of piece-goods, grain, &c. I am informed, however, it is not to trade that the majority of these people look for their returns; but in advancing money on arms, ornaments, wearing apparel, &c., or in other words, as " pawnbrokers;" and those who enter Omân penniless, after three or four years' residence, either return to Cutch with a fair competency or set up as merchants in a regular way. They appear well contented with the country in its present condition, and having neither income nor any other description of tax to discharge, are disposed to be somewhat lavish in their praises of Arab government; forgetting to attribute their freedom and security to the inadequately-esteemed fact of their being the subjects of Great Britain. The Government, on its part desirous of encouraging settlers from India of the Banyan caste, readily afford them protection; and as far as I have yet been able to observe, are inclined rather to overlook their faults than to oppress them in the slightest degree, aware, doubtless, that the sinews of war are imported to the country by these followers of peace, and precursors of plenty. Should the Banyans, as a body, ever quit this part of Arabia, I am of opinion that serious embarrassment would be the result; it being questionable if in their absence the Government could satiate the unreasonable and vexatious demands for monetary succour which are too commonly made by powerful Arab tribes.

The Battnah coast has the repute of being usually unhealthy, especially, with reference to Europeans. It possesses the advantage, however, during the summer months of being considerably cooler than Muscat, particularly at night; and the accompanying register, which has been kept with strict attention, shows that the thermometer at the hottest time of the year never indicates a high temperature, except when a westerly or hot wind blows. Happily these are infrequent, the prevailing breezes being east and north-easterly. The Arabs consider the months of May, June, and July to be healthy (which I found to be the case), but immediately the dates are plucked from off the trees, fever commences. A practical illustration of the latter statement also, both as regards myself and servants, unhappily convinced me of its truth, and the repeated applications for quinine from the town's people demonstrated the prevalence of the malady. Rain, during the summer, seldom falls; but its absence is compensated for by the extremely heavy dews at night, and the frequent fogs, dense mists, and humid atmosphere during the greater part of the day. I observed the latter commenced about the beginning of August, or just as the dates were being gathered, which may probably account for the wild native idea attributing the presence of fever to the act of plucking the fruit.

Meteorological Register for the Month of June on the North Eastern Coast of Arabia, 1861.

SEEB { Latitude 28° 41' N. Longitude 58° 15' E.

Days of the Month	Moon's Phases	SUNRISE Dry bulb	SUNRISE Wet bulb	SUNRISE Difference	NOON Dry bulb	NOON Wet bulb	NOON Difference	SUNSET Dry bulb	SUNSET Wet bulb	SUNSET Difference	9 P.M. Wet	9 P.M. Dry	9 P.M. Difference	Aneroid Barometer at Noon	RAIN	WIND Direction	WIND Force	APPEARANCE OF ATMOSPHERE, &c.
15	☽	96	75	21	99·30	74·30	5	93	83	10	82·30	93	10·30	inch. 29·78	None.	N.Wly.	1	Dry hot wind, cirri, Ther. at noon in sun 110°, wet bulb 76°.
16		90	82	8	98·30	71·30	7	96·30	78	8·30	75	93	18	·76	„	W.	1	Dry hot wind, cirro cumulus, heavy dew at night.
17		87	77	10	97	79·30	17·30	96·30	81	5·30	82·30	92·30	10	·70	„	W.	2	Ditto do.
18	☽	86	81	5	91·30	86·30	5	89	84	5	83	89	6	·70	„	E.	1	Cirri, heavy dew, P.M. clear.
19		86	81·30	4·30	89·30	83	6·30	89	81	8	81	88	7	·76	„	„	2	Ditto.
20		85	80	5	90	83	6	89	82	6	82	88·30	6·30	·75	„	„	1	Cumulostratus,
21	○	86	80	6	91	84	7	91	83	7	83	89	6	·66	„	„	1	Ditto.
22		88	83	5	92	85	7	90	84	6	84	90	6	·68	„	„	2	Cirri, dew.
23		89	84	5	91	86	6	89	84	6	84	90	6	·74	„	NE.	3	Do. heavy dew.
24		85	82·30	2·30	90	85	6	87	83	4	84	89	5	·78	„	„	3	Do.
25		87	83	4	87	83	7	87	83	6·30	83	89	6	·74	„	E.	2	Do.
26		87	81·30	5·30	88	82	6	87	81·30	4	82	86	4	·72	„	„	2	Cirrostrati, dew.
27		84	80·30	2·30	87	83	6·30	87	83	4·30	83	87	4	·74	„	„	2	Ditto.
28		85	80	5	84·30	83	6	85·30	81	5	82	85	3	·72	„	„	2	Cirrocumulus, dew.
29		83	80	3	88	82	6	86	81	6	82	85·30	3·30	·74	„	„	2	Ditto P.M. clear, heavy dew.
30	☾	83·30	80·30	3	89	84	5	89	83		83	89	6	·74	„	„	2	Cirri, Do.

Meteorological Register for the Month of July.—Seeb, N.E. Coast of Arabia, 1861.

Seeb { Lat. 23° 41' N. / Long. 58° 15' E.

Days of the Month	Moon's Phases	Temperature in the shade and depression of Moist Bulb.												Aneroid Barometer at Noon.	Rain.	Wind.		Appearance of Atmosphere.
		Sunrise.			Noon.			Sunset.			9 P.M.					Direction.	Force.	
		Dry bulb	Wet bulb	Diff'nce.	Dry bulb	Wet bulb	Difference.	Dry bulb	Wet bulb	Difference.	Dry bulb	Wet bulb	Difference.					
1	◁	87	84	3	93	85	8	95	83	12	94	82	12	inch. 29·72	N.ne	N. E.	2	Cirri, heat very oppressive.
2		93	79	14	98·30	83·30	15	96	81	15	96	83	13	·72	„	W.	1	Do. observed a large comet bearing N. 28° E. near Polar Star.
3		95	79	16	98·30	83	5·30	100	80	20	99	83	16	·72	„	„	2	Cirri, from 10 P.M. to 2 A.M. very strong westerly breezes with parching atmosphere.
4		95	78	17	98·30	85	18·30	100	80	20	99	80	19	·71	„	„	4	Clear.
5		97	78	19	102	83	19	97	84	13	100	82	18	·74	„	„	2	Do. hot wind.
6		98	80·30	7·30	102	85	17	102	83	19	100	83	17	·74	„	„	4	Cirri, very hot at night.
7		98	81	5·30	102	85	17	100·30	84·30	16	100	82	18	·70	„	„	4	Do.
8	●	97	81·30	5·30	101	84	17	98	88	10	96	88	8	·70	„	„	2	Do.
9		94·30	82	2·30	97	88	9	98	84	14	97	82	15	·70	„	„	2	Cirrocumuli, dew at night.
10		95·30	79·30	16	100	85	15	95	88	7	95	88	7	·68	„	„	2	Do. do.
11		92·30	87	5·30	89	84	15	88	83	5	88·30	83·30	5	·68	„	E.	3	Do. very cool at night.
12		93·30	80	3·30	87·30	82	5·30	87·30	81·30	5·30	85	81	5	·70	„	„	2	Do. heavy dew.
13		84·30	81	3·30	87	83	4	89	80	9	88	79	9	·70	„	„	2	Do.
14	▷	85	81	3·30	87	83	4	85	81	4	86	80·30	5·30	·68	„	„	4	Cumuli, agreeable weather.
15	P	83·30	80	5	89·30	81·30	8	83·30	80	3·30	85	80·30	5·30	·70	„	„	4	Do. do.
16		86	81	5	89·30	83·30	6·30	87	80	8	87·30	81	6·30	·66	„	„	3	Do. do.
17		84·30	80·30	2	87·30	83·30	4	89	82	7	84·30	82·30	6	·66	„	„	2	Do. heavy dew.
18		85	83		89·30	84	5·30	89	82	7	89	82	7	·66	„	„	1	Do. do.

Meteorological Register for part of July and August.—Burka, N.E. Coast of Arabia, 1861.

Burka. { Lat. 23° 44′ N. Long. 57° 55′ E.

Days of the Month	Moon's Phases	Sunrise Dry bulb	Sunrise Wet bulb	Sunrise Diffence	Noon Dry bulb	Noon Wet bulb	Noon Difference	Sunset Dry bulb	Sunset Wet bulb	Sunset Difference	9 P.M. Dry bulb	9 P.M. Wet bulb	9 P.M. Difference	Aneroid Barometer at Noon	Rain	Wind Direction	Wind Force	Appearance of Atmosphere
19		86	84	2	89	85·30	3·30	88	84	4	88·30	83	5·30	inch. 29·66	None	E.	2	Cumuli, heavy dew.
20	◯	86·30	84	2·30	87	85	2	86	82·30	3·30	85·30	83·30	2	·60	,,	N.E.	3	Do. do.
21		84	82	2	85	83	2	84	82·30	1·30	83·30	82·30	1	·66	,,	E.	3	Do. do.
22		84	80·30	3·30	83·30	80·30	3	84·30	82	2·30	83·30	81·30	1·30	·66	slight	,,	3	Do. do.
23		83	81	1·30	84	83	1	84·30	82·30	1·30	84·30	83	1·30	·66	,,	,,	2	Do. do.
24		83·30	82	1·30	87	83	4	84·30	82·30	2	84·30	83	1·30	·66	None	,,	2	Cirri
25		83·30	82·30	1	86	84	2	85	83	2	84	82·30	1·30	·72	,,	N.E.	2	Cumuli and nimbi, heavy dew.
26		83	82	1	85	83	2	84·30	81	3·30	84	80	4·30	·74	,,	E.	2	do.
27		83	80	3	85	83	2	87·30	83	6·30	84·30	83	1	·71	,,	,,	2	do.
28	☾	85·30	83·30	2	85	83	2	84·30	83	1·30	84	83	1	·70	,,	,,	4	do.
29		83·30	82	1·30	85	82·30	2·30	84·30	82	2·30	84	82	2	·73	,,	,,	2	do.
30		84	83	1	85	83	2	85·30	83·30	2	86	84	2	·72	,,	N.W.	4	Cirri
31		84	83	1	86	85	1	90·30	74	6·30	92·30	76·30	6	·70	,,	,,	1	Do. hot wind.
Aug. 1		86	72	14	90·30	86	4·30	92	88·30	3·30	92	88	4	·70	,,	E.	2	Do.
2		86	81	5	89	87	1·30	89	87	2	86·30	86·30	2	·72	,,	,,	2	Cumuli, dew at night.
3		87	85·90	2·30	87	84·30	2·30	86·30	83	3·30	87	83·30	3·30	·74	,,	,,	3	do.
4		85·30	82	3·30	85·30	82·30	3	85·30	83	2·30	85·30	83	2·30	·78	,,	,,	2	do.
5		86	83	2·30	86	84	2	87·30	84	3·30	87	84	3	·76	,,	,,	2	Cirri, heavy dew, at Sohar.
6	●	86	83·30	2·30	86·30	84	2·30	85·30	85·30	3	86	85	3	·80	,,	,,	2	Cumuli do.
7		87	82	5	86	83	3	86	83	3	85·30	83	2·30	·80	,,	,,	3	Do. do.

Meteorological Register for part of July and August.—Burka, N.E. Coast of Arabia, 1861.

BURKA. { Lat. 30° 44' N. long. 57° 55' E.

Days of the Month	Moon's Phases	SUNRISE Dry bulb	SUNRISE Wet bulb	SUNRISE Difference	NOON Dry bulb	NOON Wet bulb	NOON Difference	SUNSET Dry bulb	SUNSET Wet bulb	SUNSET Difference	9 P.M. Dry bulb	9 P.M. Wet bulb	9 P.M. Difference	Aneroid Barometer at Noon	RAIN	WIND Direction	WIND Force	APPEARANCE OF ATMOSPHERE
8	P	85·30	81·30	4	86	82	4	86	82·30	4	85·30	83	2·30	·80	None	E.	3	Cumuli } Sohar { Lat. 24°22' N. Long 56°58 E.
9		85·30	81·30	4	86·30	82·30	4	86	82	4	85·30	82	3·30	·80	”	”	3	Do.
10		86·30	81·30	5	86	81	5	84·30	80·30	5	84	80	4	·80	”	”	3	Cirri.
11	☾	83	79·30	3·30	84·30	80·30	4	85	81	4	85	81	4	·84	”	”	2	Do. heavy dew.
12		83·30	80	3·30	85	80·30	4·30	86	83·30	4·30	87	85	2	·84	”	”	2	Do. at Burka.
13		84	80	4	87	83	6	84	81	9	83	81	2	·78	”	”	2	Do.
14		84	81	3	87	81·30	5·30	84·30	82	2·30	84	82	2	·70	”	”	2	Cirri, very heavy dew.
15		84·30	81·30	3·30	86·30	81	5·30	85	81·30	3·30	84·30	81·30	3	·72	”	”	2	Do. do.
16		84	80·30	3·30	87	81·30	5·30	84·30	81	3·30	85	81	4	·74	”	”	2	Do. do.
17		84	81	3	85·30	82	3·30	84·30	80·30	3·30	85·30	80·30	5	·72	”	”	2	Cumuli and nimbl.
18		83·30	80·30	3·30	85	81·30	3·30	85	81	4	84	82	2	·70	”	”	2	Ditto.
19	O	83·30	80·30	3	85·30	81·30	4	74·30	82	2·30	84	82	4	·70	”	”	2	Cirri.
20		83	80	3	85	78	6·30	84	80	5·30	84	80	4	·70	”	”	2	Do.
21		83·30	78·30	5	84·30	81	6·30	84	80	4·30	84	80	4	·72	”	”	2	Do.
22		83	78	5	85	82	4	84	80·30	3·30	85	80	5	·74	”	”	2	Do.
23		84	80	4	85·30		3·30	84	81	3·30	85	81	4	·74	”	”	2	Do.
24		83·30	80·30	3														

Abstract of Meteorological Registers.

	MEAN TEMPERATURE at SEEB, from 15th to 30th June 1861.	MEAN TEMPERATURE at SEEB, from 1st to 18th July 1861.
Mean temperature (by Fahrenheit Thermometer)	90° 04'	92° 46'
Mean depression of wet bulb	89 19	80 02
Mean of difference	6 27	11 00
Mean of Aneriod Barometer at noon	29 in. 72	29 in. 70
Maximum of temperature	99° 30'	102° 00'
Minimum ditto	83 00	83 30
Extreme depression of moist bulb	71 30	78 00

	BURKA, from 19th July to 4th August.	SOHAR, from 5th to 12th August.	BURKA, from 13th to 24th Aug. 1861.
Mean temperature	85° 49'	86° 05'	84° 54'
Mean depression of wet bulb	79 45	82 13	78 39
Mean of difference	2 52	3 30	3 49
Mean of Barometer at noon	29 in. 70	29 in. 80	29 in. 72
Maximum of temperature	92° 30'	87° 30'	87° 30'
Minimum ditto	83 00	83 00	83 00
Extreme depression of wet bulb	72 00	79 30	78 00

(True Copies)　　　　　　　(Signed)　　W. M. PENGELLEY,

A. KINLOCH FORBES,　　　　　　　British Agent, Muscat.

Acting Secretary to Government.

With the advance of meteorology, we have been able to obtain more accurate and long-term statistics to supplement those given (in Fahrenheit) by Pengelley. These are the 1984 temperatures recorded at Seeb International Airport in degrees Centigrade, and approximately valid for anywhere in the Capital Area, though I am not alone in believing that Muscat's bowl always seems distinctly hotter than anywhere else.

	Minimum	Maximum
Jan.	18	26
Feb.	17	26
Mar.	22	32
Apr.	27	37
May	29	39
Jun.	30	40
Jul.	30	38
Aug.	28	36
Sep.	29	38
Oct.	24	34
Nov.	21	31
Dec.	19	27

Dhufar is significantly cooler with humidity at its highest from June to September. These are the 1984 C° temperatures recorded at Salalah International Airport, which is much hotter than the mountains behind the coastal plain.

	Minimum	Maximum
Jan.	17	28
Feb.	17	28
Mar.	20	30
Apr.	23	33
May	26	32
Jun.	27	31
Jul.	25	28
Aug.	23	27
Sep.	23	28
Oct.	21	30
Nov.	20	30
Dec.	20	29

The tropical monsoon area touches Dhufar, and can be identified by the ubiquity of coconut palms in fertile districts, whereas hundreds of miles, almost unbroken, of date palms assign the Batinah coast neatly to the North African and Middle Eastern climatic zone. Musandam can be cold and wet in the winter, as low cloud and mist affect the barren mountains. Rainfall again divides the Dhufar region from the Jabal Akhdhar and northern coast. The Capital area enjoyed virtually no rainfall during the whole of 1984, and only 25.6 mm. and 46.7 mm. respectively in February and April 1983.

400

Al-Khaburah and Saham

Between Suwaiq and al-Khaburah we kept to the main road, enjoying distinguished, even elegant contemporary domestic houses, typically light brown and white, each with its walled garden. Boys in their uniform *dishdashas* and satchels dashed pell-mell from green school gates on to the sandy track alongside the main road. We passed a camel jogging along beside a car whose owner was holding its halter with one hand while steering with the other. At the Rija' Restaurant I ordered chicken and rice biryani, eating in oriental tumult of 'Ya, rafiq! 'Ataini khubz! Dajajah! Wain ash-shahi? Kaif haluk, ya Salim!'

Al-Khaburah Agricultural Development Project, recently taken over by the Ministry of Agriculture and Fisheries from the original sponsors, Petroleum Development (Oman) Ltd., has been an unqualified success since it was established in 1976, in the spheres of animal husbandry, rural handicrafts such as weaving, and research into irrigation, fertilisers, and plant protection, working in conjunction with dozens of agricultural extension centres located strategically throughout the Sultanate. Local goats have been cross-bred with imported Anglo-Nubian goats, to produce an animal 50% larger than the local specimen, improving the birthrate from an average of 1 per litter to 1.5. The Chios sheep from Cyprus was selected to crossbreed with local sheep, the resulting animal being larger, healthier, and with a litter averaging four lambs. Research into fodder requirements indicated that Rhodes grass is best suited to the Batinah, and could grow there year-round despite high levels of salinity and drought conditions.

The first weaving expert arrived in 1977, reintroducing the declining craft of spinning, then weaving. This women's project was taken over in 1981 by the Ministry of Social Affairs and Labour, who now look after more than forty women's centres all over the Sultanate. Women of al-Khaburah, highly motivated and skilled in weaving, have produced two intricate giant wall hangings for students' residences in the new Sultan Qaboos University, which opened in 1986. The women weavers of al-Khaburah have taken over a previously male-dominated craft, and in a *barasti* hut twenty feet long they built themselves they weave on looms made of four sticks set in a rectangle in the ground with two horizontal beams across top and bottom. The women are accustomed to a rug-weaving technique enabling them to produce camel bags, halters, girthstraps, and even tapestries.

Beyond the experimental farm, on the way to the sea, stands the massive old castle of al-Khaburah, a new mosque and a *suq* parallel with the seafront, obtaining the greatest benefit from sea breezes. The air, soft as eiderdown, wafted in purity past a boys' school darkened to benefit from cool shadows, and alongside The Northern Assurance Co. Ltd, shuttered for prayers.

From al-Khaburah the road north enters Saham in 31 km. I headed straight for the milk-foam sea, Canaletto-bright, Bellotto-clear. Youngsters rushed towards me in a crowd, elbowing me in their

Saham. Boy beside shashah

keenness to discover as much about me as I was to discover about them
and their village. They showed me great lobster nets, new boats with
Yamaha outboard motors, decrepit wooden boats beyond repair; but I
was mainly interested in a parade of *shashahs*, lined up as if for
numbering on the higher sands. The *shashah*, that marine equivalent
of a *barasti* hut, is so small and so primitive in appearance that I could
not help thinking that (despite absent evidence) it must have been used
by the first Stone Age denizens of these coasts when they first ventured
on the waters. Nothing more simple could have been devised. Palm
stems are bound with coir into pointed bow and pointed stern. Even
when lined internally with palm-bark, coconut fibres, and the bulbous
ends of palm-branches, these craft no more than ten feet long look as
though they will sink on their first encounter with heavy surf. Not so:
they can last up to three years, though the tougher *huri* has generally
replaced it elsewhere, and the *shashah* never took hold in Dhufar,
Musandam, or any rocky shores. Oddly enough, Yamaha outboard
motors have been used by some enterprising Batinah boatmen as
successfully on the palm-branch *shashah* as on the dug-out *huri*, and by
replacing palm-butts with polystyrene, and coir by nylon thread or
rope, the *shashah* may well have a future as long as its past.

I was invited to a charmed circle of fishermen taking tea and oranges
on a square, informal *sablah*, protected from the sun by palm-branches
running perpendicular to stouter poles.

'Ja'far has just been given another daughter by his wife,' a man in his
forties was saying, squatting opposite me. 'Two daughters', mused a
bearded man, poking his stick into the sand. A young boy reached over

for a segment of peeled orange, goggling at me like a herring flapping against murder. 'And no sons', reflected a third man, smoothing his *dishdashah* with a dark-brown hand, an index-finger distorted out of true.

Daughters may be an affliction to a man when he has to feed her, clothe her, and prepare her for marriage to another man, but when she comes of age she brings with her a bride price proportionate to the income of the husband or his family, or perhaps according to the degree of consanguinity: you ask more from a stranger than from someone within the family. It has always been usual for a man to marry his *bint 'amm*, that is the daughter of a paternal uncle, and either he or his father will pay the bride-price. He can divorce her by saying 'Halaytak wa baraytak' following which he will expect the return of some of the bride-price. An Omani asks a relative to propose a marriage to a virgin girl, but if she has been widowed or divorced he may propose to her himself. The prophet Muhammad (*salla Allah 'alayhi wa sallam*) has said, 'A widow shall not be married until she be consulted; nor a virgin until her consent be obtained, by her silence.'

I spoke of the castle of Saham. Yes, it is open every day in the morning, but this is 3.20, and you must take more tea and oranges. And young Ramadan will fetch some cakes for the guest. I closed my

Woman wearing a birqa in the Batinah

eyes in the peace and joy that comes from letting an active mind loose in a lagoon of passive receptivity. The owner of the Shuruq Pharmacy nudged me to accept a tiny cup of coffee as the copper pot had boiled. Crows on the beached croaked in disapproval.

Women passed us, each masked in a mysterious *birqa*. The more expensive masks are handmade, but you can (and I did) buy much cheaper examples. The better class are usually made of black silk or cotton dyed deep blue-black with indigo. It covers the eyebrows, nose and mouth, and is fastened to the head by two tiny elastic bands at top and bottom. The types of mask vary widely, from the large *badu birqa* covering most of the face, which is particularly practical in a great heat or sandstorms, to the tiny *birqa* of Musandam, where sand is not a problem. You also see on the northern Batinah (and on visitors from there to the shops of Matrah and Ruwi) the black veil called *lahaf*. Women of Sur and the Sharqiyah traditionally wear a *shatha*, two cloths joined together with lace forming a colourful head-shawl. Women's dresses or *khandurahs* tend to be much more colourful in Oman than in neighbouring countries, especially in Jabal Akhdhar. Visitors often do not realise this because black *abayahs* are nearly always worn over these dresses in public. A *khandurah* is worn over long trousers, or *sirwal*, which are fastened at the ankles with zips. The old custom of decorating hands, feet and face with henna is still widespread, as is the antimony-derived *kohl* around the eyes. Women's jewellery can be observed in all cities and towns in a special jewellery *suq*.

Suhar

On the way to Suhar, I stopped at Sur al-Balush, a typical Batinah village with its *sur* and its closely-knit community, in this case long-established immigrants from Baluchistan. The *sur* in question has been razed to make room for a new private walled compound.

A great dam has recently been completed between Saham and Suhar, trapping up to a million cubic metres of water a year on and near Wadi Salahi. The Government has spent 38 million rials in the second Five-Year Plan (1981–5), the most impressive of a chain of dams being the Wadi al-Khawdh dam, built between March 1984 and April 1985 at a cost of 6 million rials. Eight metres high and nearly 5 km long, it will bring 400 more hectares to the Sultanate's existing 41,000 hectares of cultivated land. We should also remember that it is the six thousand *aflaj* which supply more than 70% of the water used in Oman, about 900 milion cubic metres of water each year.

When you arrive in Suhar, your first impression differs markedly from that of Istakhri, who wrote of Suhar in his tenth-century *Kitab al-masalik wa 'l-mamalik*:

'It is the most populous and wealthy town in Oman, and it is not possible to find on the shore of the Persian Sea nor in all the lands of Islam a city more rich in fine buildings and foreign wares.'

At that time, we know that Oman exported dates, dried fruits, copper, horses, and ambergris, but much of Suhar's prosperity lay in its function as an entrepôt for trade between Europe and the Middle East to the north, and India and the Orient to the East.

In the sixth century and after, if not before, large merchant ships plied from the head of the Gulf over seven thousand kilometres as far as Guangzhou (Canton). Their cargoes included gems, cloth of all descriptions, spices and drugs, and indeed any commodity that possessed rarity or luxury value. Basra's importance as the port of Baghdad may be gauged from the fact that Guy Le Strange (in *Baghdad during the Abbasid Caliphate*, 1900) carefully demonstrates that Baghdad in the tenth century covered an area greater than that of modern Manchester. Siraf, on the Persian shore, imported aloes, amber, camphor, gems, bamboo, ivory, ebony, paper, perfume, sandalwood, drugs and spices. Qais (Kish) and Hurmuz possessed equal commercial significance, if not greater

For outward-bound vessels, Suhar was the last landfall before India, and if they sailed with the north-easterly monsoon, they would lay in supplies here for the month's voyage.

Palgrave's 'sheltered anchorage' at Suhar is such an arrant invention that it vitiates any plausibility his account of Oman might have in later readers' eyes, and it is astounding that the scholarly *Encyclopedia of Islam* should rely so heavily on Palgrave. The strange thing is that Suhar prospered as it did, its open beach being partly exposed to the *shammal*, the Gulf's dominant wind, and its lee shore to the *nashi*, or nor'easter. The only explanation must be the plentiful food and water provided at most seasons by the position of Suhar at the mouth of Wadi Jizzi, which is not only fertile, but provides excellent communication through the mountains to Buraimi.

Karen Frifelt has shown extensive habitation by identifying thousands of burial cairns in the hills behind Suhar dating to the third millennium. The site of Suhar itself is not recorded until the early centuries B.C., when Nestorians name 'Mazun', specifically identified later by both Mas'udi and al-Bakri with Suhar, and a Nestorian bishopric was centred on Mazun as early as 424 A. D., and Nestorians specialised in the cargo trade with India before Islam.

The first Muslim missions to Oman were warmly welcomed by the Julanda leaders, as Baladhuri tells us; Yaqut adds that the Persian governor with his men were driven out of Suhar in 629, very shortly after the Hijrah, and after seizing control of their own country the Omanis' commercial expertise took them not only to Basra, but to East Africa, following a rebellion against Caliph Abdulmalik towards the end of the seventh century under Sulaiman and Sa'id, followed by a succession of later waves.

Tadeusz Lewicki, in 'Les premiers commerçants arabes en Chine', appearing in *Rocznik Orientalistyczny* (1935) has shown that in the eighth century an Ibadhi Omani named Abu 'Ubaid 'Abdullah ibn al-Qasim made the first known voyage by any Arab to China. The implication is that many other Omanis were making less arduous voyages to India, and to South-East Asia, at the same time, before and after.

Yet it was in the tenth century that Suhar became, as Istakhri put it, 'the greatest seaport of Islam.' He, as a national of Persia, travelled throughout the Gulf in the tenth century, and if he did not know Suhar at first hand, he definitely knew Siraf, then almost as large as Shiraz. Muhammad ibn Ahmad al-Muqaddasi, in *Ahsan at-Takasim fi Ma'rifat al-Akalim*, proclaims Suhar 'the hallway to China, the storehouse of the East and Iraq, and the mainstay of Yemen.'

By the fourteenth century everything had changed, and Suhar, as Abu 'l-fida' Isma'il ibn 'Ali notes in *Taqwim al-Buldan*, was reduced to a ruined village. Pottery sherds from the tenth century, and traces of kilns for baked bricks of the same period, cover the burial grounds and occupied area of the period, to an extent of over seventy hectares: roughly four times the size of the present town. The agricultural system covered in a similar fashion nearly four times the area cultivated today, yet Suhar is one of the most intensively-cultivated zones in modern Oman. Pottery finds prove that the decline of Suhar was rapid: from the tenth century to the twelfth.

Between 1405 and 1433 the Ming emperors sent out no fewer than seven major seaborne expeditions to the west. The later Portuguese voyages were intended to spread Christianity, capture slaves from 'infidel' countries, plunder all the coasts in their way, establish colonies to increase the trade and if possible monopolise it; in this they differed hardly a jot from British expeditions. The Chinese by contrast were more interested in equitable trade, the creation of diplomatic missions, and in obtaining recognition of nominal suzerainty by the Son of Heaven over the barbarians abroad. The great Admiral Zheng He, himself a Muslim (as are 30 million inhabitants of the People's Republic today), took 317 ships to Java, Sumatra, Sri Lanka and India in 1405, a total of 27,870 men. His final expedition from 1430–33 took him past India to Hurmuz (and probably Suhar), Salalah, Aden, Jiddah, and the East African coast including Malindi.

The French archaeologist Monique Kervran courteously showed me over her work, excavating near Suhar fort. The earliest fortress was discovered by the Mission Archéologique Française à Bahrain et Oman in 1980, under the present fort, where the Wali of Suhar used to live and work. He now has a new office, between the fort and the seashore, and the fort is being restored as a showpiece.

The first fortress was built of baked bricks, surmounted by mudbricks, forty-five metres long on each of its four sides, with a massive tower at each corner. Date-store, armoury and arsenal: all were partly uncovered, but much of the structure had been completely obliterated by later work. The earlier fortress dates to the 13th or possibly to the early 14th century, and Monique Kervran attributes it to the Princes of Hurmuz, or the Al Tibi. The Portuguese occupied the fortress with 300 men, reducing the boundary wall in 1621, and replacing the old bricks with stone. Two towers were built on the shore side, one being destroyed 200-300 years ago, and the other in 1982. Many questions have still to be answered in later seasons of excavation.

I wandered round the splendid fort, restored with such pride by the

Suhar. Fort, from the north

Wali of Suhar; then the new mosque with its green dome. Crows pecked and gawked officiously at a safe distance. I found the Sea Lord Restaurant in the *suq*, but could obtain no food because the oven was not working. The heady sea air blew a curtain wildly from side to side; it managed to keep out some of the fresh air, but not much. Salim 'Abdullah, at his shop in the old *suq*, about to disappear when the new new market arises from the sand, showed me heaps of dried fish and ropes, but I was more interested in buying an incense burner from Liwa, and made off with my trophy wrapped in an old issue of *Al-Watan*. 'Pragji Purshottam' and 'Toyota Genuine Parts' I read on the *suq's* bilingual signs. 'Laxmidas Tharia Ved Exchange Co.' (a money-changer), 'Bengal Hairdresser.' 'Al–Mulla Electronics,' 'Ratanshi Trikamdas Textile', 'Ismail Salim Carpentry Works'. The fort and the old and rising new bazaar alike are all in the old heart of Suhar: the quarter called al-Hajra. But I wanted to explore all the other quarters described by Frederick Barth in *Sohar: culture and society in an Omani town* (1983). To some extent, Suhar is just like any other coastal town of the Batinah, with its fish-market, its fort, general market, and settlement straggling like a vanishing tail north (to Hadirah quarter) and south (to Shizaw, Sabara, Harrat ash-Shaikh, and Ghail Shibul). *Barasti* houses outnumber more elaborate cement-block houses in these peripheral areas and *shashahs* more sophisticated boats. Date-palms in plantations two to three kilometres wide parallel the shore, as do fields dominated by animal fodder. Twenty to thirty dhows would have called each day on average up to 1960, but today two at most ride at anchor, loading or offloading by smaller boats. Trade and traffic is now by lorry or private car congesting the main road north to Abu Dhabi and Dubai, or south to Matrah. In 1960, up to twenty-five dhows were owned by Suharis; now the figure appears to be none at all, if one excludes a couple of *sambuq*s operated by villagers of Za'fran, a little to the north.

The covered *suq* I saw in 1986, which must be vanishing even as I write, had about 120 shops in a grid of sandy lanes, each open-fronted and more or less uniform in size, with many manufactures such as plastics and tinned goods from China, but also products from the U.S.A., India and Pakistan, Japan, Taiwan, Korea, and both Western and Eastern Europe. Each rectangular shop in the market mimics night in its secluded hinterland, especially when emptied for the hours of prayer. Everything is imported, and even the native pharmacist buys herbs and simples from other places in the Middle East. A couple of bakers, a few *halwah* cooks, two blacksmiths and half a dozen tailors make up the town's productive population, yet Suhar remains an emporium not only for its suburbs and outlying villages, but for all the foothills of Wadi Jizzi and its mountains for many miles around. An open market sells vegetables and fruit, while the fish market is divided into an auction area for retailers, and a direct sales area for consumers.

Suhar is religiously tolerant, as elsewhere in Oman, with Shi'a and Sunni, the latter represented by all four orthodox schools. There is a significant Ibadhi community, to which the Al Bu Sa'id family belongs. Hindus predominate in the merchant class. You hear not only Arabic but Baluchi, Farsi, Zidgali (a Sindhi language akin to Kutchi) and Kutchi. Intermarriage is common across 'barriers' of sect, occupation, language, and ethnic origin, class and tribe; business partnerships likewise prove social catalysts. When, in 1947, the Luwatiyah of Matrah opted for Omani citizenship on the independence of India, the Luwatiyah of Suhar, who held British passports, opted for Indian citizenship and left Suhar. The Banians, as Indian passport-holders have, however, remained. The Pakistani labour migrants have arrived too recently to integrate into the community.

Shinas and the Border

I stayed at the Suhar Hotel, 10 km north of the city on the road to Liwa, which opened in 1985. Dawn displayed its effect not as rubies deftly conjured, winking red on orange as at Nizwa, but as a pearly-white translucence borrowed from one of Turner's misty distances. Dew lay heavy on the grass, as haughty crows pecked and stalked, angrily squawking their rivals out of their way.

Majis, the next coastal village, may once have exported copper from the mines flanking Wadi Jizzi, but it is now simply a beautiful oasis set like brilliant white orchids in a quiet bouquet of green. Childlike, I threw aside my sandals and ran to the glinting sea, swimming like a dolphin first released back the water after capture. Exhausted, shivering, I then sat at the tideline, wet sand drying on me like sugar, allowing the sea to lap its transparent sleeves up and down my ankles.

Green nets lay on the sands to dry, beside powered fishing boats. Low rocks provided some little shelter from prevailing winds, but not much: the contrast with the towering cliffs of Musandam could hardly have seemed more startling.

Three km. beyond Majis the turn-off to the New Sohar Port indicated 4 km, then in 9 km more the turn-off to Liwa read 2 km: we took the latter. Some distance from the quiet fishing village we came to the castle on an eminence, its main gate strategically open on the sea-wall, round towers on three corners and a square central donjon. Several *aswar* are situated nearby; dilapidated *shashahs* paled on the beach like camel skeletons in the distant desert.

Between Liwa and Shinas, no settlement is marked on the map, a distance of 32 km, but there are at least half a dozen small communities straggling along the narrow beaches between Dawanij and Sur Bani Khuzaimat, such as Sur Bani Hamad, a name like so many others revealing the existence of community enclosures for civil defence. Almost every man and boy waved or nodded as we passed, perpetuating the courtesy legendary throughout Oman even in these motorised days.

The castle of Shinas, close to the shoreline, was being restored, scaffolding climbing up it like rigid ivy. A hardtop road now approaches the sea. The minaret of the Friday mosque harpoons the empty blue sky above the Western Hajar hills as they encroach even nearer to the sea. I lunched off chicken, rice and bread, with Zahor white chocolate from Spain at the Gharifah Restaurant beside the fortress, simple fare attuned to the plain pleasures of this dreamy fishing village, *shashahs* beached for another long day of sunshine and salt breezes. Seven km from the cross roads at Shinas a left turn takes

Shinas. Village, 1809

you on the highway to Dubai and Sharjah, Ajman, Umm al-Qaiwain and Ras al-Khaimah; ahead is the dual carriageway to the border at Khatmat Malahah and the U.A.E. (Fujairah). We took instead the rough road 2 km to the right, to the seashore at al-'Aqr, where an ancient mosque faces the sea, 500 metres from the fort, amid beautiful palm-groves.

Back on the main road, I saw another of the many signs posted just before declivities in the road: 'Stop if the Water is at Red'. At the lowest point of the dip in the road, vertical posts are painted white below red; if flood water rises above the white mark, your car risks being waterlogged and carried away across the road towards the sea.

The road across Wadi Hatta, a modern dual carriageway, lures one through dream mountains of rough crags rearing above poor tufts of acacia, hobbled camels attempting in bewilderment to tear up with their leathery mouths some sustenance from these nightmare thorns. The desolate beauty of a waste land induces a smile: Eliot's *Waste Land* is an indictment, not the paean that I should sing here. He refers to spiritual decay in urban society, of course, symbolism lost in these fastnesses where the Ibadhi faith grew and thrived on a harsh diet of asceticism and simplicity. Before long we had to turn back from the border post of al-Wajajah, and sought the winding, rough road of 'Ajib. The barren wastes of Castilian Spain brought out such a desire for spiritual discipline in S. Juan de la Cruz and S. Teresa de Ávila; it is responsible for the darker moods and prevailing melancholy in the poems of Antonio Machado. Ruined castles, graveyards, watchtowers on hills, *barasti* enclosures for goats: the atmosphere of Wadi Hatta will always remain with me: essentially wild but not fierce, inspiriting though not soothing.

The main road north leads in 7 km from the Wadi Hatta roundabout to Khadrawin (off right) and very quickly to al-Aswad ('The Black One') 10 km off left on a good road, winding through scrub, perfectly flat. Al-Aswad has a police post because it, too, is a point of depature for Dubai. A woman with a black *birqa* scrambled behind the castle as I approached it. One side seems stout, with a strong round tower; but the impression is illusory: the tower is smashed at the back, and the other walls are open to grazing goats and would-be rovers. The mosque nearby by contrast is shining new; the old mosque is, as was traditional, half covered, with the mihrab and prayer-room, and half open. Back at the frontier of Khatmat Malahah (and how final those syllables resound) I found that the mountains came down almost to the sea, and women walked unveiled. Frontier offices and officials' family houses glinted white as meringues: their tidy gardens had been planted with hibiscus and bougainvillea. Everything thrives here with enough watering, and the little community presented a delightful emblem of modern Oman.

Miles on the Route between Suhar and Buraimi

I next returned to the Wadi Jizzi roundabout, just before Suhar. S. B. Miles first visited the wadi on his way to Buraimi in January 1875, and recorded his brief journey in *The Countries and Tribes of the Persian Gulf* (1919). In November of the same year he returned, and his much fuller account appeared as 'On the route between Sohar and el-Bereymi in Oman, with a note on the Zatt, or gipsies in Arabia,' in the *Journal* of the Asiatic Society of Bengal (1877).

Having arrived at Sohár (صحار) on the 16th November, 1875, and visited the Governor Seyyid Bedr-bin-Seif Ál Bú-Sa'ídi, I requested him to be good enough to arrange for my visit to el-Bereymí, and Sheikh Ráshid-bin-Hamd, with whom I was personally acquainted, and who is a man of considerable influence in el-Dháhireh (الظاهرة), having been at one time Governor of el-Bereymí, was selected to accompany me. I could, however, only promise myself a hasty visit, as my arrangement with Captain Clayton, Her Majesty's Ship *Rifleman*, who had kindly given me a passage, was to meet again at Sohár on the 22nd. The Sheikh's preparations as regards camels, &c., were not completed until the next morning at 10 A. M., when we started with nine matchlocks of the Na'ím and Mokábil tribes, and reached about thirty miles by nightfall, encamping for the night at Sahílah, (سهيله) a village in the Wádi Jezze (وادي الجزى) belonging to the el-Kunúd. The road, after leaving the belt of palm groves and cultivation outside Sohár, ran N. W. for an hour to 'Auhí, (عوحى) a little patch of date groves and gardens irrigated by a *felej*, and then turned west over a stony, gradually rising plain, covered with thin acacia jungle and underwood towards the hills. The Wádi Jezze, which we came to soon after, is here neither broad nor deep, being but a few inches lower than the plain, and barely distinguishable from it, showing that no great torrent ever rushes down it, but that after rainfall, which in 'Omán is rarely heavy, the water that is not absorbed by cultivation is sucked in by the porous soil on the way. Another hour brings us to the site of an ancient ruined town, attested by heaps of fragments of black rock lying in squares and ovals, which mark the foundations of houses, and by parts of ruined walls and towers on adjoining hillocks, covering altogether a considerable extent of ground. From the appearance of the foundations, the houses must have been on a small scale and of rude construction. No vestige of any edifice of architectural pretensions remains. At the present day the locality is uninhabited, and a place of more dreary and complete desolation I have rarely seen. My companions could not tell me the name of the site, their only traditional knowledge was that it belonged to the Persians in the time of ignorance, and that it was destroyed by God on account of the refusal of the inhabitants to embrace the blessed truths of Islám.

A little further is a dried up *felej* leading from the hills, called Felej-el-Súk, (فلج السوق), and also ascribed to the Persians. At 2½ P. M., we came in sight of Sehlát (سهلات), a village picturesquely situated on the top of a hill, and having at a distance a somewhat imposing appearance, though a closer view dispelled the illusion. It belongs to the Bení Gheith (بنى غيث), a petty Ḥinawí tribe, subordinate however to the Na'ím. Here we rested for a short time while the camels were fed. The next village we come to an hour later is Mileyyeneh, (المليذنه) and our road henceforth lies in the bed of Wádí Jezze as far as Ḥail, (حيل) for we have now reached the foot of the hill range and commence a more steep and winding ascent. Just below Mileyyeneh, where the Wádí narrows considerably, is an arched aqueduct, of solid masonry that supplies a village, called el-Ghorák (الغراق), belonging to the Bení Gheith. I could learn nothing precise about this aqueduct, which is evidently of very ancient construction, from the Sheikh, the tradition as usual being limited to the fact of its having been constructed by the Kátirs before Islám. Here we were overtaken by a heavy thunder-storm, which soon drenched us through and made the rocky path too slippery for the camels to venture out of a walk; the sight, however, was very grand from the picturesque scenery around us; the dark blue hills of the back ground, streaked by deepest black ravines and gorges, and with ridge upon ridge of lower hills in front being lighted up here and there by the rays of the sun, now near setting, glancing through a rent in the dark heavy masses of clouds above us and showing a strange contrast of light and shade. It was some time after dark when we reached the groves of the little village of Sahíleh, (سهيله) and here the Arabs having lost the path and nearly brought us to grief among the water channels and low walls of the plantations, we were fain to wait for a villager to light us with a torch to our camping ground. This place belongs to the el-Kunúd, and has two small towers for defence. The next day, starting soon after sunrise, we pushed on more rapidly and reached el-Bereymí at $7\frac{1}{2}$ P. M. Travelling at first in a south-westerly direction, we pass after an hour an affluent on the right bank, called Wádí el-Súfán, (وادي السوفان) up which is a village of the same name, and then the Wádí Jezze, here forming a loop, we cross the 'Aḳabat Ḳumáshí (عقبه قماشي) to join it again. A few miles further S. S. W. brings us to Burj el-Shikeyrí, (برج الشكيري) just beyond which lies the village of Kán, (كان) memorable for a conflict between the Wahhábís and the Ḥadhramí troops of Seyyid Sa'íd-bin-Sulṭán some sixty years ago, in which the latter were signally defeated with great loss, and which paved the way for the onward progress of these fanatics to Shinás (شناس), where they again destroyed His Highness' forces. A cemetery of the slain in the bed of the Wádí near Kán attests the severity of the contest there. The tower of Shikeyrí is on a peak some 200 feet high on the right bank, and is joined to another tower below by a stone wall. It was built for the special purpose of barring the progress of the

Wahhábís. After another hour and a half we pass Wádí Wásiṭ, up which lies a town of that name. Sheikh Ráshid was very anxious I should visit this place, as the Sheikh Suleimán-bin-Saʻíd el-Shámisí was a great friend of his and a man of some importance. It lay, however, too much out of the road, and time was of consequence. The next place we reached was el-Khoweyrej, (الخويرج) a village of the same clan as Wásiṭ, the Showámis, (شوامس) a sub-division of the Naʻím, and having the protection of a fortlet and two towers. The cultivation here was extensive and very refreshing to the eye in contrast to the drear and arid rocks around. The fields were neatly arranged in terraces on the right bank, advantage being taken ingeniously of every available spot of ground capable of production. They were well kept and evidently received much attention, irrigation being carried on by means of channels leading from the copious stream above. Contiguous to this is Ḥail, another large village, the two forming the most considerable settlement I met with between Bereymí and Sohár. Ḥail has several towers, and on the opposite bank is el-Rabí, (الربي) a pinnacle rock about 200 feet high, on which is perched a tower surrounded by a low wall now crumbling away. The work is ascribed to the Persians, and the position is well chosen for the purpose for which it is said to have been constructed, namely, to serve as an outpost to protect the maritime plain from the inroads of the Bedouins. This outpost not improbably marks the limit of the grasp of the Persians in the age immediately preceding the introduction of Islám, when they are related to have held the sea-coast of 'Omán, the Arabs maintaining themselves in the highlands and interior. Two petty clans of the Naʻím, the Rashídat and Ḥadídat occupy Ḥail, and are at enmity with their neighbours the Showámis of Khoweyrej and Wásiṭ. As may readily be imagined, quarrels among such very close neighbours are very bitter, and they are said to ripen every four or five years into a free fight, which is not terminated without bloodshed. In such cases the people of Ḥail have the advantage of being able to cut off the water-supply of those below by damming up the stream, which is a very effective punishment, and is generally instrumental in bringing hostilities to a speedy termination. At Ḥail the Wádí Jezze is joined by its confluent, Wádí el-'Abeyleh, (العبيله) up which the road now leads. The course of Wádí Jezze above the junction is short, and lies W. S. W. towards Ḳábil and Seneyneh (سنينه). A few miles up it is a steep pass, marking, as my Sheikh informed me, the boundary of el-Dháhireh. We rested a little while at Ḥail, and then continued our route, soon reaching more open and level ground with woody ravines and scattered herbage. The highest point of this pass is called el-Nejd, where the aneroids showed an elevation of 1,860 feet, the peaks of the range on each side rising above us 1,000 feet or more. On descending the other side, we reached after an hour the Wádí 'Ain, which runs like all other water-

courses on the southern side of the range in a south-westerly direction until
its waters are absorbed by the thirsty desert. We now continue W. N. W.
over an unbroken plain sparsely studded with acacias direct on to Bereymí.
On our right lies el-Maḥdhah, (المـضـحـل) the habitation of the Bení Ka'b
tribe, while to the S.W. the lofty isolated range of Jebel Hafít, (جبل حفيت)
looming some 20 miles away, alone breaks the level expanse before us, and
we stand on the border of that inhospitable sea of sand and waste that
stretches without break or interruption for nearly 800 miles across the peninsula, and forms the greatest sand desert of Asia. The Bedouins of our party
having remained behind at a watering place we had passed, we found we
had been jogging on for some miles without them, and the Sheikh professed
some apprehension lest we should encounter any of the 'Awámir Bedouins,
who, he said, were constantly prowling about. It is customary in 'Omán, when
moving from one part to another, to take a man or two as Khafír, or pro-
tector, from each of the more important tribes through whose country one
has to pass. This applies not only to strangers, but also to any Arab passing
through the territory of another tribe with whom his own are not in " saff "
or league, when it is of course unnecessary. Sheikh Ráshid had not been
able to procure an 'Amírí at Soḥár before starting, and hence his anxiety.
We had another thunder-storm this evening, but not much rain. On
arrival at Bereymí, I went to the house of Selím-bin-Mohammed, whose
father, the Chief Sheikh of the Na'ím tribe, resides at Dhank (ضنك). Our
arrival was the signal for a general assembly of visitors, whom I soon tired of,
and I therefore beat a retreat to another house outside, which was cleared
for my reception.

Early the next morning, I set out to visit the fort, which at present is
in the hands of Sheikh Ḥamd. I saw his house on the way, and as he was
laid up with a dislocated shoulder from a fall off a camel, he sent to invite
me in to take coffee. The house consists of two lofty rooms separated by
an arch and with no furniture, but a carpet or two and an array of coffee
pots ; two or three damsels bundled out as I entered, but the fowls and
goats that seemed to make up the complement of inmates, being less bash-
ful, remained. The walls of the courtyard are loop-holed for musketry, and
a rusty iron gun lies half-buried in the ground inside. It took some little
time to get coffee ready for the assembly that had crowded into the house
and filled both sides of the room, but when it was over, I took leave and
continued my way to the fort. On reaching the gate, I received a salute
of three guns, which put the ordnance *hors de combat* by dismounting them
from their rickety carriages, and thus prevented any further expenditure of
powder. I was then taken over the fort, and the objects of interest generally were pointed out to me with great readiness and evident pleasure by
the Sheikh's nephew and his people. I was gratified by their civility, and

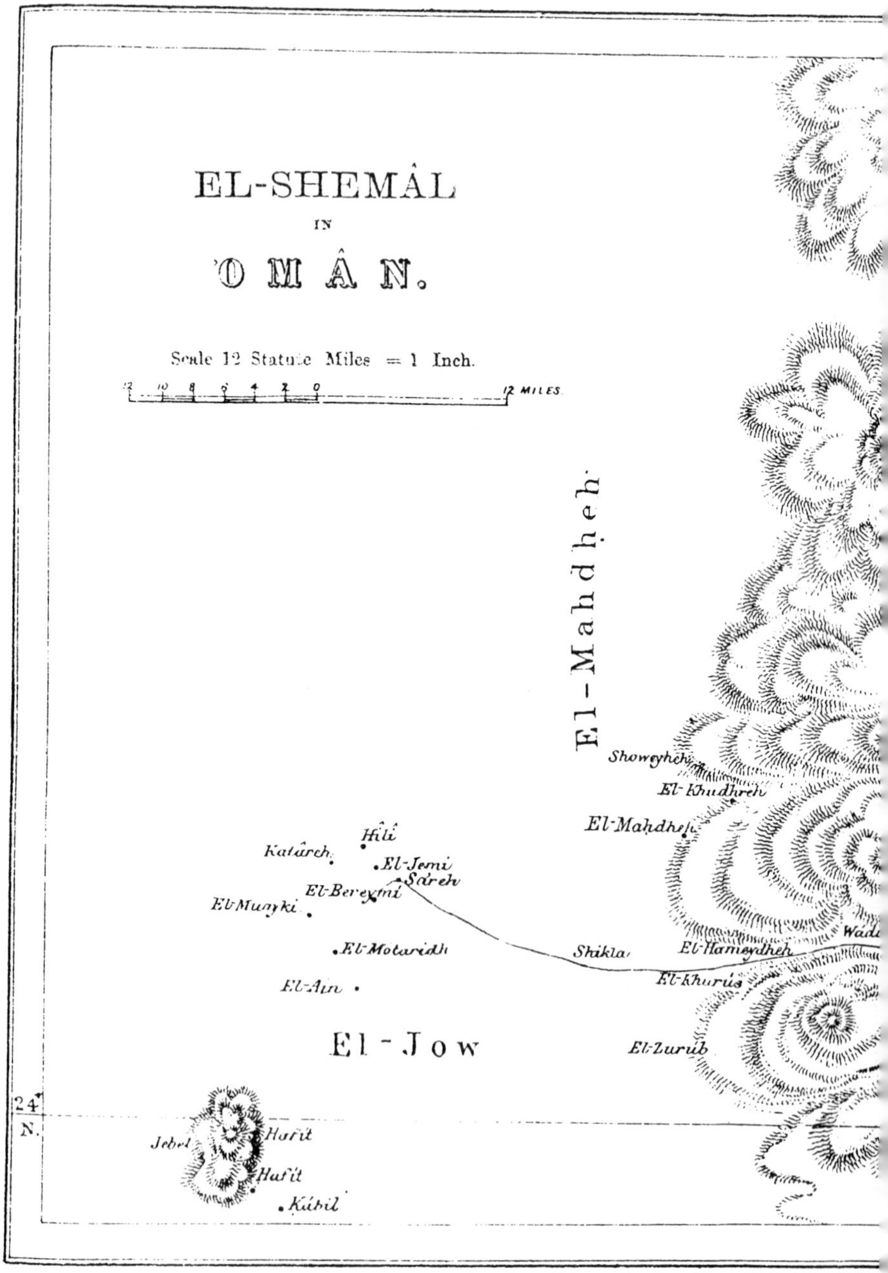

EL-SHEMÂL

IN

'OMÂN.

Scale 12 Statute Miles = 1 Inch.

12 10 8 6 4 2 0   12 MILES

El - Mahdḥeh.

Showeyheh
El-Khudḥreh
El-Mahdḥeh

Ḥili
Katâreh.
.El-Jemi
.Sâreh
El-Bereymi
El-Munyki .

.El-Motaridh
Shikla
El-Hamedheh
Wadi
El-khurús
El-Ain .

El - Jow
El-Zurub

24°
N.

Jebel .Ḥafit
Ḥafit .
.Kúbil

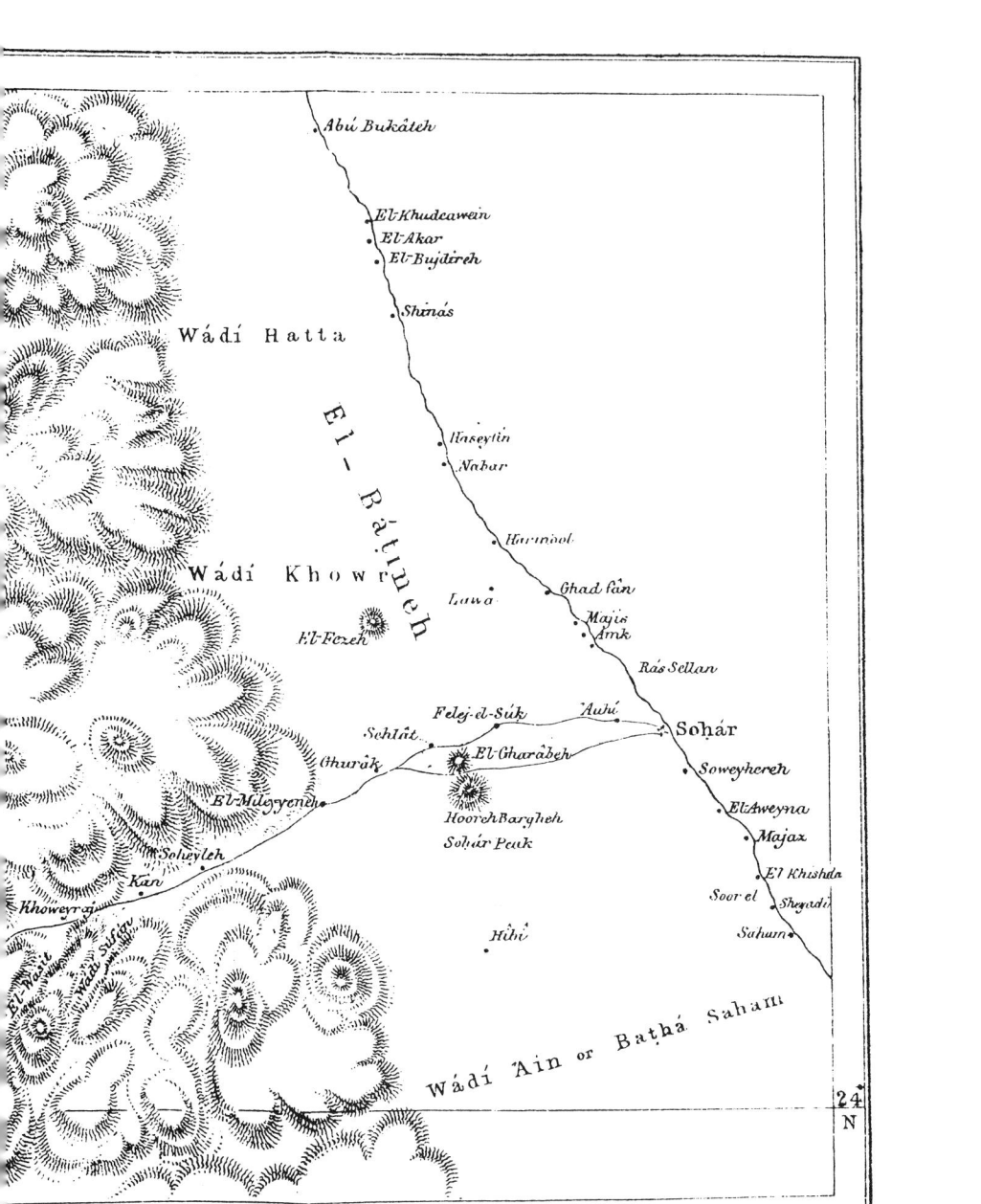

Map to show Miles' route from Suhar to Buraimi, 1875

spent some time in looking over the place, the strength and importance of which in their eyes are by no means undervalued. The plain in which el-Bereymí stands being so level, the view from the upper towers is extensive and interesting, embracing as it does the whole of el-Jow, and enabling one to get a tolerably clear notion at a glance of the topography of the settlement. To the S. W., at a little distance off, lies Sedeyrí's house, erected many years ago by that Wahhábí Chief for his own residence. It was solidly built of stone, but was destroyed by Seyyid 'Azán in 1870, who disapproved of seeing so strong a fortified house so near the fort, and who required the stones of which it was built to repair and strengthen the latter. After leaving the fort, I spent some time in walking through the section of the settlement more particularly known as el-Bereymí. The dates had nearly all been gathered, and the fields were being just freshly sown for the spring harvest, except a few late fields of jowárí, and the fruit season was over, so I did not see the place to full advantage, but the fields were neat and regular, and the orchards well cared for. Indeed, great attention is paid to horticulture throughout 'Omán, and at all the centres of population the 'Bostáns' are the great objects of interest. The houses in these settlements are seldom grouped together, but are scattered among the date groves; they are principally of mat and date leaf construction, and form perhaps the least obtrusive part of the landscape. A good proportion of the few mud houses was dilapidated and untenanted, and gave a general air of unsubstantiality to the place. The verdant appearance of this oasis, however, in which the tall and handsome foliage of the date is the predominant feature, is most attractive and refreshing to the eye of the traveller, and soon makes him oblivious of the drear and arid waste he has traversed to reach it, while the luxuriant vegetation, the sight and sound of running water, the almost entire absence of unproductive trees and plants, convey an impression of prosperity which is by no means borne out in reality. During the day several of the Sheikhs came to visit me at Sa'reh, (صعرة) and as they were communicative, I was glad of the conversation. Among those who came was Sheikh 'Alí-bin-Seif of the Bení Ka'b, who lives at el-Mahdhah; he was loquacious and lively, and was very solicitous I should visit his tribe and settlement, but it was unfortunately not convenient to do so, and I was compelled to accept his invitation for another time. He told me he had been to el-Riádh, having been taken captive by the Wahhábís in his youth and carried thither in irons. He believed they were 24 days on the road from el-Bereymí to el-Hasa, as they travelled slowly, and there was plenty of water on the way. Not having had time to visit any of the Bení Yás settlements in the morning, a messenger arrived from them inviting me to do so, but I was obliged to decline, as it was already time for me to be taking my departure. At the suggestion of Sheikh Ráshid, I requested Sheikh Selím to furnish us with an escort to the limits of el-Jow, but as he proposed to accompany me himself the first stage, I accepted his company with pleasure. It was not without strong remonstrances and pressing invitations to stay a day or two longer that we were permitted to make a start, but we succeeded in doing so about 3 P. M., and in presence of an admiring crowd of boys and Bedouins, we left el-Bereymí at a brisk trot by the same road that we had come ; the Bedoos shouting and singing and racing their camels at full speed in high spirits and evident delight at being on the road again, and already oblivious of the fact that half an hour before they had been sulky and disgusted at being told to collect the camels for the journey. After some miles a peak to our left, named Ḳaṭar, (قطر) was pointed out to me as having on the top the remains of a Persian settlement with trees, dwellings, &c., and said to **have been** occupied by them after their repulse at el-Bereymí. We passed Khaṭmet el-Shikla, (خطمه شكله) a small village at 12 miles, and soon after observing a low reddish coloured hill some distance off very conspicuous among the other dark rocks, I told a man to bring me a specimen. Nearly the whole party started off, and returned with sufficient stones to fill a portmanteau. It was a red compact nummulitic limestone, and was the only block of the kind I noticed on the journey. It was just dark when we reached el-Khurús, (الخروس) a deep cleft in the rocky bed of the Wádí 'Ain, which holds a perennial supply of rain water. The Arabs were anxious to push on to Ḥail, another 15 miles or so in the dark, but I objected ; wood and water were sufficiently abundant, and we soon had a good fire under way and suppers cooking. I noticed the men obtained fire from a spark by rolling the tinder into a ball with dry Arabs grass and swinging it round until it burst into flame. The dew to-night was very heavy, and my blanket was drenched by morning. The elevation of el-Khurús is 1,630.

20th.—Before starting I took leave of Sheikh Selím-bin-Moḥammed, whose protection was no longer required, giving him a suitable present, and I may mention as an instance of the entire want of shamefacedness in the Arab in begging, that he came up to me quietly, after he had received his *douceur* from my *factotum*, to whisper a request for two dollars more. I insinuated he had already received what I thought was proper for him, on which he said, " Well, give me one, only one more dollar, and I will be satisfied." This man's father is Chief of one of the largest tribes in 'Omán, and he himself a man of much influence and consideration. We reached the Nejd two hours after leaving el-Khurús. The ground here, which for some extent is comparatively level, is well wooded with acacias, rhamnus, &c., and green with low brushwood and grass. It is visited in the season by the Na'ím and Ka'b Arabs for pasturing their camels and goats. The mountain range we were crossing is, except in the valleys and water-courses, where the *detritus* from the hills forms a little soil, remarkably arid and sterile, and is everywhere from Jebel Akhdhar to Ruús el-Jebál entirely unclothed with verdure, presenting one of the bleakest ranges it is possible to imagine. The large valleys contain a good deal of herbaceous, but very little ligneous vegetation. Among the trees and plants in the Wádí Jezze and in the plain beyond the range were noticed the rhamnus, screwpine, *samar* or acacia vera, acacia Arabia, two oleanders, calotropis G., castor-oil, and colocynth gourd, two or three euphorbias, the wild lavender, a rush much used for making mats called *rasad ;* the *maranneh*, (مرنّح) the narcotic plant noticed by Palgrave ; the rose-scented *shirkeh*, and much coarse grass in tufts.

At Ḥail, which we reached in three hours from el-Khurús, and which is perhaps 35 or 40 miles from Bereymí, we halted for a short time for breakfast and to feed the camels, which were beginning to show symptoms of fatigue, for since the commencement of the journey we had been keeping up a jog-trot of from five to seven miles an hour, whenever the road rendered it at all practicable. Our dromedaries were all pretty good, and were as sleek and well-shaped in appearance as they were swift and easy in going. 'Omán camels are acknowledged the best in the world. The Sherif of Mecca rides none other, and last year he received a present of six from Seyyid Turkí, which were sent by sea to Jedda. They fetch from $100 to 150, if very superior, but the ordinary ones run from $30 upwards. Depth of chest is considered one of their chief points. The Bedouins of 'Omán sometimes take their camels from el-Bereymí across Arabia to Nejd and el-Hejáz for sale. In 'Omán, Arabs ride behind the hump, the saddle being very small with a felt and sheepskin over it. The camel's nose is not pierced, but a headstall with a chain nose-band and a thick camel hair rope forms the bridle. In Yemen, they sit in front of the hump, resting the feet on the neck of the animal, the saddle having a high pommel to give support. On leaving Ḥail, we entered the political boundary of Soḥár, and proceeded down the Wádí Jezze by the side of running water, flowing at intervals as far as Mileyyeneh ; owing to the alternate porosity and rockiness of the bed, the stream rushes along at one place for some distance, then suddenly disappears to re-appear again as suddenly further on. Kán and Shikeyrí, Sahílah and Mileyyeneh, are successively passed and left behind, until at nightfall we reached our halting place not far from Sehlát, where the road branched off to Jebel Gharábeh, a spot I was anxious to visit. During our march to-day we had been joined by two Sheikhs of the Na'ím tribe, who were taking two horses for sale to Soḥár. One of these, a chestnut mare, was a very casty, well-formed, and pretty animal, and from the price put on her the Sheikhs seemed to have a very exalted idea of her value. We passed on the road several small parties of donkeys and camels laden with dried and salt-fish going up to el-Bereymí, where the consumption of these articles is very considerable. The elevation of this place was found to be 850. The road we have been travelling lies almost entirely along the smooth sandy bed of the Wádís or torrents, and presents no difficulties of any kind to communication. For the first 30 miles or so from Sohár the road winds up Wádí Jezze, and for 10 miles more its confluent Wádí 'Abíleh. We then cross the ridge of the chain, here very low and not exceeding 1,900 feet, while the peaks on each side reach about 3,000. On descending the other side, we find ourselves on the plain of el-Jow, which borders on the great desert. Though heavy in places, the road is quite practicable for guns, and I remember no place that would be likely to cause artillery more than an hour's detention. The mountain range that has been crossed is part of the chain leading from Rás Mosandim (راس مسندم) to Jebel Akhdhar, and divides the provinces of el-Dháhireh and el-Bátineh. The hilly district between these two provinces is styled el-Ḥajar (الحجر) by the Arabs, but there is no general name for the range that connects Jebel Akhdhar with Ruús el-Jebál. The water-sheds towards Ruús el-Jebál lie east and west, while lower down towards Jebel Akhdhar, they lie N. E. and S. W. Running water was met with by me only in Wádí Jezze from Ḥail to Sehlát, and then never more than a few inches deep, but there is no scarcity of water anywhere. The inhabited spots are irrigated both by canals drawn from the stream, and by wells. The rocks met with were mostly sediment-ary, the principal being a very dark limestone and an argillaceous slate, the latter lying in great angles. The bed of Wádí Jezze is cut through a breccia containing fragments of granite, green limestone, quartz, and a beautifully variegated sandstone, and the breccia or conglomerate is pene-trated in some places by masses of slate or shale. The range is very peaked and sharp ridged, and here and there the strata were in waves, appearing as if the rock had been at some period subjected to pressure. Throughout the whole route, the aspect of the country is extremely barren and sterile, and, as might be expected, is unable to sustain much animal life. The only wild animals, I noticed, were a few ravine deer and foxes, and birds were everywhere extremely rare. On starting the following morning, we pro-ceeded down the bank of Wádí el-'Aweyneh (وادى العوينه) of the el-Mokábil, who occupy chiefly the upper part, where it is well wooded, and where run-ning water is abundant. In an hour and a half we reached Jebel Gharábeh, (جبل غرابه) where, as I had already learnt from Sheikh Ráshid, who had been regaling me throughout the journey with the traditionary lore of 'Omán, were situated the ruins of the citadel of 'Omán, the pristine name of Sohár, at one time the capital of the whole country. I determined here to take the opportunity of exploring these interesting and ancient ruins, which are probably not paralleled in 'Omán, and accordingly made a short halt for the purpose. The hill is not high, perhaps 250 or 300 feet, but it is very steep and inaccessible, and there being no semblance of a road, I was glad to accept the assistance of the Bedouins, who are as agile as cats and clamber about the most difficult places with ease. Around the summit, which is irregular, are traceable the ruins of these fortifications extending perhaps for half a mile. The wall still stands in places, from two to six feet high, and it is possible to trace the outline of part of the buildings at the highest point, though the greater part are an undistinguishable heap of ruins. Along the line of fortification at intervals were small circular towers, several of which are still conspicuous. The thickness of the walls was uniformly about three feet, and they are constructed entirely of rough fragments of the rock of which the upper part of the hill is composed, *viz.*, a white oolitic limestone, cemented with clay, and I could detect only three places in which mortar had been used at all. These were an arch in the wall, the curbstones in the path that led down the hill and the water cisterns. One of these cisterns, of which there are two, is quite at the summit, and is in shape an oblong, constructed of round pebbles cemented and plastered with mortar. The other is oval-shaped and of similar construction. It is lower down the hill. Both these tanks are small and shallow, and in such good condition, that, if cleaned out, they might still be serviceable. There are no signs of wells having been sunk that I could see; indeed, the quality of the rock precludes the idea of such an experiment. Somewhat below the highest point is a low arch in in the wall, built of selected stones cemented together, and was not improbably that of the gateway, as it is just over the road. There is one other arch in a tower still lower down, but it is made with long slabs placed together uncemented. Only a few yards of the roadway are traceable near the top, the rest is entirely obliterated, but it probably wound round the greater part of the hill. It was very narrow, and is faced at the edge with curbstones. It was at the extreme summit where the outline is best preserved, that the residence of the Chief or Governor probably stood, but to judge from the heap of stones *in situ*, the building was apparently of no great extent; perhaps a small stone house for the Chief and rude shelter for the garrison were all that was needed. From its position and strength, however, the importance of the castle is sufficiently apparent, and it was doubtless considered quite impregnable in those days of slings and bows, while it of course commanded the whole plain of Sohár from hence to the sea.

According to the tradition of the Arabs, as related to me by Sheikh Ráshid and confirmed by Seyyid Turkí and others, the castle at Jebel Gharábeh, as well as the city of 'Omán, were founded by Julandá-bin-Karkar, (جلندا بن كركر) under whom the city covered a great part of the maritime plain lying between Jebel Gharábeh and the sea; but there is no doubt that both the ruins of el-Gharábeh and the city are much anterior to the time of the Julandaites. Another legend related to me was of the daughter of Julandá, whose hand became diseased and withered, while the physicians were unable to apply any remedy. She remained thus for a long time, and at last decided to try the benefit of the sea air and bathing, after experiencing which for a short time she entirely recovered. The above tradition, it may be remarked, deriving the Julandaites from the 'Amálekite Bení Karkar is opposed to the more generally received account, according to which they descend from the el-Azd of 'Omán. The 'Amáleka properly derive from the Ishmaelite or Nejdean stock, as does also Sohár, the brother of Tasm and Jadis and the Eponymus of the ancient town. The Julandaites were a powerful dynasty in 'Omán, and for some time previous to the second advent of the Persians before Islám held dominion over el-Bahrein and the whole of the Persian Gulf. According to Ross' Annals of 'Omán, the Persians at the time of the introduction of Islám had possession of the sea-coast of 'Omán, while the Arabs had the interior. Mohammed sent messengers to the two sons of Julandá, who then ruled in 'Omán, and these messengers alighted at Damsetjerd near Sohár, a fortified place built by the Persians. Julandá's sons, 'Abd and Jeifar, and the Arabs agreed to accept Islám, but the Persians, refusing, were attacked in Damsetjerd by Jeifar and compelled to quit 'Omán. This Damsetjerd I am inclined to identify with the ruins at Felej el-Súk described above, and this is indicated, I think, both by their position and the Arab tradition attaching to the spot. The extreme antiquity of Sohár as one of the principal emporiums of 'Omán is shown by its identification with the ancient city of 'Omán, the capital of the country, which depends not alone on Arab tradition and authority, but is accepted by European writers, who see in Sohár the Omana or Omna of Pliny and Ptolemy.

The exact period, however, at which the town changed its name is a question more difficult of solution, and regarding which history and tradition afford no clue. The ancient history of Sohár is very obscure, as, though frequent references to it may be found in all ages, no good description or account of it exists that I am aware of. A short notice, however, by Ibn Mojáwir is worth quoting. He says :—" Sohár had 12,000 houses, and every nákhodá dwelt in a separate house, and the people used to draw their drinking water from the aqueduct. Some one told me there were 192 steel-yards for the weighing of merchandize between vendors and purchasers. The town was built of bricks, mortar, and teak-wood, and it became ruined, and the jinn haunted the castles around. Abú Bekr el-Bisrí informed me that the country belonged first to the kings of Kermán of the Seljúk dynasty, then it was ruled by the el-Ghozz, and afterwards it became deserted and was destroyed by the Arabs." Some cultivated spots are still to be seen near Jebel Gharábeh, but the only outward and visible sign of its former greatness remaining is the aqueduct Felej el-Mo'taridh (فلج المعترض).

This work, which is of good stone masonry, leads along the surface of the ground, with a gentle declivity from the Wádi Jezze in the vicinity of Húreh Bargheh, (حورة برغه) or Sohár Peak as we call it, to the shore, a distance in a straight line of 14 or 15 miles, and is still distinctly traceable for the greater part of the way, disappearing at the outskirts of the present town. I noticed by the side of it more than one small cistern of exactly the same pattern and construction as those on Jebel Gharábeh. The modern town of Sohár has for some years been in a gradually declining state. It has been described in the bright pages of Palgrave, but has still further decayed since his visit. In Sohár proper, which lies between Rás Sellán (راس صلان) and the village of Soweyhereh, (سويحره), the population is now only about 4000, including 400 Persians, a dozen Jews, who have been gradually decreasing in numbers year by year, and half-a-dozen banians; the bulk of the inhabitants here as also along the coast from Sohár to Majis being of Persian and Belúch descent. The citadel, in which the Governor Seyyid Bedr resides, is a lofty, square, plain building, with a strong entrance and well defended by a moat; next to Sohár Peak it is the most conspicuous object seen from seawards. From the roof an excellent view is obtained of the surrounding country, and a lovely landscape it is; the sea-shore being fringed with a belt of stately palm gardens and cultivation about three miles broad, while behind, the plain rises gradually, until broken up by the lower spurs of the lofty, dark, serrated range in the back-ground. In the second story of this house is the tomb of Seyyid Thoweyní-bin-Seyyid, but the room in which it stands has been bricked up since my last visit. It is remarkable that of the number believed to have been associated directly or indirectly with Seyyid Selim in his parricidal act, some nine persons, only two are alive, the rest having all, with one exception, met violent deaths. The town wall is very dilapidated, especially the front towards the sea, and is now fast crumbling down, and there are no towers or bastions to it, but the moat is still kept clear on the land side. The circuit of the wall is about a mile, but a very small extent of the area inside is covered with houses, the rest of the ground being bare or occupied with date and other fruit trees. The market contains about forty shops, and is good and well filled, the fish market particularly; the fisheries on the Bátinah coast being abundant almost to a miracle. The custom duties are 5 per cent., and the farm this year has been sold to a Persian for $175 per month; but this is no gauge for the amount of imports, as foreign goods are obtained from Muscat, where they have already paid duty, and are consequently free from assessment here. Trade is said to be decreasing annually, and to be transferring itself to

Shargah, which is almost as easy of access to the principal customers as
Sohár, the Arabs of el-Dháhireh and el-Jow, and where goods are cheaper,
being imported thither direct from Bombay, and thus saving Muscat dues
and re-shipment. Sohár has no harbour, nor even the slightest shelter for
native craft, and is dependent entirely on its position in being able to tap
the trade of Upper el-Dháhireh and el-Jow for existence, and should this
source of prosperity be in part drawn off by rival ports, it must sink in
time to the level of other towns in the Bátinah. El-Jow, in which el-
Bereymí lies, is the smallest of the six provinces of 'Omán, and is situate
between el-Dháhireh and the Shemál. It is bounded on the south by
Jebel Ḥafít, on the east by Khatmet el-Shikla and el-Maḥdheh, on the
north by el-Shemál, and to the west by the desert. El-Jow is inhabited
by several tribes, both Gháfirí and Ḥinawí; the former having been in
the ascendant since the accession of Seyyid Turkí. The most powerful
and the predominant Gháfirí tribe at present is the Na'ím, which is divided
into two distinct and about equal sections, each having numerous sub-
divisions, and numbers on the whole some 20,000 souls. They occupy el-
Bereymí Proper and Su'areh, (صوعره) and their possession of the fort enables
them to overawe the whole of the settlement. Since the time of Seyyid 'Azán,
they have been practically uninterfered with by the Muscat Government,
but of course own allegiance to the present Sultán. The Na'ím are at feud
with the Bení Yás, who occupy part of el-Bereymí, and their hostility is
interrupted only by occasional truces; collisions frequently occurring between them. Of the two sections of the Na'ím one inhabits more particularly el-Jow and Bereymí, the other el-Dháhireh. They are of the more
orthodox or Sunní persuasion, unlike the generality of 'Ománís who are
Ibádhiya. The Chief Sheikh of the tribe is Moḥammed-bin-'Alí-bin-Ḥamúd,
who lives at Dhank, his representative at el-Bereymí being his son Selim.
The principal Ḥinawí tribe at el-Bereymí is the Bení Yás, who formerly
gained so much notoriety by their piratical exploits. The Chief of this
tribe is Sheikh Zaid-bin-Khalífah, a man of strong character, and perhaps the sole individual in these parts possessing any real personal power
and authority. He resides at Abúthabí, and there are four smaller
Sheikhs subordinate to him residing at el-Bereymí. This tribe takes the
lead on the Ḥinawí side in all dissensions between the Ḥinawís and Ghá-
firís at el-Bereymí, and during Seyyid 'Azán's reign held the predominant
position here. The Bení Yás occupy the villages of Jemí, Kaṭáreh, Ḥeylí,
(هيلي القطاره الجيمي) and the Wádi Mes'údi (وادى المسعودى) at el-Bereymí,
and are said to have formerly out-numbered the Na'ím, but this state of
affairs has become reversed of late. The Gháfirí tribe next in importance
to the Na'ím, is the Bení Ka'b, which numbers some 15,000 souls, and occupies the district of el-Maḥdheh, which includes the mountain range and val-

leys between Wádí el-Jezze and Wádí Hatta. There are about 20 villages in this district, the principal of which is el-Mahdheh, where the Sheikh 'Alí-bin-Seif resides. They are all irrigated by conduits drawn from the hills, and as the soil is the same, Mahdheh produces the same kinds and quality of grains and fruits as el-Bereymí. There is no direct pass through the range from el-Mahdheh to Sohár between Hatta and el-Jezze. Other Gháfiri tribes are the Bení Kattab (بني كتب) and el-Darámikeh (الدرامكه) ; and Hinawí tribes are the Dhowáhir, (الظواهر) which occupies el-'Ain, el-Dáúdí, el-Kharais, el-Mareyjib, Sa'neh and Mo'taridh, (الخريس الداودي العين المعترض سعنه المريجب), and rank next in power to the Bení Yás and the 'Awámir. This last is a very large nomadic tribe, widely scattered over 'Omán, but occupying chiefly the desert outskirts from Kooria Mooria Bay to the Sabkha, (صبخه) and roaming about with their flocks and herds in a state of semi-savagedom. The 'Awámir are genuine Bedouins, and no wilder or more predatory race exists, I believe, in Arabia. One of their clans, the 'Affár, (عفار) are popularly supposed to feed upon the bodies of animals that have died naturally, but this is denied by the tribe who, however, admit that they are not unfrequently reduced to devouring their skin clothing. A large portion of this tribe has settled down, particularly in the province of 'Omán, where they occupy a district of twelve villages called the ' Buldán el-'Awámir' and follow agricultural pursuits. The wandering 'Awámir do not even respect the members of other clans of their own tribe unless they are acquainted with them, but plunder indiscriminately all they meet. In August 1874, a party of this tribe arrived at Muscat from the neighbourhood of Wádí Rekot in Kooria Mooria Bay to assist His Highness Seyyid Turkí in an expedition he was engaged on, and returned afterwards, as they had come, by land. They professed to have no difficulty in making their way over the great desert.

El-Bereymí is the appellation usually applied to a collection of seven villages or settlements, of which the one, specially bearing that name, is the largest and most important. The others are Su'areh to the N., Jemí, Katáreh and Heylí to the N. W., and 'Ain and Mo'taridh to the S. E. ; and the population of the whole may be estimated at 12,000 to 15,000. From the outside the appearance of these settlements is very pretty and refreshing, the date palms and orchards forming a green-setting to the low palm leaf huts, which are scattered throughout, and which just peep through the foliage. They have a striking similarity to the " ábádís" or settlements in Mekrán. The general condition of the people is low, and there is a noticeable equality of property throughout, but this is owing probably more to the want of good government and the chronic state of warfare and insecurity they live in, than to the natural disadvantages of the land. Agriculture is in rather a mediocre state as regards cereals and vegetables, the principal object of culture being of course the date. They are not dependent on the annual rainfall which is small, but are able to irrigate with certainty by means of their valuable aqueducts drawn from the hill range as well as from wells, water being abundant and at no great depth. Each settlement has at least one of these canals, that of el-Bereymí Proper being brought from a perennial spring in the hills distant about twenty miles. The water in this canal was quite warm to the touch, but I forget what they told me about the source. The grains grown are wheat, jowari, maize, barley, and bajri, the spring crops being wheat, the autumn, jowari, and bajri. This is sometimes succeeded by a crop of beans or pulse, but the latter are never sown intermingled with cereals. The stubble is always ploughed in and never burnt, and the only other manure used is cattle dung. The vegetables grown are sweet potatoes, radishes, cucumbers, egg-plants, onions, and pumpkins. Tobacco, cotton, red and white, and lucerne are also grown, the last for the use of cattle. Eight or nine crops of this are obtained in the year, showing the quality of the soil, which is fertile but thin. But more care and attention are bestowed on the fruits than on anything else, and they consequently arrive at considerable excellence. All the best kinds of dates are cultivated, *fard*, *maseybilí*, *khalas*, &c., though they are not considered equal to the same varieties in Bedíeh (بديه) and Semáil (سمايل). The other fruits are peaches, mangoes, custard-apples, limes, sweet-limes, oranges, mulberries, pomegranates, melons, guavas, figs, and grapes. There are only a very few horses at el-Bereymí belonging to the Sheikhs; cattle too are scarce; camels are abundant and cheap; and asses are used extensively for burden and riding. The food of the people is chiefly dates and coarse bread or rice, varied by salt-fish, camels' and goats' flesh. Milk is abundant, and a hard sort of cream cheese is made, the juice of an euphorbia being sometimes used instead of rennet for coagulating the milk. On the sea-coast the intestines of fish are often used for this purpose. There being no banians or other regular traders, there is no general bazar at Bereymí, but every afternoon a market is held where the Bedouins assemble with their produce and animals for sale or barter with those who can supply their wants. Money is little used on such occasions where cloth, articles of food, camels, donkeys, goats, and all the miscellaneous articles of an Arab household, are exchanged. The most trifling things change hands, and the scene is, as may be imagined, a lively and picturesque one. The ladies here, I observed, did not wear the tinselled mask seen in Muscat, but covered their heads with a black cloth veil, which is still more unbecoming. I must not omit that like their European sisters they wore high-heeled shoes. Their occupations, besides household affairs, are spinning, mat-weaving, felt-making, and tending goats and kine.

El-Bereymí formerly possessed two forts, only one of which is now standing, the other has been demolished, and lies a heap of ruins. Both are said to have been built by the Showámis, a strong clan of the Na'im occupying chiefly the Wádi Jezze, but the fort still standing was improved and strengthened by the Wahhábís during their occupation. It consists of four towers joined by curtains and surrounded by a deep ditch. It is of square form, built entirely of mud or unburnt bricks, and carries eight guns of sizes. The breadth of the ditch is about 25 feet, and both scarp and counterscarp are quite steep and faced with brick work. The rampart is eight feet high and two thick, and there is an open space of 20 paces between it and the towers. These towers rise perhaps 40 feet, the curtain somewhat less than half way up, and each side of the square formed by them is about one hundred and fifty feet. The gate is the weakest part of the structure, there being only a single small wooden door standing half way across the ditch, which is here bridged with the trunks of two date trees. Inside the fort is a residence for the Sheikh with accommodation for the men, and some godowns. Water is abundantly provided by two wells, which would yield sufficient for a large garrison. I tasted the water of one, and it was perfectly sweet and good. Near the outer gate is a brass 24-pounder, mounted as a field-piece, having the name of Seyyid-bin-Sultán, A. H. 1258 in Arabic, and the English date 1842. It is one of a batch of 20 that Seyyid Sa'íd procured from America at that time for his corvette the *Sultan*. This gun was brought from Sohár by Seyyid 'Azán, in 1870, in his expedition against Bereymí, and was used against the fort it now defends. With unusual energy and forethought for an Arab, Seyyid 'Azán brought spare carriage wheels, harness, and tents, all of which are carefully stored up in a godown. The harness did not look as if it had ever been used, and they told me the gun had been dragged thither entirely by manual labour. The fort is fairly well situated, and stands out on the plain, but on the N. W. side the houses and cultivation encroach somewhat close upon it, and on the other side lie the ruins of Sedeyrí's fort at no great distance, which would afford capital shelter for an enemy. Both as regards strength and position it is the most important fort in this part of 'Omán, and is generally regarded as the key of the country towards the west. Its reduction, therefore, would be considered necessary by any force approaching from that side.

I endeavoured to gather information respecting the route between 'Omán and Nejd, but the accounts were somewhat discrepant. According to some the first district beyond el-Jow is Beinúneh, in which is el-'Ankeh, a hamlet of the Bení Kattab, with a small date grove, the Sheikh of which is Saí'd-bin-Aweydimí. Next to Beinúneh lies el-Dhafreh, inhabited chiefly by the Menásir, and where there is a watered grassy vale called Da'ús, visited in season for pasture by the Menásir, 'Awámir, Bení Yás, Bení Kattab, el-Mizaniyeh, and el-Ghafaleh nomads. Further on between el-Dhafreh,

Katar, and el-Ḥasa is the district of el-Ja'fúr. Through these districts lies the route from el-Bereymi to el-Ḥasa, from whence the road continues to el-Riádh. There is no tract that can be followed, as the sand is blown about by the wind, but there appear to be two general routes, one of which is used more in winter, the other in summer ; the first is straighter and shorter, the other passes near the sea, is more winding, and after leaving the Sabkheh turns north for three days. The journey is not considered dangerous or difficult, as water is found in a great many places, though usually very brackish, and they seldom have to carry a supply for more than two days. Caravans very rarely make the journey, and travel only at night, taking about thirty days from el-Ḥasa to el-Bereymí. Troops as a rule travel by day only, their pace being a gentle amble, and they cover the distance in twenty to twenty-five days. A ḳásid takes ten days. The Menásír and Bení Yás chiefly hold possession of the eastern part of the route, the Ál Morra of the western. No hills are met on the way, and the only Wádís are el-Sabkheh and el-Soḥba. I give in a tabular form the halting stations of the two routes, but as already observed, water is procurable in many other places. The Sabkheh, or Sabkheh Mattí, as it is sometimes termed, is a marshy tract or Wádí about forty miles in breadth, commencing from the vicinity of Wádí Jabrín and entering the Persian Gulf between Long. 51° 50′ and 52° 20′, lat. 24°. In some parts it is a treacherous morass, only to be crossed at the beaten tracks, and it is said that should the camel miss the path, he becomes engulphed in the mud. The Sabkheh, according to the concurrent testimony of all the Sheikhs and best informed persons I have spoken to on the subject, both in el-Jow and Muscat, including His Highness Seyyid Turkí, is the boundary line between Nejd and 'Omán, and has been so considered from time immemorial. The water-shed of el-Aarid and Yemámeh appears to lie S. E., the Wádí Hanífeh and all other Wádís converging towards el-Randha, where they unite in the Wádí el-Soḥba, which falls into the Persian Gulf just above the Sabkheh Mattí, probably at Khor el-Dhoan. The Bedouins in the great desert rear great numbers of camels, the sale of which constitutes their chief support. The Ál Morra and 'Awámir are said to traverse it extensively, as it is not entirely destitute of water, which can be obtained of brackish quality in places by digging. Palms and other large trees are not met with, but dwarf acacias and herbaceous vegetation, suitable for camel fodder, are sufficiently abundant. Besides two species of gazelle and the oryx, numerous ostriches inhabit the more northern and western portions, and are hunted for the sake of their feathers, which eventually find their way to Mecca, there being no sale for them in 'Omán. There is said to be a route running direct S. E. from Nejd to Mahra that takes twenty-five days. Water is procured every three or four days, and is carried on in skins, the

Bedouins finding their way without difficulty ; a light camel-load of dates and flour enabling them to traverse a long distance. In 1870, Sa'úd-bin-Jelowí came straight across the great desert from Nejrán to Abúthabí in fifty-six days, travelling leisurely, but for the last fifteen days he and his followers were greatly pressed for food, their store having become exhausted. His purpose was to meet Seyyid 'Azán, which he did at Burka, and then accompanied him in his expedition against el-Bereymí.

South of Yemámeh and three days from el-Hasa lies the fertile and well-watered valley of Jabrín, whose groves of date palms are said to extend for several miles. It is situate entirely in the desert, and does not form part of Nejd. It was formerly a large and flourishing settlement, but it subsequently became so malarious and unhealthy, that the inhabitants were driven away, and it is now almost entirely destitute of permanent residents. The Arabs claim an antiquity of 800 years for it, but it has long since fallen to ruin, though I believe the fort and some of the walls of the houses are still standing. It is also said that after heavy floods gold coins are sometimes picked up by the Bedouins. The dates belong to the Ál Morra and Dowásir tribes, who visit Jabrín in the summer to collect the harvest, which is carried for sale to Nejd and el-Hasa. It is also extensively resorted to by the neighbouring nomads with their flocks and herds for the sake of the luxuriant pasturage.

In el-Bereymí I found a small colony of Arab gipsies, Zaṭṭ (زط) or Zaṭúṭ, as the Arabs call them, settled and I have since had further opportunity of observing these people. In his 'Alte Geographie Arabiens' Dr. Sprenger has identified the Zaṭṭ with the Jats of India, and though, as he shows, they have been in Arabia upwards of 1,000 years, they are at once distinguishable from the Arabs as a distinct race. They are taller in person and more swarthy, and they have that cunning and shifty look stamped on their physiognomy so observable in the gipsies of Europe. The Zaṭṭ are spread over Central and Eastern Arabia from Muscat to Mesopotamia, and are very numerous in 'Omán. Everywhere they maintain themselves as a separate class and do not intermix by marriage with strangers. It occurs sometimes, I believe, that an Arab takes a Zaṭṭiya to wife, but no pure Arab girl would be given to a Zaṭṭ, though daughters of Arabs by slave mothers may occasionally be obtained by them. It is probable, too, that the race is continued to some extent by adoption as well as procreation, as they do not seem to be a prolific people. In 'Omán, besides those who have been permanently settled in the country, are to be found many who come across from Persia and Belúchistán in search of employment or to visit their kinsfolk, but their stay is seldom prolonged. The Arab Zaṭṭ are divided into numerous clans or families, for which they have adopted Arab nomenclature, such as Wilàd Maṭlab (ولاد مطلب), Wilàd Ḳabâl (ولاد قبال), Wilàd Shaghraf (ولاد

شغرف), Musandé (مسندى), Ḥarímal (حريمل), Ḥaik (حيك), 'Ashorí (عشرى), &c., and each of which is in a state of clientship to some powerful Arab tribe, generally that of course with which it has most trading connections. The Zaṭṭ all profess the Musalmán religion, but no doubt they retain many of their own customs and usages. The levirate law obtains among them, but should there be no brother, the nearest male relative can take the widow to wife. They are looked down upon by the Arabs as an inferior race, but they are valued for the useful services they perform; and as their persons and property are always respected, they usually go about unarmed. In Nejd, I hear, the Zaṭṭ women are considered to be very handsome and dance publicly for money, but they are reputed to be chaste and moral; they are a necessary ingredient at private festivities, as they set off the assembly by their beauty and the party is not thought complete without them; they are consequently also more sought after by the Arabs there as wives. In 'Omán the case is different. The 'Omání women are more highly endowed by nature than their Nejd sisters, and the Zaṭṭ are not thought so favourably of by comparison. They appear to lead a semi-nomadic life, and move about from village to village with their families and chattels, working as occasion requires, but a few families may be found permanently established in most of the large towns and settlements. Their little mat hovels are the smallest and wretchedest human tenements I have ever seen, being merely a couple of mats arranged round three or four sticks tied together at the top, and the whole concern not usually exceeding 4 or 5 feet in height. They are accomplished handicraftsmen, being farriers, smiths, tinkers, carpenters, weavers, and barbers. They manufacture also guns and matchlocks; indeed most of the trades and manufactures seem to be in their hands, and they are to the natives of the interior what the Banians and other Indians are at the sea-port towns.

The Arabs assert that the Zaṭṭ speak among themselves in a dialect unintelligible to strangers, and they call this 'Rattíní' or 'Fársí'; but it is my belief that the original tongue of the Zaṭṭ has become almost entirely obliterated through long and intimate intercourse with the Arabs, and that what they speak among themselves is a jargon or gibberish of their own particular manufacture, composed of a corrupted Arabic mixed with the few Jat words they have retained. To effect this they have invented a simple and ingenious system by which they are able to transmute any word required into their own jargon without the slightest hesitation. The plan is to prefix the letter m and to suffix an additional syllable eek, while lengthening the first or second syllable of the word itself. Thus Bard (cold) becomes Mbardeek; Ḳamar (moon), Mḳámareek; Ghol (غول snake), Mgholeek. I subjoin a few words that appear to be of their own vocabulary, as specimens:—Father—Bweieekee; Mother—Mahiktee; Brother—

Mánas; Son—Kashkáshee; Man—Fseyil; Woman—Fseyileh; Slave—
Daugeh; Head—Kerrâ; Body—Kerrásh; Bread—Kshayim; Rice—Fidá-
mah; Knife—Jerrâha; Water—Tsammee; Donkey—Gyadoor; Go—Batûs;
Child—Towâtneck.

The Arabs do not of course trouble themselves with speculations as to
the origin of this people, but have a traditionary belief that they immigrated to Arabia from Persia at some remote period. I may remark, in
conclusion, that resemblance between the Zaṭṭ and the Gipsies of Europe
in character, appearance, habits, and profession (I have no means of comparing the languages) is so striking and complete, that the hypothesis
of their identity of origin must be regarded as, at least, highly probable.

A Route from el-Bereymí to el-Ḥasa.

Names of places.			
El-Dhafreh	..	..	الظفره
Khotem	..	..	ختم
El-Serádíḥ	..	..	السراديح
Bedú el-Moṭowwa'	..		بدو المطوع
Beinúmah	..	..	بيذوه
Bedú Jerash	..	..	بدو جرش
Sabkheh Maṭṭí		..	صبخة مطي
El-Sala'	..	..	السلع
Salwah	..	..	سلوه
El-Ghodha	..	..	الغضي
El-Sakik	..	..	السكك
El-Ḥemrúr	..	..	الحمرور
El-Ṭaraf	..	..	الطرف
El-Ḥasa	..	..	الحسه

From El-Ḥasa, i. e., Hefúf, the usual
road to Nejd is followed.

Another Route from el-Bereymí to el-Hasa.

Names of places.			Quality of water.	R E M A R K S.
El-Johar	..	أكجهر	Wells of good water	Acacia jungle, but no cultivation. Belongs to-el-Dhowáhir tribe. Half a day from Bereymí.
El-'Ankah	..	العانكة	Good, from shallow pits..	Date trees and cultivation. Hamlet of the Bení Kattab.
El-'Aweyneh	..	العوينة	Plentiful, but brackish; wells very deep.	Acacias and camel fodder. Two days from the sea.
Bedú Showeybí	..	بدو شويبي	Brackish; wells deep	Barren country with scrub. Limit of Beinúnah.
Dafis	..	دفض	Brackish; wells deep	Low ground with acacias, scrub, and grass. Visited in rains for pasturage by el-Manásir, Bení Yás, Bení Kattab, El-'Awámir, Mizárieh, and el-Ghafálch nomads. Two days from the sea.
Bír el-Motowwa'	..	بير المطوع	Sweet; wells very deep..	Acacia jungle. This place is so-called from one Mohamed el-Motowwá, a noted character, having been slain here. His son Hamd is now one of the most noted freebooters in 'Omán, and a constant terror to the Bátinah.
Ghadír el-Lál	..	غدير اللال	Sweet; but wells deep ..	Scanty scrub. The Sabkha lies between this and Bír el-Motowwa'. Road level, but winding among sand hills.
El-Sala'	..	السلع	Springs of sweet water on surface.	Acacia jungle. Belongs to el-Menásir. Half a day from the sea. Limit of el-Dhafrah.
'Aklet el-Nakhlet	..	اخلة النخلة	Sweet, near surface	Among sand-hills. Belongs to el-Menásir and Al Morra.
El-Sakik	..	السكك	Sweet, from springs	Woody. Lies a little S. W. of el-Katar.
Salwah	..	سلوه	Sweet.	
Faïj	..	بجيع	Sweet; wells five fathoms deep.	
Bajásh	..	بجاش	Sweet.	
El-Menáïot	..	المناعية	Sweet.	

432

To Buraimi

At the crossroads, al-Buraimi is marked 100 km left, and Falaj al-Qaba'il ('The Water-Channel of the Tribes') 1 km off right. Buraimi is shown as 100 km away, and Magan 5.5 km. The latter is the location of a recent French archaeological mission, and off right is the Oman Mining Company town, with a splendid new housing complex, situated about 25 km from the modern mining sites, which may be visited only with a prior permit. 3 km from Magan town is a poor graded road leading in 2 km to Falaj as-Suq, the old Persian settlement of Damsetjerd confirmed by ancient walling on the ridge overlooking the plain. Semi-sedentary *badu* were camping nearby, their goats disapproving as young kids gambolled ever farther off from parental observation. Sirhan tells us in the *Annals of Oman* how the Persian Shah Anushirvan appointed one Dad Fairuz Hashmushfan, called al-Muqabir, to govern Oman and Bahrain from Damsetjerd (or Jamsetjerd), where he built and founded a town. Fully four thousand Persians were stationed there. The Persian fleet, having conquered the southern Gulf as far as Suhar, then sailed to Dhufar and Hadramaut, finally conquering even Aden, to add 1,500 miles of coastline to the glory of the Persian Emperor. But after Anushirvan's death in 579, the Arabian possessions began to diminish, and the first waves of Islam in the seventh century removed Persian armies and Governors for ever.

I stopped at Hail al-'Ada, where traces of an aqueduct and castle surmount a transverse wadi beside the main road; numerous artificial stone circles can be seen hereabouts. The land is barren now, with tough-looking acacia making a goat's life a grim tenacious struggle against starvation, but clearly this region was quite thickly populated in earlier centuries. The scenery hereabouts is fantastic and terrible in

Magan. Oman Mining Company housing complex

Damsetjerd, near Falaj as-Suq

a way that Loutherbourg would have understood, and painted with delighted *frissons*. Past the signs off to Thaqabah, we came on the settlement of Suhailat ash-Sharqiyah: modern flat-roofed houses gleamed white and blue and brown in the morning sunshine, distant mountains enhancing the sense of community in the huddle of

Ha'il al-'Ada. Aqueduct and dam foundations

buildings, where only goats stirred. A little farther on you see a bright new creamy white mosque, intimate with a low minaret, and a chain of watchtowers behind it like giant fingertips appearing above the horizon.

This is all copper country, as you can read in Andreas Hauptmann's excellent study, *Die Entwicklung der Kupfermetallurgie vom 3. Jahrtausend bis zur Neuzeit* (*Der Anschnitt*, Beiheft 4, Bochum, 1985). Secondary copper minerals were smelted in the Early Bronze Age (third millennium B.C.) and in the Umm an-Nar period (third to second millennia) to a total of 2,000 to 4,000 tons from 150 ore deposits in Oman. In early Islamic times (9th-10th centuries A.D.), the figure rose to between 40,000 and 60,000 tons, proving that extensive and well-organized factories existed to smelt copper ores. There is a total break in the record after two hundred years, then the mines begin production again on a very much smaller scale, producing only between 3,000 and 3,700 tons between the 12th and the 19th centuries. Evidently the colossal amounts of wood required for mining supports and production caused catastrophic devastation of tree cover throughout the northern mountains and probably the northern coastline to a lesser extent.

It must be pointed out that copper sites are by no means confined to Wadi Jizzi, but occur in a crescent linked with the mountains beginning west of Fujairah (U.A.E.) towards the hinterland of Suhar, and then swinging eastward again down to al-Maisar. However, as Paolo Costa has stated in 'The copper mining settlement of 'Arja: a preliminary survey' (*Journal of Oman Studies*, Vol.4, 1978, published in 1981), 'the richest copper deposits were identified in the area of the Wadi al-Jizzi'. The most important archaeological remains, with substantial buildings, were found at Samdah, al-Baida', the linked sites of 'Arja and Tawi 'Arja, al-Asail, and Mulayyinah. Tawi 'Arja had a lithic culture, with microlithic jasper tools inducing Maurizio Tosi to suggest that the quarrying of jasper not only predated the quarrying of copper here, but might even have generated it. Al-Asail lies just south of the modern road, and 'Arja and al-Baida' farther to the north, both of these last benefitting from the watercourse of Wadi Bani 'Umar al-Gharbi, cascading from Jabal Ghashan (1265 m high) and discharges a little north of Majis town. The Oman Mining Company has resumed mining at 'Arja and al-Asail, and work by archaeologists and mining experts attached to the Ministry of National Heritage and Culture began in 1978 at 'Arja. Tawi 'Arja (*tawi* means 'well') has one of the only two wells in the area, used by the Shawawi shepherds, but there is a very elaborate system of ancient *aflaj*, while no evidence for channels or wells was found near the mine of 'Arja. Dr Costa concludes that the 'distribution of freshwater resources and the location of cultivable land are certainly the two basic factors of settlement in Oman, but the exploitation of mineral resources has quite often led to the formation of settlements in areas ill-suited for the life of a community, even a small and perhaps seasonal one. The smelting of the ore on the very spot of the extraction, which we have seen was the most common practice in Oman, determined at 'Arja the formation of a small mining settlement clustered around the copper-bearing outcrop, while a 'supporting' village grew up at Tawi 'Arja where the physical setting was adequate for the development of agriculture.'

It is extraordinary beyond words to encounter, deep into the virtually uninhabited desert, mounds of copper slag left here a thousand years ago. And 'Arja claims our attention too for a square two-tier structure, some twenty metres square at the lower tier, with remnants of a ramp in the northwest side, leading up to the second tier. Dr Keall of the Royal Ontario Museum sees similarity to a Mesopotamian ziggurat, which, if true, would date it to the third millennium B.C.

After signs to al-Jahili (8 km) and al-Wasit (10 km), we took a turn off left to the village of al-Khan, a rough track winding along the wadi floor, with palm gardens on both sides. The *Handbook of Arabia* of 1916 estimated the number of houses held by the Bani Hina and

Khan, Wadi Jizzi

Maqabil at eighty, a figure not dissimilar from that I estimated seventy
years on. Healthy children in bright clothes ran in tumult to greet us,
and played among the pebbles as I chatted to the two local men about
their goats and the hope for a good date harvest.

Beyond Khan the main road passes turns to Sahban (6 km off left)
and Saya (2 km off right), the grandeur of the mountains taking our
breath away, their barren crags relieved only by the occasional tuft of
acacia tormenting a few scavenging goats.

The checkpoint of the police in Wadi Jizzi stands at Hail bin
Suwaidan, just beyond Khuwairij; together they form the largest
settlement between Suhar and Buraimi. The only village still to be
negotiated is the tiny Khurus, perennially supplied with water from a
cleft in the rocky bed of Wadi 'Ain, and then, after driving between the
peaks of Khatmat ash-Shiklah and Khatmat as-Suwwad, and then over
the plain of Muzailah to the plain of Jau, its main town being Buraimi.

The town of nine villages, once united, is now divided into two parts.
One is governed by the U.A.E. from the centre called Al-'Ain ('The
Spring'), the traditional base of the ruling family of Abu Dhabi, in the
United Arab Emirates. The other five Abu Dhabi-ruled villages are
Hili (site of numerous successful excavations), al-Jimi, Qatarah,
al-Mu'tarad, and al-Muwaiqi. Oman administers the other three:
Buraimi proper, Sa'arah and Hamasah. The fort of sun-dried brick,
originally occupied by the Na'im, is being restored. Buraimi, the first
Omani city to become Muslim, was invaded by the Wahhabis from
al-Hasa in 1800, and they occupied Buraimi for some seventy years and
then again from 1952–5, spurring the Imamate revolt in Oman from
1957 to 1959. The Saudi troops were evicted from Buraimi and al-'Ain
in 1955 by Trucial Oman Scouts, and eventually in 1974 a Saudi
communiqué accepted the *de facto* arrangement still in force.

Hamerton on Northern Oman

The first European in Northern Oman was Captain Atkins Hamerton
of the Bombay Army, whose journey in January–February 1840 took
him from Sharjah to Buraimi, about 160 km in all. He checked its
defences on behalf of the British Government, and then traversed
Wadi Jizzi, on the track of the Wahhabi invaders who had frequently
raided the Batinah, to the coast at Suhar. Buraimi, as the key to Oman
from Eastern and Central Arabia, had twice been occupied by
Wahhabi troops, who had been expelled in 1839, after the conquest of
Hasa and Najd by Muhammad 'Ali. The British had received
intelligence that the Egyptians in Eastern Arabia might intend to
invade Oman, and the British therefore determined to discover the
latest position in Buraimi, and to counteract Egyptian expansionism in
Oman by reopening the Political Agency in Muscat. Hamerton was
appointed Agent within a matter of months after this journey was
completed. In this abridged form it originally appeared in *Selections
from the Records of the Bombay Government* (n.s., no. 24, 1856),
reprinted in 1985 by The Oleander Press under the more accurate title
Arabian Gulf Intelligence.

Captain Hamerton's Route (Abridged) in January 1840 from Sharjah to Buraimi ['Brymee']

Names	Hours	Direction	Remarks
Sharjah to Fallah	3	S. by E.	A spot on the desert, having a well with good water, and three large trees, but no houses or tents whatever. No trace of a road.
Bir Muhafidh	10	S.S.E.	The ascent the whole way was very gradual, over sandhills. No forage whatever. A fine well.
Ghuraif	3½	S. by E. & S.S.E.	The road tortuous, An old ruined fort, in the midst of thick babool jungle, having several wells of good water. Formerly belonged to the Shuamis of Buraimi, who were driven out about fifty years ago by the Beni Kaab or Chaab. At present occupied by the Beni Kuttub Bedouin Tribe. Good forage in the cold season for camels.
Gibul Yiff	6½	Ditto.	A high peak, so called in the country of the Beni Kaab. The track from Bir Mohafiz over hard sandhills: on its right babool jungle; some of the trees of considerable size.
Brymee	13	S.	The road very heavy, over and winding round the base of high hills of sand, in many places so steep that the camels could not ascend or descend. At seven and a half hours (from the time of starting) entered a third range of hills, and an hour after descended into a beautiful valley (called Hurmullioh), covered with wheat fields, just coming into ear (the property of Beni Kaab). In the valleys were two large towers, called Koheel and Jiburee, for the protection of the cultivation.

Brymee is a town of considerable size, built of sun-dried bricks, and surrounded by a wall constructed of similar material; but the greater part of the town is represented to be in a dilapidated state, and the wall a perfect ruin. On the south side of the town, however, in an open plain, is a fort, nearly square, surrounded by a dry ditch, about twenty-four feet wide, inside of which is a wall, about eight feet high, for the protection of matchlockmen while defending the ditch. About thirty feet distant, and inside of this wall, is the fort wall, about fourteen feet in height, and five in thickness at the base, and at the top only eighteen inches or two feet. It has round towers at the angles, but ill constructed, and the whole built of sun-dried bricks. The length of the fort, inside, Captain Hamerton found to be sixty-one paces, and the breadth sixty. On the north side, about three hundred paces distant, is another and smaller fort, about thirty-five paces in length, and fifty in breadth, the wall about fifteen feet high, and loopholed. In time of emergency Brymee could muster about 800 men for its defence, but under two chiefs, not always on the best terms with each other. The fort might offer an effectual resistance to undisciplined Arabs, with their matchlocks, but Captain Hamerton is of opinion that it could not for an hour withstand the attack of disciplined troops, with artillery.

The Arabs who hold the forts of Brymee, however, are only a branch of a tribe which occupies the adjacent district, and goes under the general appellation of the Naeem of Brymee; but the person who is considered the principal chief resides at a place called Zuneh, about eighteen miles distant from Brymee, and, if all united under him, might number 2,600 fighting men; but there are at least four chiefs, who pretend more or less to independent authority, whose submission to the Chief of Zuneh must be viewed as in some measure voluntary, and whose united power is subject to be greatly weakened by the jealousies and misunderstandings apt to spring up among them.

The principal value of Brymee would appear to be derived from its groves of date trees, easily reared, and brought to great perfection by the plentiful supply the plain on which it stands receives of water drawn by aqueducts from the adjacent hills. It is not known by whom these valuable conduits were constructed, but the Natives assign the merit to Solomon the son of David, possibly Suleiman the Magnificent; but more probably they are the work of the Persians, who conquered and held Bahrein and the Arabian Coast in the time of Nadir Shah. This latter supposition receives support from the circumstance of similar aqueducts being in common occurrence almost everywhere in Persia.

The Brymee dates are considered superior to any produced in the province of Oman. Wheat grown in the valley is of a fine description, but does not appear to be much cultivated. Fruit, such as oranges and lemons, grapes, figs, mangoes, olives, and pomegranates grow in great perfection. Coffee, too, was formerly cultivated on the hill Hafeet, but from the indolence of the inhabitants, or other causes, its growth has been abandoned.

The fort of Brymee, as far as has been ascertained, was built by the Wahabees, who, shortly after they had established their authority on the Arabian Coast, compelled the adoption of the tenets of their creed, and the payment of the Zukat or the fifth of all property. They are said to have had their attention first directed towards it by the inhabitants attempting to rebel, and the Wahabee General finding it necessary to summon the chiefs on the Arabian Coast to besiege and capture the place, which was soon effected, and the defences, which consisted merely of a wall of mud or sun-dried bricks, were improved, and entrusted to the charge of a Wahabee garrison. This person, who was called Bin Gendrik, was shortly afterwards succeeded by another of Soud, the Wahabee Chief's General, the famous Mootluk, to whom is principally attributed the obstinate resistance and brave defence of Shinas against the British in 1809; and it was under his superintendence that the fort received its present shape, and was surrounded by the ditch, now forming its most formidable line of defence.

Brymee has been generally held by the Wahabees to the date of the fall of their power and influence, when the Naeem tribes assumed possession in independence. Being from its position and advantages a place of considerable importance, its capture would be deemed essential by any force invading Oman.

ROUTE FROM BRYMEE TO SOHAR
Traversed in January-February 1840

Names	Hours	Direction	Remarks
Brymee to Bir-ool-Humeeza	6	E. by N. ½ N.	Ambek, a place on the road in the hills, having wells, and abundance of good water. At Bir-ool-Humeeza (off the track) is a good well. The road all the way very good, and water in several places.
Wadi-ool Tizzee	3½	NE.	The bed of a mountain torrent, near a fine date grove, called the Jrukhl-ool-Heil, with a fine stream of water. Two hours further, another large date grove, the Nukhl bin Kuttub, with also a stream of good water, and two hours further Shigeeree.
Shigeeree	4	NE.	A pass between two high peaks in the Wadi-ool-Tizzee. There is here a round tower (Heerinal Shigeeree), built by the Imaum of Muskat, but now belonging to Hamed bin Syf (Ool Bokhee Banee), and a spring of good water. Two hours further an aqueduct upon arches, crossing a ravine called El Molana, belonging to the Kurnide Arabs, and two hours further Saleet.
Saleet	4	NE.	A small town, with large date groves, and a fine stream of water. From Ool Humeeza to Saleet (eleven and a half hours) the road very good, winding through high mountains. Plenty of water all the way, but no forage, except for camels. Shortly after leav-

ing Saleet the sea becomes visible towards the north-east, and the descent from the hills commences to the plain at the Kemaats, called Felligh-ool-Goball, aqueducts, which extend from the hills to the date groves of Sohar.

Al Ohei	4½	. . .	Al Ohei, a small town.
Sohar			

CONCLUDING REMARKS

The two tribes who reside in the part of Oman visited, are the Beni Kaab and the Beni Kuttub; the latter a Bedouin Tribe, without any fixed place of residence, numbering about 600 men. The Beni Kaab inhabit that part of the country between Gibul Yiff and Hurmullioh; they are estimated at 600 men.

The road from Shargah to Brymee is difficult, while that from Sohar to Brymee may be considered good. Plenty of water the whole way; forage for camels abundant, but none for horses.

Buraimi Today

Although Buraimi is in the desert, as an oasis it possesses a microclimate distinctly cooler than that of its surroundings; the average daytime soil temperature in oases is some 15% lower than in surounding deserts, with relatively high humidity. Buraimi's climate falls in fact between the extremes of the hot, dry Rub' al-Khali desert, and the maritime desert climate of the Batinah. There is a distinct winter and also a distinct summer, with strong wintry prevailing winds; but the extremes of diurnal temperatures are not as marked as in the hot, dry desert. Water is supplied both by underground and surface *aflaj* systems, with their consequent limitations as to irrigable area. Agriculture is thus confined to *falaj* zones, and most houses are built on land unsuited to cultivation, so that forts and *aswaq* are built around the limits of the date gardens. The few houses found in the groves are raised above the *falaj* to assist smooth irrigation and avoid flooding during winter. A change in the last fifteen years has seen the growth of vegetable gardens in the penumbra round the date groves, and new houses are being built near these new gardens.

In 1982 a Royal Directive established a Regional Development Committee to co-ordinate and accelerate infrastructure in the wilayats of Buraimi and Mahdah, an area touching the western border from the U.A.E. in the north to southwest of Sunainah and 'Ibri in the south. The basic objective of the new Committee, in conformity with the known wishes of local people and Omanis who have migrated to the Buraimi zone, is to accommodate growth and immigration in the urban areas, but at the same time to encourage residents of smaller communities to remain within their rural settlements by introducing

Buraimi. Husn al-Khandaq

basic services. The objective has been achieved by maintaining gravel roads linking all settlements to the main blacktop road system; repairing all feeder *aflaj* affecting existing agricultural areas; distributing water to all villages inhabited by fifty persons or more, and smaller groups along the delivery network; connecting centres inhabited by 250 persons or more to the main electricity grid (or provide them with independent power supplies); allocating plots for replacement houses in all communities, including Buraimi and Mahdah themselves; and finally developing new agricultural zones by providing Government land where potential exists.

Buraimi. Vegetable suq

442

In practical terms, the road south from Buraimi to al-Qabil was opened before National Day 1984, Husn al-Khandaq's restoration in authentic mud-brick with teak doors was completed for National Day 1985, and in 1986 electrical services illuminated Hafit, al-Qabil, Sunainah, Wasit, Hail, and Rabi'. A major new source of municipal water supplies has been identified in the Zarub Gap east of Buraimi. A new airport, aligned roughly parallel with the Al-'Ain International Airport of the United Arab Emirates side of the frontier, will replace the present airstrip during the Third Development Plan. The 2,000 m runway will cater for Twin Otter and occasional Fokker Friendship movements, and permanent terminal facilities for sixty passengers will be provided in Stage 1.

Buraimi is benefitting from the same tidal wave of progress that has affected every other facet of life in the Sultanate of Oman, yet the rate of change has luckily not been so rapid that it has radically altered the essential Omani personality there.

Eccles on Northern Oman

J. G. Eccles travelled with the D'Arcy Exploration Company's geological survey in Northern Oman in November–December 1925, with the eminent geologist G. M. Lees and the well-known Muslim convert William Richard (Hajji Abdullah) Williamson. Born in 1872, Williamson had made his first pilgrimage in 1895, his second in 1898, and was to make a third, in 1936. W. E. Stanton-Hope has written Hajji Abdullah's biography as *Arabian Adventurer* (1951), and a fascinating tale it is, satisfying the plea in Eccles' third paragraph.

D'Arcy Exploration Company's Geological Survey of 1925.— The party was composed of: Messrs. G. M. Lees and K. W. Gray, geologists; Mr. Joseph Fernandez, deputed by the Bombay Natural History Society; Mr. A. F. Williamson (al-Hag 'Abdullah); and myself.

I was in charge of the social and political branches; Williamson organized the transport and messing; the remainder were free to carry out their strenuous technical work.

It is a great pity that Williamson, or someone for him, does not write the story of his life. Some twenty-five or more years ago he was in Aden. There he became a Muhammadan and went up to Sana'a and the interior of the Yaman. Since then he has wandered over the greater part of Arabia. He has performed the Hajj—twice, I believe.

At the outbreak of war he was acting as agent for the Shaikhs of the great Muntafiq tribe, and during the war he was attached to the intelligence branch in Iraq. His knowledge of Bedouin life and customs is very profound, and it was strange to see 'Omanis, who are famed for their camel-breeding, bringing their camels to him and consulting him as to treatment for wounds and illness.

As the principal object of the expedition was a geological survey, the main scarp of the Jabal Akhdhar naturally offered most attraction. But, as I have already mentioned in describing the political situation, the whole of this region is controlled by Shaikh 'Isa bin Salih in the name of the Imam. The former was at that time (October, 1925) making preparations for his advance into the Dhahirah, and on being approached wrote politely regretting his inability to control his tribesmen should the party enter his territories. Although in consequence all the land to the south and south-east of the Wadi Hawasinah was barred to us, Shaikh 'Isa's advance, which coincided with our journey, was undoubtedly a help to us, as many Shaikhs who might otherwise have proved refractory received the party with open arms, hoping that exaggerated reports would reach Shaikh 'Isa of the presence of Muscat officials in their territories, and perhaps also seeing a vague chance of getting some help from the Muscat Government.

Williamson for the greater part, and I for the whole journey, travelled in Arab dress, and we found that it had a great effect in making our various hosts and visitors more friendly and in loosening their tongues. Every day after we had camped, between the afternoon and the sunset prayers, he and I and the Shaikhs and their followers would sit round chatting, and many were the amusing conversations we had. Another helpful factor in overcoming the shyness and suspicions of the Arabs was that one of my personal following was a slave of the Sultan's household famous for his coffee-making. He would carry all the apparatus on his riding camel, and wherever there was a long halt he would hastily prepare coffee for the party.

On November 3 all the transport and stores had been collected at Bait al-Falaj, my headquarters, two miles inland from Matrah, and on the following morning we made a start. Passing through Boshar (the Besheyr of Palgrave), famous for its hot springs, where we were hospitably received by Shaikh 'Ali bin 'Abdillah al-Khalili, brother of the Imam, we entered the Batinah plain at Sib, and followed the coast to Khaburah, where we arrived on the 10th. The Wali of the place, Mudhaffar bin Sulaiyim, is the son of an African slave who rose to a position of great influence in 'Oman, becoming Wali of Sohar and controlling more than half of the Batinah. Mudhaffar accompanied us throughout the whole trip. He inherits much of the character of his father—is forceful, active, much liked by the Arabs, and became our general manager and trusted adviser.

THE COUNTRY OF THE TRIBES AL-HAWASINAH AND BANI 'UMR.—
Khaburah lies at the mouth of the Wadi Hawasinah, which takes its
name from the tribe which inhabits it, and up which we proposed to climb
to reach and cross the watershed. Here we were joined by Shaikh
Saif bin Muhammad of the Hawasinah, a Hinawi tribe, and Shaikh
Ghussun bin Salim of the Bani 'Umr, a Ghafiri tribe. The two tribes
are generally at feud with one another, and the two Shaikhs are quite
dissimilar in character and disposition. The former proved himself a
miserly, weak, shifty obstructionist; the latter a tactful, firm, and
helpful disciplinarian.

Our path led us over the Batinah plain toward the mountains.
Leaving on our right the deserted village of Qasaf, where the bare
stumps of the palm trees bore mute testimony to the repeated failure
of the rains during the past ten years, we reached Ghaidhain
(555 feet), wrongly placed in Hunter's map of 1908. A description of
the village will suffice to portray all the villages of this area, which are
alike in essential details. It is well built of stone and local cement
and stands on a terrace on the left bank of the Wadi about half a mile
inside the foothills. Extensive date groves and gardens lie in a re-
entrant behind the village and are watered by a *falaj* or water
channel which begins from a spring underneath the surface shingle of
the Wadi, a mile above the village, and continues underground for
some distance before emerging into an open cemented channel. Those
who are familiar with the Persian Qanat, or Kariz, will recognize the
same type. In this case the open channel is divided into two streams
just before it reaches the gardens, each stream watering a half of the
date groves. Hence the name Ghaidhain, "The two groves." Pursuing
a leisurely course up the Wadi we passed through Suddan, described
by Wellsted, whose route we now joined, Bida'ah, and Falaj al-
Hadith. Here we came upon running water, and the valley bed was
sprinkled with oleander and ziziphus shrubs, many of which reached
a height of 10 feet. As we approached the junction of the Wadi
Hawasinah and Wadi Dhillah the valley opened out and we passed a
large tower built for the protection of the caravan route and occupied
by three men of the Hawasinah. Two of them came down to meet us
and to receive the customary *douceur*, which we gave them and passed
on. But the third man, either not realizing that we had paid, or
thinking it not enough, ran to a point directly overlooking the Wadi, fired
over our heads, and started screaming in the sharp, high-pitched
feminine tone used by men in this country to warn their friends of the
approach of hostile or strange parties. His own companions ran to
him from the fort, and our following started loud explanations from
below. Williamson and I went steadily forward and left them to
settle it.

During our passage up the Wadi Hawasinah we had suffered much

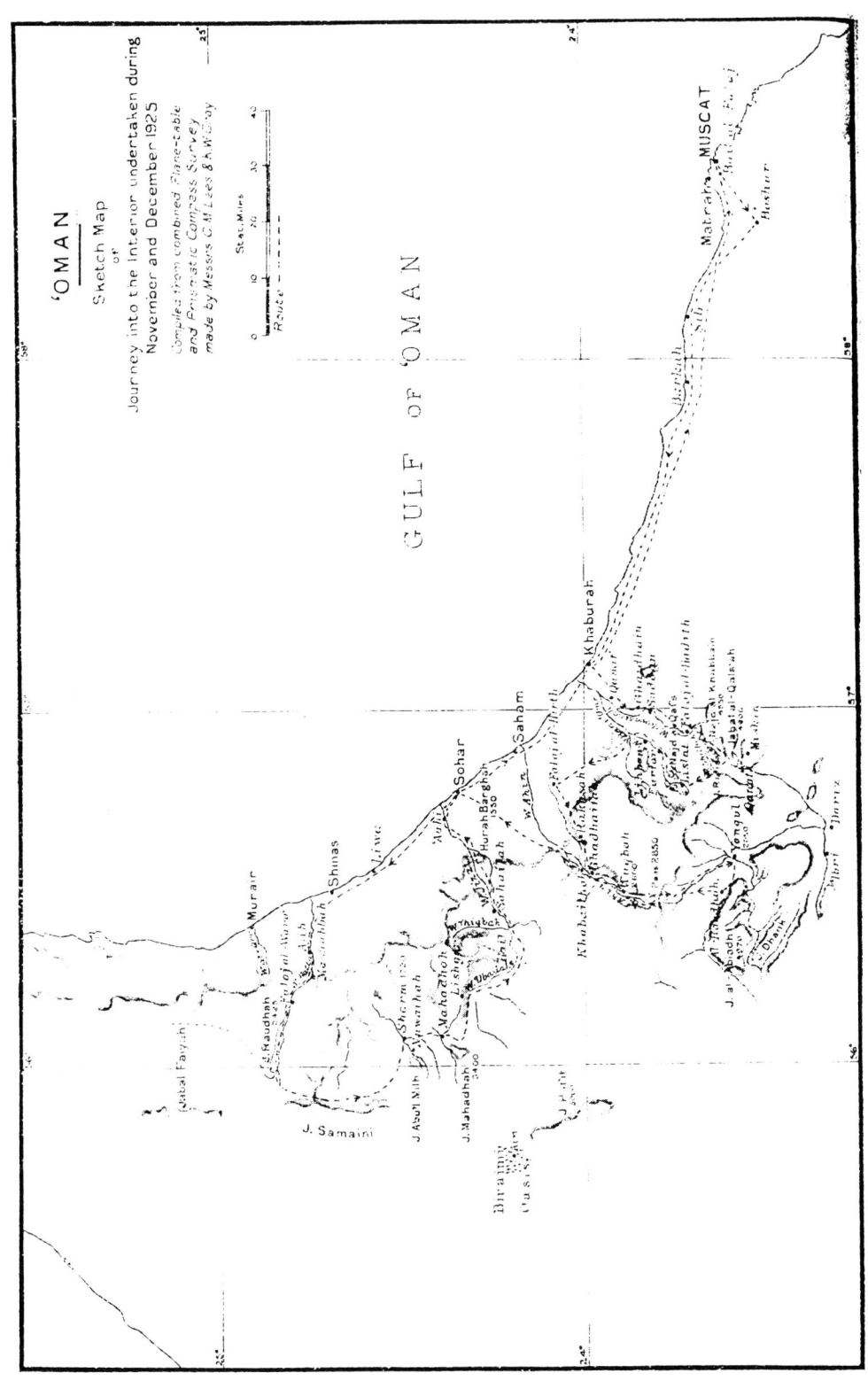

Map to show Eccles' route in the Batinah and Interior, 1925

trouble and obstruction from Shaikh Saif, and so we were relieved to turn into the Wadi Dhillah, which belongs to the Bani 'Umr. From a high hill Lees gained a good view of the Wadi Ḥawasinah above its junction with the Wadi Dhillah. The hills appeared to recede some-what from the river bed and to allow of small plots of cultivation on the terraces. A number of villages could be seen, including Hajajah, 'Abailah, Tawi, Suwairiq, Washah, and Harm 'Ali. The Wadi Dhillah is narrow with vertical cliffs rising sheer from the valley bed. It is uninhabited except for some Bedouin squatters. As we proceeded the valley became steeper and more narrow and the going became very bad, until we reached a zigzag path known as Najd al-Khabbain (2,970). Here we were forced to dismount and lead our camels to the top of the pass over the range which divides the Dhahirah and the Batinah.

We found a pleasant camping site on the further slope and determined to stay here a few days, as I had been having attacks of malaria, and the transport required some reorganization. From Jabal al-Qala'ah (4,900) Lees gained a magnificent view of the Dhahirah up to the edge of the great desert. To the south-east the great scarp of Jabal Akhdhar was plainly visible. The Wadis, draining west and south-west, run out into extensive plains broken here and there by groups of small hills. The villages of Qanat, Dariz, and 'Ibri could be seen, but Miskin was hidden by a range of low hills.

Two miles along the scarp west of the camp we were shown a wonderful gorge, called en-Naqs, 10 to 20 feet only in breadth, with vertical walls rising to a height of 400 feet. From half a mile away it is quite invisible. Breaking through the knife-edge ridge of Jabal Ra'is it joins the Wadi Dhillah and provides for good mountaineers, such as all the local tribesmen are, an alternative track to the Najd al-Khabbain. In the narrowest part the torrent bed drops a sheer 40 feet, and over the precipice thus formed a chain has been hung. Well forged with long, narrow links it is firmly secured between two rocks, but as it is not long enough to reach the bottom a rope has been attached to the further end. The face of the cliff is concave, so that no foothold is obtainable. The tribesmen said that up to twenty years ago there had been only a rope and many had lost their lives by its breaking. They could not name the public-spirited Shaikh who had substituted the chain.

On November 20 we packed up reluctantly and set out for the Wadi Bani 'Umr, by which we had decided to return to the Batinah. During the loading up Williamson tried to make one of the tribesmen take on his camel some hens which had been bought for our messing. But the prejudice against the carrying of any kind of fowl by a man proved so strong that insistence would certainly have resulted in a general strike. In the same way we could only obtain eggs if we sent one of our own servants to fetch them. Our way led along a wide valley under the outlying spurs of the main ridge, which falls in far gentler slopes on the Dhahirah side. On a small hill in the centre there stood a strong but dilapidated fort dominating the valley. A short but difficult pass brought us over the watershed into the Wadi Bani 'Umr, which we followed past the village of Suwaidah to Lislat (or Hail Islat, 2,300 feet). The latter had belonged to the Hawasinah, and was protected by a strong fort built on the summit of a solitary hill which rises abruptly from the valley bed. A well had been dug in the Wadi against the perpendicular side of the hill and enclosed by the semicircular wall built against the cliff right up to the fort 100 feet above. The place was captured by the Bani 'Umr some forty years before. They have abandoned the old village and built a new and miserable one of palm branches on the opposite bank. When I asked Shaikh Ghussun the reason, he said that his and his father's deliberate policy had been to discourage the use and building of strong forts and villages, which proved only a source of weakness to the tribe, as each petty village Shaikh comes to rely more on his own fortifications than on the unity of the tribe, in which lies their real strength.

A little beyond Lislat the road left the Wadi, and passing the small village of Qisah on the right, climbed over the northern shoulder of Jabal Ra'is by a pass called Najd Bani 'Umr or Najd al-Qafs (" the pass of the cage," 3,110 feet). This was the roughest stretch we experienced throughout the journey, and almost the whole day's trek had to be done on foot. We re-entered the Wadi Bani 'Umr at Furfar, and continued along it to the low foothills, passing Lihban, a prosperous and well-built village, with extensive date plantations on both sides of the Wadi. Another day's march brought us to Falaj al-Hirth in the Batinah, seven miles inland from Saham. During the past few days several misfortunes had overtaken us. Lees and Fernandez had both had severe attacks of fever. Gray had upset a bottle of Indian ink over the plane table, and Williamson one day had lost the way, taking with him our special cook, and had not turned up till eight o'clock at night after darkness had fallen.

THE COUNTRY OF THE BANI 'ALI.—We had been joined in the Hawasinah country by Shaikh Muhammad bin Hilal of the Bani 'Ali, an Arab of the finest type, quiet, unassuming, yet naturally dignified. He now came to me and invited us to visit Yanqul, the capital of the tribe and residence of his brother Khalifah, the paramount Shaikh. By this time Shaikh 'Isa bin Salih in his advance from 'Oman proper had reached Dariz. 'Ibri was still holding out, but its fall was imminent. The Bani 'Ali had always remained loyal to Muscat, and had turned a deaf ear to Shaikh 'Isa's enticements. But now, when town after town in the Dhahirah was falling, Khalifah's position was very precarious, especially as there were in Shaikh 'Isa's camp two rival claimants to the Shaikhly office. He had twice appealed to Muscat, but without avail. In inviting us therefore his motives were twofold. Firstly, Muscat might be induced through our influence to give him some support; and secondly, if he made a great show exaggerated reports would reach Shaikh 'Isa, who might really think that Muscat and even the British Government were behind him.

But whatever the motive it was a golden opportunity for us, and we determined to start for Yanqul the following day by way of the Wadi Ahin. A rapid journey over the plain soon brought us into the foothills again, and we entered the Wadi at its junction with the Wadi Hibi, which enters from the south, and contains the village of Hibi, wrongly placed on Hunter's map. During the march Lees dismounted and turned aside to climb a hill. His guide immediately started to take the camels on, so that Lees was forced to turn back and stop him. An old woodcutter then appeared, and started chatting with the guide. Lees made another move to climb the mountain, whereupon the woodcutter ran towards him shouting abuse, and, as Lees continued on his way, began to throw stones at him. Lees, a tall and doubtless terrifying figure in his battered panama, turned and took three paces toward him, brandishing his geological hammer. It was quite sufficient. The old man turned and ran, followed by a shower of stones from Lees.

As we approached Wuqbah the river bed narrowed until we reached a part where there was a considerable stream with deep pools. In order to avoid this stretch, which probably in rainy weather became a series of rapids, a road had been built over the side of the Wadi, with steps made of stone and local cement (saruj) at either end from the river bed. These steps were a source of great pride to the tribesmen, who had several times warned me to look out for them. The valley starts to widen as it approaches Wuqbah, a village with very extensive gardens, wrongly placed in the Wadi Dhank in the " Handbook of Arabia," and gradually broadens above the river into a plateau well covered with jungle bushes and grasses. A short and easy pass brought us over the watershed, which is here not very clearly defined. A mile beyond there stretched from north-west to south-east a wide, well-wooded plain, bounded on the east by the irregular and serrated black peaks of the serpentine hills which form the watershed, and on the west by a long regular limestone scarp, known as Jabal al-Abiadh. We sent the caravan to camp at ar-Raudhah, a mosque and well immediately below the centre of the scarp, whilst we turned aside and made for Yanqul, which lies at the southern end.

I will not weary you with a description of our welcome, as it was almost identical with that given to Sir Percy Cox at 'Ibri, and described by him in his paper to the R.G.S. last year. As soon as politeness allowed we proceeded to the house set apart for us. It was a strong two-storeyed building overlooking a pleasant garden. The upper rooms were high and well-windowed. The beams were painted red, and decorated with verses from the Quran in white. The staircase, as in all Arab buildings, was very narrow, steep, and low. The baths constituted a pleasant feature, which I saw in no other town. These were formed by irrigation channels running from the main *falaj* directly under the houses, where special bathing-pools were constructed. Strong gratings were placed over the channel where it entered and left the house. I noticed that Shaikh Khalifah could hardly speak, and wondered if he had a sore throat, but when he came to see me privately he told me the true reason. The night before our arrival an attempt had been made to murder him by three of his cousins, who would have succeeded but for the loyalty of the doorkeeper of the fort. There must also have been a disaffected element among the townspeople, as he had lost his voice in haranguing them after the failure of the plot. The three brothers were imprisoned in the fort, and he and Mudhaffar started to discuss their fate. Mudhaffar was all for killing them. "Wait," he said, "till the captain and his friends have departed, then slay them all." "Idhrib bi'l-saif wa kul 'asal." But no decision was made, and I have never heard what happened to them. On the following day news came that Shaikh 'Isa had captured 'Ibri. It was too late for Khalifah to expect any help, and two days later he set out for 'Ibri to make the best terms he could.

Yanqul stands in a strong position between the southern end of Jabal Abiadh and a high conical peak named after the town. The fort is in good condition and is a large rambling building with a low tower. The gardens lie mostly to the west of the town. Whilst Williamson and I were occupied with interviews and conversation, Lees and Gray climbed the scarp of Jabal Abiadh and gained a splendid view of the Dhahirah up to the Biraimi oasis. Bearings were taken on 'Ibri and Dariz which gave intersections with those taken from Jabal Qala'ah.

COUNTRY OF THE BANI KA'AB.—We returned to the Batinah by the same route and entered Sohar on December 7. Our intention was to follow the coast to Murair at the end of the Batinah plain and work back to Sahar under the foothills. But a fortunate event occurred which changed our plans. Between Liwa and Shinas Williamson and I turned aside to visit a Baluchi Shaikh famous throughout 'Oman for his hospitality. I had only intended to drink coffee, but when I told him so he seized my beard (by now quite a respectable one) and insisted on our staying the night. We compromised on a midday meal, during which I was introduced to Shaikh Ma'adhad, brother of Shaikh Salim bin Diyan, Tamimah, or paramount Shaikh of the Bani Ka'ab, whose district in the Dhahirah extends from Jabal Raudhah to the Wadi Jizzi in a long narrow strip among the outlying spurs of the main mountain ridge. Ma'adhad invited us to visit his country and made himself responsible for us in everything. We naturally jumped at the opportunity. Matters were quickly settled and that night it was decided to work up the Wadi Hatta. Shaikh Ma'adhad gave me the impression of a man of very strong passions repressed by an equally strong will, giving him outwardly a quiet and reserved demeanour. In durbar he never spoke unless appealed to, when he would answer as laconically as possible. But his judgment was always direct and his followers held him in respect. Only when on the trek would he break through his reserve and talk and chant continuously.

For the third time we turned to re-enter the hills. Between Liwa and Wadi al-Qor they are of less elevation than those further south, but north of Fujairah they are reinforced by the Shimailiyah ridge. Our first halt in the Wadi Hatta was at 'Ajib, where, amongst other things, tobacco was being extensively grown. A little above the village the river bed narrows to a gorge, called al-Wajajah, with running water and deep pools which the path skirted. At the near end on a shelf over-looking the gorge stands a strong tower whose only method of ingress is by a rope thrown from a window. This tower marks the administrative boundary of the Wali of Sohar. Emerging from the gorge we passed the village of Mashabbah, the gardens of which have been almost entirely killed by drought, and entered an enclosed plateau dotted with small hills. As we advanced, passing several villages and continuous gardens, this plateau gradually merged into a wide plain covered with desert bushes, whilst in the distance a line of sand-hills gleamed red and gold in the sun. There was no marked watershed or pass. The stony plateau marking the head of the Wadi Hatta simply loses itself in the plain which slopes down towards the Trucial coast. The reason was explained by Lees, who found recent marine shells on the plateau at a height of 1,050 feet. This level plain is in reality a raised beach, and the Shimailiyah country to the north must have been an island when the sea was at this level.

We were now very close to Sir Percy Cox's route, which passed to the west of Jabal Raudhah. Lees climbed this mountain and over-looked Jabal Faiyah and Dubai to the west and the Gulf of 'Oman to the east. Here we turned to the west and passed under the main ridge of Jabal Samaini, keeping to the east of the scarp. Shaikh Ma'adhad had tales of panther to be found, which he said took toll not only of the flocks but also of grazing camels. The southern end of the ridge terminates in a peak known as Jabal Munfarid (3,700 feet). Some miles further on the road climbed a small pass and dropped into the plain, which here has the effect of a backgammon board, as long wedges of sand run right up to the black serpentine hills, leaving patches of stony, well-wooded plateau dovetailed between them. We were forced to cross one sand dune, and then turned again into the first range of hills, and leaving Jabal Abu 'l-Milh on our right reached Sharm, Ma'adhad's home, and the second in importance of the Bani Ka'ab settlements. The welcome given us was on a scale second only to that of Yanqul, and the dancing, which was of the kind seen by Wellsted at Suwaiq, continued throughout the rest of the day. On the same night letters came announcing the murder of Shaikh 'Isa bin Ṣaliḥ at 'Ibri, and the hurried retreat of all his forces toward 'Oman. This was most fortunate for us, as Shaikh Ma'adhad had apparently begun to regret having invited us to his country and tried to persuade me to return by the road by which we had come. But now the path was smoothed, for although the news of the murder was false, it was true that 'Isa and his troops had retired in confusion.

We therefore set out with light hearts for Mahadhah, where we arrived after an hour and a half's riding over a stony plain within the first ridge of hills. The town lies at the side of a broad, bare " baṭḥah " (wide torrent bed), which may possibly be the same as that crossed by Sir Percy Cox on the outskirts of Biraimi. The fort stands on a small hillock in the centre of the torrent bed. On the further side and to the west of the town rises Jabal Mahadhah, blocking any view of the plain and sand dunes beyond. Shaikh Salim ibn Diyan we found to be an older and more sedate man than his brother Ma'adhad. More frequent journeys to Sharjah and Dubai had accustomed him to European manners and town-dwellers' amenities, and he was alto-gether more polished, more intelligent, and broad-minded, though no less virile, than his Bedouin brother. Whilst Williamson and I listened to the local politics and strolled round the gardens admiring the broad and strong water channel, Lees climbed Jabal Mahadhah (3,400 feet), and reconnoitred the Biraimi oasis and Jabal Hafit. The latter, indeed, was constantly coming into view during our journey from Jabal Raudhah to Hail in the Wadi Jizzi, and we heard no other name for it, nor was any mention made of a village named 'Uqdah. It is, as Sir Percy Cox has pointed out, an isolated hog-backed hill some twenty miles in length, but he does not emphasize the difference between it and comparatively small hills such as Jabal Faiyah. Actually it rises to a height of about 5,000 feet, or nearly 4,000 feet above the surrounding plain.

Just before we left Mahadhah a message arrived from the Shaikh of the Bani Na'im inviting us to Biraimi. Lack of time forced us much against our will to forego this great opportunity. Shaikh Salim was much exercised in his mind as to the policy he should pursue in relation to the Bani Na'im. Some two days ago they had sent him a letter to inform him that they had despatched a messenger to Ibn Jiluwi, Ibn Sa'ud's lieutenant in al-Hasa, asking for help in arms and man-power against Shaikh 'Isa. They now wished the Bani Ka'ab to join them. The latter were already in alliance with the Bani Na'im, and Shaikh Salim thought that he ought to have been consulted before so important a step was taken. His grandfather (who visited Colonel

Miles at Biraimi and invited him to Mahadhah) had been taken prisoner by the Wahhabis and led in chains to Dara'iyyah, where he was kept for seven years, so that Salim had no love for them. In the earlier part of the year Ibn Jiluwi had sent his messengers to demand *zakat* from the Trucial coast and from the Biraimi oasis. Certain tribes had paid up, and it was ostensibly to prevent this and to unite the tribes against the Wahhabis that Shaikh 'Isa advanced into the Dhahirah, as we have seen.

Time pressed, as Lees and Gray had to catch the mail-boat from Muscat, so we set out for the Wadi Jizzi, accompanied for the first two miles by our hospitable host. The track crossed a level boulder-strewn plain. To the east a number of wadis emerged from the black serpentine crags, the lower slopes of which were dotted with villages and date groves. Some low foothills hid the plain on the west. By some misunderstanding the caravan had turned aside to a well other than that agreed upon. Finding it unoccupied, we went steadily forward until by sunset we were well into the hills again, and we realized that we must have lost them. After much fruitless wandering in the darkness, during which all the Arabs argued their loudest, some to go back, some to advance, and some to stop where we were, we sighted a lamp on a hill to our right. It had been placed there by Gray, who was with the caravan. When we pointed it out to the tribesmen they refused to believe it. It must, they said, be either a star or a jinn. However, we insisted on turning towards it, and after a rough ride reached the camp very tired.

On the next day we followed the Wadi 'Ubailah, which gradually narrows as it enters the hills, to its junction with the Wadi Jizzi, along which many European travellers have passed. Turning aside to visit the Persian ruins at Hurah Barghah, described by Colonel Miles and others, we reached Sohar on December 23. Here we divided. I rode back along the coast, spending Christmas Day with Wali Mudhaffar at Khaburah. The rest embarked on a dhow, and, more fortunate than Palgrave, reached Muscat in two days.

6 MUSANDAM

Ru'us al-Jibal (not 'Jabal' as Thomas would have it in the singular, for obvious topographical reasons, supported by the R.G.S. map on a scale 1 : 100,000), means 'The heads of the mountains', with the implication that here, at the northwesternmost tip of Oman, the country's mountains swerve round and come to a sudden, dramatic series of peaks, some on the mainland, and others on the rocky headlands, called the Musandam Peninsula, joined to the mainland by a narrow neck between Maqlab and Jabal Dabshun.

Khasab Hotel, in the shadow of the rocks

Bertram Thomas on Musandam

Bertram Thomas, author of *Alarms and Excursions in Arabia* (1931) and *Arabia Felix* (1932), contributed an informative article on Musandam and its people to the *Journal of the Royal Central Asian Society* (now *Asian Affairs*) in 1929, and this essay is reprinted in full by courtesy of the Royal Central Asian Society.

THE MUSANDAM* PENINSULA AND ITS PEOPLE THE SHIHUH

By BERTRAM THOMAS

HE who can visualize the map of Arabia will be aware of an excrescence
on its eastern seaboard which destroys the rectilinear symmetry of
the great Arabian peninsula. This excrescence is Musandam. A
territory belonging to the Oman Sultanate, it is separated from its
suzerain by the Qawasim corridor, which crosses the base of the
peninsula from Dibah to Sha'am. Insignificant in size—it is but forty-
five miles in its greatest length, and its average breadth is only half
that distance—it yet presents problems, ethnological and philological, of
great interest to Arabists, problems into which there has been, as yet,
little if any research. Its local name, Ru'us al Jabal, translated by
Miles " Heads of Mountains," could perhaps have been alternatively
and not less felicitously translated " Mountain Capes," for here terminates
in a forked series of seagirt headlands the great Archæan backbone of
Oman—the Hajar. This system, conspicuous farther south for its well-
defined north and south axis two to three thousand feet high, flanked
with coastal plains and threaded on the east by deep perennial wadis
and on the west by dry and shallow torrent beds, here in Musandam
takes on another character. Here is a tangled mass of mountain spurs
which sprawls out clawlike into the sea to divide the Gulf of Oman
from the Persian Gulf; the sea has closed in east and west to the main
escarpments, giving rise to an intricately tortuous mountain coastline.
As an instance of the rambling nature of this coast, at Sibi, a cove in
the innermost recess of the Elphinstone Inlet, the mountain isthmus is
only a mile and a half wide, but to reach its distant side I steamed
seven hours circumnavigating fifty miles of the deeply indented promontory. Surging round this wild, rock-bound coast are strong eddies
and conflicting currents whose provocation is increased by the waisted
entrance of the Persian Gulf. These waters, distrusted by the Arab
mariner—it is a common sight to see a becalmed dhow being swept by
some sinister force on to a rocky shore, all hands at the oars—were
dreaded even more by our merchantmen in the more remote " days of
sail," when to the natural perils were added the risks of Qawasim

---

* " Musandam " has been equated with " Anvil " by every writer since
Palgrave (see Vol. II., p. 317). In spite of a certain physical resemblance, there
is no verbal justification for this : and neither the inhabitants nor their neighbours are aware of such a meaning.

pirates (Joasmees so called). That the sterile and forbidding mountains of Musandam—a natural asylum, so it would seem, for the aboriginal retiring before a wave of invasion from whatever land direction—should have preserved some traces of an ancient race, or tongue, or cult, is in keeping with the probabilities. And today it is its strange inhabitants, the Shihuh, that have attracted the attention of travellers as the strange language of Kumzar and Khasab has perplexed them. The classical geographers are unfortunately silent about the Shihuh, as also indeed are the early Arab travellers. The term "Ichthyophagi" with which a strip of the coast is anciently labelled, while true of Musandam, can scarcely have been less true of any of the other fish-teeming shores of South-East Arabia. Economic conditions would seem to make the appellation applicable throughout the Oman littoral yesterday, today, and for ever. But the silence of the ancients is strange, having regard to the importance of the Persian Gulf as an ancient trade route, the prominence Musandam Peninsula must have attained as a landfall for mariners, and the tribal tradition of an ancient Shihuh occupation. While it is true that tribal tradition has a tendency to extravagant growth, it shares with ancient tradition—*e.g.*, the Mosaic account of the peopling of Arabia from five great patriarchal stocks of Cush, Jokhtan, Ishmael, Keturah, and Esau—the claim of reflecting, in some measure, historical facts. And here an interesting point arises. The Shihuh having escaped the notice of historical geographers, has a record of the word been preserved in the Shuhite of the Book of Job? For, on investigation, I found that Musandam and its environs—*i.e.*, Northern Oman—whether or not by coincidence, provide a remarkably close tribal setting for the Biblical story of Job.

Thus, "When Job's three friends heard of the evil that had befallen Job, they came to console with him, everyone from his own place, Eliphaz the Temanite, Bildad the Shuhite, and Zophar the Naamathite." Now three of the biggest tribes in Oman today are the Bani Bu 'Ali, the Shihuh and the Na'im. The latter two readily suggest identification with the Old Testament names ; and the Bani Bu 'Ali, the Bliulaie of Ptolemy, derive on the authority of Miles from the ancient Bani Teman. This coincidence of the "three kings of the East" with the three shaikhly comforters (requisitely necessary neighbours), while remarkable in itself, is strengthened by its fulfilment of the other condition of the story, namely, the neighbourhood's exposure to raiding by Chaldæans and Sabæans—*i.e.*, proximity to their habitats.

Now the once popular conception of Chaldæans as exclusively the people of Chaldæa in Lower Mesopotamia, and the Sabæans as the remote subjects of the Queen of Sheba in Arabia Felix, placed the land of Uz in an unenviable geographical position. That Sabæans were not only found in South-West Arabia is demonstrated by the name still attaching to a quaint people in the Lower Euphrates today, to say

456

nothing of the historical " Sabi " of the Persian Gulf, while the word
Chaldæan is not, I think, to be confined to Babylonia but to be identified
with Bani Khalid, a numerous and widespread tribe in Arabia.   Be this
as it may, the classical geographers make specific references to
Chaldæans and Sabæans inhabiting the Oman Peninsula.   Pliny places
the Gens Chaldæi under the Eblitæ Mountains on the western side of
Musandam, and Ptolemy not only places the Asabi or Sabi (the prefix
being merely the assimilated article) along both coasts, but shows that
they gave their name to the entire area—*i.e.*, Promontorium Asabo, as
he styles Ras al Musandam, and this name is still preserved in the
Shihuh village names of Khasab and Sibi.
   Thus the geographical requirements of the story of Job would seem

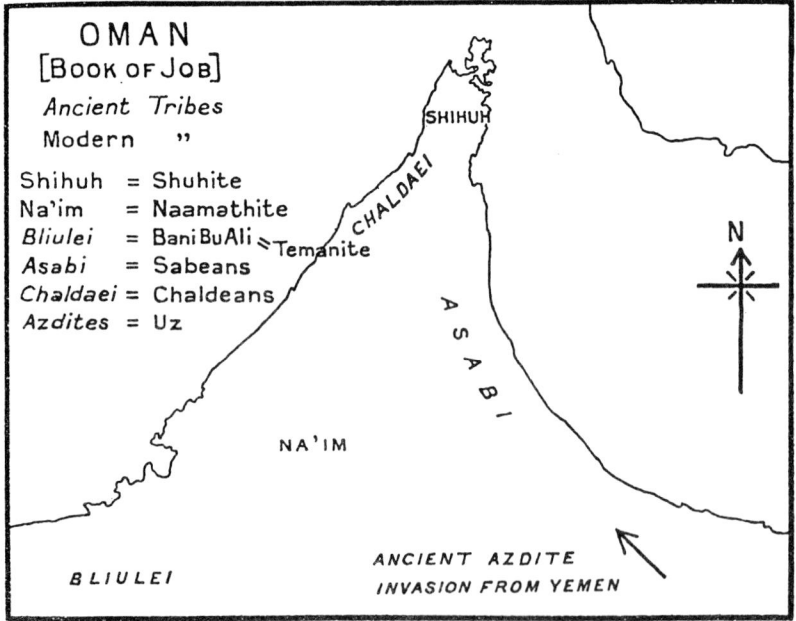

OMAN
[BOOK OF JOB]
Ancient Tribes
Modern   "

Shihuh = Shuhite
Na'im   = Naamathite
Bliulei   = Bani BuAli
Asabi   = Sabeans
Chaldaei = Chaldeans
Azdites = Uz

SHIHUH
CHALDAEI
Temanite
A S A B I
N
NA'IM
BLIULEI
ANCIENT AZDITE
INVASION FROM YEMEN

to be satisfied in an extraordinarily complete measure by tribal distribution in ancient and modern Oman.   The identification of Azd with
Uz is another remarkable coincidence if not more, but it is only fair to
add that the first historical reference I can trace of the Azdite migration into Oman is the invasion following the bursting of the dam of
Ma'rab.   Remembering, however, that Oman supplied the ships and
sailors for the ancient carrying trade between India and Babylonia,
and that from the evidence of its theology the Book of Job is held by
many to have been written not earlier than the time of the Babylonian
captivity (see " Cambridge Companion to the Bible," p. 59), it is not, I
think, unreasonable to suppose that it was to Oman that Job's author

came for his local colour. This theory would give the Shihuh a seventh century B.C. local antiquity.

The late Dr. Hogarth, taking part in a discussion in early 1927 following Sir Arnold Wilson's lecture to the Royal Geographical Society " A Periplus of the Persian Gulf," stated in connection with the people of Musandam promontory: "There is no more interesting survival; they are almost certainly a part of what is left of the pre-Semitic people of Arabia." I was able later to send to Dr. Hogarth an account of my slight researches, and I was to receive in Oman from him, alas! a posthumous letter on this same subject. The Shihuh is not racially a single unit; it comprises three elements physically and linguistically distinguishable and localized in their settlements of Musandam Peninsula. These are:

1. The Interior Mountain Badawin, belonging to one or other of two confederations—Bani Hadiyah and Bani Shatair.

2. The Kumazara.

3. The Dhuhuriyin.

For convenience I will take them in the reverse order:

1. *The Dhuhuriyin.*—It is doubtful whether the Dhuhuriyin should ethnologically be included as Shihuh at all. They are to be identified, I think, with the Darrhæ of Pliny, in ancient times a powerful and numerous Nizarite tribe of Dhahirah. They have a tradition that at one time they were masters of Musandam, and that the Shihuh dispossessed and absorbed them. Today their remnants number fewer than a thousand, all fishermen living in the coves of the Elphinstone and Malcolm Inlets (there are a few at Larek Island, who are sailors besides) and subject to one or other of the two rival Shihuh political factions. In summer there is a general exodus of them to Dibah, where they spend the hot months in their few poor garden possessions. Although subject in Musandam to the Shihuh, in Dibah they pay tribute not to the Shihuh Shaikh there, but to the Qawasim Shaikh. This may be a legacy of non-Shihuh origin, when they would naturally have been in "suff" with the Qawasim against the Shihuh, the hereditary enemy of them both. The Dhuhuriyin are a poor, spiritless people, inoffensive and amiable. Their heads—I measured six or seven at Sha'am and Dibah—though round and short in appearance, with close-cropped hair sometimes growing well down over their foreheads, proved to be unmistakably dolichocephalic. Of comparatively fair complexion, they lack the big bone of the Shihuh Badu. Arabic is their language, and they know no other (only the insignificant Larek element which has become mixed has acquired Kumzari). Except for one or two dialectical peculiarities—words generally pertaining to fishing—the Dhuhuri fisherman's Arabic is as pure and unaccented as any I have met in South-East Arabia.

2. *The Kumazara.*—The name derives from Kumzar, the promontory's

northernmost village, of which they are exclusive inhabitants. They have also a third share of the date groves of Khasab, where they are to be found in summer, but they are primarily fishermen. They are regarded throughout Oman as Shihuh, and they claim themselves to be Shihuh, a claim which is not questioned by their fellow-Shihuh tribesmen, over half of whom, indeed, in the south, they have established a complete ascendancy; for one of their Shaikhs habitually resides at Dibah, is the *de facto* Shaikh of the Bani Shatair confederation, and claims to be the paramount Shaikh of the entire Shihuh tribe. The Kumazara are physically peculiar in their lack of Semitic features characteristic in some degree of their fellow-tribesmen. It is they, and they alone, who speak the strange tongue which has baffled and confused strangers. Of this tongue I have made a vocabulary of some 400 words, and discovered its chief simple grammatical rules as a result of many visits during the past three years. It is a compound of Arabic and Persian, but is distinct from them both, and is intelligible neither to the Arab nor to the Persian nor yet to the bilinguist of both. Nor has it any resemblances, as suggested probable by S. Zwemer, to the Hamyaritic languages. I gathered from a local "Mutawwa" that some of its Persian words are archaic, deriving from the classical "Farsi" of remoter Persia rather than the spoken "Ajami" of the Persian Gulf seaboard, a point that may suggest a certain antiquity for it. Of my word list, less than 20 per cent. cannot be traced either to a Persian or Arabic origin; and in this connection Sir Arnold Wilson, in his "A Summary of Scientific Research in the Persian Gulf," has suggested a possible Sumerian origin at least for such words as pertain to the sea. If Sumerian words are looked for in support of a Persian Gulf origin for the Sumerians, no support for this theory can, I consider, be looked for in the physical features of its people. Word coincidences may be found, of course. One interesting one I encountered in South Arabia—namely, the Mahri and Shahari word for "mountain pass" is a slight modification of the word "ziggurat," which suggests the Babylonian temple tower—both dwelling-places of the gods. But this is a diversion.

The strange tongue used by the Kumazara elements of the Shihuh has given rise to the pardonable though, I think, erroneous belief that its users are aboriginal and pre-Semitic. They are, in my opinion, of Persian or some kindred South Asiatic origin.

3. *The Interior Badawin—Bani Hadiyah and Bani Shatair.*—These are the inhabitants of the mountains. They monopolize the interior and form, according to local estimates, a clear majority of the inhabitants of the Peninsula. All are shepherds. Men of big bone, Semitic in appearance, a handsomer type than either of their fellow-tribesmen Dhuhuri or Kumzari, they are, I think, the true Shihuh. Their only language is Arabic, but an Arabic quite distinct from that

of Oman, and they know not a word of the strange coastal dialect of the Kumazara. To a foreigner who has perhaps of necessity acquired Arabic from textbooks or at the feet of a literary townsman of Baghdad or Damascus, and comes with fixed ideas of what is Arabic and what is not, this dialect will at first be almost unintelligible, but that it is Arabic and no other language there can be little doubt. This is, of course, true of most Badawin speech, including that of the Awamir, distant neighbours, with whose supposed "language of grunts" I have had recent occasion to obtain a slight familiarity. But the Arabic of the mountain Shihuh has many peculiarities which are definitely suggestive of foreign influence: a detailed analysis is given in Dr. Jayakar's "The Shahee Dialect of Arabia." How such influences could have penetrated to the remote insular Shihuh it is difficult to understand, bearing in mind that the Kumazara elements, probably Persian in origin, and the Dhuhuriyin, who, living along the coasts, are therefore exposed to foreign influences, speak an Arabic which is comparatively free from such contamination. Whether it can be traced to the ancient Persian occupation of the Arabian seaboard is an interesting speculation. That influence must have been considerable, existing at the time of the Great Cyrus in the sixth century B.C. and lasting intermittently for more than a thousand years up to the rise of Islam. At the same time, such a view does not receive support from those parts of Oman where Persia seems to have dug herself in —witness the localities of her great "falaj" systems—for here no vestige of Persian influence now colours the locally spoken colloquial Arabic. Is it possible that the exponents of this Musandam dialect are descendants of the original Persian colonists of Ma'zun (the pre-Islam name given to the area corresponding probably with the territory lying between Sahar and Rostaq and perhaps the Magan, from whence the Sumerians got their copper), and were driven north before the Qahtani invasion of the second century to inaccessible Musandam, where, unmolested, they have preserved the frills of their ancient speech?

Against this otherwise plausible theory there would appear to be three possible objections: (1) Their Hinawayih partisanship, (2) their definitely Semitic type of countenance (their heads—I measured two —gave a dolichocephalic index), and (3) the "bait al jahl," a circular tomb (?) of black lava pebbles which, as I have observed elsewhere, is the most typical and common archæological feature of Ma'zun, is unknown in Musandam. What of Job's Shuhite? Are the Shihuh a survival of Israel, and are their peculiarities of pronunciation of Hebrew and not of Persian derivation? Jews in Oman there have been throughout historic time—the last of them is recorded at Sahar within living memory—and it is not impossible that their original home was the Persian Gulf. In the Dhahirah province there is a universal tradition that its earliest known people were "Yahud," and today two tribes

460

found there, the Bani Kalban and Al Yaqib, the latter a powerful
settled element possessing Ibri, the first town of commercial importance
in the province, are by a unanimity of Omani opinion descendants
of Bani Israel. Incidentally the Persians are supposed never to
have penetrated to the Dhahirah where the falaj systems enjoy the
exclusive name of Da'udiyat—i.e., of David. These considerations
would account for Hebrew survivals in Oman.

Does this bring light to the problem of the mountain Shihuh?

The Shihuh composing these three diverse elements are split up
into two rival political factions :

1. The Bani Hadiyah, comprising the confederation of that
name in the northern end of the Peninsula, and perhaps half of the
Dhuhuiryin under a Bani Hadiyah coastal Shaikh with headquarters at
Khasab.

2. The Kumazara, consisting not only of the purely Kumazara
elements but of the Badu confederation of Bani Shatair to the south,
and the remaining Dhuhuriyin, under a Kumzari Shaikh with headquarters at Dibah.

The former shaikhship is hereditary, the latter elective—elective not
by the elements under its jurisdiction, but by the ruling families of the
distant fishing village of Kumzar. The two factions, Bani Hadiyah and
Kumazara, normally hostile to one another, unite in the face of a
common danger—e.g., the Qawasim. In politics they call themselves
Hinawi (except the Dhuhuriyin, who make a secret boast of their
Ghafiri origin). These two political labels, Hinawi and Ghafiri, are, in
practice, though not in theory, the Omani equivalents of Yemeni and
Nizari, which in turn are the two great Abrahamic families of Qahtan—
i.e., Jokhtan of Genesis and Adnan (descendant of Ishmael) respectively—into which the tribes of Arabia divide themselves. The word
Shihuh is believed, both by Bani Hadiyah and Kumazara, to be patronymic, that the tribal progenitor was Shih, son of Malik bin Faham,
the Azdite leader of the Yemeni invasion of Oman, probably in the
second century A.D. If this genealogical claim were established it
would vitiate the theory of an earlier Shihuh occupation of the promontory than the century in question.

The claim, however, is not borne out either by historical record or
tribal tradition of neighbours. Thus, " S'abaikh adh Dhahabi," the
work on Arab tribal genealogies by Shahab ad Din al Abbas, makes no
mention of one Shihi amongst the sons of Malik bin Faham. Their
internal accounts are also conflicting, for the Bani Hadiyah element,
while conceding that the Kumazara are Shihuh, do not admit a common
ancestor. I gathered from a shaikhly informant of the Qawasim tribe,
their more erudite neighbours, that the Malik from whom the Arab
Shihuh spring was, according to Al Siyuti, one bin Hazam, not a
Yemeni at all, but a Nizari. This Nizarite origin would, I suggest, be

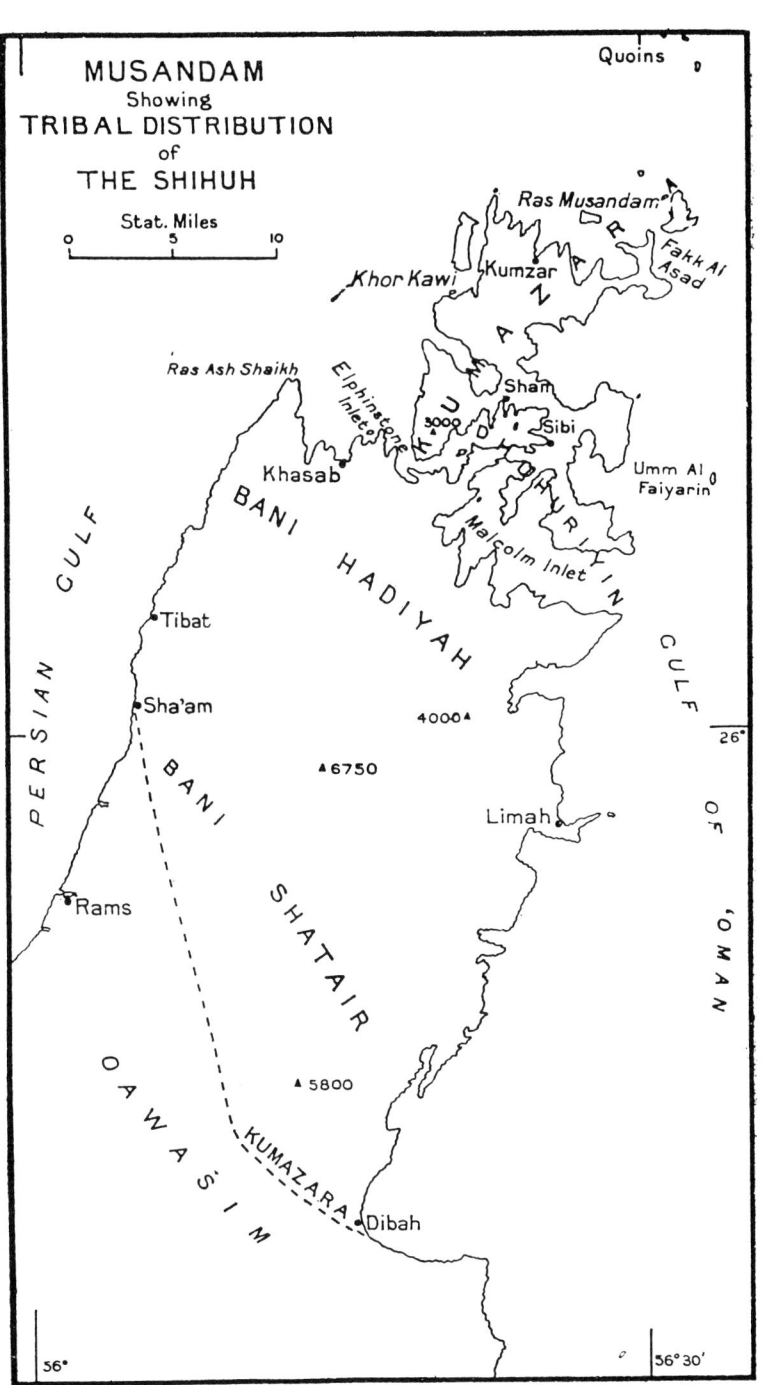

Map of Musandam, 1929 (Thomas)

more in accordance with their remote location, and some of their linguistic variations from the rest of the more or less uniform-speaking Hinawi tribes of Oman.

The traditional feud, which normally lies dormant, between Hinawi (Yemeni) and Ghafiri (Nizari) loses none of its pristine fervour in Oman when kindled into flame—Ghafiri tribes range themselves with the Ghafiri disputant and Hinawi with Hinawi. To the Shihuh the word Ghafiri is anathema. It arouses, I suggest, the especial zeal of the proselyte or the renegade. Political expediency has accounted for tribal coat-changing in the past, and such a notable tribe as the Bani Yas, today avowedly Hinawi, belongs historically and ethnographically to the opposite side. It would seem to me that Shihuh Hinawi partisanship is of this order, and is to be accounted for by Ghafiri persuasion of their neighbouring hereditary enemy—the Qawasim. This change would, in some remote past, have been facilitated by adhesion to the Shihuh of Kumazara elements, which, if as I believe, are of non-Arabian origin, would have had no racial claim to, or sentiment for, either denomination, and would have been guided purely by considerations of expediency. Failure to claim either would have invited degradation to the status of Baluchis, who in Arab countries of the Persian Gulf are sometimes to be found as slaves. Consideration of these factors inclines one to the belief that the Shihuh—exclusive of the Kumazara—are of Nizari origin and of great local antiquity. But today the Kumazara are an integral part of the Shihuh, and all call themselves Hinawi.

The Shihuh are primarily shepherds in the interior, fishermen, boat-builders, and to a limited extent date-gardeners along the coast. They are by far the most primitive tribe in Oman, the most superstitious, and, as regards the mountain folk, the most difficult to cultivate friendly relations with. Nowhere in the interior is water sufficiently copious or localized to support large village settlements or allow cultivation beyond a strip of wheat and barley sufficient for local needs. Inaccessible mountains forbid any animal transport whatever, except for a few hours' trek up the wadi behind Khasab. There is, it is true, a camel route along the southern boundary between Dibah and Rams, but strictly speaking it is, I think, in Qawasim territory. The interior mountain folds, where they form natural catchment areas, support acacia jungle. Here the horse is unknown, the camel less met with than in any other part of Oman; even the hardy ass can scarcely cope with the mountain conditions, and man lives on his forbidding hill-tops with a few herds of nimble goats. The traveller who wishes to penetrate this country must do so on his flat feet, and be prepared to climb 3,000 feet to reach small inhospitable settlements where water is scarce, food is unobtainable, and he is unwelcome. But he will probably be dissuaded by a coastal shaikh from hazarding the perils of the journey. Few, indeed, desire to share with the Shihi his mountain secrets. The Indian trader who for hundreds of years has established himself amongst the Arabs of Trucial Oman, or in the Sultanate, has never obtained a footing in Ras Musandam. The chief port of the Shihuh is Rams, probably the Regama of Ptolemy, which is not now Shihuh territory, though preponderantly Shihuh in inhabitants, but belongs to the Qawasim. Through this port the simple wants of the Badawin find their way from Dibai—rice, coffee, sugar, tea, piece goods, and spices. Ports next in importance and belonging to the Shihuh are Dibah, Khasab, and Lima.

The mountain Shihi is often a troglodyte. Natural caves exist in the mountains and provide him and his flocks with shelter. Besides this, he quarries and builds an underground house called "Bait al Qufl." Here are his winter living quarters, and here his few worldly possessions are kept locked up throughout the year. First a pit is dug twelve to fourteen feet deep, some fifteen feet in length and twelve in breadth. Side walls are built of stout stone, and the whole is roofed over first with timbers of the "samr" and then with a covering of loose earth or sand. All that is visible above ground to the approaching stranger is a foot or two of roof. There is no ventilation except for a small hole in the roof, a condition of things which does not, however, prevent him from having his fire there in winter. The dark interior has to be lighted by a lantern. Entrance is by means of a doorway at the side and thence down a stairway. This doorway, made exceptionally strong, opens inward to a long key. A feature of the Bait al Qufl, as this laborious creation is called, is the enormous "kharas" or water tank, much too big to have been introduced through the doorway, a large piece of pottery capable of accommodating the contents of fifty water-skins—a three months' supply, replenished during the rains— which went into the house with its foundations. Ancient graves exist in some parts of the mountains, which are said to have an orientation towards Jerusalem and not Mecca.

The superstitions of the Shihuh Badawin are notorious, but though they build no mosque in the mountains and their knowledge of religion may not be profound, they are Muslims. Palgrave was mistaken in his statement that all here are staunch Ibadhis. Ibadhism does not exist in Oman much north of Sahar, nor has it ever flourished in Musandam. The Shihuh are avowed Shafis. This does not preclude their belief in jinns, janns, afarit in general, and umm subiyan in particular. The first named, despite diminutive corporeal proportions, keep the Shihiyin in their houses after dark; janns are monstrous creatures to be distinguished by the light that shines from their foreheads; the afarit have the terrifying faculty of going upon their bellies; umm subiyan is a woman of ordinary appearance save that she casts no shadow. A she-devil, she walks in rags amidst waste places carrying a basket, and her mission seems to be to cause miscarriages to pregnant women. The Shihuh set much store by visits to shrines, those at Sha'am and Aida and Ras ash Shaikh being the most famous. At Ras ash Shaikh—Shaikh Mas'ud, to give the saint his earthly name —not many years ago there landed from a ship some foreign workmen in search of stone to build the Quoins lighthouse. Seeing a suitable specimen close to the shrine, they attempted to carry it away, but found it too heavy. My local informant, whose faith was strong, told me that, surprised though they were by their failure—the stone was not a big one—the truth never occurred to them that it was the saint who was thwarting them. Here the sterile woman comes to intercede for a son, and the sick man for his recovery. But the "Mazar," as the shrine is called, is not the only means of combating misfortune or disease. There is a widespread belief in the curative effects of Quranic Holy Writ. Thus a sick man will come to the mulla for a sacred script, and this he wears as an amulet. For a snake-bite the remedy is a reading from the Quran, and the patient is not permitted to sleep. Sometimes the mulla elects to read over a glass of water, which is then taken to the invalid to drink. The relative of the man sick unto death will be found making what to the unsophisticated may appear an irreverent business proposition to Providence, in that he publicly announces that, conditional upon the invalid's restoration, he will kill the fatted calf and make a feast for the poor.

For some inexplicable reason among Shihuh Badawin the cow is anathema. They will neither eat its flesh nor drink its milk. In every other part of South-East Arabia the bull is met with and respected. Well fed, he labours unflaggingly yoked to the leathern hoist which he lowers into the well and hauls up to refresh the thirsty gardens, while his wives and family are met with browsing peacefully in the country, or singly drawing the plough, or in due time crowning the banquet. Here, in interior Musandam, the mountainous nature of the country may in part be responsible for discrediting the worthy beast, but cannot be wholly so. Be that as it may, a hungry Shihi from the interior is alleged to flee from the feast where beef is exposed. In the hills the plough is drawn by women, but Shihi women are not otherwise degraded; they all wear the veil.

Another peculiarity of the Shihi Badu and fisherman alike is the "nadabah" or "kubkūb." This is a kind of tribal war cry, the use of which, however, is not restricted to martial occasions, but is freely indulged at all times of rejoicing, a feast, a marriage, or a circumcision. My first experience of it was at Fujairah, where the Dibah Shihuh were guests of their ally, the local shaikh. Following an Arab feast of princely proportions, the Shihuh shaikh rose, and, standing a little aside with hands still unwashed—this, I gathered, was a necessary observance—placed his left arm across his chest and his right arm bent above and behind his head; then, straightening and bending the raised arm, he set up a curious howl not entirely unmusical, ascending and descending the scale over a compass of perhaps nearly an octave. This noise, which may best be likened to a vocal imitation of a swanny whistle pitched two octaves lower, was carried on for some few minutes. Meantime, a dozen or so of his tribesmen standing close together in a ring about him with their hands to their mouths, "muedhdhin" fashion, broke in at intervals with a curious, sharp, dog-like bark. I could detect no words, but was told that " Shih al Mahyub "—*i.e.*, " Fearsome Shihi "—was the formula used preparatory to the "howl." This " nadabah," as it is called, characteristic of the Shihuh tribe alone, is common alike to the mountain Badu and the Kumzari fisherman. A sword dance is said sometimes to accompany it, but the sword dancing of the Shihuh is disappointing to one who has seen the set order and combination of such displays in Oman. The Shihi performance is entirely individual, every performer careering around with trembling blade, oblivious of everybody else, in much the same uncouth manner as practised at Murbat in South Arabia. In Musandam music is at a discount. No virtuous woman would raise her voice in song ; the drum is tolerated, but limited to "Ids"; the simple " rababa " and honest " zimmur," beloved by Transjordan tribes, are abjured ; while the lute of Muscat is bracketed with the vice of great towns and is only to be mentioned in a whisper. Tribal law has here its own orientation. The almost universal Arabian law of asylum, or " dakhala," has but a nominal significance. Thus, if a Shihi arrives upon a strange scene in pursuit of a refugee from some blood feud and finds " his man " the guest of other Shihiyin, he has no compunction about killing him on the spot. Such a course would cost him his life in any other part of Oman. To attempt it would be an affront to his hosts, who would be in honour bound to protect the refugee, or, in the event of his blood being shed on their soil, to avenge themselves on the newcomer. But it is almost unthinkable for such an incident to arise in Oman or elsewhere in tribal Arabia, where the law of sanctuary is sacred. Such are the customs and behaviour of the Shihuh of the mountains. Wild and untamable men, they are generally well armed with modern rifles, but will be met carrying the spear or straight, double-bladed sword for preference and the inevitable Shihi club the " yurz," a straight stick with a small iron axe-head.

But even the remote Shihuh are not exempt from the summer migratory habit to which the entire inhabitants of South-East Arabia are subject. The pearl banks, of course, attract a Shihuh quota, though incomparably smaller and less representative than the Batinah coast and Hajar contingents. The wives and families of all divers, including those who habitually live along the Trucial coast, come to spend the season amid the date groves of the Oman Sultanate. In summer the whole world gravitates to the date grove. The Kumazara desolate their fishing village in favour of Khasab. The Dhuhuriyin of the Elphinstone

Inlet betake themselves in a body to Dibah. The Shihuh Badawin in large numbers move down to Rams, Khasab, Lima, and Dibah. It is remarkable how strong is the attraction to the Arab of the " ruttub "— the ripening date. In spite of the fierce heat and scorching "simum" wind, here called " gharbi," that are necessary to the most satisfactory condition of the ripening fruit, the date grove makes an irresistible appeal at this time. Entire populations of villages and towns move out to " qaiyidh," as it is called. I have ventured an " Al hamdu l'illah " when, after a scorching day, a cool moist wind has sprung up in the July evening and just made life tolerable ; but the response showed few sympathisers. Indeed, the cool wind is deplored for its effect on the quality, not only of the " ruttub," but of the entire date crop—i.e., the whole of next year's supply. Only the wind with the breath of the furnace is welcomed at this season.

And so the traveller coasting northwards along the steep shore from Sha'am past the coves of Tibat and Bakha will find in summer little of human interest. Palgrave pictured it under a winter sky in his delightful and inimitable way—viz., " Now the granite wall went sheer down into the blue ocean ; now it spread out into cliffs down which winter torrents ran and where little villages niched themselves like eagles' nests ; close by them patches of green sprinkled on the mountain ledges." Thence he rounds Ras ash Shaikh and enters Khasab Bay. Here is a fine natural harbour five miles long and three miles across surrounded by sharp and naked mountains. The cove itself, in which the village stands, is about three-quarters of a mile wide and long From the sea at high water Khasab presents the appearance of a palm grove skirting the water's edge with a fort midway along it. But when the tide ebbs, a normal fall of six and a half feet, this scene is fronted by a thousand yards of gently shelving beach. The village for the most part lies tucked away behind the grove in a " bahta," which debouches on the west side of the harbour, though the temporary reed huts of summer visitors, chiefly from Dibai, are conspicuous amidst the gardens and along the front. The fort, now ramshackle with age, is the residence of the Wali, a Sultanate nominee. But the ignorant and uncouth Shihuh, never very amenable to authority, are here virtually masters in their own house, the revenue they pay being scarcely sufficient to meet the cost of the Wali and his garrison. The fort itself is an early seventeenth-century Portuguese erection consisting of a square of four bastions connected by curtain walls. It is chiefly famous for its employment in 1624 as a base by the Portuguese Admiral, Ruy Frere da Andrada, in the attempted recapture of Hormuz, which lies across the Straits fifty-five miles away on a bearing a little to the east of north. In Khasab the cow comes into its own, as the creaking rig of the wells testifies, for the villagers do not share the prejudices of their Badawin brethren of the mountains. Here in the date gardens an unsavoury method of fertilizing the soil is met with. The "haudh" or well reservoir is packed to the brim with metoot, the local sardine, which is caught in vast quantities, and for the most part is exported to the Trucial coast and Bahrain for cattle fodder. Here in the Khasab "haudh" the freshly caught metoot is allowed to rot under a three days' scorching sun without more effective disguise or discouragement than a covering of old sacking. The well is then set going, and the water, before flowing into the irrigation ducts which cross and recross the garden, is first washed through this fish-pit. The effect on the surrounding air may be left to the imagination. In the harbour, conspicuous amongst native craft of all kinds, is the small but picturesque "batil," recalling somewhat in the beauty of its lines the Venetian gondola. Combining a "badan" stern, characteristic of the small Batinah-built coasting craft, with a "bum" bow, the feature of the Kuwait-built "dhow," this small fishing boat is built here at Khasab and at Kumzar of teak wood imported from Malabar. At one season of the year when the harbour becomes lively with the caperings of a kind of tunny, the fisherman, standing in the bow of his batil, will be found employing curious measures. His coiled hand-line carries a hook baited with nothing more than a cotton wad, sometimes with two side-feathers to give it the appearance of a "Jarad al bahar"—i.e., the flying-fish, here called "locust of the sea." This line is flung out lasso-fashion to its greatest length and is hauled in rapidly hand over hand to the undoing of the tunny. Most of the fish of these waters after satisfying local needs is salted and sent to Linga, from whence it is exported to India. Sharks, which are numerous, are invariably exported, because the Shihuh, unlike their fellow-Omanis, have little fancy for shark meat. I am inclined to sympathize with them, having myself dined off young shark on some occasions. Other exports are firewood, charcoal, and sheep, chiefly to the pearl coast at the end of the diving season. A feature of Khasab harbour is the extraordinary purity of its light. The scenery, grand in itself, seems here to take on some ethereal quality. Stratified cliffs reflect great coloured patches in the clear still water, and the high hills are often lit up early morning and late afternoon in rose-coloured splendour. On the east side of the harbour and to the north is Elphinstone Inlet, a remarkably fine fiord seven miles long, with Ras Sham, a sentinel mountain of 3,000 feet, marking its otherwise deceptive entrance. Throughout its serpentine course mountain ridges, scarcely less imposing than Jabal Sham, flank it on both sides. On their desolate slopes appears here and there an incongruous tree, or a flock of goats may be seen swarming along precarious ledges a thousand feet above a Dhuhuri village. But for such slight indications of life the visitor will find the inlet desolate in June. A reputation of being the hottest place in the world suggests that it was of Elphinstone rather than of Muscat that the Persian poet quoted by Bent wrote, " It gives to the panting sinner a lively antici-pation of his future destiny."

Telegraph Island Buildings, abandoned since 1867, still stand, and the shade temperatures taken there on the forenoon of June 30 last year read : Dry bulb, 96·8; wet bulb, 84·4. I landed on the opposite shore on the same day at the village of Sham to examine a Dhuhuri village. The houses, if they deserve the name, clustering about the beach, were all deserted for the summer ; only one old man and his son had been left behind in charge of the village flocks. The Dhuhuri house is a tiny rude stone skeleton affair, a single room, ten feet by six feet, and four feet high, the stones locally quarried, and just sufficiently dressed to require no cement or similar binding. A doorway is formed by a gap in the long wall, and there is no ceiling. The only interior decoration is a small mud fireplace for the coffee dallal. On the fore-shore at Sham is a mosque in more ambitious style and proportions. Beyond Elphinstone there is no variation in the wild and rugged shore until Kumzar Cove is reached. Here, shut in on three sides at the base of bleak, precipitous, windswept hills is the mysterious village already mentioned. From seawards—there is no land approach—Kumzar presents a line of a hundred fishing batils dismasted and drawn up with singular uniformity along the yellow sandy beach, and behind and above them a mass of closely crowded, tiny, flat-roofed mud or rude stone houses. The narrow lanes between teem with fisherfolk who all use the strange language to which they have given their name. They live principally on fish. Their fish diet and prolific birth-rate are facts which they imagine, unscientifically I suppose, to be in some way related. The Kumazara bury their dead in their own houses, under their living-rooms. True, there is a small graveyard in the middle of the village which must have been full to overflowing long centuries ago, but the villagers have never turned elsewhere as they easily could have done to bury their dead. This practice of house burial, observed in ancient times in South Persia at Anau (see the " Cambridge Ancient History," i., 87, and therein ascribed to Dravidians or some south Asiatic race), has not, I think, been observed elsewhere in Arabia. Is this another clue to their mysterious origin? Kumzar possesses one copious well under the cliffs at the back of the village, the scene of a constant stream of fishwives. A legend attaches to it that whenever a dog shall bark in the village the well will dry up, and dogs are in consequence rigorously excluded. An interesting marriage custom is that the bridegroom on the night of the nuptials is brought down by his friends to the edge of the sea and there must immerse his right foot to the ankle. He is then conducted back to his house to consummate the marriage. A feature of the dress of the coastal Shihi —I have never seen it affected by other Arabs, though it is common in India, of course, and less so in the Dhufar province of South Arabia—

is the spare, neatly folded "wazar" (skirt) which is carried thrown over one shoulder : and he generally carries in his hand the "yurz" or else a stout straight walking-stick. Unlike his fellow-Shihi, the Kumzari will occasionally marry with freed slave blood ; the interior mountain Shihuh, on the other hand, are almost unique among Badawin in possessing no slaves at all.

And thence the traveller passes on through the troubled waters off Ras Musandam. Soon after changing course south, opens the treacherous passage between the mainland and Musandam Island—a strait named in charts "Fakk al Asad," but known to local Arabs as "Al Bab" or "Lahio" (Kumzari dialect). The shallow waters of the Persian Gulf here give place to the deepening ones of the Gulf of Oman. Soon comes the Island of Umm al Faiyarin, standing stark 360 feet in height, four and a half miles off the shore. This island is known to the Kumazara as Ko Kaig, or the egg mountain, in reference to its popularity with sea-birds in the nesting season. And so on to Dibah, the limit of Shihuh territory—a village set in a considerable date grove along a magnificent sweep of bay, with the blue Hajar for the first time receding slightly from the coast. An important port today, Dibah belongs in part to the Qawasim and in part to the Shihuh. Here is probably the mart which Alexander the Great's admiral Nearchus noted, here is indubitably Pliny's Dabanegoris Regio, and here was fought during the Caliphate of Abu Bakr the battle (A.D. 633) which gave Oman to Islam.

Musandam Today

Subsequent writings on Musandam include Walter Dostal's 'The Shihuh of Northern Oman: a contribution to cultural ecology', and two papers on the Musandam Expedition 1971–72, all published in *The Geographical Journal* (respectively March 1972, February 1973, and October 1973).

The Musandam Development Committee was founded in 1976, the Khasab Hotel opened to guests in 1981, and the 2 km-long runway and airport began functioning in 1982, replacing a gravel strip of 600 m.

The logistical problems are immense; it is not only that by a freak of political geography Musandam is separated from the rest of Oman by the United Arab Emirates, but that the Ru'us al-Jibal rise almost everywhere so steeply from the sea that access even by boat is hazardous in the extreme, except at a few isolated bays such as that of Khasab, which has become the administrative centre of the zone.

Strategically, Musandam is situated at a sensitive narrowing of the Gulf, close to the Iranian shore and offshore islands. Shipping, such as international oil tankers, must be protected and a new radar post is currently in course of erection at the highest point in Musandam to strengthen security.

Access to Musandam by road (Dubai-Sharjah-Ras al-Khaimah-Bukha-Khasab, or Muscat-Shinas-U.A.E.-Bayah-Khasab), depends on possession of a multiple exit and re-entry visa which must be obtained beforehand. Oman Aviation Services operates a five-day-a-week link from Seeb to Khasab, sometimes via Limah or Bayah. The local people have always depended on their precarious boats to move from coastal village to village; the Shihuh of the interior by contrast travel from inland village to village on foot and with pack animals such as donkeys, as many of the tracks used to be too steep for camels.

A costly roadbuilding programme opened a graded road first from Bukha and the border post at Tibat (U.A.E. and Omani immigration and customs); the latest map I have shows no road between Tibat and Ras al-Khaimah, but as a matter of fact one can reach Dubai in less than three hours. Later in 1982, a graded road was hacked out of the almost perpendicular mountains, enabling a motorist with four-wheel-drive (and the inevitable exit and re-entry visa) to pass the Royal Oman Police checkpoint at Wadi Salhat, Bayah, Khor Fakkan, Fujairah, and Suhar, reaching the Capital Area in about seven hours of hard driving. The latest spectacular achievements in road engineering are the hairpin track along the cliffs to Khor Naid, on the east coast, and the new tracks giving site access to the dams of Wadi Shariyah and the Wadi Mawa (protecting the airport from flooding) and Wadi Khasab farther south; many other graded roads are planned according to a list of priorities.

Round the Coast to Bukha

My flight from Seeb to Khasab departed at 6.45 a.m., in a Fokker Friendship with 17 passengers and 2 crew. It followed the shoreline of Oman, with the familiar pattern of sea, sandy beaches, villages with date plantations and cultivated fields, and the coast road of the Batinah leading straight as honesty to Suhar and the border point of Khatmat Malahah. By 7.50 the plane was overflying the Gulf, where tankers were ferrying from terminals on both shores some of the oil to lubricate world industry: more than 50% of all oil exports to North America, Europe and Japan passes through the Strait of Hurmuz, sixty kilometres wide.

Bayah could be made out in the clear morning air far below us, a tiny green patch indistinguishable from its close neighbours, Dibba Hisn (belonging to Sharjah) and Dibba Muhallab (belonging to Fujairah). At 8.15 I sighted Limah on the left, nestling between rocks and sea, and then the awe-inspiring jagged mountain-tops appeared to rise to meet us as we began the descent to Khasab, landing at 8.25. Most of my fellow-passengers were administrators working for Musandam Development Committee, or contractors with one of the many firms involved in the creation of a new Musandam, with an infrastructure of electricity, roads, water-supply, communications and consumer goods that might seem to threaten the future of traditional Musandam. But the quality of life in the interior is hardly likely to be affected: harvests must always be limited by the pocket-handkerchief size plots, dug with tenacity in the few level patches between barren rocks. And by the climate of baking suns and heavy storms.

Without a multiple-entry visa for Oman, or an entry visa for the United Arab Emirates, my journey around the west coast road would have to terminate at Tibat. To go there, I had hired in advance a four-wheel-drive Toyota pick-up from the Musandam Development Committee. After a quick shower and changing into my toughest shirt and jeans at the Khasab Hotel (which can be booked through MDC in Ruwi), I drove out of Khasab, which Bertram Thomas has described, and in 8 km encountered the tiny village of Qidah, in the sandy cove at the head of Khor Mukhi. Mukhi itself is a characteristic village of the Musandam coast, sheltering from the excesses of tempest, sun, and pounding sea below rugged cliffs. So few travellers have ever been able to drive along this road (since it has never been opened to tourism, and is only very recently traversable by saloon cars) that it remains virtually unknown. Yet in its way it rivals the

Al-Harf, hilltop village on the Khasab-Bukha road

spectacular drive from Amalfi to Sorrento, or from Jiddah to Taif. Battling between overhanging cliffs, with landslides perennially imminent, and sea-spray so effervescently insistent that on even the hottest day you have to keep your seaward windows closed to avoid drowning or at least drenching, you enter the elemental world that Hemingway evoked in *The Old Man and the Sea*, and Melville in *Moby Dick*. The waters of the Gulf threaten to carry you away in winter storms, and although the roadmakers have seemingly protected the narrow track with stout walls and huge rocks to withstand hammering, one knows that sooner or later the roaring Gulf waters will win.

The boats at Mukhi could not, surely, stand a chance in the high tides, but Omani seamen have engaged the Gulf for millennia, and are equally determined now. A cold store and fish-freezing unit have been built, enabling local fishermen to sell their catch locally, without the need to travel beyond Musandam, at standard prices assuring adequate payment independent of market fluctuations.

The sea glittered green from the vantage-point of Hana village, and in a few minutes I entered al-Harf, a cluster of houses atop the mountain saddle, surprising as a cock's egg.

Then down again to the coast perilous, steep rocks to landward bearing the great scars of mechanical incision. Rubble has been piled along the dirt track to seaward. I could not hear myself shout as the fearsome waves battered and roared. Mile after mile of winding track, rounding rock after rock, led to the village of al-Jadi, with modern houses that clearly owed much to the roadbuilding programme. No longer accessible only by boat, al-Jadi is now within easy reach of Khasab, and even easier reach of Bukha, a pleasant village of 571 people, the largest settlement in Musandam outside Khasab (3,267), Bayah (2,226), Kumzar (1,174) and Mudha (1,038).

Bukha's main attraction is the fort commanding the beach, but I found another fort on a hill 2 km beyond this point. There is a Shell petrol station, a football pitch, and regular daily spraying and refuse collection. Power plants were built here in 1976, as at Khasab and Bayah, and the distribution was extended in 1980 throughout the west coast, a total of 457 electrical connections. The anticipated drift of population away from Musandam to the Capital, familiar elsewhere in the Sultanate, has been halted at 11,153 (1984 figure) due to the rapid expansion of development-based manpower, linked with the efficient spread of adequate fresh water supplies by regular boats, and subsidies for fishermen and farmers alike. I took tea at the Sea Coast Restaurant with a homesick waiter from Kerala who regaled me with extra cups of tea, sticky cakes, and tales of his large and demanding family back home. 'The more they ask you to send them home, the more you love them,' he nodded, his black eyes becoming bright with tears that he would not allow to fall. There was no other person in the Sea Coast Restaurant: he was the only cook and waiter; I was the only customer and guest. We shook hands on parting. I could not bring myself to say 'Look around you and rejoice

Bukha. Fort

in the wedge-tailed shearwaters and Wilson's storm petrels, the salt spray flicking, gleaming into the air, the fine yellow unpolluted sand, with its angular crabs, the air warm and soft as eiderdown.' His eyes were fixed on a wife who longed for his return, and children part bereft, part proud, who would look in amazement at the money transmitted from a father they could scarcely remember.

The latest official figures available for citizens and migrants in Oman appear in *Abhath wa dirasat nadwa al-istikhdam al-amthal al-qawa al-'amala al-wataniya* (Ministry of Labour and Social Affairs, 1984), showing an estimated population of 900,000, 70.6% being Omani citizens (635,000) and 29.4% expatriates (265,000). The workforce consists of 45.7% Omani citizens (116,500), and 54.3% (137,200) expatriates. Of the total number of expatriates, 9,200 are Arab migrant workers, 127,500 are non-Arab migrant workers (chiefly from the Indian subcontinent), and 128,000 represent the balance of the non-national population.

Omanis too migrate to find work: 10,000 in Saudi Arabia and 33,450 in the other Gulf Co-operation Council States (Bahrain, Kuwait, Qatar, and the U.A.E.), according to Birks and Sinclair in *International Migration in the Arab World* (Beirut, 1982).

I motored on past Fudgah and Ghumdah to the customs checkpoint at Tibat; veering left before it to explore for the first time that hinterland so mysteriously connected in my imagination with 'the avaricious' (for that is apparently how the Shihuh got their name), the silent people of the interior who abominate cows, almost as much as they distrust outsiders. In the winter they tend to live in stone houses in a *bulaidah* or *harah*, between about November and April.

474

A *bulaidah* is a settlement comprising just one extended family, while a *harah* has several, each comprising a quarter. Winter quarters have terraced fields, mill-houses, water cisterns, and stone granaries. In the summer they live in a *musaif*, characterised by *barasti* huts made of palm-fronds, stone houses built to let the wind through, date-palm gardens, and *aflaj* for irrigation. Low rainfall, leaving their cisterns empty, forces the Shihuh down towards the coast in the summer, and they migrate to their date-palms, for their barley and wheat crops are unreliable and generally insufficient.

The drive back to Khasab was uneventful in that I managed to avoid punctures and other catastrophes, and once back in the capital of Musandam I took lunch at an Indian restaurant named 'Musandom'. The morning newspaper in Arabic reported heavy snow all over Western Europe, while I was enjoying spicy fried fish with chapatis in a thriving town bathed in brilliant sunshine, a sea-breeze fluttering the vertical strips of plastic that did service as a door. I paid my bill at the counter and sauntered out to the Toyota pick-up, whistling 'Stranger in Paradise', which for some reason came into my head at that moment.

Across the Mountains to Bayah

The road up the wadi runs parallel with the airstrip then veers left into Wadi Khasab. Eleven km outside the town a turn off right is signposted to Mahas, then a sudden 'Mountainous Road' warning brings a smile recognising understatement. I had seen some of these 'roads' from the air and they seemed about as adventurous as driving straight up Kanchenjunga. Indeed, changing gear down to second,

Isolated oasis near the mountain road Khasab-Bayah

Wadi Salhat from the Khasab-Bayah road

then down to first to rise above the clouds, I positively welcomed my
slow progress, as it gave me a chance to stare in astonishment from
side to side at this miraculous feat of road engineering, made possible
by generous oil revenues. People scrambled in the back of the
pick-up as I slowed down, then signified their wish to get off by
banging on the roof. I have canoed along the rivers of Borneo,
trekked in the mountains of New Guinea, and explored the Maya
ruins in the jungles of Yucatán, but I never imagined that one day I
should operate a free, unscheduled bus service among the cloud-
tracks of Musandam. Down in a wadi 6 km from the warning,
bursting into a high fertile valley out of James Hilton's *Lost Horizon*
9 km later, and after 15 km more down again into Wadi Salhat, past a
wrecked vehicle that never made it thus far, I stopped at the Royal
Oman Police checkpoint to talk to two young policemen sitting in
their four-wheel-drive vehicle on the lookout for whom? Illegal
emigrants? Not many people come down this road to Bayah, and they
were glad of conversation. No, they did not find the scenery
remarkable; all mountains look alike (they were from Nizwa and
Bahla in Jabal Akhdhar). No, they didn't miss their families because
they were still unmarried, but they hoped to marry local Jabal
Akhdhar girls soon. Shihuh girls? No, they laughed, the Shihuh only
marry among themselves, they are not true Omanis like the rest. One
of them tried to stamp on a lizard, calling it 'Shaitan!' (the Devil) but
I managed to intervene and save the creature's life by a kind of mad
show. After all, all Englishmen are mad, aren't they? So I never got
to Bayah: I turned round and saw the lavish spectacle of Ru'us
al-Jibal, drawing my breath in awe at every turn, rolling away
boulders freshly fallen since the last vehicle had passed, greeting

Shihuh as they turned from tending their goats, revelling in an afternoon burnished with sunlight, darkened with shadows on the dolomitic limestone twisted, faulted, and folded as if a crazy giant had played with clay until it had assumed the most fearsome shapes, then baked it for millions of years in the hottest of all ovens. The few terraced fields, insolently green in contorted aridity, are overlooked by Jabal al-Harim, the tallest peak in Musandam at 6,848 feet. The army engineers even plan to bulldoze a track up here, for a radar station!

Khasab and Kumzar

Back in Khasab, I met John Dymond, Director of Operations in Musandam. His office was awash with maps, plans, and reports, but he was clearly full of enthusiasm for the changes that will transform Musandam from an arid waste into a lively, healthy and prosperous community in the shortest possible time. Musandam Development Committee was established by Royal Decree in 1977 as an advisory body, without executive powers or funds. Then, in winter 1979–80, it took over local executive responsibility from the Omani Ministries of Agriculture and Fisheries, Commerce and Industry, Communications, Electricity and Water, Land Affairs, Municipalities, Post and Telecommunications, Information, Youth Affairs, and Public Works. Two examples show very clearly the vast improvements made in a very short time.

Before M.D.C. took over responsibility for agriculture in Musandam, Khasab had nine staff, including one tractor driver and one veterinary assistant, who lacked co-ordination; Bayah had two men,

Khasab. Fisherman

who merely issued a few seeds sent to them. In the first two years, 28 new staff joined, including a qualified veterinarian and an agronomist. In one ten-month period in 1982, tractors spent 758 hours, ploughing fields in Khasab, Bukha and Bayah, and the officials presented a thousand trees, 200 kgs of seeds, and 12,000 kgs of fertilizer. An Agricultural Centre has been built at Bayah, with a four-hectare experimental plot to test the suitability of seeds locally.

If we consider water supply, probably the most crucial factor of life in Musandam throughout the centuries, the position in 1979 was that well-sites existed at Khasab, Ghumdah, Bayah, Qidah and Limah, and two small desalination plants at Kumzar and Shaisah. Only Khasab and Bayah had piped water distribution; outlying villages depended on the delivery of water by tankers and bowsers to water-tanks, from which supplies were drawn.

In Khasab, new wells doubled the supply in 1980, and a new town supply tank was completed in 1982, with a capacity four times that of the old tank. Between January 1981 and January 1983, the average water production per month rose from 26,000 cubic metres to 36,000. The Khasab-Bayah road permits delivery by truck to mountain villages as well as to those in the plain, and Shihuh in the mountains are being assisted with building materials flown in to repair their cisterns. Now the Shihuh crops, such as alfalfa, wheat, carrots, green vegetables, radishes and onions, can be watered to provide a more reliable harvest.

Khasab has a new port and breakwater with cargo berths of 90 and 60 metres, and facilities for roll-on roll-off landing craft. A separate harbour for small fishing boats and coastal dhows incorporates a floating pontoon jetty connected by a ramp with the shore, aiding quicker on- and offloading.

I explored the *barasti* huts near Khasab town centre, and the Portuguese fort, still in use today. The *suq* was busy in its sparkling new pedestrian precinct, and a crescent moon far above in the satin-black sky echoed the crescent on the roof of the new mosque.

Apart from the occasional helicopter, Kumzar is accessible only by boat from Khasab, an awesome voyage of two hours that reminded me more than once that Sindbad is thought to have been a sailor from the shores of Oman. Tucked neatly in a little bay like a claw into a lobster, the village of 2,000 or so people now has a desalination plant (1976), electricity, health clinic and school. The inhabitants cluster in flat-roofed houses of mud and stone, divided roughly into two to allow the wadi, if in spate, to rush its course from mountains into sea, without lifting homes with it.

Kumzar has been propelled abruptly into the modern world with a telecommunications system that for the first time has permitted the introduction of the telephone, radio and television. Nevertheless, most Kumzaris still migrate to Khasab in the summer to tend their date harvest.

7 DHUFAR

If mountainous Musandam seems like a different world after the metropolitan bustle and flyovers of the Capital Area, and the wide green valleys sloping from Jabal Akhdhar, then the visitor to Dhufar will catch his breath again, for this is yet another distinct climatic and botanical zone in the Sultanate of Oman.

On the flora and fauna, *The Journal of Oman Studies* Special Report no. 2 (1980) published 'The Scientific Results of The Oman Flora and Fauna Survey 1977 (Dhofar)', based on six weeks' work in the comparatively short period immediately after the lifting of the monsoon cloud when the scene is one of spectacular verdant beauty, after heavy rain and before autumnal desiccation set in.

To the north lies the Empty Quarter, that fearsome Rub' al-Khali lying partly in Oman and partly in Saudi Arabia which for so long resisted the efforts of travellers to explore its fastnesses. The fringes of the desert had been touched by G. M. Lees in 1925–6, by R. E. Cheesman, by Bertram Thomas in 1930–1 from the south, H. St. John Philby in 1932 from the north, and Wilfred Thesiger firstly in 1947 and again in 1947–8. Their records are *Arabia Felix* (1932) by Thomas, *The Empty Quarter* (1933) by Philby, and *Arabian Sands* (1959), and must be considered three of the greatest travel books ever written.

Haines on Dhufar

Dhufar south of the Empty Quarter has been distinctly less fortunate in its chroniclers. Stafford Bettesworth Haines, Captain of the survey ship *Palinurus*, landed an officer in 1834 to report on the situation in Dhufar, and the relevant section from his *Memoir of the South and East Coasts of Arabia* is reproduced from the *Journal* of the Royal Geographical Society (1845).

'The soil of the district or province of Dhafár (for there is no *town* of that name‡), is abundantly luxuriant, well irrigated by mountain-streams, enabling the inhabitants to employ their industry in cultivation if they choose, and abundantly repaying the farmer for his labour. Still, though nature has been thus bountiful, the people are extremely indolent, generally contenting themselves with what the soil yields spontaneously, in preference to improving the crops by tillage. In some parts which I shall hereafter mention, the little labour they have bestowed on cultivating the ground has amply repaid them, and has, in fact, been one means of making them more industrious.

On the lofty mountain range of Subhán, 4000 feet high, which runs parallel with the coast at a distance of about 16 miles, and has a luxuriant Tehámah, or belt of low land between it and the sea, the soil is good; wild clover growing in abundance and affording pasture for cows and immense flocks of sheep and goats, while in many places the trees are so thick that they offer a welcome shade impervious to the scorching rays of the sun. Mr. Smith, an officer of the vessel which I commanded, was deputed by me to examine the whole of the Subhán range. He traversed it entirely in perfect safety, and, under the name of Ahmed, became a great favourite with the mountaineers. He was everywhere hospitably entertained by them, and they would not even permit him to drink water from the numerous clear mountain-streams that were meandering in every direction. "No," they said, "do not return, Ahmed, and say we gave you water while our children drank nothing but milk." In every instance they gave him the warmest place at the fire, and invariably appointed some one to attend to his wants. They even extended their generosity so far as to offer him a wife and some sheep, if he would only stay and reside among them. On Mr. Smith's expressing a wish to see some of the numerous wild animals whose footsteps were everywhere visible over their park-like mountains, they immediately despatched a party, who returned with a splendid specimen of an ibex,* a civet-cat, and a very fine ounce. He himself saw plenty of smaller game, such as antelopes, hares, foxes, guinea-fowl and partridges.

These hospitable mountaineers are handsome, well-made, active

---

‡ Perhaps Captain Haines was misled with respect to the present existence of that place by the strange pronunciation of the natives of Mahrah. A passage in M. Fresnel's paper on the Geography of Arabia (Jour. Asiat. iii. x. 192) seems to justify such a supposition: – "Remarquons ici," he says, "que la position assignée par Ptolémée à la métropole de Sapphar (*Zhafâr* ou Dhafâr), le Sephar de la Genèse, le *Tsfór* des modernes Homerites, cadre parfaitement avec celle du promontoire Syagros, supposé Râs Saugra (Saukirah). En effet, la longitude Orientale de ce cap surpasse d'environ deux degrés celle de Zhafâr dans nos meilleures cartes. Or, je vois dans Ptolémée la longitude de Sapphar marquée 88 degrés, et celle de *Syagros extrema* 90 degrés, ce qui nous donne précisément la différence voulue de 2 degrés dans le sens voulu. Je ne puis donc comprendre pourquoi d'Anville a mis Sapphar du côté d'Aden, et rejeté le promontoire Syagros à Râs-al-Hhadd."

* I have the horns by me, as a fine specimen; they are 3 feet in curve, with 21 knobs.—S. B. H.

men, and always well armed, their weapons being the same as those used by the Mahrahs. They are of the Gharrah tribe. Their women are handsome, and much fairer than any seen on the coast. I have seen as many as 200 at a time, who came down to barter their cattle, butter and gums, for dates, at Morbát. Curiosity induced me to ask them how they accounted for being so fair, and their reply was, that it was owing to their drinking nothing but milk from their childhood; little dreaming that they were indebted to the renovating breezes and temperate climate of their native hills, on the summit of which in February the thermometer ranged from 49° to 72° Fahrenheit.

The dress of these women consists of a coarse cotton petticoat, with a blue robe over it; their dark hair, as usual, is artificially lengthened and arranged in long narrow twisted tresses.

The plants found by Mr. Smith in the Subhán mountains were the same as those in the more elevated parts of Sokotrah; dragon's blood, frankincense and aloës were seen in abundance.

We now return to the Tehámah or low land. The first village near the sea, to the E. of Rás el Ahmar, is called Audád, being about 1 mile S.W. of the principal village of Sallálah, and having a population of 300 or 400 souls. This village is protected by a fort, and has its " Jámi'," or mosque, in which the service on Friday may be performed.* It is surrounded with gardens, date-trees and millet† fields, with some wheat, cotton, and indigo; and the soil is abundantly irrigated either naturally or by artificial canals from the neighbouring lakes.

The next village near the sea-shore, S. E. of Sallálah, is Haffer, in 16° 57′ 30″ N. and 54° 11′ 00″ E., about 1½ mile distant, containing a population of about 100 men.

Two miles and a half E.N.E. of Haffer, there is a fresh-water lake, formed by a copious spring, near which there are extensive ruins. This lake is deep and thickly covered with bullrushes, where we here found abundance of wild-fowl.

About 1½ mile inland, and 2½ to the N.E. of Haffer, is the village and white mosque of Robát, with a population of 100 or 200 souls. The whole country surrounding the above-mentioned villages is cultivated, producing cotton, indigo, millet,‡ and other kinds of grain, a few vegetables, but no fruit. They apparently

---

* *Mesjid*, whence our word mosque, signifies " a place of worship;" *Jámi*, " a place of assembly," a meeting-house. The latter only has a pulpit (minber), whence the Khotbah (prayer for the Sultán) is pronounced and sermons are delivered by the Khatib (preacher), and where the sacrifices and services of the great festivals (ʿid-el-Kurbán, &c.) are performed.—Murádjah d'Ohsson, Tableau de l'Empire Ottoman, ii. 453, 8vo. ed.

† Dhurrah, Sorghum vulgare.

‡ In the original, Jowárí, Sorghum vulgare (the Dhurrah of the Arabs), and Bájrí, *i.e.* Pennisetum typhoideum, probably called Te'ám (food) by the Arabs.

care little for either of the two last named articles, their accustomed diet being milk and millet-bread, with meat occasionally.

Three miles to the E.N.E. of Haffer is the fort and village of Diríz, having a population of about 150 souls. The village has a salt lake immediately eastwards of it, and from thence, proceeding in an easterly direction, towards Morbát, all traces of cultivation are lost till we reach the village of Thagah (Thákah), which has a small population, with a date-grove and some cultivated ground west of it. There are also several ruined forts near the hills, which at Thákah approach the sea. Thákah is in 17° 00' 40'' N. and 54° 30' E.

The extensive plain of Dhafár is bounded on the W. by the high mountains of Seger, and to the E. by Jebel Subhán. To the N. each of these mountains gradually decrease in elevation, while towards the sea they are skirted by a low sandy beach, having regular soundings and good holding-ground, from 10 to 4 fathoms. During the north-easterly monsoon, the gusts off-shore from the N. and W. are at times very violent.

The sea-coast continues low and sandy till within 17 miles of Morbát, when it is terminated by a dark precipitous bluff of moderate elevation.

Trading-boats now frequently touch at the villages along the shore of Dhafár, and barter dates, rice, and cloth for gums, butter and grain; and, as this coast forms the shore of the gum-country, it might, with a good system of government, and an industrious population, be rendered a most flourishing tract. This fact did not escape the notice of Sayyad 'Akíl, a celebrated chieftain on this coast; and, had Providence ordained him a longer life, the now neglected plain of Dhafár would, doubtless, have presented the same appearance of opulence and bustling activity as characterized it in former ages.

The frankincense and gum-arabic annually exported from Morbát and Dhafár vary from about 3000 to 10,000 maunds,* which is nothing to what might be procured, the trees being exceedingly numerous on the mountain-declivities and in the valleys inland, and attaining a height varying from 15 to 25 feet. The bark is of a greyish colour, easily pierced, and the leaf large. In this neighbourhood is found the aloë-tree† of Sokotrah, growing out of masses of primitive limestone, apparently without any earth to sustain it. Its height averages from 3 to 15 feet.

The inhabitants of the villages in the plain appear to have but

---

* Mans; but do the Arabs use this Indian measure?

† Sabr, the name given by Captain Haines, is the Arabic word for the aloë. His drawing shows that it must be the variety called "arborea" by Förskal (Flora Arabica, p. cx.). The officinal aloë is called Aloë Socotrina.

little intercourse with the Bedowins of the interior, who only visit them for purposes of trade.

The people of the plain are of mixed blood, owing to the influx of settlers during the time of Sayyad Mohammad 'Akíl. They are (as most town-bred Arabs) timorous, indolent, and much addicted to the use of tobacco. The dress of the higher orders is that commonly worn by all respectable merchants, viz., a white robe, bound round the waist with a shawl, and a " fótah," or waistband. Their heads are shaven, and protected by the customary 'amámah,* or turban. The poorer classes wear merely the " fótah," secured to a neatly-plaited leather belt, the workmanship of the Bedowin girls, called " 'akab," which is tightly secured round the waist: when out of doors they wear the yambe'.†

The Gharrah Bedowins, who are the roving rulers of the country, prefer their glens and mountains to the hotter Tehámah, and wander from spot to spot, as the pasture serves for their cattle and flocks. They employ themselves during the S. W. monsoon in collecting gum, and frequently reside in the cavities of their limestone mountains.

They are a fine, athletic race of men, dressed in a blue, glazed waistband, which is, in general, their only covering. Their arms are the matchlock, yambe', and short, straight sword; but some, who cannot afford to purchase these weapons, arm themselves with a piece of very hard, heavy wood shaped thus,—

which they throw with great precision as far as 100 feet—at that distance, indeed, they could kill a man. This weapon is thrown so as to rebound along the ground, and every lad carries one in his hand. They allow their hair to grow long, and it is then gathered up behind, like the Mahrahs', which gives them a wild appearance.

Immediately before the fast of the Ramazán‡ both males and

---

* Or 'Immámah, from its fulness or size; properly a large, official turban—such as judges and public officers wear.

† Spelt Jambea by Captain Haines; but that Yanbe' is the proper spelling appears from Niebuhr (Arabia, p. 62), who, as a German, spells it Jambea. The final *a* is used to express the letter *'aïn*. The word is probably a colloquial term introduced in modern times. In this sense it is not found in Arabic lexicons: it means literally, " it flows, spouts out," as blood from a wound, or water from the earth, hence *Yanbu'*, a spring, is the name of a town in Arabia, on the Red Sea, from the abundance of springs in its neighbourhood. Janbíyah might mean in Arabic a " side arm;" but no such word appears to have been ever used, therefore it must be supposed that Niebuhr's orthography has been adopted by Captain Haines.

‡ Ramadhán, in Arabic.

females visit the Tehámah for the purpose of barter, and it was then that we had an opportunity of seeing them.

It struck me that their women (who are modest, though they wear scarcely any covering), and their young men, have a Jewish cast of countenance. Their faces are longer than Arab faces generally are, their eyes large and bright, and they have figures that would have delighted the eye of Canova, could he have seen them. They are much fairer than the Arabs of the coast, and were apparently pleased to see men stouter and fairer than those of their own tribe. Indeed, they were frequent lookers-on at my crew when playing at cricket; and I then had forty fine extra Europeans on board, having saved the crew of the Reliance whaler, which had been wrecked on one of the Curia Muria [Khuryán Muryán]* Islands.

The Gharrah Bedowins seldom eat meat, excepting on festivals; not that they dislike it; as their favourite dish is young camel's flesh, but they value the milk too highly to slaughter the females of either camel, cow or goat. The males of the two latter, they frequently dispose of on the coast for dates, cloth, &c.

As Sayyad 'Akíl, formerly ruler of Dhafár, was at one period conspicuous and much dreaded, I shall add a short account of him, to show how from being an object of detestation, he at last commanded respect, and even veneration.

The 'Akíl† family were merchants. The brothers Sayyad Mohammad and 'Abdu-r-Rahmán were in the habit of trading in a large bugalá belonging to their father, which gave them a predisposition for a roving life; and, as Fortune favoured their speculations, they added to the number of their vessels, and purchased 500 slaves from Mozambique. In one of their voyages Sayyad Mohammad visited Dhafár: the luxuriant appearance of the country tempted him to settle there, and he gradually rose to be master of the place. With a large retinue of slaves, assisted by his own ability and bravery, he defeated the Gharrah tribe in every engagement, and was latterly much dreaded by them. Under his just rule, the district flourished, and trade and population increased. He extended his conquests as far as Morbát, and there built a fort for the protection of the town.

Ambition and avarice, united with his predilection for a roving life, led him to commit piracies on the high seas; and his vessels,

---

* Curia Muria, the name introduced into European maps by the Portuguese navigators of the sixteenth century, shows what the vowels of these names should be, while the Khartán Martán of Idrísí and other Eastern geographers prove that the fine nasal was, as is often the case, dropped in cômmon parlance, and that through the ignorance of transcribers the letter before á was written with two points above instead of below, and made *t* instead of *y*.

† 'Akíl is the name of a distinguished Bedawí tribe (Burckhardt's Notes on the Bedouins, p. 232).

among other prizes, captured in the Red Sea an American ship, of which all the crew were murdered, with the exception of one boy, whom he carried to Dhafár, and educated in the tenets of the Mohammedan faith. When we arrived at Dhafár, this young man had nearly forgotten his mother tongue. He was a Mohammedan, and had a wife and several children, and seemed perfectly contented with his lot.

After some years of cruelty and plunder, the Sayyad's conscience smote him, and he suddenly gave up the sea, and settled quietly in Dhafár, anticipating the comforts of a quiet life, and anxious to make others happy; but in this he was disappointed. The Gharrah tribe deceived him, and for a time led him to imagine that they were contented with the justice of his government. They traded freely with the Tehámah, and apparently all animosity between them and him was buried for ever. This calm lasted from 1806 to 1829. The district still improved, and even Morbát could number a population of perhaps 2000 souls. This bold rover, with his mode of life, had changed his habits also. He became devout, and averse to shedding blood; was loved by his subjects for his mild and impartial rule, and dreaded by his enemies. Treachery, however, had long been at work, and opportunity alone was wanting for the Gharrahs to take their revenge for the many acts which they deemed tyrannical and oppressive. Moreover, there were many others between whom and the chief there existed a mortal feud, on account of relations who had been slain by his followers; and all these persons eagerly joined the cabal against him.

The long wished for opportunity occurred after the month of Ramazán,[*] in 1829. The Sayyad, returning from Morbát with a smaller retinue than usual, was mortally wounded by a matchlock-ball, fired from the low brushwood. When he fell, his slaves immediately fled, and the Bedowins, who were lying in ambush, dispatched him at once. His body was afterwards found, by a strong party sent out to recover it, pierced with numerous wounds from their daggers, or yambe's.

The Imám of Maskat, hearing of the death of Sayyad Mohammad 'Akíl, sent a force to take possession of the territory for the brother of the deceased, Sayyad 'Abdu-r-Rahmán, who was still a merchant, and at that time in Bombay. But, when he heard the particulars, he prudently declined the proffered honours of so unsatisfactory a sovereignty, and preferred the more peaceful and profitable calling of a merchant, which he still exercises at Mokhá, where he is distinguished for his intriguing disposition, as well as his great wealth.

The Imám of Maskat requiring troops for the settlement of

---

[*] Ramadán, or Ramadhán, in the mouth of an Arab.

his southern possessions, the force at Dhafár was withdrawn, and the district once more fell under the rule of the Gharrah tribe, who soon drove away the greater part of the inhabitants by a system of plunder and monopoly, and thus their villages have dwindled away almost to nothing.

Immediately E. of the cliffs to the W. of Thákah the soundings on the coast become deeper, with alternate cliffs and small sandy beaches. About 7 or 8 miles W. of Morbát there is a small rock, called Jawàní (Husein), having some ancient ruins of hewn stone on its summit. It is distant about 50 yards from the mainland. Its length is about 300 feet, by 200 broad. Tradition says a bridge formerly connected it with the mainland.

Morbát, or Merbát,* is a small village,† in 16° 59' 15" N., and 54° 47' 40" E. (reckoning from Bombay, as before stated), situated in the centre of a small but well-sheltered bay, named after it, containing about 50 houses, and a population from 150 to 200 souls, who may be divided into three classes:—1st, a few Arab merchants not born there; 2nd, Arabs who are either descended by their mother's side from individuals of the Gharrah tribe, or have married Bedowin wives; and, 3rdly, slaves, the females of whom are not celebrated for their morals. The head man, or Sheïkh, when I was there in 1835, was Ahmed of the Makyat branch of the Gharrah tribe, a strong, well-made man, 5 feet 7 inches in height, and 35 years of age, with good features, and a benevolent countenance. I received great civility and politeness from him. He was true to his word, and extremely obliging, which much facilitated my work. In return for his kindness, I presented him with a rifle, thirty German crowns, and some cloth. The population we found extremely indolent, addicted to smoking, and lolling at their ease. They possessed no vessels, not even fishing-boats, and were too lazy to make rafts.‡ One of the younger merchants purchased a boat from a bagalá, while we were there, with the money he had amassed to pay for a wife, which speculation turned to good account, as I employed him to supply the ship with water.

The houses in the village are miserable hovels; those that are inhabited are erected on a rising ground, immediately S. of the landing-place, having to the S.E. a small, square, ruined fort; and to the N. one of much larger dimensions, built by Sayyad Mohammad 'Akíl, surrounding which, are the relics of numerous houses in ruins.

---

* Mirbát, commonly pronounced Merbát, is the spelling fixed by Abú-l-feda, Geog. i. 98.

† Its position was derived from numerous observations. By 30 azimuths in 1831, and the variation determined at 3° 12' W. In 1836, variation by 62 observations on shore was 2° 27' W. High water at 8 or 9 hours, rise and fall 6 feet 10 inches.

‡ Catamarans.

There are the remains of another village near the base of Jebel 'Alí (a red granite hill near the beach at the head of the Bay), which apparently surrounded a tomb called Ḳubbat * Sheïkh ibn 'Alí, dedicated to the patron-saint of the place. On the extreme point, forming the south side of the anchorage, are the ruins of another tomb called Ḳubbat Sheïkh Hidrús [Idrís?].

Both the inhabitants and vessels are supplied with water from holes dug in the sandy soil of a small valley near the hill called Jebel 'Alí. This water is brackish and unpalatable at first, but becomes tolerable after a time, and we never found it to possess any pernicious quality.

Morbát affords but few supplies. All we obtained were goats and bullocks brought from the interior, and a few radishes and onions from Ḍhafár: wood is brought from the mountains.

Morbát or Merbát Bay is a small, secure, and well-sheltered anchorage for 24 points of the compass, but from S. to W. it is open. The low and rocky point to the S., called Rás Morbát, has a sunken rock off it, at 300 yards' distance. The bay turns suddenly from the pitch of the point, in a northerly direction, having two or three small points and bays, ere that upon which the present village stands is reached, and from thence the deepest bay of any, forms the landing-place; and after passing the watering-place, the shore turns gradually in a western direction towards Ḍhafár.

During the N.E. monsoon the water is as smooth as a mill-pond. The soundings extend but a short way off-shore; and a vessel will quickly shoal from 30 to 10 fathoms, between which and 5 fathoms, from 500 to 600 yards off-shore, there is the best anchorage. I generally anchored in 6 or 7 fathoms off the village.

A leading mark for making Morbát, used by native navigators, is Jebel Dekan (or Jebel Morbát), as they term it, being nearly true N. from Rás Morbát. This peak is nothing more than an elevated part of the Subhán range, from which the mountains rapidly decrease in height in a westerly direction, thus rendering it a conspicuous object from the sea.

The revenues of Morbát are trivial; but the Sheïkh receives a present from most vessels anchoring in the port, which enables him to pay the annual stipend of 70 dollars to his tribe, and to live respectably himself. He also levies a small anchorage-fee (nominally), in proportion to the size of the vessel, and the will and liberality of the Nákhodá.† I have known one, two, or three

---

* Kubbah, or Kobbah, whence our word alcove, signifies a sepulchral chapel—a saint's shrine, frequented by the devout—a place of pilgrimage. K is pronounced in this part of Arabia, and in Egypt, commonly like *g* in *geese*. Ghubbat, as Capt. Haines spells it, means a bay or creek.

† Master of the ship; an Indian term from náó, ship, and khodá, which here signifies not God, but lord or master.

bags of dates given, and sometimes a bag of rice. The power of the Sheïkh extends nominally from Thákah to Rás Nús; but I doubt whether he would attempt to inflict fine or punishment upon any offender except one of his townsmen.

While surveying and examining this part of the coast, I took an opportunity to ascertain the number of vessels that annually supply the S.E. and Southern coast of Arabia with dates, thence deducing an estimate of the immense quantity brought from the Persian Gulf and Maskat. This also shows that any strong naval power could almost cause a famine among the inhabitants of that tract.

Some of the more intelligent merchants, when I mentioned this to them, were much astonished at my remark, as to the ease with which the inhabitants of the south coast might be punished for any offence they had committed by a blockade, which would almost reduce them to starvation, as the growth of dates on their coast would not supply one-twentieth part of the quantity needed for their support. When they clearly understood me, one of them exclaimed—" That is not the idea of a man, but of the devil; for into man's imagination such a thought for the wholesale destruction of his species could never enter. Say no more about it; for dates are bread, and bread is the staff of life."

The season for the run of the trading-boats down the Arabian Coast from the Persian Gulf is from the beginning of November to the end of December. From the 21st of November to the 10th of December, 40 boats anchored in Morbát Bay, all laden with dates, and varying in size from 30 to 150 tons; and 121 boats passing the port were hailed, varying from 30 to 300 tons, which is about one-half the number for the season; so that the whole may be nearly as follows :—

	Tons.
In 18 days, 40 boats anchored with dates, average 80 tons	3,200
In 18 days, 121 vessels passed with dates, average 80 tons	9,680
Total	12,880

This amount I witnessed; but believe that the remaining days in the two months above mentioned would make the annual supply little short of 25,000 tons.

The larger class of boats return before the S.W. monsoon sets in; but others, well equipped, with a navigator on board, return with the " Tadhbírah " * in June, or after the first blast of the S.W. monsoon has been felt upon the coast, their cargo being

---

* *Tadhbír*, as it should be written, signifies a certificate: it is here probably used technically for the first indication of the monsoon.

principally coffee. The smaller craft, called bedans, bakárahs, batillahs, and tránkís* of the Moṣeïrah and Ṣúr districts, make a coasting voyage, and employ themselves in fishing along-shore, and then return with the current in March or April. I have met them in fleets of fifty or sixty boats, with from eight to ten men in each, and do not hesitate in saying that they plunder whenever an opportunity offers without personal risk. As a proof of this I may mention, that while carrying on a trigonometrical survey of the coast below Cape Isolette, I had left the ship in my launch and cutter at 3 A.M., accompanied by Lieutenant Sanders and Midshipman Fleming, with the view of commencing my work about eight miles to the N. in Jinzerah Bay, by sunrise. When we were about four miles from the beach, and it was still dark, we crossed a large bakárah on the opposite tack, and spoke her in passing. The cutter being some distance astern, with only Lascars in her, my attention was naturally attracted to her, as I doubted the honesty of these traders much: nor was it without reason, for the bakárah wore round and stood for her. We immediately bore up to the assistance of the Lascars, and when close, received a volley of matchlocks from the bakárah, which we returned, and stood for her. Finding that we were well armed, and not inclined to be intimidated by her fire, she took to her heels. I ordered the cutter to keep on her off-shore side, while I pulled and sailed in the launch in her wake, keeping up a fire of musketry. My object was to keep her in-shore close to the high breakers, and, as the day dawned, for the surveying-vessel to open fire upon her, and cut off her retreat, as she was too fleet for us.

As daylight dawned, the nákhodá of the boat found himself in a most awkward predicament. On his larboard bow was the Palinurus within half a gun-shot; on his starboard bow and beam heavy breakers; close astern the launch, firing at him; and on his larboard quarter the cutter. He was so hemmed in that his only alternative was to run his vessel ashore, which the second 9-pounder shot from the Palinurus compelled him to do, and all hands swam on shore. I afterwards sent the launch with a gun to destroy her; and complained to the Imám of Maskat, whose subject owned the boat. He immediately took notice of it, and imprisoned its nákhodá and owner for life.

Prior to quitting the subject of Morbát, I would observe that during the prevalence of the sudden and dangerous blasts from the N. and W. (called by the Arabs belát†) which a vessel will sometimes experience in Curia Muria Bay, a strong south-

---

* Taránkí is probably an Indian term from táran, " to cross over, to swim."

† Belád, *i.e.* country, land ; the final *d* is often pronounced like *t*. El belád here signifies the land, provinces, countries in the interior.

easterly breeze will be found blowing over the point of Morbát during the day, and light and variable airs during the night with smooth water.* I account for this change of wind by the extensive precipitous wall of Subhán, which forms a barrier on its S. and E. face, varying in elevation from 3000 to 5000 feet, and running in a N.E. by E. direction from Morbát to Rás Nús; so that on rounding Rás Nús for Morbát the wind diminishes in strength, and gradually blows parallel with the line of Subhán, until the Valley (Wádí) of Dhafár is opened, through which northerly and westerly winds rush down with violence. Owing to the same cause but very little rain falls during the year upon the rocky belt of land at the base of Jebel Subhán, and Morbát rarely has the benefit of a shower, while to the W. the sides and summit of the Subhán range are covered with verdure.

Rás Morbát is a low rocky point forming the southern part of Morbát Bay, and the S.W. point of the low belt of land which extends in breadth from 6 to 12 miles from the Subhán mountains. Its extremity is very low, and a rocky reef extends from it about 400 yards. Caution, therefore, is requisite in rounding it, as the soundings are very bold—10 fathoms being close off the pitch of the reef, and 20 fathoms not 300 yards from it. It is in 16° 57' 50" N., and 54° 47' 26" E.

Ten years afterwards, the *Palinurus* returned to Dhufar, under Captain J. P. Saunders, whose landing parties found the *badu* more hostile than on the ship's last visit. Surveyors reported being fired on, or warned away by serried ranks of matchlocks. Assistant-Surgeon H. J. Carter was of the party, and his 'Descriptive account of the ruins of El Balad' appeared in *Transactions of the Bombay Geographical Society* (1846) but is not reprinted here because the scientific content has been superseded by Paolo Costa's 'The study of the city of Zafar (al-Balid)' in *Journal of Oman Studies* (Vol. 5, 1979), 1982. Carter was also responsible for a pioneering 'Memoir on the geology of the south-east coast of Arabia' in the *Journal of the Bombay Branch of the Royal Asiatic Society* (1852); brief 'Notes on the Gharah tribe, made during the survey of the Southeast Coast of Arabia in 1844–1845' and substantial 'Notes on the Mahrah tribe of Southern Arabia', again in the same journal (1845 and 1847 respectively).

Miles on Dhufar

S. B. Miles, Political Agent, visited Dhufar, including al-Balid (the Jibbali name for the Arabic al-Balad ('the town') in December 1884, aboard Her Majesty's ship *Dragon.* At Damkot, according to the *Administration Report of the Persian Gulf Political Residency and Muscat Political Agency for 1884–85,* the Muqaddam, 'a decrepit old man, said he remembered the boats of the *Palinurus* being engaged in surveying the coast half a century ago, and had seen no ship here since then. None of the people had ever seen a steamer before, and they had in fact shown evident signs of alarm at the *Dragon's* approach, some clustering together, and others scampering away up the hills at the back; but they were soon reassured.' The account by Miles continues with his arrival at Salalah, and is reprinted in full from this point, to his cursory stay at the Kuria Muria, returning to Muscat eighteen days after he had set out, in the cool season.

'Passing Ras Sajar, which marks the southern limit of the territories of the Sultan of Muscat, the *Dragon* arrived at Sallala and anchored off there on the 19th December. The district of Dhofar, according to some, includes the whole tract lying between Ras Resoot and Ras Noos, consisting of a maritime plain 85 miles long, enclosed by the lofty range of Jebel Samhan, which touches the sea at these points. But the application of the name is more usually restricted to the rich alluvial plain between Resoot and Thakah. This plain, which is of half-moon shape, is 30 miles in length, with an extreme breadth of about 14, and is formed by the curvature of the hill range; the coast-lines subtending this arc lying due east and west and having a sharp turn or bay at Resoot and Merbat, which form sheltered harbours for vessels in the south-west and north-east monsoons respectively. It is one of the most fertile and favoured districts on the southern coast of Arabia, and its chequered history shows that it has ever been a coveted possession. Extensive ruins of towns and forts scattered over its surface are an attestation of a former populousness and importance that have long since passed away. Dhofar contains at present five separate towns, all of them near the sea – Okad, Sallala, El Hafah, El Dahareez, and Thakah. The Governor and his garrison reside at Sallala. Numerous wadies or water-courses intersect the plain, some of which are well wooded and grassy, while others form small fresh-water lakes. The range about Merbat rises to an elevation of 3,000 or 4,000 feet, and is thickly wooded to the summit with tamarind and acacia, frankincense and bdellium, dragon's-blood and other gum-resinous trees, and affords pasturage to herds of cattle and camels, and flocks of sheep and goats. But the western portion of the range behind Dhofar is neither so high nor so well clothed with vegetation, the upper part being almost bare of trees. The soil of the plain is light and rich, and excellent water is found everywhere at a few feet from the surface. Cotton, jowari, bajri, pulse, lucerne, and cocoanuts are grown, but to a limited extent only. The trade of Dhofar is chiefly in the hands of Khoja merchants, who are agents for houses in Bombay and Moculla. The imports are rice, grain, dates, sugar, cotton cloth and oil, the cloth being mostly indigo-dyed stuffs from Bombay, and may amount to $50,000 annually. The exports are frankincense, ghee, hides and skins, wax, &c., brought down from the hills by the Gara Bedouins, besides cotton, sardine oil, and shark-fins, &c. The

Sámhan hills are known to be rich in balsamic, rubber-producing, and other useful trees, but they have never been properly explored. The rubber tree grows to the height of 15 or 20 feet; it is called Isbak by the Arabs and Tishkot by the Garas. Specimens of the product have been sent to Muscat, but it has not yet become an article of trade.

The inhabitants of the plains of Dhofar are mostly Katherees. This Hadhramaut tribe, under Sultan Bedr Ba Towarek, invaded and overran Dhofar about 300 years ago and made El Dahareez their capital. They were in turn dispossessed by others, but subsequently regained supremacy; they still form the bulk of the inhabitants and number altogether about 1,500. The total populaion of the plain may be 2,000. The headmen of the towns and the Kazis all came to Sallala to visit the Political Agent; they are on good terms with the Governor, and expressed themselves as being well satisfied with the rule of His Highness Seyyid Toorkee.

The Garas are an extremely interesting and peculiar race, and are but little known. They are allied to the great Mahra tribe of whose language they speak a dialect. They are taller than, and of a different physique and physiognomy from, all the Arab-speaking tribes of Yemen and Omán, and, though they claim to be Hymyarites, have probably a separate origin. The area occupied by the Garas is a very circumscribed one, extending only from Rakeyoot, a village 12 miles west of Ras Sajar to Hasek, and not reaching inland more than 40 miles from the sea. In habits they are pure Bedouins, but are not nomadic; they may rather be styled Troglodytes, and their cave-haunting propensity is one of their chief singularities. The hills appear to be honeycombed with these caves, some of which are of prodigious dimensions, and afford space and accommodation for a whole family with its possessions in cattle and goats. A cave near the sea explored in the preceding year was 100 feet broad, 60 feet deep, and 8 feet high, and was double-storeyed, a second smaller chamber existing over the roof of the larger. This cave was entirely natural, and was a comparatively small one. The sub-tribes of the Gara are 10 in number and are thus distributed: At Rakyoot – Beyt Shemasa and Beyt Elsa; at Dhofar – Beyt Saeed, Beni Kattan, Beyt Jesjyon, Beyt Tebbook, Beyt Keshoop, Beyt Jahbool, and Beyt Maashinee; at Merbat – Beyt Makheir. The strength of the whole Gara tribe does not probably exceed 3,000 souls.

Among the ruins with which the plain is interspersed the most extensive and interesting are those on the shore between El Hafah and El Dahareez, covering a space 2 miles in length. These ruins, now known as El Baleyd, are believed by Sprenger to be the remains of the ancient Mansoora, but this name is unmentioned now in local tradition. The citadel, towers, and mosques are still standing in part, and the town wall and ditch can be distinctly traced. They have been measured and fully described by Carter. According to local tradition this city was founded by the Mainjooi or Nejui dynasty, which rose to the height of its power in the fifth century of the Hijra. The existence of this dynasty has been discredited by European orientalists, but without reason. The tombs of the Sultans near El Robat, a few of them exquisitely worked and inscribed by Persian or Sanaa artists, have been examined and copied. The prosperity of Mansoora was doubtless owing in great measure to the existence of a copious stream of perfectly sweet water which encircled the town on three sides. This stream, which is 4 or 5 fathoms deep, formerly communicated with the sea and formed a most excellent creek or harbour for dows and boats. It is now closed by a sand-bar, but this only requires to be removed to render the port again available for native vessels.

About half a mile from the ruins of El Baleyd lies the principal Moslem shrine at Dhofar, the tomb of the Zamorin, known as Abdulla-el-Samiry. He was the Raja of Cranganore in Malabar and was converted to Islam in the beginning of the third century of the Hijra, circa 210. Being compelled to leave his kingdom, he embarked in an Arab dow and came to Dhofar, where he died four or five years afterwards in the odour of sanctity. He is reputed to have first brought rain to Dhofar by his prayers, and his tomb is still visited by numbers to beseech his intercession in time of drought. The tomb is enclosed by an unroofed wall of mud and stone 25 feet by 10; it is 18 feet long by 4 broad, and lies north by south, with a broken headstone of black basalt. The inscription is imperfect and there is no date.

The recent history of Dhofar may be said to commence from the seizure of power by the famous pirate Muhammad Akil, who established order and ruled with a strong hand until his murder by the Garas in 1829. On hearing of his death, His Highness Seyyid Saeed, the Imam of Muscat, immediately sent a force and took possession of the place, although Muhammad's brother, Abdul Rahman, who was then a merchant in Bombay, desired and intrigued to obtain the government. Seyyid Saeed, however, was at that time engaged in his unlucky enterprises against Siwi and Mombasa in East Africa, and could ill spare the troops required for garrisoning Dhofar. They were consequently withdrawn soon after, and the district fell into a state of anarchy, from which it did not emerge until the arrival of Seyyid Fadhl, the Moplah. In the year 1870 the Wali at Baghdad, under orders from the Porte, which at that time had designs on Southern Arabia, despatched a quasi-scientific expedition along this coast, and a liberal distribution of presents and flags was made to the various chiefs. Dhofar was one of the chief points visited by the steamer, and Turkish flags were landed here, but this tentative move was not followed up by Turkey. Seyyid Fadhl endeavoured to emulate the career of Muhammad Akil, but though he had the support of Mecca in his enterprise, he lacked the capacity for rule; and his followers, who seem to have been a band of ruffians, roused the people against them by their villainy and oppression. The general confusion became so great at last that the Katherees and Garas were compelled to unite and expel him. His Highness Seyyid Toorkee was then invited by the natives to occupy the country, and they have since remained contented and fairly prosperous under his rule.

At the time of the Political Agent's visits in Her Majesty's ship *Philomel* in November 1883, the Gara Bedouins were not on good terms with the Governor owing to a dispute about taxes, and a collision had occurred shortly before between the garrison and Sheikh Fankhor-el-Maashinee. At this visit the Political Agent found that the Garas had since submitted and that friendly relations had been re-established.

From Sallala the Political Agent visited El Dahareez and Thakah, and from the latter place the *Dragon* steamed on to Merbat.

Merbat is a town of about 300 inhabitants, situated in a little bay which forms an excellent and secure anchorage during the north-east monsoon, at the western extremity of a narrow plain 25 miles long by 7 broad, extending under the Sámhan range. There are two forts – one in ruins, constructed by Muhammad Akil in 1806; and a new one built by the order of His Highness Seyyid Toorkee four years ago, which has a garrison of 20 men. About half a mile from the town lies the tomb of Sheikh Muhammad Ali, who died in 556 A.H., now a famous shrine. The principal export of vegetable origin from Southern Arabia is olibanum or frankincense, the country producing which, the libanophorous region of the ancients, extends from the Wady Meyfa and the Himyar range in longitude 47° 30' E. to Hasek in longitude 55° 20' E., an area embracing the whole of Hadhramaut and part of Oman. Westward of Ras Fartak the tree is found but sparingly, and the collection of gum is neglected by the Arabs, the work being left to a great extent in the hands of Somális, who come across for the purpose and pay for the privilege. The trees are most abundant on the limestone summits of Jebel Sámhan, where the gum is gathered in May and December by the Garas, who call it Shlhât. The termination of the limit of the tree at Hasek and Wady Rakôt is very abrupt, and it is not found further to the east. The average annual export of the gum from Dhofar is about 30 tons, and the local value $60 per ton. Two young trees were brought to Muscat in Her Majesty's ship *Dragon* and are now thriving.

From Merbat the *Dragon* proceeded to Hasek on the 23rd. The old town here, now completely in ruins, is situated on the left bank of the Wady, where it was protected by two circular towers. The more recent inhabitants appear to have been of the Ba Malah tribe, who were attacked by the Kowasim or Beni Yas about three quarters of a century ago, and the women and children carried off into slavery. There is now no trade whatever at Hasek, and Arab craft seldom or never call here.

Hellaniyeh, the largest and only inhabited island of the Kuria Muria group, was next visited. The inhabitants of this island, 34 in number, subsist on fish and goats' milk, with a little rice and dates which they procure from passing dows by selling dried fish. Their habitations are of the most wretched and primitive description. Generally round or oval, 6 feet in diameter, and the height of a walking-stick, they are built of loose stones with a scanty roofing of mats laid over sticks and fish bones. Miserable as is their condition, their possessions are, nevertheless, coveted by others, and they are sometimes ill-treated and robbed by the Kowasim and Beni Yas Arabs.

Hellaniyeh was occupied as a station in 1861 for the Red Sea and Karachi Telegraph, but was abandoned on the failure of that cable in 1862. The guano, for the sake of which the cession of these islands was obtained in 1854, has long since been removed, and vessels now rarely touch here.

The last place visited by the *Dragon* was Ras Sharbedat. This part of the coast is inhabited by the Jenebeh tribe, who wander here during the cold weather with their flocks and herds for convenience of pasturage, and return to Jaalan in the summer. There are three fresh-water lakes in this neighbourhood.

The Political Agent returned to Muscat on the 28th December 1884.'

The Bents on Dhufar

The next travellers from Europe to record their experiences of Dhufar were James Theodore Bent and his wife Mabel, and their map records travels in 1889–90. J. T. Bent was the first European to explore the mountains north of the Salalah plain, and his contribution entitled 'Exploration of the frankincense country, Southern Arabia' (1895) here reprinted by courtesy of the *Journal* of the Royal Geographical Society, is consequently of the highest importance.

EXPLORATION OF THE FRANKINCENSE COUNTRY, SOUTHERN ARABIA.*

By J. THEODORE BENT.

In revisiting the south-east of Arabia this winter, our hopes were to get from Muskat to the seat of our former work in the Hadramut, and to fill up the large vacuum in the map which this country offers. Experience has now taught us that this is impracticable, but that Arabia must be done piecemeal, and then patched together as it were; and, indeed, this is the most profitable way of doing any country, for the traveller on a great through journey generally loses the most interesting details. Probably no country in the habitable world is at present so little known as Arabia. Arab fanaticism, waterless wastes, piracy and brigandage, have all combined to exclude Europeans, and now that we know so much of the history of Greece, Egypt, Palestine, and other Oriental countries from monumental records, Arabia is about the only place left that will afford us new and startling discoveries in the study of primitive mankind.

Taking South Arabia from the east, we may rapidly enumerate the work that has been done. In Oman we have the journeys of Palgrave, Wellsted, and Colonel Miles; along the coast-line a few people have landed here and there, notably Dr. Carter and Colonel Miles, who both visited the coast of Dhofar, but no one has gone inland as far as the hills. Herr Hirsch and we ourselves penetrated into the Hadramut last year. Mr. Tate, Colonel Wahab, and Mr. Harrington have made dives into Yemen and the Yafi country, and, of course, around Aden the country is now pretty well known. Dr. Glaser has made some interesting archæological excursions to Sana, Marib, and other parts of northern

Yemen, Professor Schweinfurth and a few others have also visited
Yemen, and Mr. Walter Harris also has made a journey from Aden
to Sana, but with these exceptions the rest of South Arabia remains to
be done. This evening I propose to describe the portion we have
reclaimed this winter from the unknown, and relate our journeyings
to and from Dhofar.

We reached Muskat on December 6 last, and there, with the kind
assistance of Major Hayes Sadler, prepared to make our plans. We
learnt that Oman proper might easily be surveyed at a time when the
bellicose tribes in the interior are in a state of comparative tranquillity;
to the west of Oman proper, there are several tribes of independent
and hostile character, only nominally under the sway of the Sultan of
Oman. Of these I may mention the Jenefa tribe, noted for slave-

Muscat Harbour, 1899 (Bent)

trading and looting shipwrecked vessels; they are naturally much
opposed to the visits of Europeans, lest their sources of livelihood should
be imperilled, but if the traveller can secure his *kafeer*, or paid escort,
from these chiefs, there is every likelihood that he would pass safely
through their territories. West of the Jenefa tribe stretches the vast
desert of Oman, which connects the arid wastes of Central Arabia
directly with the Indian Ocean, and the crossing of which entails hard-
ships which even the Arabs themselves are loath to encounter.

The most interesting feature in Oman is the Jebel Akhdar, or Green
Mountain, which, with its ramifications, occupies the whole of the
central district of this country, and rises to a height of over 8000 feet,
and in the winter-time is very cold, and subject, we were told, to falls
of snow. At the foot of this mountain is Nezweh, the old capital of
Oman, and we heard reports concerning the grapes which grow there
and the wine they make, and that it was the original home of the
muscatel grape. We then proposed to pass through the Jenefa tribe
to Ghubbet el Hashish on the west, and entirely abandon the desert

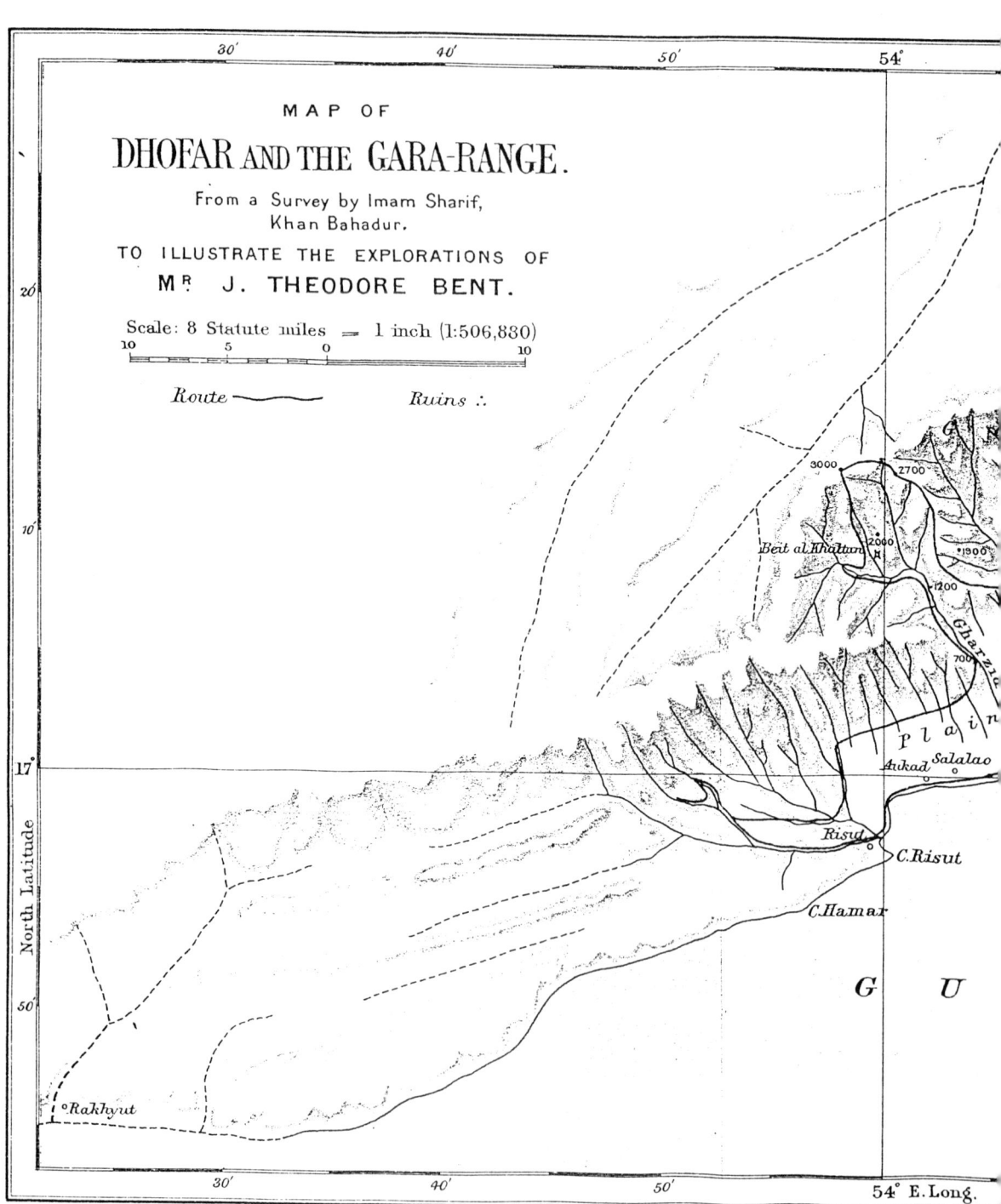

MAP OF
DHOFAR AND THE GARA-RANGE.
From a Survey by Imam Sharif,
Khan Bahadur.
TO ILLUSTRATE THE EXPLORATIONS OF
M͇ʀ J. THEODORE BENT.

Scale: 8 Statute miles = 1 inch (1:506,880)

Route ———   Ruins ∴

Map of Dhufar coastal plain and Jabal Qara, 1900 (Bent)

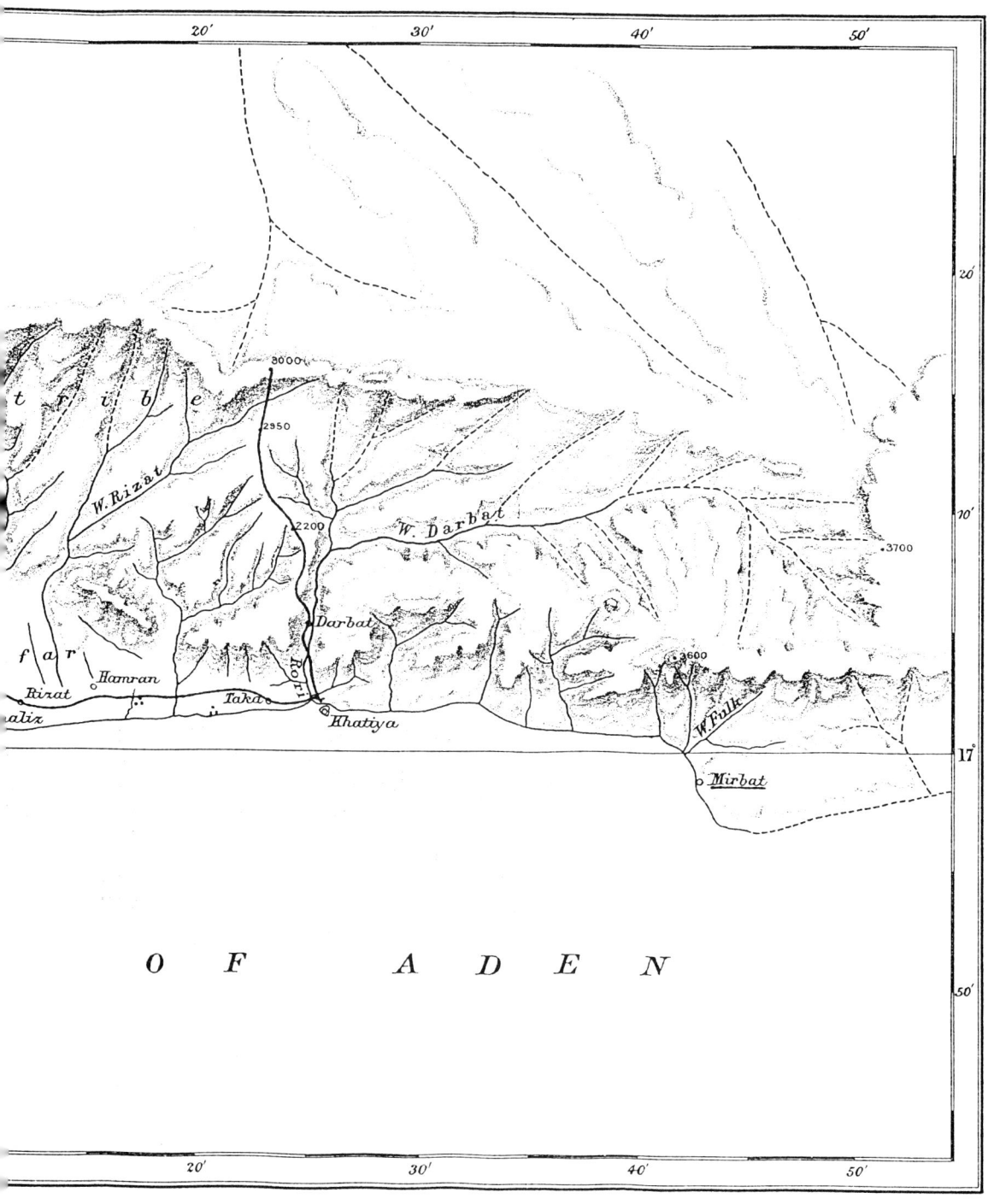

20'   30'   40'   50'

26°

t   r   i   b   e   3000

                                2950

W. Rizat   2200   W. Darbat   3700   10'

                                            Darbat

f   a   r
                                                              3600
Rizat   Hamran   W Falk
aliz   Taka   Khatiya   17°
                             Khatiya   Mirbat

                                                                        50'

O   F   A   D   E   N
                                                                        50'

20'   30'   40'   50'

route as impracticable for Europeans. To arrange for this journey, the Sultan of Muskat and Major Sadler took infinite trouble; camels were hired, and a horse for Mrs. Bent, and the sheikhs of the various tribes through which we should pass were summoned to escort us. Owing, however, to the illness of some of our party, we were at the last moment obliged to leave the expedition for another season, and, as events proved, it was fortunate we did so, for the insurrection broke out almost immediately afterwards, and in all probability we should not have returned alive to relate our experiences.

Before entering upon a description of the Dhofar district, which formed the principal part of our winter's work in Arabia, a few words may not be amiss on the present condition of Oman, as it has lately been the scene of considerable local agitation. Muskat, the capital and seat of government, has many points about it which remind one of Aden. It has a good harbour, and is surrounded by arid volcanic mountains. It bears the same relation to the Persian Gulf that Aden does to the Red Sea, and it is now practically British as far as the environments of the town are concerned, and the Sultan occupies much the same position as the native independent princes do in India. If it were not for British protection, and the presence of our gunboat in the harbour, Sultan Feysul, the present ruler, would have long ago succumbed to the attacks of the chiefs from the interior, and the ambitious projects of his brother Mohammed. Though on paper the territory of Oman is exceedingly large, extending to the north as far as the Turkish limits beyond Bereymi, and to the west beyond the confines of the great desert, including Dhofar, a distance of over 700 miles, nevertheless the authority of the present Sultan is almost entirely confined to the two towns of Muskat and Mattra, and the small strips of fertility immediately contiguous to these places. His great-grandfather, Sultan Seyid Said, was a man of considerable power, but on his death Zanzibar was separated from Muskat, and Sultan Toweyni's power was much circumscribed by the ambitions of his nephew Tourki, who ultimately succeeded in getting his uncle assassinated, and in ascending the throne under British protection; his son Feysul now reigns in continual dread of the machinations of his immediate relations. He and his half-brother Mohammed never meet without an escort to protect them from each other, and under this state of affairs the power and audacity of the Bedouin chiefs outside Muskat have increased to an alarming extent, and this has been the cause of the recent disturbances.

The present dynasty has been on the throne since 1741, when the Said family rose from obscurity, and the founder of the dynasty was elected Imam by popular acclamation; he shook off the Persian yoke, and established Oman as a power in the East. His son, Said bin Ahmed, was the last of the elected Imams. The Imam, as head of the Church and army in Oman, had to take an oath to fight against all infidels, but the next heir protested that this position was too expensive to maintain, so the Imamate, or spiritual lordship of Oman, has been in abeyance ever since. The result of this course of action has been a great laxity of religious opinion in Oman, fostered also by their natural antagonism to their fanatical neighbours, the Wahabi, who occupied the country and oppressed the inhabitants for a brief period at the commencement of this century. Nowhere in Arabia is Mohammedanism so lax as it is in Oman, offering a striking contrast to the bigotry of the people of the Hadramut, amongst whom our lot was thrown the previous winter.

The Omani form a third sect, as distinct from the Sonnites and Shiites, amongst whom the rest of the Muslim world is divided, and are known as the Ibadhiyah sect, or the followers of Abdullah bin Ibadh, who lived in the years 685–705 after the Hegira. The Sonnites recognize four successors of Mohammed as Imams—namely, Abubekr, Oman, Othman, and Ali. The Shiites reject all but Ali and his family, whereas the Ibadhiyites recognize none of these, but say that an Imam, if required, must be elected by the voice of the faithful. In Muskat one may enter the mosques without let or hindrance, and wretched neglected buildings they are, with only an apology for a minaret in one corner, about three feet high like an inverted bell. The chief buildings of Muskat are of Portuguese origin—namely, the two forts at either extremity of the town, and the ruined cathedral, which is now used as a stable for the Sultan's horses. Muskat has but few architectural attractions, a few carved balconies and doors, a few palaces, notably those of the Sultan and grand-vizier, and its aspect is one of squalor and dirt; few more unhealthy places could be found in the world. The narrow, dirty streets are the hotbed of disease, fostered by the moist heat to which it is subject. Just outside the walls by the fish-market is a malarious pond, into which the inhabitants throw their dead fish and other refuse, and the mephitic vapours from this, when the wind is from that quarter, spread fever broadcast through the town.

At the time when piracy was rife in the Persian Gulf Muskat was a great trade centre, but the overthrow of piracy and the introduction of steam has reduced Muskat to its normal condition, namely, that of a date-exporting harbour; nevertheless its strategical position and harbour will always ensure it a certain amount of trade and importance. The population of Muskat is strangely cosmopolitan for an Arab town. The merchants are chiefly Banyans from India and Persians; outside the town, in numerous encampments of reed huts, dwell colonies of Belouchis from the Mekran coast, African negroes from Zanzibar and Somaliland, and Bedouins from the mountains, by whom all the active labour of the town is carried on, whereas the Omani Arab is essentially lazy, and does as little for his livelihood as he can. Mattra, the commercial centre, from which all the roads to the interior start, is about

3 miles from Muskat round a headland; the journey thither is chiefly performed by water, the camel-track taking a long round through the mountains to get there. There is another road used in rough weather, partly by water, and then across a pass at the back of the headland. The scenery around Muskat is particularly fantastic and weird, but absolutely arid and unproductive, except where a few gardens are maintained by irrigation. The fishing village of Sadad is built on an inlet of the sea 4 miles south of Muskat, and the view over this, with its palm gardens belonging to Said Yusuf, and its fantastic background of tossed and tumbled volcanic peaks, intermingled with inlets of the sea, is one of the most striking I have ever seen. The coast-line, too, is very fine, with its beetling cliffs of black and green tufa rocks, the home of countless sea-birds. In one of the retired bays near Muskat, approachable only by sea, is the European cemetery, in which secluded retirement lie buried many men from the British gunboats, and the missionary

Sidab, 1900 (Bent)

Bishop French, who is probably the only man who has ventured to make an attack on the religion of Southern Arabia.

Dhofar, the ancient frankincense country, at which point we elected to commence our winter's campaign, is 640 miles by sea from Muskat; this distance we were prepared to traverse in a boat lent us by the Sultan, called a *bateel*, rather smaller than a dhow, when luckily one of the Turkish pilgrim steamers on its way to Jeddah put into Muskat, and the captain consented to drop us for a consideration at Mirbat, the first point of Dhofar, where the desert ends. Thus we were saved an uncomfortable sail of doubtful duration along an inhospitable coast, where the inhabitants are few, and celebrated for their marauding tendency.

Dhofar is nominally under the Sultan of Muskat. I may here emphatically state that the southern coast of Arabia has absolutely nothing to do with Turkey, and from Muskat to Aden there is not a single tribe paying tribute or having any communication with the Ottoman Porte.

Eighteen years ago the inhabitants of Dhofar were in such a state of internal turmoil, and in such dread of the Bedouins of the Gara tribe, that they applied to Sultan Turki for a ruler. He sent them Wali Suleiman, a man of remarkable strength of character and determination, who has gained for himself a great reputation for bravery amongst the neighbouring Bedouin tribe, who all respect him and his authority. Suleiman was the son of a slave of Sultan Toweyni, and a most intimate friend and adherent of Sultan Turki, whom he greatly assisted in coming to the throne of Oman. He had only a hundred Arab soldiers with him when he reached Dhofar, but his skill was such and his powers of organization so efficient, that he defeated the Bedouins in several encounters, and now he boasts that he has twelve thousand Bedouins devoted to him, and told us with pride that two years ago he had sent two thousand rupees as tribute to Muskat, last year he had only sent one thousand, and that this year he had sent none. His next step will probably be, when a favourable opportunity offers, to declare himself independent under British protection. For this reason he was exceedingly polite to us, entertained us in his castle by the coast at Al Hafa during our stay, and arranged with the greatest possible assiduity for our safety during our exploration of the interior.

Dhofar and the Gara mountains which encircle it form a quite abnormal feature in this otherwise arid coast. From Cape or Ras Risut on the west to Mirbat on the east we here find a long narrow stretch of flat alluvial soil at the foot of the mountains, very little raised above the level of the sea. This plain is never more than 9 miles wide, and at the eastern end, where the mountains come down nearer to the sea, it is reduced to an exceedingly narrow strip. Water is here very near the surface. Streams making their way to the sea are of constant occurrence; consequently the plain is very fertile, and capable of producing almost anything. Along the whole line are many groves of coconut palms. Tobacco, cotton, Indian corn, and various species of grain grow here in great abundance; in the gardens we find many of the products of India flourishing, namely, the plantain, the papya, mulberries, melons, chilis, brinjoles, and fruits and vegetables of various descriptions. In fact, Dhofar and the Gara mountains may be termed one large oasis by the sea, bounded on the north by the Nejd desert, on the east by the Oman desert, and the gradual tendency to the west is towards the arid hills and sand-choked valleys which we met in the Hadramut last year. As we shall presently see, the Gara mountains are full of water, forming itself here and there into small lakes. They are decked to their summits with rich vegetation, and this will account for the fertility of the plain of Dhofar, and the strange contrast it forms to the rest of the coast-line of Arabia.

The one drawback to the progress of this country is its harbourless condition. During the north-east monsoons dhows can find shelter at

502

Mirbat, and during the south-west monsoons at Risut, but the rest of the coast-line is provided with nothing but open roadsteads with a rough surf always rolling in from the Indian Ocean, and we had considerable difficulty in landing ourselves and our goods at Al Hafa in small hide-covered boats specially constructed for riding over this surf.

We traversed the whole of this plain between Capes Risut and Mirbat in various directions, and found thereon the sites of ruined towns of considerable extent in no less than seven different points, though at the two capes where now is the only anchorage there are no ruins to be seen, proving, as we afterwards verified for ourselves, that anchorage of a superior nature existed here in antiquity, and which has since become silted up, but which anciently must have afforded ample protection for the boats which came here in the frankincense trade. At Takha, as we shall presently see, there was a very extensive and deep harbour,

Coastal scenery west of Salalah, 1900 (Bent)

running a considerable distance inland, which with a little outlay of capital could easily be restored.

After a close examination of these ruined sites, there can be no doubt that those at spots called now Al Balad and Robat, about 2 miles east of the Wali's residence, formed the ancient capital of this district. We visited them last Christmas Day, and were much struck with their extent. The chief ruins are by the sea, around an acropolis some 100 feet in height. This part of the town was encircled by a moat still full of water, and in the centre, still connected with the sea, but almost silted up, is a tiny harbour. The ground is covered with the remains of Mohammedan mosques, and still more ancient Sabæan temples, the architecture of which—namely, the square columns with flutings at the four corners, and the step-like capitals—at once connects them architecturally with the columns at Adulis on the Red Sea, those of Koloe and Aksum in Abyssinia, and those described by M. Arnaud at Mariaba in Yemen. In some cases these are elaborately decorated with intricate patterns, one of which is formed by the old Sabæan letters ⊙ and X, which may possibly have some religious import. After seeing the ruins at Adulis and Koloe, the numerous temples or tombs with four isolated columns, no doubt can be entertained that the same people built them all. As at Adulis and Koloe, there is unfortunately an absence of epigraphy which would materially assist us, but the subsequent Mohammedan occupation and the conversion of the temples into mosques may account for this.

This town by the sea is connected by a series of ruins with another town 2 miles inland, now called Robat, where the ground for many acres is covered with ancient remains; big cisterns and water-courses are here cut in the rock, and standing columns of the same architectural features are seen in every direction.

With the aid of Sprenger's ' Alte geographie Arabiens,' the best guide-book the traveller can take into this country, there is no difficulty in identifying this ancient capital of the frankincense country as the Μαντεῖον Ἀρτέμιδος of Claudius Ptolemy. This name is obviously a Greek translation of the Sabæan for some well-known oracle which anciently existed here, not far, as Ptolemy himself tells us, from Cape Risout. This name eventually became Zufar, from which the modern name of Dhofar is derived. In A.D. 618 this town was destroyed and Mansura built, under which name the capital was known in early Mohammedan times. Various Arab geographers also assist us in this identification. Yakut, for example, tells us how the Prince of Zufar had the monopoly of the frankincense trade, and punished with death any infringement of it. Ibn Batuta says that "half a day's journey east of Mansura is Alahkaf, the abode of the Addites," probably referring to the site of the oracle and the last stronghold of the ancient cult.

Claudius Ptolemy is certainly very clear in his geography of this coast. East of the Hadramut is the promontory of Syagros (Ras Fartak), and east of Syagros, on the Sachalites Sinus, first is mentioned the city of Manteion Artemidos (Μαντεῖον Ἀρτέμιδος), and then Abyssapolis (Ἀβυσσάπολις). Between this last-mentioned town and Rasal Hadd (Cape Coradamus), a long distance then as now of desert coast, he gives us no name. Sprenger sums up the evidence by saying that the town of Zufar and the later Mansura must undoubtedly be the ruins of Al Balad. He also associates Abyssapolis with Mirbat, and the existing evidence, as we have seen, quite confirms this statement. Thus, having assured ourselves of the locality of the ancient capital of the frankincense country—for no other site along the plain has ruins which will at all compare in extent and appearance with those of Al Balad—we shall, as we proceed on our journey, find that other sites fall easily into their proper places, and an important verification of ancient geography and an old-world centre of commerce has been obtained.

Wali Suleiman gave us of the best he had to offer, and placed the best rooms of his castle at our disposal. The residences of the Arabs in Dhofar will not compare with those of the Hadramut, which surprised us so much last year. Wali Suleiman constructed his own when he came to the country, equi-distant from a cluster of villages which contain the chief Arab population of the plain. It is strong and substantial, and stands in an isolated position close by the coast; a fine gateway leads into a long dark passage, lined on either side by stone benches, where the Wali's soldiers recline, and where sheikhs from the mountains are regaled with coffee out of a huge coffee-pot with a long bird-like beak when they come to visit him. This passage leads into a spacious courtyard with a well in it, near which dwells the white she-ass, Wali Suleiman's only steed, on which he pays his state visits to the various villages in his dominions. Also there are here kept a considerable number of state prisoners, Beduins convicted of rapine, and Mahris captured in war. They are all bound with iron fetters, and the worst characters are chained to blocks of wood. Every night these prisoners said their prayers in a corner, led by an imprisoned mollah, and bewailed their misdeeds into the small hours of the night. This fact alone attests to Wali Suleiman's power over his neighbours. We never saw such a sight as this amongst the Hadramut sultans.

Wali Suleiman's ordnance is not of a high order; his soldiers mostly fight with the antiquated matchlock guns, and a few rusty old cannons stand near the doorway; but these are sufficient to overawe the Beduins, who have but little in the way of firearms. Outside the castle there is a large enclosure or bazaar, where the frankincense trade is carried on, and there is also a long palm-thatched shed which serves as a sort of parliament-house, where the chiefs who visit the Wali sit during the day, and a fire is always burning there to provide these guests with coffee. Almost directly you leave the castle you step into cultivated land extending in every direction for miles, and dotted with coconut groves around the villages, the houses in which are plain but substantial, being mostly built of stones brought from the ruins. The mosques here, as in Oman, are most insignificant, and the inhabitants do not appear to be the least fanatical.

After a few days' delay at Al Hafa, under the hospitable roof of Wali Suleiman, our arrangements for an expedition into the interior were made. To secure our safety, the Wali had summoned seventeen sheikhs or heads of families of the Gara tribe, who own the mountainous district behind the plain up to the confines of the Nejd desert, and in accordance with his summons they all arrived, bringing with them their camels with which to convey our party and our luggage. The whole convoy was placed under the guidance of one sheikh, Sayel, an old grey-haired, wiry man, dressed in nothing but a loin-cloth, but the possessor of so great wealth in flocks and herds that all the others treated him with great respect, and termed him the sheikh of all the

Garas. He owns seventy camels, worth 500 rupees apiece, and cows, sheep, and goats innumerable.

We never had to deal with wilder men in our lives than those who constituted our escort; they wore long unkempt hair, tied down with a leather thong like a boot-lace. Each man carried his wooden shield called a *gohb,* so constructed with a knob that he could turn it round and use it as a stool to sit upon; his wooden spear pointed at both ends, called *ghatrif,* a weapon peculiar to the Gara tribe, which they hurl with wonderful precision; and his flat iron sword called *saif.* Very few also carried matchlock guns. We found these wild men in most respects friendly during our wanderings, but of most independent spirit. If we asked them to do anything for us, they would reply, "We are sheikhs, not slaves;" and when once away from the influence of Wali Suleiman, they paid no heed to the orders of the soldiers sent by him, and during the time we were with them we had the unpleasant feeling that we were entirely in their power. They would not march longer than they liked; they would only take us where they wished, and they were unpleasantly familiar; with difficulty we kept them out of our tents, and if we asked them not to sing at night and disturb our rest, they always set to work with greater vigour. They got hold of our Christian names, and were for ever using them, to our great annoyance. They affected indifference for money, and absolutely the only hold we had over them was medicine; they positively loved my medicine-chest, and during our journey consumed an incredible quantity of pills, quinine, and other dainties. At first they would chew the pills, with disastrous results, but we soon taught them to swallow them in the orthodox fashion. Every night we had a row of them wanting to be doctored, and with this feeble weapon we ruled. Certainly they did well by us on the whole, and eventually we were satisfied that they took us to see everything in their country, but at first we doubted them greatly. They would chat pleasantly with us as we went along, but were ready at the slightest provocation to fly into wild incoherent rages, and the information they gave us about the country was never twice the same. Altogether the Garas are different in character and physique to any natives I have met elsewhere; in type they are akin to the Hadramut Beduin, small, active, and with finely cut features, but they are much wilder and less accustomed to contact with civilization. They live chiefly in caves and under trees, only using reed huts when they come down to the plain to encamp with their flocks during the rains.

Not a whit less wild than their masters are the camels of the Gara; they danced about like antelopes, and made hideous noises when loaded, and we had the greatest difficulty in getting our goods fastened on to them. The Beduins were totally ignorant of camel-loading; they brought no ropes or thongs, a supply of which we were luckily able to obtain

506

from the Wali, and during the first days of our journey most of our baggage was thrown off and damaged, and it became quite a common sight to see a camel scampering away across the plain in terror, dragging its fallen load over rocks and through thorns. In the Hadramut we were surprised to see the camels eating fish; here in Dhofar we were still more surprised to find them consuming bones with avidity, and if they saw a bone on the side of the track, no power of ours would prevent them from stepping aside and appropriating it. As for the riding camels, they too were painfully wild, and we all in our turns had serious falls, and counted ourselves lucky that no bones were broken. The fertile highlands of the Gara country are celebrated for breeding camels, but they do not use them themselves except for bringing frankincense to the coast, as there is no trade route or communication with the interior through their country. These were our chief difficulties in travelling through the Gara country, but as we had no fear for our personal safety and the attacks of hostile tribes, we felt infinitely happier than in the Hadramut.

We left Al Hafa on December 30 last, and our first day's march took us close to Cape Risut, past several ancient ruined towns of minor importance. Near Cape Risut a large tract of country is covered with frankincense trees, with their bright green leaves like ash trees, their small green flowers, and their insignificant fruit. The frankincense, the old staple trade of this district, is still gathered in three places in the Gara mountains. The best is obtained at spots called Hoye and Haski, about four days' journey inland from Mirbat, where the Gara mountains slope down into the Nejd desert. The second in quality comes from near Cape Risut, and also a little further west, at a place called Chiseri, frankincense of a marketable quality is obtained, but that further west in the Mahri country is not collected now, being much inferior. The best quality they call *leban lakt*, and the second quality *leban resimi*, and about 9000 cwt. are exported yearly and sent to Bombay. It is only collected in the hot weather, before the rains begin, in the months of March, April, and May, for during the rains the tracks on the Gara mountains are impassable. They cut the stem, and after seven days return to collect the gum which has exuded; this they do three or four times a month, and in the cool weather, as the gum comes but slowly, they leave the trees alone. The trees belong to the various families of the Gara tribe; each tree is marked and known to its owner, and the product is sold wholesale to Banyan merchants, who come to Dhofar just before the monsoons to take it away.

One could not but feel interested in the existence still of this old-world trade on the very spot which was once so celebrated for it, when the odoriferous gum was much more prized for temple-worship and household consumption than it is now; and as we rode through the groves of this incense-bearing shrub we thought of the cunning old-world legends of the dragons which were supposed to guard these trees, and of the death-giving odours which they were believed to exhale; for the old Sabean frankincense merchants were jealous guardians of their treasures, and sought by awe-inspiring anecdotes to keep off competition.

Lake in Wadi Jarziz, 1900 (Bent)

From Cape Risut we went inland to the base of the mountain range, and spent several days in visiting lovely little gorges, which run a short way into the mountains; but there is no approach by them to the

heights above, the wall of rock being here abrupt and impassable. They
are lovely ideal little spots, with running streams and ferns and trees, bulrushes, reeds, and tropical vegetation. Bednin families live in the caves
around, finding here ample fodder for their cattle. We originally understood that Sheikh Sayel was going to take us up to the mountains by a
valley still further west, but for some reason, which we shall never
know, he refused; some said the Mahri tribe was giving trouble in this
direction, others that the road was too difficult for camels. At any rate,
we had partially to retrace our steps, and following along the foot of
the mountains, found ourselves encamped not so many miles away from
Al Hafa.

The next day we entered the mountains by the Wadi Ghersid, the
regular Gara track between the coast and the interior; the entrance to
this gorge is about 9 miles from the Wali's castle, and on entering it a
great surprise was in store for us. Instead of the sand-choked valleys of
the Hadramut, arid except where irrigation is carried on with immense
labour, we here were plunged into a valley covered with the richest
tropical vegetation. Just above our camp, on the second day, water
coming out of three holes in the mountain-side forms itself into a small
and exquisitely beautiful lake, well stocked with duck and other water-
birds, photographs of which we took; the encircling rocks are overhung
with creepers, and covered with maidenhair and other ferns; huge fig-
trees block up the valley, the lower branches of which are full of *débris*,
showing how in the rainy season this gorge must be a raging torrent;
limes, cactus, aloe, and mimosa form on all sides a delightful forest,
whilst the mountains rising above the lake are clad almost to the summit
with timber. Such a scene as this we never expected to witness in Arabia;
it reminded us more of the rich valleys leading up to the tableland of
Abyssinia, and never shall we forget the delightful evening spent by
the lake of Ghersid. It is doubtless probable that a knowledge of such
valleys as these gained for Arabia its ancient reputation for floral
wealth. Passages in Theophrastus, Strabo, Athenæus, allude to this, and
more especially to its wealth in aromatic plants. Aristotle calls Arabia
εὐώδης, sweet-smelling, and Pliny more especially gives us a list of the trees
and herbs grown in Arabia, and it is highly likely that the frankincense
merchants who visited Dhofar in pursuit of their trade knew of these
valleys, and not unnaturally brought home glowing accounts of their
fertility, and gained for Arabia a reputation which has been thought to
be exaggerated.

Next day we pursued our way up the gorge of Ghersid, climbing
higher and higher, making our way through dense woods, often
dangerous for the camel-riders, and obliging us frequently to dismount;
sweet-scented white jessamine hung in garlands from the trees, and the
air was fragrant with the odour of many flowers; above us towered
grey rocks, and the hill-slopes were clad on both sides with trees.

We had one more camp in this lovely valley, almost at its head, where it is very narrow, and then on the following morning we commenced the ascent up a rugged path exceedingly difficult for the camels. The highest point of this range of mountains is not more than 3000 feet, and at our camp that night we registered 2600 feet. From above the aspect of this country is very curious. On the side towards the sea the mountains are cut by several deep gorges, similar to the Wadi Ghersid, full of vegetation, and all the hills around up to the summit are covered with grass and clusters of trees; as it was the dry season, this grass was converted into hay, which no one cared to harvest. Here and there in the brown expanse were isolated groups of fig-trees, of which we counted three varieties, and the thick foliage of these trees was full of birds; these groups of trees give a very park-like aspect to the country, and dotted over it we saw numerous herds of camels, goats, and cows grazing, which belonged to the Garas. We constantly came across their homesteads, which consisted of deep caves in the hillside, in which the families and the flocks live together in happy union; the calves and kids are penned in holes in the rocks, the milk is churned by shaking it in a skin attached to a tripod, and all their implements are of the rudest kind.*

We found the Gara women exceedingly shy and retiring—so different from the bold hussies in the Hadramut, who tormented us with staring into our tent; they fled, if we approached them, like timid gazelles. They have but poor jewelry—silver necklaces, armlets, and nose and toe rings; they love to join their eyebrows with antimony, and stick some black sticky stuff like cobbler's wax over their noses and foreheads; they are very small, and like Japanese; they do not cover their faces, and are very lightly clad in dark blue homespun cotton garments.

After proceeding along these highlands for two days, we decided to halt for two nights, nominally for a rest, as we had now been on the march for many consecutive days; but there came on the most frightful hurricane from the north, which blew steadily for two days and nights, and effectually prevented us from getting the rest we required. Our tents were with difficulty kept erect, and the cold was very trying. Our Beduins lay in an inert mass round wood fires, and our whole camp was for the time being thoroughly miserable.

From this point, however, we were able to take excellent observations around us, and form a clear idea of the configuration of the Gara range. The oasis-like nature of this range is here very marked; in all directions beyond it is desert. As it slopes down to the north it gradually becomes more and more arid, vegetation becomes more and more sparse, until it ends altogether in the yellow desert of Nejd, stretching as far as the eye could reach, and ending in the horizon in

* It is interesting to read in the 'Periplus,' § 32, a description of this coast, and of the high mountains behind, "where men dwell in holes."

a straight blue line as if it was the sea. Sheikh Sayel promised to take us across the Gara border into Nejd if we wished; but as it would have entailed a considerable delay and parley with the sheikhs of the Nejd Beduins, and as we could see from our present vantage ground that the country would afford us absolutely no objects of interest, we decided not to attempt this expedition.

On leaving our very exposed and nameless camping-ground, we pursued our course in a north-east direction, still passing through the same park-like scenery, through acres and acres of lovely hay worth nothing a ton. It is exceedingly slippery, and dangerous foothold for the camels; consequently numerous falls were the result, and much of our journey had to be done on foot.

On the second day we began again to descend down a hideous road, and a drop of about 1500 feet brought us to a remarkable cave just above the plain, and only about 10 or 12 miles from Al Hafa. This cave burrows far into the mountain-side, and is curiously hung with stalactites, and containing the deserted huts of a Beduin village, only inhabited during the rains. Immediately below this cave in the Wadi Nefas are the ruins of an extensive Sabæan town, in the centre of which is a natural hole 150 feet deep and about 50 in diameter; around this hole are the remains of walls, and the columns of a large entrance gate. We asked for information about this place, but all we could get in reply was that it was the well of the Addites, the name always associated with the ruins of the bygone race. In my opinion this spot is the site of the oracle mentioned by Ptolemy and others, from which the capital of Dhofar took its name. It much resembles the deep natural holes we found in Cilicia in Asia Minor, where the oracles of the Corycian and Olbian Zeus were situated. It is just below the great cave I have mentioned, and, as a remarkable natural pheno-menon, it must have been looked upon with awe in ancient days, and it was a seat of worship, as the ruined walls and gateway prove; further-more, it is just half a day's journey east of the city of Mansura or Zufar, where, Ibn Batuta somewhat contemptuously says, "is Al Akhaf, the abode of the Addites," and there is no other point on the plain of Dhofar where the oracle could satisfactorily be located from existing evidence. Sometime, perhaps, an enterprising archæologist may be able to open the ruins about here, and verify the identification from epigraphical evidence.

From this point we rode across the plain for about ten miles to the coast, at a place called Rizat, and were entertained for the night by Wali Suleiman, at a house he has built here, some 12 miles west of his permanent home at Al Hafa; but as his accommodation here is limited, we remained in our tents. Here he has utilized a running stream to fertilize several acres of ground, in which he grows tobacco, sugar-cane, jowari, and various other products; his garden is well stocked with fruit and vegetables, and is delightfully shady, and we enjoyed a rest under a mulberry tree after our hot ride, and ate quantities of the small fruit. Wali Suleiman spends much of his time here, getting away from his troubles both domestic and political.

From Rizat to Takha is an uninteresting ride of 13 miles along the plain, past the mouths of several streams with plenty of water, clear but brackish, and with dense growth of mangrove down by the sea.

At Takha are extensive remains of an ancient town, which must have been second only in importance to the capital at Al Balad; similar columns, only without decoration, are standing here and there, marking the sites of tombs and temples, and we passed by several large stone sarcophagi and other remains. As the mountains come down here very near to the sea, the position is considerably more attractive, and the modern village is one of the largest of those scattered over the plain of Dhofar.

It at once occurred to us that this must be the site of the town which is alluded to by Claudius Ptolemy and Arrian as Ἀβυσσάπολις. Yakut tells us (iv. 481) "it is the harbour for the town Zufar, and is distant five parasangs from it. The harbour is good, and is often mentioned by traders, whereas the town Zufar has only a roadstead but no harbour." Ibn Khaldun, in his geographical account of the countries beyond Yemen, says, "Zufar was the seat of empire of the Tubbas, and Mirbat was situated on the seashore; both cities are now in ruins." Again, Bunbury, in his work on ancient geography, says, "The port of Moscha, which appears to have been a place of considerable trade, must probably have been situated in the district now known as Dhofar, a little to the west of the modern town Mirbat." Dr. Glaser, in his Arabian Geography, says, "If the Moscha of the Periplus must be sought for about 10 miles west of Mirbat, then I have no hesitation in saying that it must be identified with Abissa Polis and Tafa."

These accounts, if they at first sight offer a little discrepancy, can on inspection be easily reconciled. The ruins of Takha are exactly as Yakut says, 5 parasangs or 20 miles west of those of Zufar at Al Balad, and as there are none at the modern town of Mirbat, and only indifferent anchorage during the north-west monsoons, it would suggest itself that the Mirbat of antiquity was situated here.

On the following day this somewhat puzzling question was settled for us by finding the only thing wanted to identify the spot, namely, a commodious harbour. An hour's walk from our camp near Takha took us across a promontory where the estuary of a river forms quite a large lake, separated from the sea only by a narrow sand-belt over which the water flows at high tide. Around this lake are the ruins of several ancient buildings, and what is now a headland connected to the mainland by a neck of sand is surrounded by an ancient wall and fortification, and bears the appearance of once being an island protecting the

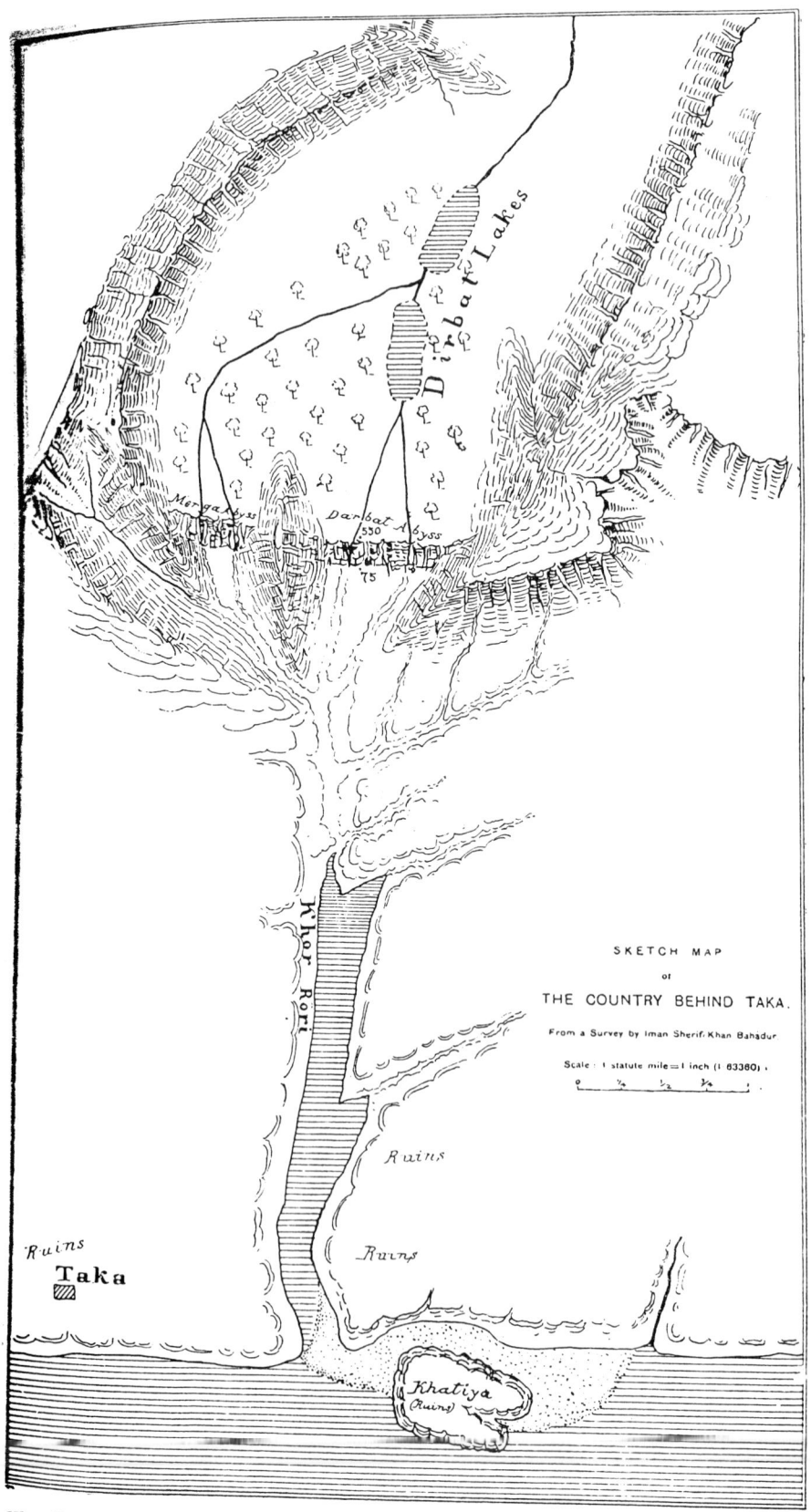

Khor Rori and Darbat Falls, 1900 (Bent)

entrance to a harbour. The similarity to some of the ancient Greek harbours is here very striking, and the lake, when connected with the sea by a proper channel, as it must have been until quite a recent date, must have formed a most spacious and commodious harbour.

Here we had the one thing wanting to identify the site, namely, the harbour of which Yakut tells us, where the ships which came to Dhofar in the frankincense trade found anchorage. The Abyssapolis of Ptolemy, like Manteion Artemidos (Μαντεῖον Ἀρτέμιδος), is evidently the Greek equivalent for some Sabæan name, or merely called from the existence near here of a remarkable abyss which we shall presently visit. The name given us by Ibn Khaldun of Mirbat is still attached to the village and anchorage 12 miles to the west, where the modern dhows go, and the term "Moscha" or "Mocha," which Arrian here introduces, is one frequently occurring on the Arabian coast, and apparently means, as Dr. Glaser tells us, an inlet or harbour, and consequently we have no difficulty in deciding that the ruins and harbour near Takha are those of the ancient town and harbour known to the Greeks as Abyssapolis (Ἀβυσσάπολις), or Moscha, and to the natives as Mirbat.

We skirted along the lake, which is called by the natives Kho Rowri, for a mile or more as we rode inland from the coast, until it dwindled into a narrow stream, and then lost itself in the dry rocky bed of a torrent coming down from the valley we were about to enter. A ride of 4 or 5 miles brought us up to the higher range of mountains again and the opening of the valley, and after following this for another mile, we pitched our camp at the foot of one of the most stupendous natural phenomena we have ever seen. The valley leading down to the sea has been filled up in the course of ages by a calcareous deposit, collected on either side of an isolated hill in the middle of the valley, about 1000 feet in height. This deposit has taken the form of a perfectly straight and precipitous wall 550 feet in height and three-quarters of a mile long on the eastern side of the hill, and about a quarter of a mile long and 300 feet high on the western side.

Over these walls feathery waterfalls precipitate themselves something in the style of the Staubbach, adding perpetually to the chalky secretions of which these walls are constructed. During the rains the falls must be magnificent, but, as I have said before, it was the height of the dry season when we visited it.

The general appearance of these walls is white and whitish-grey, with long white stalactites hanging down in tumbled confusion; it is streaked here and there, where the water perpetually falls, with patches of green, and below it plantains 20 feet high, enormous castor-oil plants, daturas, and many other plants flourish, and the Beduins have utilized the stream before it loses itself in the rocky channel to make small gardens.

The rocky channel below is itself very curious, being a flat surface about 50 yards across, of perfectly white calcareous rocks, and just below the wall where the water comes down there is an enormous amount of white calcareous deposit, quite soft and springy to walk upon.

The general aspect of these two walls is exceedingly striking from below, they are so sheer and straight; and it was curious to see the Beduins, who live above, like tiny dwarfs looking over the dizzy height in wonder at the first Europeans who had invaded their wonderful abode.

As we looked at this precipice, there seemed to be no doubt as to why Claudius Ptolemy had given the name of Ἀβυσσάπολις to the town on the coast. The merchants who came there for frankincense must have known of it quite well, and marvelled like we did at this great natural phenomenon in the mountains just behind the town. Similarly, another town in Arabia is called Abisamapolis (Ἀβίσαμαπόλις) by Ptolemy, which has a steep mountain behind it 4000 feet high, up which a road leads to Marib (Sprenger, § 96), and it is also clear that Greek names were given by foreign traders to the places they visited from local peculiarities; for example, the μαντείον Ἀρτεμιδος must have been the seat of the oracle of some female Sabæan goddess corresponding in attributes to Artemis. Be this as it may, the stupendous abyss stands there still as one of the world's wonders, constructed by nature on the same principle as the pink and white terraces in New Zealand, and the calcareous deposits of Yellowstone Park.

To thoroughly explore the vicinity of this wonderful spot, we stayed in our camp at the foot of the abyss for three nights.

On the first morning we set out early to climb the hill and see what is to be seen at the top of the abyss. A rough camel-track has been made to climb this hill, which has been partly swallowed up by the calcareous deposit, and forms a spur covered with thick vegetation, with strange peeps through the branches at the wall and the waterfalls on either side: from this path we got an excellent view and photo of the smaller abyss, which the natives call Merga, whereas the larger abyss and the water at the top of it is known as Dirbat, which looks as if it had some close connection with the name Mirbat which we have just alluded to. More water falls over the smaller abyss than over the larger one, and when we were there two considerable falls were precipitating themselves over it.

On reaching the top of the abysses, we found ourselves on a lovely grassy plain, as flat as a billiard-table, and grazed over by quantities of cows belonging to the Gara herdsmen. We walked along for about a mile under the grateful shade of large trees, and when we had rounded the jutting hill, we found ourselves by the side of a long, narrow lake, which feeds the several channels which fall over the abysses; and when the Beduins want to water this extensive meadow, they dam up these channels, and thus have a natural source for irrigation provided for them even in the driest weather.

The banks of this lake are adorned with very fine timber, principally fig-trees and mimosa, amongst the branches of which a fine convolvulus with large pink flowers was creeping. The lake is full of bulrushes, and quantities of birds live on it—wild duck, herons, and waterhens; in some parts it is very deep, and it is divided into two parts connected by a narrow running stream; it is not broader in the dry season than a wide river, and in length it is about 2 miles, and the source which feeds

Abyss of Darbat, 1900 (Bent)

it comes out of the mountains behind, and can be traced up a narrow gorge for about two days' journey.

The Beduins have many superstitious connected with it, and, indeed, it is a most fairylike-looking spot; they affirm that jinnies live in the water, and that whoever wets his feet here is sure to have fever. Sheikh Sayel assured us he had actually seen the jinni, as also one which dwells in a cave hard by, where some Beduin herds have their farmstead. The mountains on either side are very curious, being honeycombed with caves and fissures, and in one place there is a large hole in one of the mountains through which daylight can easily be seen. The slopes of the hills on either side of the valley are covered with rich timber and vegetation, and the possibilities of this valley, if brought under proper cultivation, struck us as very great. Altogether, the abysses, with the lake and flat valley above it, struck us as quite the

516

most weird and fascinating spot we had ever visited in any of our wanderings.

After going up to the head of the lake, we partook of some refreshments under a wide-spreading fig-tree by the side of the lake. We then went down to the end of the valley to peep over the abyss. The Beduins had been damming up the channels for irrigation, and we had to cross much swampy ground, and got our feet wet without catching the inevitable fever; and after pursuing our way for about a mile, we reached the edge of the large abyss and peeped over into the dizzy depth below. Needless to say the view is a most striking one, and the sun's rays made lovely rainbows in the feathery falls, and far below us our tents looked like tiny specks in the trees.

Every November a fair or gala is held up here by the side of the lake, to which all the Beduins of the Gara tribe come and make merry,

Above the abyss, Darbat, 1900 (Bent)

and the fair of Dirbat is considered by them the great festival of the year. A round rock was shown us on which the chief magician sits to exorcise the jinni of the lake, and around him the people dance. There is doubtless some religious purport connected with all this, but, as I have said before, it is extremely difficult to get anything out of the Beduins about their religious opinions; like the Beduins of the Hadramut, they do not observe the prayers and ablutions inculcated by the Mohammedan creed, and the Arabs speak of them as heathen, but beyond this we could not find out much. Their language, too, is different from anything we had heard before. They can understand and converse in Arabic after a fashion, but when speaking amongst themselves none of our party, Arab or European, could make out anything they said, and from such simple words as we were able to learn—such, for example, as *ouft* for *wadi*, a valley, *shur* instead of *yom* for " day," and *kho* instead of *nahr* for a river—we were led to believe that they speak an entirely different language, and not a dialect as in the Hadramut.

Sprenger (§ 449) supposes that the tribal name Gara or Kara corresponds to the ancient Ascites whom Ptolemy places on this coast; but as the Ascites were essentially a seafaring race, and the Gara are a pastoral tribe of hill Beduin, the connection between them does not seem very obvious. It is more probable that they may correspond to the Carrei mentioned in the campaign of Aelius Gallus as a race of Southern Arabia, possessing, according to Pliny, the most fertile country.

After another day spent in sketching, photography, and measurements, we felt that we had thoroughly explored the neighbourhood of the abyss, so we started back to Al Hafa, which we reached in three days, to prepare for our departure from Dhofar, and the interests which centred in this small district—the ancient sites, the abyss, and, above all, the surprising fertility of the valleys and mountains, the delicious health-giving air, and the immunity from actual danger which we had enjoyed—combined in making us feel that our sojourn in Dhofar had been one of the most enjoyable and productive of any expedition we had hitherto undertaken, and that we had discovered a real Paradise in the wilderness, which will be a rich prize for the civilized nation which is enterprising enough to appropriate it.

It was with regret and manifold misgivings that we now entered upon another phase of our Arabian journey, and faced a series of disappointments, which, had we foreseen, I think we should have tarried far longer under the favourable influence of Wali Suleiman.

He did all he could to assist us in the further progress of our journey. When we told him that our object was to go across the Mahra country by land to the Hadramut, he wrote and sent a messenger to the neighbouring Mahri Sultan of Jedid to ask his permission for an escort through his country; the reply was unfavourable, so we had nothing left to do but to hire a *bateel* and set sail along the coast for Kishin, to the Sultan of which place I had a letter from the British political agent at Muskat. A long sea-journey in an Arab bateel is exceedingly uncomfortable. We had a cabin in the stern open all round, and with a sail in front to secure some privacy, and so low that we could not sit up in it, and for six days Mrs. Bent remained in her camp-bed without getting up, for the simple reason that there was nowhere else to go to. Then the smell of bilge-water was horrible. Every silver thing we had about us turned black, and our clothes will probably never recover from the sulphureous vapours to which they were exposed; the water in our butts was the home of cockroaches and scorpions, and our cooking was done in a square wooden box with ashes in it. For the first two days we had little wind, and made very slow progress, our compensation being that we passed slowly by most exquisitely beautiful coast scenery. From Cape Risut for about 100 miles the coast is exceedingly rugged and precipitous, rising sheer out of the sea to a height of 1000 feet. Vegetation here and there adorns the

518

only available spots where earth is to be found, and now and again
appeared outcrops of the calcareous deposit, like waterfalls down the
cliff—small replicas of the abyss which we had visited behind Takha.

We anchored for one night off Rakhiut, the most western fortress
of Wali Suleiman, which he has erected to ward off the attacks of the
hostile Mahri tribes; another night we were in slight alarm from some
Mahri Beduins, who called to us from the shore. Our sailors told us
they were trying to beguile us ashore with a view to plunder; consequently we turned a deaf ear to their cries, and anchored about 100
yards from the shore. As slave-traders and wreckers the Mahris have

Patterns on a capital, Dhufar, 1900 (Bent)

the worst possible reputation along this coast amongst the owners of
Arab buggalows, dhows, and bateels, and no wise captain ever ventures
to land about here if he can help it. At length the longed-for breeze
arose, and carried us in two days past Ras Fartak to the harbour of
Kishin.

Between Jedid and Ras Fartak the land is low and recedes, and as
we sailed along we decided that it was the mouth of some big valley
from the interior, and after careful cross-examination of the Sultan of
Kishin and our sailors, we gathered that this was actually the mouth
of the great Hadramut valley, which does not take an extraordinary
bend as is given in our maps, but runs in almost a parallel straight line

to the sea from west to east, and the bend represents an entirely distinct valley, the Wadi Mosila, which comes out at Saihut.

At Kishìn I landed to interview the Mahri Sultan Salem, and see if he would conduct us up to the Hadramut valley; here again we met with a positive refusal. "No one goes up that way," he said; "it is full of robbers." The town or village of Kishìn is a truly wretched place, with a few houses and reed huts scattered about in a dreary waste of sand; very different from the fertile plain of Dhofar, and very like the environs of Sheher, the sea-coast city of the Hadramut. Sultan Salem is very old and decrepid, and he is one of the chief sheikhs of the Mahri tribes, the head of a gang of thieves. Of course, he was civil enough to me with my letter from Muskat, but I think his people were rather sorry to see so likely a prize depart unmolested. Sultan Salem is nominally the lord-paramount of the island of Socotra, but he does not trouble his island dependency much, and from what I could see I should say he and his tribe were fast succumbing to senile decay; he is the brother of the Sultan of Saihut, another robber-chief of the Mahri tribe, who is greatly averse to admitting Europeans into his dominions. The fact is, these tribes feel that if their present state once became the object of European inquiry, they would be no longer able to exist in their present condition.

After our futile attempts to penetrate into the Mahri country, there was nothing left for us but to start again in our boat for Sheher, and rely on the promises which Sultan Houssein Al Kaiti had given us last year of sending us under safe escort to the eastern portion of the Hadramut valley, which must contain much of interest, and has never yet been visited by Europeans. He undoubtedly would have done it for us last year, having made all the necessary arrangements, but the season was so advanced that it was impossible to undertake it, as we must have waited till Ramazan was over. This year, however, to our great disappointment, our reception at Sheher was cold and inhospitable; the Sultan refused to let us go out of his town, and quite negatived all his previous promises. So far as this year was concerned the further survey of the Hadramut had to be abandoned. Our plan was to go up the Wadi Mosila, which is said to contain many interesting sites of ancient towns, to cross over the tableland again east of the hostile Katiri tribe, visit the tomb of the Prophet Hud, the Mecca of this portion of Arabia, and the volcano of Bir Barhut, which has been described to us as a large hole in the mountain-side, out of which issue volumes of smoke; it is similarly described by the old Arab geographers such as Hamdani, Idrisi, and others, and is the place described in the Koran as the abode of infidel souls after death. This neighbourhood should be replete with natural and historical interest, and we fondly hope that, with adequate support from the Aden Government, we may be able to reach it another year.

In conclusion, I must add that Imam Sharif Khan Bahadur, who accompanied us last year in the Hadramut, was again kindly placed at our disposal by the Indian Government, and was as before exceedingly useful to us in our various difficulties. To him we owe the maps of the Hadramut and Dhofar districts, and I am sure his regrets were equal to our own at not being able to join these two important surveys of Southern Arabia, and thereby place before the world a consecutive map of this unknown region.

NOTE FROM KEW ON THE COLLECTION OF PLANTS.

"The botanical collection, although more numerous in species than that made in the Hadramut, comprising about 250 species as against 150 from the latter country, has less of a local character, and, as might have been expected, exhibits a larger proportion of species having an eastern extension. Apparently it contains no new generic type; but as the plants have not yet been thoroughly examined, some one of the obscure-looking things may prove to be such. Certain it is, however, that there is no very striking novelty of this rank, and the number of new species will not be very great. The most interesting feature in the collection is the presence of specimens of several plants of more than ordinary importance in relation to their products or distribution. Noteworthy among these are the wild cotton, *Gossypium Stocksii*, and excellent specimens of the frankincense, *Boswellia Carteri*. This species of cotton was previously only known from Sind; and the specimens of frankincense prove Carter to have been right about the species from this region, though the fragments Kew previously possessed were insufficient to place the fact beyond doubt. The present specimens are exactly like those collected by Playfair and Hildebrandt in Somaliland. *Balsamodendron Opobalsamum*, one of the myrrhs, is also in the collection. *Acridocarpus Orientalis*, a very conspicuous member of the Malpighiaceæ, was found at the foot of the Dhofar mountains. It is interesting, because it was originally collected by Aucher Eloy, but the locality was uncertain. He probably collected it in the Muskat district. *Gypsophila montana*, a wiry herb of the Caryophyllaceæ, is an illustration of that element of the flora common to Aden, Sokotra and Somaliland.

"W. B. H."

Before the reading of the paper, the Chairman, Admiral WHARTON (Vice-President), said: It requires no words from me to introduce to this Society Mr. Theodore Bent, who has so often delighted audiences here, and I hope he is going to do so again to-night. Mr. Bent has had many difficulties to contend with, and he has been disappointed in not being able to accomplish all he has wished, but I feel certain his paper will prove very interesting, and I will now call upon him to read it.

After the paper, Sir WILLIAM FLOWER said: I am asked to say a few words, but I am sorry to say that Mr. Bent has not given me the slightest opening to say anything about his paper. He has been to a most interesting country, and he has told us he has collected plants, and I congratulate Mr. Thiselton Dyer, of Kew, for having the opportunity of seeing something of the flora of the country. But he has not mentioned a single creature—except I think I heard him speak of a wild duck—in the animal kingdom, about which I can have anything to say. Of course it is impossible that a man can do everything, and such explorers as Mr. Bent must first go and open out these unknown countries, and naturalists will follow afterwards. I am sure that in a country so full of beautiful vegetation as this, there must be a variety of interesting animals. In Mr. Bent's previous journey he had a collector of zoological specimens, and he brought back many of great interest; but I am afraid that the experience of taking the zoological collector was not on the whole, so far as the geographical results of the expedition were concerned, very satisfactory, and so Mr. Bent informed me, to my regret, before he went on this journey, that he should not impede himself with such a person again. Well, I suppose he was right, because, as I say, no man can do everything at once, and, although I have had no opportunity of saying anything in reference to my particular subject, I think I can say, as no one else has got up to do so, that we all here feel great satisfaction and pleasure at the safe return of Mr. and Mrs. Bent from this interesting journey. We all feel that they have done a very great work of geographical exploration, because they have brought before us features of the country which are entirely new to very many of us. I had no idea that behind the arid rock-bound coast, as it always seemed, of Southern Arabia could be found these beautiful green and watery vales, the photographs of which remind us at times of some parts of our own land. I therefore feel that Mr. Bent has done excellent work. Moreover, he and Mrs. Bent have performed this work under circumstances of considerable danger. They have happily surmounted all these dangers. And we hope this is not—I will not say the beginning, because they have begun already a long time ago—but we hope that this is only a stage of progress reached in a work which may go on for many years yet, and I am sure that geographical science, and with it the progress of humanity and civilization, will be very much forwarded by their exertions.

The CHAIRMAN: Mr. Bent has not taken us over a very large extent of country this time, but he has taken us to a very interesting one—to the home of the old frankincense trade, of which history scarcely knows the beginning. He is very much to be congratulated, I think, although he had difficulties that prevented him going in other directions where he wished to, that he found this country open; because the last account we had of it was, that you could not venture more than half a mile beyond the beach, and the inhabitants were always internecinely engaged in killing one another, and two men of different villages had to get an intermediary to allow them to talk to one another. Mr. Bent has happily come after the beneficent rule of Wali Suleiman, who has made his strong hand felt there, and who has opened the country up. I think there is nothing more interesting than going through a country with a history and endeavouring to recognize the old sites, and I think Mr. Bent has unearthed the harbour that must have existed in ancient days when the frankincense was more sought after than now, and when the trade was very much greater. Whether it could be made of any service now is another question. Mr. Bent seems to think it might be, and it it is very possible it may be so. I am sorry that on the maps we have here, Al Balad, the ruins of which he has discovered, is not marked. It is to the westward of Al Hafa, and I think there is no doubt that that was the old centre of civilization in this region. I think, ladies and gentlemen, you will all join with me in heartily thanking Mr. Bent for the very interesting information that he has given us in his paper.

Bertram Thomas on Dhufar

Bertram Thomas retraced Bent's route to 'Abyssapolis' in the course of a journey in 1928, described in 'Among some unknown tribes of South Arabia', here reproduced by courtesy of the Royal Anthropological Institution from their *Journal* (1929).

AMONG SOME UNKNOWN TRIBES OF SOUTH ARABIA

By BERTRAM THOMAS

FROM a point of view of exploration, South Arabia is divided by the line of longitude 55° E. What little exploration has been done by Europeans is to be placed west of that line. Eastwards no European has at any time before penetrated the hinterland (except for a small journey made in the Ja'alan by Lieutenant Wellstead more than a hundred years ago). South-Eastern Arabia has thus remained *terra incognita* and its map a blank. Commercial intercourse with South Arabia, chiefly in connection with frankincense, was carried on probably by the Greeks, and, certainly later by the Romans, and this led to a knowledge of local geography which has been preserved for us by Ptolemy, Pliny, Dionysius, and others, but that intercourse was of necessity confined to the western half of the south coast, and we turn in vain to the classical geographers for any considerable light on the eastern half of the southern littoral. The Arab geographers add little, and that, except for Ibn Batuta when he speaks of Dhufar, on the evidence of others. The truth would seem to be that sterility and barbarism, the two chief characteristics of the south-eastern borderlands of Ruba al Khali, have throughout the ages been prohibitive alike to Europeans and to Arabs.

In January and February of 1928 I accomplished a journey through country in which no European had previously set foot. Fifteen camels and an escort of twelve Arabs (a changing personnel except for three guides[1]) constituted my caravan, and the journey of 650 miles from Suwaih to Dhufar occupied forty-eight days. This included two halts, one of seven days at Khaluf and the other of three days at the entrance of Wadi 'Ainain made for the purpose of changing camels, and this, in effect, divided the journey into three approximately equal stages.

I should say at the outset that I travelled under no Government auspices— in fact, no auspices but my own, and the success of the journey depended upon my living as a Badu and mixing freely with the natives and being passed from Shaikh to Shaikh. It is country that owes allegiance to no government, and where every man goes armed, and three times my innocent caravan came under hostile fire for no better reason than that we were mistaken for raiders with whom it would be suicidal to do other than shoot first and enquire afterwards.

[1] *See* Pl. V, Fig. 1.

Space does not allow of a narrative of the whole of my journey, and I propose here to limit myself to a description of a previously unknown group of tribes (except for the Dhufar elements), occupying a part of the second and the whole of the third stage of the journey, their customs and dialects, and the archæological remains found in their habitat.

The area in question stretches from Wadi Sarab (lat. 20° 10′, long. 57° 45′ E.) to Salala, the capital of Dhufar (lat. 17° 00′, long. 54° 6′ E.). This region forms the habitat of a group of five tribes, which it would appear are racially distinct from the Semitic Arab, and perhaps from one another, and they speak four separate dialects which are not understood by Arabs, and which have closer affinities with Ethiopic than with Arabic. The names of these tribes in the order in which I passed through them from east to west are Harasīs, Bautāhara, Mahra, Qara and Shahara, and, judging from their appearance, traditions and customs, I think there can be little doubt that they constitute a block of non-Arab tribes of great local antiquity ; and I further assume, in the light of the facts which I shall set forth, certainly as regards the last-named four, that they are Hamitic. It would thus seem that their ethnological and linguistic affinities are at variance.

Before I describe these tribes in particular I should point out that there are two hostile political factions, Hināwi and Ghāfari, to one or other of which every tribe in South-East Arabia, including these tribes, owes allegiance. Superficially the terms Hināwi and Ghāfari would appear to date from a dynastic squabble over succession in Oman in the early eighteenth century ; but, as I have observed elsewhere, they are of much deeper significance, for, generally speaking, the Hināwi label coincides with the tribes of avowed Qahtani descent and the Ghāfari label with those of Maadic or Nizari origin, and all other non-Yamani stock. Within limits, therefore, the division is in origin racial. It is not surprising then that this group should claim to be Ghāfari, except for the Harasīs, who, moreover, acknowledge their defection in times past to the Hināwi party through political expediency. Throughout Oman and South-East Arabia this Hināwi-Ghāfari feud still smoulders, and no man dare travel through country used by a tribe belonging to the rival faction without a *rafīq*.[1] To do so would be to court attack and capture of his camels, if not his own death, except during the brief and intermittent period of truce. To pass safely through a tribe it is not always essential to have as *rafīq* a member of that particular tribe, though that is preferable, but it is essential that a *rafīq* should be taken of that political faction, Hināwi or Ghāfari, to which the tribe belongs. Only amongst the Qara is a slave acceptable as a *rafīq* ; elsewhere it is the custom to kill the free man and take captive the slave. In truth it would appear that the slave has a commercial value, and as such shares the privilege of the camel taken in the raid.

---

[1] Generally a representative of the tribe, whose presence guarantees the traveller and should ensure safe conduct.

524

The *Harasīs* are a small, dwindling nomad and pastoral tribe, perhaps numbering not more than 200 men, whose habitat is not regarded as their own exclusively, but belongs in name to the western Janaba tribe in the steppe land bordering the desert and extending from the hinterland of Wadi Sarab as far west as the 56th line of longitude. Although clients of the Janaba they belong to the rival political

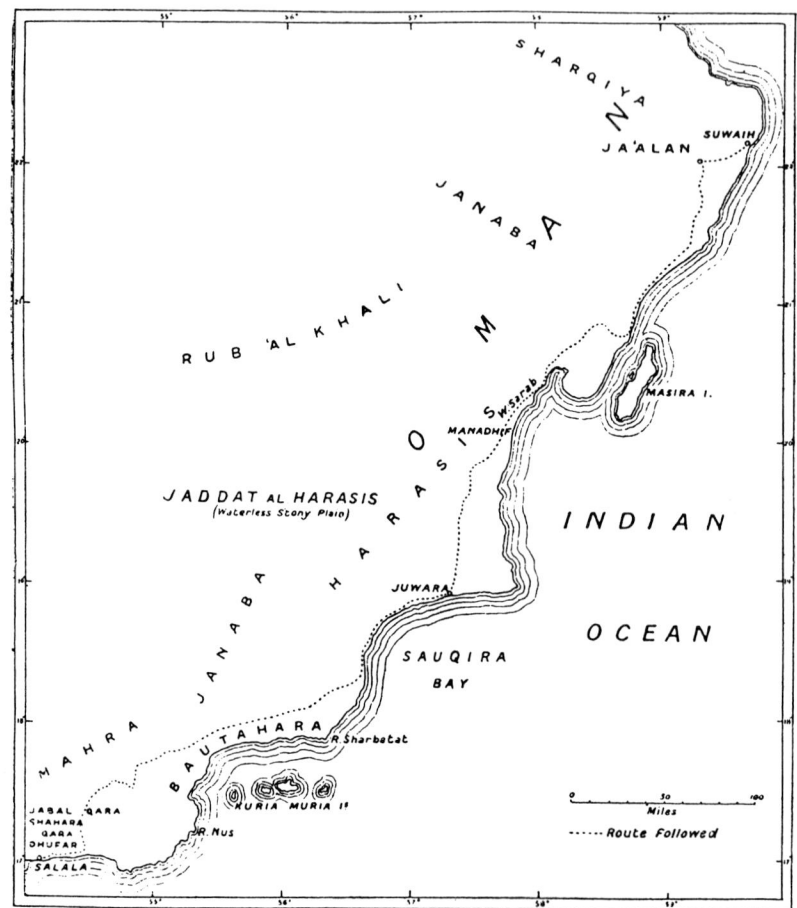

Map to show Thomas' route from Suwaih to Salalah, 1928

faction of Hināwi, but pay a nominal *zakat* to the Awamir tribe, by which tribute they are indirectly brought under the ægis of Ibn Saud. Rude and primitive they appeared to me, as distinct from their Hamitic neighbours to the west as from their Arab neighbours to the east. They are, generally speaking, bilingual, though

Badu of the Janaba and Yal Wahibah, 1928 (Thomas)

A AND C, BADAWIN OF JANABA TRIBE.   B, BADU OF YAL WAHIBA.

Chipped figure on rockface, 1928 (Thomas)

their Arabic has a slight accent, the noticeable feature of which is a flat " a " sound, suggestive of that vowel-value as met with in Syria, but otherwise having no resemblance to the Syrian Arabic, being a virile Badawin dialect with a rather pleasing inflexion of voice. They have short broad faces, low sloping foreheads and large heavy aquiline noses. The youths, unlike any others I have met, could almost be taken for a type of southern European. A tribe with noble traditions, they are now a scattered collection of seven sections, acknowledging only a nominal hereditary Shaikh, by name Ibn Akīs.

The *Bautāhara* is a still smaller tribe, primarily engaged in shark-fishing, which they carry on in a primitive way, swimming on inflated skins. They have a few camels, and their habitat extends from Ras Sharbatat to Ras Nūs. They are held by their neighbours, with whom they cannot marry, to be of ignoble origin, and until recently were so wild and disreputable that no traveller could pass in safety even accompanied though he might be by a *rafīq*. Ghāfari in politics, the Bautāhara are now subject to the western Janaba. The only Bautāhari I could get hold of, passing as I did on the desert side of their habitat, was a fisherman from the coast. He appeared to be a very low type, though his complexion was comparatively light brown for South-East Arabia ; he had comparatively straight hair and high cheek-bones. A greater local antiquity is allowed to the despised Bautāhara, who were once largely pastoral, than to any other local tribe save the Shahara.

The *Mahra* habitat extends through the desert westward of Jabal Zāūlāūl behind the Qara mountains. Of Ghāfari adherence, the Mahra do not claim descent from a single ancestor as is usual with other tribes ; the eastern element claim descent from one Bīr Bāuki, and the Arabs refer to these Mahri neighbours as Hasarīt. These purely nomadic Mahra had noses which might almost be regarded as European, round brown eyes, short under-lips, jaws wide at the base tapering to pointed chins, and long curly hair. The Mahra exercise suzerainty over a depressed tribe, the Bait Al Hafi, a similar relationship to that subsisting between the Janaba and Bautāhara. .

The *Qara* are a numerous settled and prosperous mountain tribe occupying the Jabal Qara and Jabal Samhan, the most prosperous tribe of all the Hamitic group, possessing innumerable camels, herds of cattle, and the richest frankincense country. They resemble the Bisharin tribe of the Nubian desert, to judge from the photographs of the latter. Men of big bone, they have long faces, long narrow jaws, noses of a refined shape, long curly hair and dark brown skins. (Pl. VI, Fig. 2.)

The *Shahara* are a small and disunited tribe living in groups amongst their Qara overlords, for whom they are mere hewers of wood and drawers of water. They have comparatively fair brown skins, less curly hair and more noticeably high cheek-bones. Weak and degraded to-day, they claim to have been possessors of the land before the Qara dispossessed them. (Pl. VI, Fig. 1.)

The entire group, brown in colour, perhaps early Ethiopic in speech, non-Arab and non-negroid in appearance, and occupying almost exclusively the central south of Arabia, have an ancient tradition of a North African origin *via* Hadhramaut —an origin, as regards those of Dhufar, noted by the Arab traveller Ibn Batuta.

These Hamitic tribes have many interesting and perhaps unique customs. The women are not veiled, except in the case of those Badawin elements, *e.g.* Hasarīt and Harasīs, who move amongst Omani tribes. They have a custom of tattooing the chin with a short vertical line with a dot on either side—still practised in modern Egypt (*vide* Lane's *Modern Egyptians*), and I met instances of a bracelet-like design around the wrist. In Dhufar they paint their faces red, black, and green on occasions of religious and other festivals, marriages, circumcisions, etc., generally a line along the edge of the cheek, another under the cheek-bones, and one bridging the nose across the eyebrows. The men are not tattooed. The Qara women of Dhufar have a custom when young and almost invariably before marrying, of having a half-inch strip of hair with its attendant flesh shaved in a centre-parting along the top of the head. The hair never grows again and sometimes the operation has fatal results. Shaving the hair around the forehead to show a large expanse of brow is another observance of feminine beauty. Throughout the whole of these tribes it is customary for a boy's hair to be cut short, except for a strip an inch wide, which is allowed to grow long down the centre of the head—suggestive of a certain Hindu caste, or perhaps of the Horus-lock affected by young Egyptian princes in ancient times in honour of the god. Men either shave the upper lip wholly or leave the slightest suspicion of a moustache not more than an eighth of an inch wide ; they shave the side face both downwards from the cheek-bone and upwards from the throat, leaving a slight beard around the edge of the jaw, but they never shave their chin-tuft, though sometimes it is trimmed with scissors. It would be shameful to shave it, for by it a man swears. The growth of face hair is not strong, but the chin-tuft is coarse. Male head hair, characteristically long and curly, particularly with the Mahra and Qara, is generally worn like bobbed hair, and in the case of the Qara it is sometimes bunched up and tied with a bun on top—in both cases generally presenting a greasy appearance from generous treatment with coconut oil. Men go bareheaded, except for a narrow leather thong called a *mahfif*, which is merely a coil wound round the widest part of the head nine times and resembles an Arab *aqal*. Their body-dress consists of a short black shirt to the knees, with a leather girdle around the waist which is looped as a cartridge-belt. They go at all times barefooted and barelegged, except that the Badawin element in summer, when travelling over hot sands, sometimes affects a woollen sock.

Circumcision by these tribes is universal—the boy on approaching adolescence, sometimes after reaching it ; the girl on the day of her birth or the second day. This system of adult-male and infant-female circumcision is the reverse of that employed in Oman, where the practice is infant-male circumcision (six years) and circumcision of the girl when approaching the age of ten. In both regions with the male the whole of the foreskin is removed, but as regards the female, while the Arabs of Oman merely incise the top of the clitoris the Hamitic tribes of the central south perform clitoridectomy. In Dhufar there are sometimes elaborate rites attending male circumcisions, and batches of youths undergo what is a severe public test of their fortitude on the same day. Large numbers of men and women assemble around a large open space. On a rock in the centre sits the boy of fifteen, a sword in hand. This sword, which has been blunted for the occasion, he throws into the air to catch it again in its descent, his palm clasping the naked blade. Before him sits the circumciser, an old man ; behind him stands an unveiled virgin, usually a cousin or a sister, also sword in hand. She raises and lowers her sword vertically, and at the bottom of the stroke strikes it quiveringly with the palm of her left hand. The stage is now set. The boy sits, his left hand outstretched palm upwards, waiting for the actual operation. This made, he has promptly to rise bleeding and run round the assembly raising and lowering his sword as if oblivious of pain, the girl running after him. He must complete the circuit three times without being caught by his fair pursuer, and his manliness will be judged by his performance. If he fails, he will be regarded as a weakling. Success is attended by singing, and firing of rifles.

The wife in these tribes may not betray grief on the death of her husband. Mothers, daughters and sisters may weep and raise their voices, and amongst the Qara they let down their hair, beat their heads, and pour dust upon them ; but in a wife it is held to be shameful publicly to show pain at the loss of her man, and she hides herself. The Mahri buries his dead without much ado. With the Qara it is a time for wholesale sacrifice, and on the death of a rich man his family will kill as many as ten cows, four or five camels and twenty sheep, the flesh being given to the poor. The cow is held in great esteem, and it is of interest that the virtue of slaughter-ing it at this time is held to be seven times greater than that of slaughtering a camel, though the camel is intrinsically the more valuable animal of the two. The sacrificial value of the camel and the sheep would appear to be the same. This special reverence for the cow is noticed in the custom governing its milking—it is a male prerogative. The Shaikh of the tribe will milk a cow, but it is held shameful for a woman to touch its udders, though she may water and tend it. Curiously enough, this reverence for the cow by the Hamitic tribes is completely reversed in the neighbouring Arab tribes of Oman. There no man would demean himself to milk a cow ; it is an undignified labour, fit only for women. Regarded almost as an unclean animal, the very word for it, *buqara*, is never to be used in polite society. The words chicken and egg suffer from a similar taboo in Oman, where no man would dare in the presence of the Sultan, a Shaikh or 'Alim, mention these things by name. He would probably use *banāt al muedhdhin*, i.e. " daughters of the muezzin," for the former, and *rasās*, i.e. " bullets," for the latter, and the conventional idea of their vulgarity is shared by the Hamitic tribes. To prepare a lunch of chicken and eggs for a Qara Shaikh would inevitably cause offence.

The burial of a slave in Central South Arabia is attended by much ceremonial, which may conceivably be negroid in origin. A procession of slaves advances singing and playing drums. Men constitute the vanguard, then follow the musicians and corpse, and last of all come the women. Those of the vanguard dance, wearing over their knees a rattle called *khish-khish*, presumably an onomatopœic word. Those in the body of the procession sing, and the responses are taken up by the ladies behind. After the burial everybody, male and female, dances frenziedly around the grave until tired, and the procession retires again with music and dancing, the participants happy in the belief that the departed will now rest unmolested by evil spirits.

A curious method of spirit exorcism and snake-bite cure is practised by certain sections of these tribes, notably the Bait Shaitāna, Bait Qawās and Bait Al Bilhaf. The sick man is laid upon his back. Four of his fellow-tribesmen, two a side, kneel over him. Bending and straightening their backs, so that their faces touch his naked body in turn, they chant, or bark rather, some magic formula, working themselves and their patient up into a state of hysterical emotion. After a time they add to their activities that of gently spitting on his body, and finally they apply their lips to the patient's abdomen in the region of the liver, and draw up his flesh an inch or two, letting it flick back as they raise their heads. In time the patient faints or gets into a state approaching exhaustion. He is then left to come round free of pain or fever or cured of his snake-bite.

With the Harasīs and Mahra another curious custom obtains—that of never milking their sheep into a cold receptacle. A hot stone, heated in a fire, must first be introduced. The idea it suggests, that the warmth thus applied to the udders encourages a facile milking, cannot be wholly satisfactory, because the practice is observed only in respect of sheep, not of camels. I have seen milking taking place from behind—a habit depicted on the early limestone inlay from Tal al Abaid near Ur. In the Dhufar mountains, when lactation is ceasing in the cow, the tribesman blows down the animal's vagina. The Qara, amongst whom this practice is common, believe that the irritation causes an increased supply of milk. A shepherdess of the Bait Ash Shaikh tribe, from whom I purchased a sheep in Wadi Dhikûr, would not agree to its being slaughtered in sunlight because of the fear that it would bring misfortune to her family, and this was a common belief. A Harsusi of my escort informed me that in no very distant times past the Harasīs would not only not slaughter, but not milk their flocks in sunlight, and to this day there are two breeds of sheep, " Banat Al Murtal " and " Banat Al Maqtuf," which no tribesman of whatsoever tribe would dare slaughter until after dark. This may, I think, conceivably hark back to some ancient sun-, moon- or star-worship cult.

Ordeal by fire is also in common use. There are several centres, the chief perhaps being Ghaidha in Mahra country. The process is for a red-hot iron to be placed on the suspect's outstretched tongue. To prove innocence he must promptly, at the conclusion of this operation, be able to command sufficient saliva to expectorate. My guide, Luwaiti, in avowing its fairness and respectability, told me of a case then pending where a young man who had killed his father, accidentally he affirmed, had volunteered to submit to the ordeal. Belief in witchcraft is general. Old men are particularly suspect, and are sometimes killed on the grounds that they could never have attained so ripe an age without communion with supernatural powers. Death is often attributed to the spell of some suspected witch, who is forthwith persecuted. A tribesman of Bait Ash Shaikh, who incidentally fired on us when approaching his camels in Wadi Afar, had as a young man killed his widow cousin for being a witch—a murder which had public approbation, if indeed the murderer was not actuated by public opinion. A recent case occurred within a month of my arrival, where an alleged witch had been murdered by no one knew whom, and no one was ever likely to tell. It appeared that she had been accused for long, but had proclaimed her innocence and had betaken herself to the exponent of the fire ordeal. There she emerged from the test vindicated ; but even this failed to convince her tribe.

Steeped in superstition and clinging to many pagan customs, these Hamitic tribes of South Arabia are all avowedly Muslims of the Shafi sect. In diametrical opposition to Wahabi tenets they have much veneration for the shrines of saints, which they periodically visit. That of Muhammad Bin Ali, known also as Mazar Al Wali, near Murbat, is the most famous of them. The annual pilgrimage to it takes place on the 15th Sha'aban, when pilgrims from the far mountains assemble at the shrine, walk around it, and salute it with the nose-kiss, and here read the Qur'an. Amongst the credulous three consecutive journeys to this shrine are accounted to have the virtue of a pilgrimage to Mecca. Other shrines are Salih Bin Hud at Siddih, Haddad at Dahariz, and Bin Arbait at Raisut. Few of these tribesmen make the pilgrimage to Mecca ; in fact, no Harsusi, I was told, had ever made the Mecca pilgrimage, but on the death of a relative a Harsusi will often defray a large part of the expenses of an Arab pilgrim on behalf of the departed. Disputants of these tribes are not always satisfied with an oath on the Qur'an ; an oath upon a shrine is more acceptable. They have a special veneration for the Shurafa Saada, descendants of the prophet, after shaking whose hands they will raise their own reverently to their nostrils and take a few sniffs, apparently inhaling virtue thereby.

The languages of this region are four in number. Their distribution is as follows :

Shahari is spoken by the Qara and the Shahara tribes,
Mahri by the Mahra, and
Bautahari and Harsusi are the dialects of the Bautahara and Harasis, respectively.

The region occupied by these tribes is to the Arabs of Ja'alan known directionally as Hadra; by them the Indian Ocean bordering these coasts is called Bahr Al Hadri; finally, the block of tribes occupying Central South Arabia, including all these Hamitic tribes except the Qara, is known as Ahl Al Hadāra; I therefore venture to give to these dialects the name of Hadāra group. The identity of the word with Hadoram of Genesis is suggested, also with the Adramitæ of Pliny. During my sojourn amongst these tribes I collected vocabularies of some 500 words of each of these dialects (fewer in the case of Bautāhara for lack of opportunity, passing as I did through the desert in their rear), and deduced a few simple grammatical rules. The four dialects would appear to be classifiable into two distinct categories :—

  (a) Shahari.
  (b) Mahri, Bautāhari, Harsusi.

(a) and (b) are mutually unintelligible, whereas members of the three dialects of (b) can with difficulty understand one another, though less so in the case of the Bautāhara than of the other two. It is interesting to observe that the rich and important Qara tribe employ Shahari, the speech of their dispossessed underlings. As regards the affinities of the three dialects of group (b), Bautāhari is considered by the others to be a depraved tongue, and is disparagingly referred to by the oft-quoted Arabic tag "language of birds." There is no written form of any of these Hadāra dialects, but I suggest that one or more of them must be of great local antiquity, for I discovered traces of rude inscriptions of an early alphabet, which seems to me to resemble Sabæan or Ethiopic characters, on boulders of monuments of an archaic kind over an extensive area; but these I will describe later. What strikes the ear of the listener to these dialects is the frequency of the *ll* sound met with in Welsh. In words akin to Arabic it seems to take the place of س ش ص ض ذ ز ظ. But not invariably so, as the Arabic sound-values of these characters are also met with in the Hadara dialects. Indeed, all Arabic sound-values exist, except the guttural ع, and in addition to them ب (*p*) گ (*g*), and in the case of Shahari a nasal " n " or " m." This, and the fact that in words common to both Arabic and the Hadara group there are variations in the vowel-sounds, account for the fact that most of the geographical names have two slightly varying pronunciations, only one of which I have been able to record on my map. The affirmative is sometimes expressed (almost invariably by the Bautāhara) by drawing in the breath while gently nodding the head upwards; the Shahari tribesman, if he does not understand, puts his tongue out; the time-honoured Arabic expression *Inshallah* is here replaced by *Am Katīb* or *Ham Katīb*, a fatalistic reference to " that which is written." Two of these dialects, the Mahri and Shahari, at least, as they are spoken in the Western Hadhramaut and Socotra, have been extensively written up by the philologist Dr. Maximilian Bittner, working on the material brought home by Dr. Müller's

Arabian Expedition of 1902 and Count Landberg's expedition of 1898 and 1899.
It is curious that Bittner calls part of his work " Studies in Shauri Language in the
Mountains of Dhofar on the Persian Gulf." The mountains of Dhufar are not on
the Persian Gulf, but on the Indian Ocean, and the term " Mountains of Dhufar "
would itself seem here to be misleading. The region of activities of the Viennese
Expedition was, I understand, the neighbourhood of Ba'l Haf and Nakab Al Hajar,
and is not coincident with the mountains of Dhufar of this paper. While I have
no pretensions to a knowledge of philology, and my word-lists have been made on
the spot as one only among many activities, I hope a philologist will look into my
material. A comparison of my word-lists with those of the South Arabian Expeditions,
of the existence of which I had no knowledge at the time mine was made, may have
some interesting results, especially as the fields of investigation would appear not
to be the same. So far as my own uninstructed researches go it seems to me that the
pronouns, verbs and language structures are the same in both sets of lists. But
whereas the plurals of nouns I noticed to be almost invariably of the " broken-
plural " variety, the dialects further west as analysed by Bittner seem generally
to use an external termination approximating more to the regular Arabic forms.
There are also differences in many substantives, and it is possible that the dialects
of the tribes I visited who occupy the mountains or roam the southern confines
of the Rub 'al Khali differ from the coastal dialects investigated by the philological
expeditions. My Bautāhari and Harsusi dialects, which would appear to be related
to Mahri, have never, I think, been previously noticed.

In contrast to the coastal plain and parched sands of the earlier part of my
journey, the terrain of the Hamitic tribes passed through consisted of bleak and arid
limestone hill and rolling country crossed by Wadi beds, shallow in the early stages
and trough-like as the greater altitudes in the later stages of the journey were reached.
Here, in these Wadi beds alone a sparse nourishment for man and beast was to
be found in the water-hole, the acacia and gum-arabic trees, and the various
camel thorns known as *warikh, thurmad, haram, ghussab, dha'a, thidiya, rimram,
al ithal* (tamarisk), *thamama* and *hamiyat*. The course from Manadhif (lat. 20° 00',
long. 57° 40') to Juwara in Sauqira Bay lay over the plain of Al Dhahir, reaching a
maximum elevation of 470 feet. Everywhere on our right hand the vast empty
spaces of Jaddat al Harasis spread out northwards, eventually to lose themselves in
the sands of Rub 'al Khali. The term Rub 'al Khali, incidentally, is nowhere
used by the inhabitants of these Southern Borderlands to mean the entire desert
in its geographical sense ; when used by me in talking to the inhabitants, however,
it was sufficiently literal as to admit of no misunderstanding. Beyond Juwara
we hugged the coast of Sauqira Bay through the Jazir littoral below what appeared
to be low igneous formation. Leaving it by way of Wadi 'Ainain, we ascended to
the plain of Jaddat Arkad about 1,000 feet in height, the Wadis throughout
descending south-eastwards towards the coast. From Wadi Muqarrad, the western

Shahara tribesmen, 1928 (Thomas)

Qarah tribesmen, 1928 (Thomas)

limit of Arkad, we climbed again, threading our way through the northern fringes of Rakibīt, the great steppe borderland of Rub al Khali coming into view on our right hand, as now and then our tortuous course took us towards the edge of the mountain system on the desert side, and hence we continued gradually climbing to 2,000 feet, in Wadi Dhaghaub, the deep troughs of the Wadi beds now trending everywhere to the north.   For six days prior to reaching Wadi Andhaūr the camels went unwatered, and we ourselves were reduced to the meagrest ration.   The herbage was scant, and our camels carrying loads for six or seven hours a day through this difficult mountainous country were in a sorry condition.   The horse cannot live in such conditions, and I saw no horse west of Ja‘alan, and only a few dogs at Khaluf, throughout the whole of this journey through the southern borderland till reaching Dhufar. Goats will go longer without water than camels. Our camels, which were faltering on the sixth waterless day, could, I gathered, had they been without their loads, have gone ten days or more, and if grazing in the rains would survive a month without watering.   During the rainy season these Badawin wander with their flocks of sheep and goats for six weeks without bringing them to water.   They take no water for themselves, subsisting on the milk of their flocks.   The oryx gazelle and rīm, and sand lizards must, of course, be practically non-drinking animals.

The archæological feature I met with throughout this journey which points to a great local antiquity for the non-Arabic speaking tribes, was a class of crude ground monument sometimes bearing inscribed characters of a possibly Sabæan alphabet (Pl. IX, Fig. 3).   This monument consisted of a system of triliths, three elongated blocks of undressed stone (or sometimes round boulders with a naturally smooth surface), about 1½ feet high, standing on end and leaning inwards with their tops touching to ensure stability.   These triliths were set up in series along one alignment, each pile standing equidistant from its neighbour about one and a-half paces.   Some-times the trilith had a fourth and smaller boulder superimposed, and occasionally a series of triliths was enclosed by an elliptical line of small pebbles.   The series varied in number.   I found them of 5, 7, 9, 14 and 15 triliths.   Running parallel to each series at about three paces distant was a smaller series of large conical rubble heaps, modern equivalents of which I have seen elsewhere used for the kind of cooking, known to Arabs as _mashuwa_, a method of grilling flesh on heated stones.   These, I suggest, had some sacrificial significance.   Some of the smaller series of triliths, _e.g._ those of five in number, were without them; the longer lines had these sacrificial piles in the proportion of one to three or four triliths.   Between the two were sometimes small square boulders which had no obvious function, unless they were used as seats.

These monuments had no common directional orientation, but wherever found were aligned with the axis of the Wadi.   Most were entirely without inscriptions, certainly all the Eastern ones.   Inscribed boulders were first met lying near the monuments in Wadi Andhaūr, but in Wadi Dhikur they were found as the head stone of one of the terminal triliths.   That the rude scrolling was humanly made was a discovery that the illiterate locals and my escort would scarcely believe. The inscriptions were generally separate characters $1\frac{1}{2}$ inches in size, rudely done, and having a dotted superficial impression which suggested that the implement used was a nail-headed flint hammer. Transport limitations prevented my bringing back more than one specimen of this work—presumably representing the picture of a camel—which I have presented to the British Museum (Pl. V, Fig. 2). On account of weathering of the stone, a great many of the inscriptions were unrecognizable, but in almost all of them a character or two showed up clearly, and amongst these I noticed ᚺ ψ ⌣ ☉ ൦ ⋎ ധ (see Text-fig. 2). Those inscriptions that were better preserved I copied in full : of one I made a squeeze (No. 6) and another I photographed. The places where these monuments were found are as follows : Wadi Sarab, entrance to Wadi 'Ainain, Wadi Banat Ar Raghaif, Wadi Haradha, Wadi Andhaūr below Khunghari Pass, Wadi Dhaghaub, and Wadi Dhikur. It is in the two last-mentioned places, however, that they occur with great frequency, and are inscribed, particularly in the Wadi Dhikur, where there are long lines of them in parallel groups. It is, I think, beyond doubt that they are graves. Not only is this suggested by their appearance, but the Wadi Dhikur preserves a continuous burial tradition, as is witnessed by the following facts :—(i) It contains a large Muslim cemetery, and was a favourite burial-ground up to quite recent years ; (ii) two other cemeteries of a pre-Islamic period are marked by two different types of non-oriented grave ; (iii) the name Wadi Dhikur may be translated Valley of Remembrance. As regards (ii), one of these two other grave-types is a cave sepulchre. The left-hand Wadi ridge is honeycombed with cavities which, having received the body, have been bricked up with loose small stones. The other grave-type consists of a giant ovoid of large flat slabs of rock ; the largest grave was sixteen paces long and six to seven in breadth. The individual slab of rock, now black with age, seemed roughly about three to four feet square and so heavy as to require many men to handle it. Judging from the weathered condition of the stone and the rude nature of the construction, this last type may, I think, be the earliest of all. These graves, while grouped generally, with their long axis in one direction, were sometimes bunched together in rosette formation. I also saw them at Khor Ruri and Khor Suli in the Dhufar plain, whither the trilithon type did not reach.

Wadi Dhaghaub and Wadi Dhikur, the approaches by which I made my way over the Qara Mountains from the Rub 'al Khali side, involve a quick ascent from 2,000 feet in the Wadi Dhaghaub, through 2,300 feet in the Wadi Dhikur to the Divide 3,000 feet. They are the centre of the frankincense country and part of the habitat of the Qara tribe. Between Wadi Afar and Wadi Dhaghaub we crossed a large jungle of the frankincense tree, *mghur*, the most extensive of any passed. It resembled an orchard of young three-year-old fruit trees. One of my escort brought me a specimen of the sap on his dagger blade, of lard-like appearance and pleasant odour. The frankincense country, owned chiefly by the Qara, is

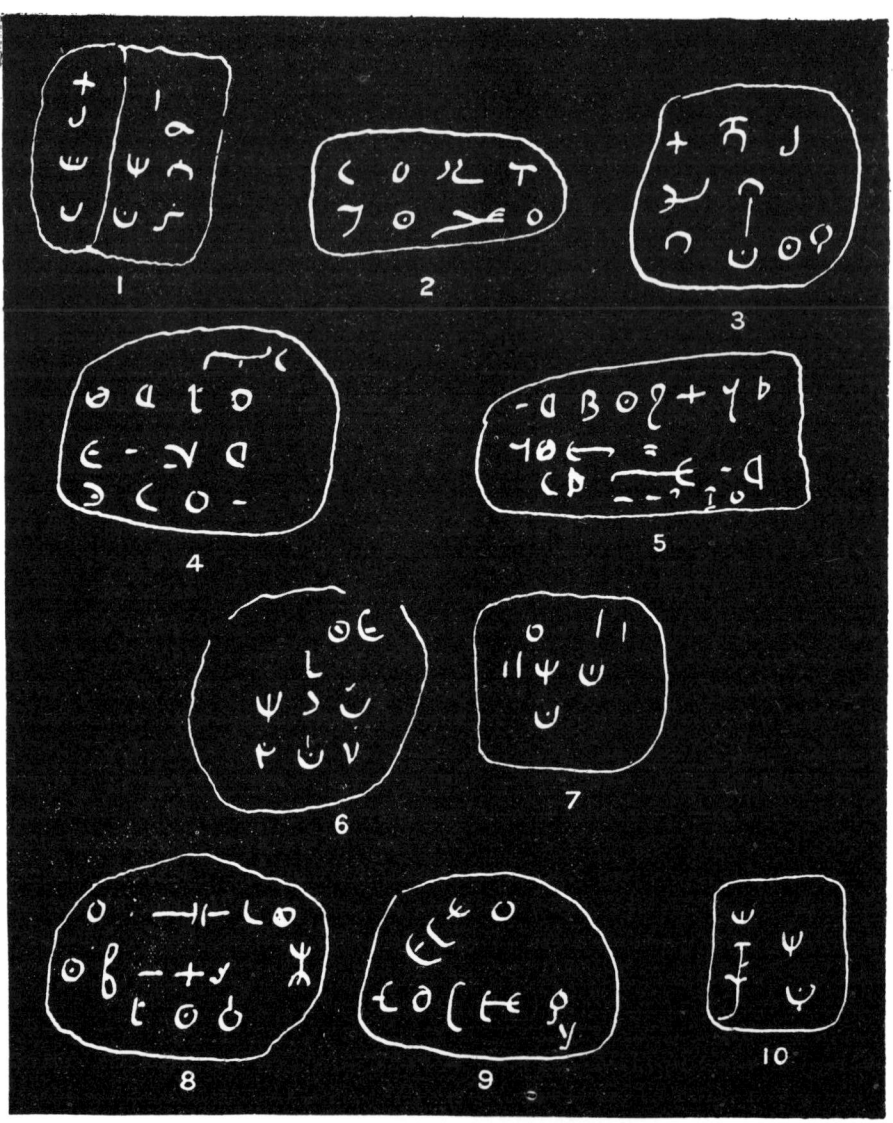

FIG. 2.—INSCRIPTIONS ON LARGE ROUNDED BLOCKS OF STONE.[1]

NOS. 1–3. IN WADI ANDHOR.   4. IN WADI DHARKABUN.   5–10. IN WADI DHIKUR.   (NO. 5 IS BROKEN
AT BOTH ENDS, AND PROBABLY FORMS PART OF A LONGER INSCRIPTION.)

---

[1] The significance of these inscriptions is uncertain, but Dr. Stanley A. Cook, who was at first inclined to treat them as South Arabian, is now inclined to regard them as camel-brands, *wusūm*, and refers to somewhat similar signs in Doughty, *Arabia Deserta*, I, 125, and Bent, *Southern Arabia*, p. 369 and Plate IV.

Camel brands on rockface, 1928 (Thomas)

worked by them and by the Mahra and Al Kathīr, women as well as men, the owners' share being one-tenth of the produce. The tree has a silver bark, of bushlike appearance, with a tiny insignificant leaf, eaten by the locally bred camel. To secure the frankincense the scaly bark is scratched with a knife (not a deep incision) and is left for a week. The sap which exudes has by then dried into a transparent green resinous substance of delicate fragrance, and this is the article of commerce which will find its way to the Indian temples. We ascended the white pebbly bed of Wadi Dhikur, and were rewarded by a magnificent spectacle. Red cliffs, 300 feet and more, towered above us on each side, their face sculped by natural action into loges throwing fantastic dark shadows. The scene reminded me of Petra. Near the top at 2,300 feet was a pool, where our animals watered while I collected one or two pink-red fossils. Thence we returned down the same Wadi to the Tomb area, with its gentler slopes clothed with frankincense trees. There three sheikhs of the Qara had arrived for me, sent by my friend Saiyid Sa'ud, the Governor of Dhufar. Bare-bodied except for the shirts about their loins, an end of which was thrown over their shoulders, and armed with swords and bucklers, they were to accompany me for the rest of the journey. On the morrow we despatched our camels by another way and climbed the steep side of the valley on the right, a hundred feet or so above. Thence we passed through rolling country thick with *tishgāut* jungle—a libaniferous shrub inferior to frankincense—gradually climbing to the Divide at 3,000 feet, the inland limit of the region which receives the plentiful rainfall of the south-west monsoon. The Qara and Shahara mountaineers are troglodytes (Pl. VII, Fig. 2), and their caves are generally marked by low hive-shaped straw huts of their own erecting. I was anxious to see the abyss which Bent had identified with Abyssapolis, so thither—the natives call it Ad Dahaq—we proceeded, descending through rolling yellow meadows where cattle grazed, and occasionally catching a glimpse of the blue sea of the Indian Ocean over Taqa, or the blue range of Samham behind Murbat, as we went. Passing the Shahari village of Shihait (Pl. VIII), we soon arrived at the brink of the right bank overlooking the valley of Darbat. The spectacle is delightful. You look down some 200 feet through a tangle of tree-tops to the stream below lined with trembling willows; opposite, the hills are equally precipitous and similarly wooded. In whatever direction you look is a wall of tropical jungle, and below it the stream which is flowing towards the abyss. We made our way down the hill-side to this stream, marked by tropical trees and luxuriant creepers and alive with heron and other water-fowl. Thence we crawled to the edge of the Dahaq and peered down its precipitous face 500 feet below. With much difficulty we clambered down to its wooded foot and followed the meandering Wadi to its mouth—Khor Ruri. The ruins hereabout—Husn Mirahadh and the entrance of Inqitat (Bent's Khatiya)—are clearly the site of Moscha of the Periplus (Ptolemy's Abyssapolis as suggested by Bent from its proximity with Ad Dahaq). Miles' identification of Moscha with Hasik, 80 miles to the east, is, I think, a less happy one. The Arab geographers give Murbat as the site of the ancient seaport and capital of Dhufar, which lasted until the fourth century Hejira. The Shaikh of modern Murbat, 20 miles to the east, who accompanied me to these ruins around Khor Ruri, volunteered the information that this was the site of the ancient Murbat, and I further elicited from a Shahari that in their dialect Murbat is called Sīk—a word which would appear to preserve the important radicals of the name Moscha. From here to Raisut is the rich plain of Dhufar with its capital at Salala. Here the coconut is the predominant vegetation as opposed to the date palm of Oman, and, besides, there grow in abundance cotton, tobacco, indigo, cereals, papya, melons, etc. This plain of Dhufar has widespread ruins (*see* Pls. IX and X). The most characteristic feature of these ruins is a plain primitive column, with octagonal shaft, square corbelled cap, and square base, generally a monolith. It is usually about 6 feet high, and this and its corbelled cap suggest that it supported arches. Arch masonry does in fact lie strewn about amidst the debris, and an arch of a later period still stands. Bent connects the Dhufar column with columns he saw at Adulis on the Red Sea, and with those of Kulue and Aksum in Abyssinia, and others he heard of at Ma'araba in the Yemen, and says that no doubt can be entertained that they were built by the same people. Bent's other descriptions of the archæological remains as given in his *Southern Arabia* are, however, superficial, and his sarcophagi are, I consider, only water-tanks (*see* Pl. X, Fig. 3). The most elaborate ruins—Balīd (Pl. IX. Figs. 1 and 2)—have, however, been very fully described by Carter, the surgeon of the Indian Naval brig *Palinurus*, in a paper given to the Royal Geographical Society in 1846. I attempted in the short time at my disposal to make measurements of the monuments I saw, and slight drawings of the decorated capitals and bases.

The word " Dhufar " is susceptible of the division Dhu Afar (Afar being the Wadi bounding one of the largest frankincense areas), though it should be mentioned that it is spelt with a ظ, not ذ. I unfortunately omitted to question the natives on its probable derivation. In the Hadara dialects modifications of this word, *e.g.* Āfur, Aufur, mean " red," and the meaning " Red Country " is satisfied by the predominant red colour of Wadi Afar and Wadi Dhikur. In the Shahari dialects the word Āfar means " clouds," and the name " the cloudy country " is satisfied by Dhufar's continuous rainfall for three months during the south-west monsoon.[1] Whether Ophir, which has been suggested on grounds of similarity of name, receives any confirmation from such speculations, is perhaps a debatable matter.

Dhufar Today

Until 1970, Dhufar was considered remote and alien by the rest of Oman, in spite of the fact that Sa'id ibn Sultan had taken over control in 1829 after the death of the local ruler Muhammad ibn 'Aqil, for each village continued to enjoy an element of independence, and for brief periods the province was governed by foreigners: 'Abdullah Lorleyd, an American-born protégé of Sayyid Muhammad ibn 'Aqil; and an Indian of Hadrami descent, Fadl bin 'Alawi (from 1876 to 1879).

By the mid-1970s, half of the population of Dhufar were living in Salalah, and most belonged to the tribe of Bait Kathir. The other half were herdsmen in the mountains of Qara, Mahra, and Shhera stock, or Bait Kathir in the coastal towns of Rakhyut, Raisut, Taqah and Mirbat. The mountain people, *Jibbalis*, probably descend from ancient South Arabian tribes, and have their own languages, with a strong admixture of Arabic nowadays, factors which tend to militate against unity with the rest of Dhufar and the rest of Oman.

The first ruler of Oman to spend considerable time in Dhufar was Faisal ibn Turki; his son Taimur spent more time in Salalah than in Muscat. But it was Sultan Sa'id ibn Taimur who took his dynasty's rôle in Dhufar so seriously that he created new agricultural estates at Jarziz and Arzat and, after spending much of the period from the outbreak of World War II to 1955 at his palace of al-Husn between Salalah and the shore, finally settled in Salalah from 1958 and never went back to Muscat. Discontent arose in 1963–4 and armed rebellion from 9 June 1965. From 1968 the communist tendency of the revolt increased with expanding aid, both ideological and logistical, from and through the (then) Democratic Republic of Southern Yemen, now the P.D.R.Y.

Sultan Qaboos came to power following a gun-battle in the palace of Salalah, and promised measures including amnesties for rebels surrendering, and a modernisation programme including new schools, hospitals, and adequate water supplies. The Sultan also increased defence expenditure to a level of 50% of the national budget, which culminated in victory over the rebels, officially declared on 11 December 1975, ten and a half years after the war had started. The visitor must be warned that mines laid by both sides cannot positively be known to have been located, and so it is advisable to keep to the well-worn roads and paths, especially in the interior.

Ways to Salalah

The preferred methods of going to Salalah from Muscat are by car (to pass the checkpoint near Thumrait you need to obtain a permit beforehand) and by plane. When Sultan Sa'id ibn Taimur made the first car journey from Muscat to Salalah in the 1950s, it took him a week across unmarked dirt tracks. When the Izki-Thumrait stretch of highway was opened in 1982, the journey took only nine hours, and now you can not only find strategically-placed filling-stations, but also top-quality restaurants and motels to break your journey, all bookable beforehand through the Ruwi Hotel in the Capital.

Oman Aviation Services operate daily flights from Seeb to Salalah Airport, which is close to the city centre. If you take your driving licence, you can hire a car in Salalah (Budget Rent-a-Car has an office in the Holiday Inn), but local taxis are quite inexpensive, in the absence of scheduled local bus services.

The most economical method of travel between Muscat and Salalah is the bus; the terminal in Salalah is beside the cemetery.

The Gulf Air Boeing 737 was packed full of European and Indian businessmen, with a few Omani families. We had left Seeb promptly at 8 a.m. and arrived at Salalah precisely on schedule at 9.25, the little terminal bustling with friendly chatter and shouts. 'Abdullah Hadid of the Bani Riyam was there to welcome me, waving from a throng of brightly-dressed men and women in the warm morning sun. The racial mix is very striking, testifying to Dhufar's turbulent history as a seaport and trading centre between the Gulf, India, and East Africa over many centuries. Indeed, one feels a distinct African pulse in the coconut palms waving in a sultry breeze, early afternoon indolence in the closing *suq*, the warm waters of the Arabian Sea, and brilliant colours in costume and ornament. It has been suggested that, apart from isolated stone age flint tools and six stone circles of 7,000–3,000 B.C. found at Bir Khasfa on the edge of Wadi Arar, and the occupation sites (5,000–3,000 B.C.) of Muqshin, Mitan and Fasad, the earliest evidence of Dhufari settlement is place-names identifying localities with others in the Tigris-Euphrates area, whose emigrants are also connected with Ethiopia, North Africa, and the Mediterranean. Himyaritic peoples from the West then absorbed, or intermarried with, these original settlers, to be followed by other immigrants from west and southwest Arabia, and from black Africa, notably from Zanzibar.

The Jibbali-speaking Qara arrived in the area about 900–1000 A.D., probably because of progressive desiccation in their grazing grounds, since they have always herded goats and cattle. They were followed about three hundred years later by the camel-herding Mahra, who speak their own language, Mahri, and are likewise nomadic, living each summer in the more fertile plains. Later on the Bait Kathir immigrated into Dhufar from Hadhramaut, and became successful merchants; their haunts are the plain of Salalah and the borders of the Empty Quarter.

Exploring Salalah must be a formative experience, like Goa, Penang, or Hangchow. There is nowhere like it in the world, and the authorities have recently recognised this by halting the demolition of historic buildings, though the policy came too late for most of the city had by then already been razed in the interest of progress, including the old fort at the junction of Shar'a an-Nahdha and Shar'a as-Salam.

The old al-Husn Palace has been restored, with a new guest wing opened in 1981, but this is closed to visitors, and your only indication of old Salalah are quarters of old two-storey limestone houses with carved wooden shutters, though these are often now interspersed with newer or restored buildings, for the age-old dilemma applies here as elsewhere: if houses are falling down as a result of their antiquity, an occupant must either find the money to restore his home himself, or rely on government aid. But if he is living in such an old house, the chances are that he will not be able to afford either to move or to restore, and if the government does not intervene as a matter of urgency, what is to be done?

From west to east, these old quarters, where traditional architecture may still be seen, are 'Uqad; the house of Shaikh Ahmad ash-Shanfari, opposite the Great Mosque in al-Hafah; a row of houses east of the Great Mosque in Salalah; and the southernmost sector of Dahariz, east of the city.

The name of Salalah is an Arabic corruption of the Jibbali *Selelt*, according to the late T. M. Johnstone's *Jibbali Lexicon* (1981), and means 'The Gleaming One', from the radiance of the tall white houses visible on the plain, so different from the one-storey brown houses and huts of the mountains. The city grew between the twelfth century and sixteenth century, relying on nearby Raisut for its anchorage and harbour. This city, the ruins known now as al-Balid, arose between two creeks connected parallel with the shore to form a rectangular island. The Great Mosque, Palace, and public open square or *maidan*, were situated in the western part, the whole surrounded not only by water, but also by a massive wall with turrets; a gated wall divided the city into two. G. Rex Smith has translated the relevant passages from the *Tarikh al-Mustabsir* (7th century A.H., 13th century A.D.) of Ibn al-Mujawir in *Proceedings* of the Seminar for Arabian Studies 1984 (1985):

'Zafar [the generic 'Dhufar', used here interchangeably with al-Balid. Ph.W.] was destroyed in the year 618 [beginning 25 February 1221] by Ahmad [bin Muhammad bin] 'Abdullah bin Mazru' al-Habudi in fear of al-Malik al-Mas'ud, father of al-Muzaffar Yusuf bin Muhammad bin Abi Bakr bin 'Ayyub. The former built al-Mansurah and named it al-Qahirah. It became inhabited in 620 [beginning 4 February 1223]. The name by which it is known is Zafar and it is on the sea shore. A wall made of stone and plaster (or baked brick and plaster) was built around it. Four gates were put into the wall: one of them the gate which leads to the sea called Bab as-Sahil. On the land side there are two, both called by the names of the gates of Zafar (previously) demolished: one in the east called Bab Harqah, leading to 'Ayn Fard; the second in the west called Bab al-Harja, leading to al-Harja, a pleasant village situated on the seashore near Zafar.'

Ibn al-Mujawir is also the first writer to mention that the coconut (native to India) was grown in Dhufar, but he does not mention when the city was first occupied. The mediaeval Yemeni geographer al-Hamdani (died 945 A.D.) notes the existence of Raisut, Mirbat and Hasik but not Dhufar/Zafar/al-Balid, leading one to assume the mid-tenth century as a terminus post quem for the city's foundation. Miles writes (in *The Countries and Tribes of the Persian Gulf*, 1919) that the ruler of Hurmuz, Mahmud bin Ahmad al-Qusi al-Qalhati, who captured and plundered Zafar in 1261, was driven off by one Muhammad Abu Bakr.

In 1278 the Habudi army ruling Zafar were routed by the second Sultan of the Rasulid dynasty (with its court in Ta'izz, Yemen), but it is a reasonable assumption that Zafar lay too far distant from Ta'izz for Yemeni rule to have been anything but nominal.

Marco Polo stated that Zafar was a thriving city, and one of the major ports of the Indian Ocean, but R. Guest has suggested that Polo never visited the city himself ('Zafar in the Middle Ages', in *Islamic Culture*, 1935), but merely reflects a view general in the 13th century.

Ibn Battuta first visited Zafar in 1329, and again in 1356. He found that 'all the people were merchants and had no other means of livelihood, that mosques were numerous, that good horses were exported to India, that rice (which formed the staple diet of the people) and the cotton of which their clothes were made was brought from there, and that good silk, linen and cotton stuffs were made in Zafar.'

Paolo Costa has stressed that Zafar's significance in the fourteenth century stemmed at least partly from the 'shifting of the political and commercial centre of Islam from Iraq to Egypt, which started with the establishment of the Fatimids and the decline of Baghdad, and continued under the Ayyubids and their successors in southwestern Arabia, the Rasulids. The consequent revival of the ancient Red Sea route determined or greatly accelerated the decline of the former leading shipping towns of the Gulf, namely Suhar and Siraf.' In its turn, Zafar decayed with the establishment of long-distance routes, transactions in bulk, and the growing complexity and sophistication of international supply and demand. The end of the horse trade marked the effective ending of Zafar's importance on sea-routes, and Zafar, nearby Ribat (or Arbat) and Raisut were gradually abandoned, with a concentration on small-scale fishing, farming, and pastoralism based on isolated rural villages that gradually merged to become modern Salalah.

Al-Balid, following centuries of despoliation as a quarry for building materials, has now been fenced off, and may only be visited with a permit from the Ministry of National Heritage and Culture, as is the case with the site of Samhar in Khor Rori to the east. A map of Zafar appears in the *Journal of Oman Studies,* vol. 5, 1979 (1982), with many photographs, including a specially valuable aerial view. Work on al-Balid was undertaken by the American Foundation for the Study of Man in 1952 but, apart from a few isolated notes and articles, there has been no satisfactory report on these excavations, 'particularly regrettable' in the words of Paolo Costa, 'because the finds have never been made accessible to other scholars'. The Foundation was interested only in pre-Islamic material, and when they found, after excavation, that the site was in fact Islamic, they abandoned work. Dr Costa notes that 'the great amount of débris and excavated soil was not removed but heaped on the tell itself. The large spoil heaps encroached on the trenches and soon after the dig began to slip,

Salalah. Al-Balid excavations

disturbing and obscuring the unearthed structures'.

The Americans unearthed the foundations of a substantial town gate, still visible today. Dr Costa's team from the Ministry of National Heritage worked in al-Balid from 1977 to 1981, and found evidence towards a complete plan of the Great Mosque, with its monolithic columns, large courtyard, and terraces on all sides except the eastern;

Salalah. New fish market

544

indications of the western moat, connected to the small western *khor* or creek, and thus to the sea, with a large bridge spanning it, later made into a causeway.

Salalah today is exhilarating because one seems to exist in a time-shuttle: the modern Family Bookshop, with British wives paying for *Woman* or *The Daily Mirror*, looks out on to a dusty road leading to a beach where athletic urchins in shorts lam into a football even in the midday heat. The old *suq* has gone, but the new one is alive and hectic with shouts and colours. A coconut was slashed open for me to drink the milk; then I asked for the meat to be sliced off, and that too, delicious, snowy white, quenches the thirst. I bought apples and oranges, and watched the trade in fish. Jum'a Ramadan Nasib, of the 'Awamir, had his bare feet swathed in polythene: he was to be my taxi-driver for a few days, teaching me Jibbali words with infectious good humour: he was enthralled and mused that anybody could actually want to use Jibbali when Arabic is the lingua franca of all Oman. The women in the *suq* congregated together, rings in their noses, and clanging, jingling jewellery on their wrists and ankles. The camel market was over, and I chatted with a certain Ibrahim, who had paid 300 rials ($900) for a young male that he hoped would grow and give him back his money soon. Female camels fetched 2000–3000 rials each.

The jewellery *suq*, near the Nahdha-Salam roundabout, repeats the age-old Arab preference for having all related shops in the same area. The best buy was the Long March suitcase from China at 5 rials or a huge trunk from the same factory at only 8 rials. A charming gold necklace at 55 rials, in the window at Ahmed Jewellery, I discovered had been imported from Italy. I looked for Salam Hotel, advertised on

Salalah. National Bank of Oman, Shar'a an-Nahdhah.

Salalah. Three styles of domestic architecture

upper Shar'a an-Nahdha, but it was closed; only al-Kalbani Laundry on the ground floor was open.

As old houses are vanishing, new houses seem to be of a much higher standard, from the aesthetic and functional viewpoints, than anyone has the right to expect in a world dominated by quick profits and high-rise apartments. A two-storey villa, with mandatory television

Salalah. A new house

Salalah. Mosque and coconut palms

aerial, looks much more than the conventional box, with a pleasing range of white, cream, light and dark brown.

I took a sweet *kunaifa* and tea at the new Nile Palace Restaurant; the place had a 'Closed' sign upside down, neatly controverting Alfred de Musset's subtle comedy 'Il faut qu'une porte soit ouverte ou fermée.'

At the Islamic Cultural Bookshop, opposite Hotel Redan, I bought the *Diwan* of Abu Nuwas, and the Arabic text of Ibn Jubayr's *Travels*. A nearby toyshop offered plastic toys from Hongkong, such as a pack of various spacemen and another of a mother-camel and two young. 'Which sells better in Salalah?' I asked the Patna-born shop assistant. He raised his hands to disclaim any knowledge of his merchandise, and simpered in wonderment. 'Everything is *very* good,' he whispered conspiratorially, very *very* nice' and wished me a fervent 'Come again soon.'.

If you get a chance to see the Salalah Raft Race held on the first Friday in March, don't miss it. Teams of four to six competitors make a raft of standard empty steel 45-gallon drums (one per team member) which must be paddled without keels or rudders. Windsurfing contests are also highly enjoyable: if it is not Bondi beach, then Salalah is certainly serene and quiet, and for the evening hours just before and after sunset the warm breeze and soft sand remind one of Seychelles or Barbados: there is nobody around, as if you were auditioning for the part of Robinson Crusoe.

Samhar and Mirbat

Next day I was to journey east, from the beach-snug Holiday Inn to

Mirbat, then north to the mountain villages of the Qara. Jum'a had said he wanted to leave at eight; I suggested seven. He countered with seven forty-five, and I proposed seven thirty and a packet of cigarettes, to which he agreed. So promptly at seven forty-five, having pocketed his cigarettes, he drove off. Have I children? 'Yes.' 'Boys?' 'No.' 'Ah,' he subsided gloomily, casting unspoken aspersions on my virility. He, by contrast, has three daughters and no fewer than five sons to continue his line. 'Did I have camels?' Regrettably not. 'A car?' 'Yes, a little old Volkswagen Beetle.' 'Zvagen very tough', he said politely. 'But not as good as your expensive American car,' I replied, feeling that we were approaching cordiality: if in doubt, praise others' cars and denigrate your own, which in my case is not difficult, for I have never liked expensive cars and their tendency to depreciate like punctured tyres.

The sea lilted brightest azure on our right, the mountains gentle fawn far left in early morning haze. Camels grazing by the roadside occasionally raised their heads to stare rudely at us. 'Too many accidents with camels,' groaned Jum'a. 'They like to run against cars: run too fast, land too flat: *shooch*', and he hamfisted the windscreen in excited mimicry as we entered Taqah. I wandered in the graveyard of the Friday Mosque, which is situated to the left of the main road, and boys followed me, explaining to each other that I was a spy *(jasus)* or an escaped criminal from the jail, come to hide in the graveyard so that the police would not find me. Fields of alfalfa were flourishing opposite the mosque in the brilliant sunshine that made colours stand clear and jewel-like; the mist had dissipated, but dew-drops still hung on tall leaves.

And then after a few kilometres we turned right to the site of Sumhur ('Sumhuram' is believed by some scholars to be merely an oblique form of the name 'Sumhur'), overlooking Khor Rori. In this I follow A. F. L. Beeston in *Journal of Oman Studies,* vol. 2 (1976) and Brian Doe in his book *Monuments of South Arabia* (1983), in preference to interpretations by Jamme or Pirenne. Its port may have been the Moscha of Ptolemy, but that has not been proved as yet.

The site we see today, covering an area of two acres, measures 68 metres × 128 metres on the east-west alignment, and is completely walled, with roughly cut stones, to a thickness averaging 2.5 metres. Sumhur (or Samhar, vowels being more fluid and less significant than consonants which form root-clusters in Semitic languages) dates from the first century B.C. (according to Pirenne) or a century or two later (according to Beeston). Alfred Beeston has described how the process of founding a daughter-city here has parallels with modes of Greek settlement. 'A party, or parties, of settlers *(hwr)* were sent out from Shabwa, the Hadramite metropolis, under the leadership of an individual named Abiyatha' SLHN, son of Dhamar'ali, whose function resembled that of the Greek *oikistes* but who had the title of 'Commander of the Hadramite *gys* in the land of Sa'kalan (= Greek Sachalitis, general name of this part of the South Arabian coast)'. In military terminology, the South Arabian *gys* was not an 'army' in the

548

Samhar. The frankincense port of the first centuries B.C.-A.D.

modern sense, but a smallish ad-hoc body made up of various elements and assigned to a specific rôle; its extended application to a body establishing a settlement would thus seem natural, and although such a body might be prepared to meet opposition, it is not necessary to infer that there was any such opposition (the site may well have been previously uninhabited), or that the settlement involved military conquest.'

Nevertheless, Samhar was fortified so strongly that opposition must have been anticipated from the start. Brian Doe sees the use of two-storey buildings with internal staircases as 'confirming the rôle of Samhar as a permanent garrison town.' It could hardly have been otherwise, as it controlled the export of frankincense from the interior, which was the main, if not the only, valuable commodity traded from Dhufar at that period.

The main gate in the northern wall consists of two separate barriers to negotiate, and must have constituted one of the most testing obstacles for potential invaders found in all Arabia, excelling even the south gate of Timna. Other important structures include a Temple to Syn or Sin, the principal deity of the Hadramites attested by a bronze tablet identified by Jamme and published in Wendell Phillips' *Qataban and Sheba* (1955). Temples to Syn have been identified throughout Hadhramaut.

At the head of Khor Rori rears the sheer limestone cliff overlooking Wadi Darbat, during the rainy season a waterfall cascading more than 300 metres to the grateful earth. This was identified by Theodore Bent as the Abyssapolis of Ptolemy, with ample reason.

Mirbat is governed by Shaikh Amr Musallam of the 'Awamir, and though he was away, we were welcomed to the fortress that overlooks the idyllic harbour, fishermen mending nets and examining their recent haul on the beach. Mirbat first occurs in the geographical literature in al-Hamdani's *Sifat al-Jazirat al-'Arab* (written before 945 A.D.):

'The people who dwell below al-Qamar were the Bani Khanzarit. They expelled the inhabitants of Raisut, who then retired to the Mahra settlement of al-Ghaith, and occupied the towns of Hasik and Mirbat for some time.' He continues, 'The Thughra tribe subsequently induced them to retake Raisut and drive off the Khanzarit. The Khanzarit chief, Muhammad ibn Khalid, then went with his tribe to ar-Radha of the Bani Riyam, a branch of the Qamr, and resided there: the Bani Riyam have an impregnable fortress in Oman. The oldest inhabitants of Raisut were the Bayasirah, near whom the Azdite tribe Hadid settled. The Hadid were overpowered by some Arab tribe who had intermarried with the Mahra, and the man who ruled after that was Musa ibn Rabi'a of the al-Udar.'

A friend in flowing *dishdasha,* eating chicken and rice beside me in a little restaurant in Mirbat, was 'Abdullah Hadid, of the same tribe that Ahmad ibn Yusuf al-Hamdani had mentioned more than a thousand years earlier. The passion for genealogy in Arab countries, with its desire to trace descendants back to Companions of the Prophet (salla Allahu 'alayhi wa sallam), must surely be intelligible in this breathtaking context. The name 'Mirbat' means a place to moor vessels, and the historical Mirbat lies about 32 km north of the present village, which probably formed an outer anchorage in the north-east monsoon. Old Mirbat was supplied with fresh water from the hills behind and benefited from three tidal creeks, described already by Haines, Carter and Bent. The Minjawi dynasty controlled Mirbat for some time from the tenth century to the twelfth; al-Idrisi tells us that in 1145 the ruler in the plain was Ahmad al-Manjuwi (or Minjawi), who held his fief from Oman. The Minjawis, who according to Carter may have come from Balkh, destroyed the city and founded al-Balid to the west. The large merchant houses of Mirbat even today constitute something of a wonder; not only for their great size in such an apparently sleepy little town, but also for the carved windows and *mushrabiyat.* Clusters of medium-sized houses one or two storeys high are separated by these larger houses, which boast plaster carvings in low relief of typical vessels, such as the *ghanjah* or *bum.*

On leaving Mirbat, we were hailed politely by a splendid figure of a man, bearded and bright-eyed, with a rifle clearly in working trim, and a dazzling *khanjar* tucked in his cartridge-belt.

Yes, of course we could give him a ride. He told us his name was Muhammad 'Ali Ahmad Muhammad, which did not stun anybody, for these are the commonest names throughout the entire Arab World.

'Abdullah Hadid informed Muhammad our guest that I was from Cambridge, near Skutlanda. He had, he recounted, a friend who spent three days in Kilarskow and had his camera, passport and all his money stolen. 'The police were very, very good. They gave him many papers

550

Scrub on the Dhufar plain near Mirbat, with Muhammad 'Ali Ahmad Muhammad against the Qarah Mountains

to write and gave him good tea. The policemen were *wonderful.*' 'But,' I interposed delicately in ensuing silence as the driver and our guest absorbed these facts, 'did your friend ever recover his wealth, and his belongings?' 'No,' he said, shrugging as if reconciled to skulduggery, 'the police say they never catch any teebs, but they are so good beeble.' I did not linger on the crime rate comparisons between Britain and Oman, for the complexity and inequalities of Western urban society are too deep to describe in a few sentences. But the Omanis in the car with me would probably never have experienced a theft in their lives, and would have had no sympathy or understanding of the criminal mentality. Muhammad courteously requested to be dropped at the side of a road which apparently led nowhere, but we all knew better than that. He would pad in his sandals towards the cliffs, where his home would be a simple hut, with grazing for his family's goats and camels nearby.

Jabal Qara

The car climbed up the zig zag bends to Tawi 'Atair, with a marvellous view down into Khor Rori; we disturbed a bird of prey (Verreaux's Eagle?) devouring lumps of meat from the ribcage of a cow hit recently by a car, but it swooped away before I could identify it with certainty or take a photograph.

Many other cows, fat and contented, raised their heads in faint surprise as the car passed them.

I was learning as many Jibbali words as possible, with or without Arabic admixture: 'Bkhairtum', Good morning; 'Alghubun', Good evening; 'Na'gid?', What do you want?; 'bira', man, child; 'ghabghat', woman; 'mizé', house; 'yi', father; 'ami', mother; 'athé', friend; 'lé', cow; 'aithá', camel; 'Inek bukhar?', What's your news?; 'Tum bukhar?', How are you?, 'ehe', yes; 'lub', no; 'arum', road; 'usé', rain; 'infid', before; 'maghari', after; 'Ghadun!' go; 'Nkya!' come; 'É luhu at'ghad?' Where are you going?.

The Jibbalis we saw bound their long hair with a leather cord termed *mahfif*; most still wear their traditional dress, like a kilt, often with a blanket over their shoulders. They comprise only twelve to fourteen per cent of the population of Dhufar, and all these hills seem extraordinarily under-populated.

Tawi 'Atair is one of those Jibbali settlements that possess a brand-new community complex, including a mosque, administrative offices, schools, a health centre with pharmacy, and a police post. Tawi 'Atair was the end of the road, so we turned back and headed for a similar administrative centre at Madinat al-Haqq ('City of Truth'), a gleaming new town in white, with a mosque at its heart. We took tea there, in a restaurant where a Hindi video film was being shown on the television to that same rapt and silent awe of the Indian audience that has made Bombay the capital of the Oriental film industry.

The clouds began to look grey and threatening, but I was assured with some amusement that this was not the rainy season: there would be no precipitation until the monsoon season starts in June, or possibly May. When the monsoon season starts in India, it starts in Dhufar.

Cattle are as crucial to the lives of the Qara tribesmen as they are to the Dinka of Sudan or the Masai of East Africa, and it has been estimated (by V. C. Peterson, in 1960) that there are 25,000 head of cattle in Dhufar, each with its own name. Peterson notes: 'The origin of the Dhofar cattle is not known but they are similar in conformation and colour to the early Spanish and Portuguese cattle. They are very small. Adult females stand about 3' 4" at the shoulder and weigh approximately 300 pounds. Bulls are somewhat larger.' Exceptional cows may yield a gallon at each milking at the hands of men, not women; elsewhere in Oman only women milk the cows.

Eating habits are also curious: the Qara consider the carrion-eating hyena permitted meat, treating the jaw muscles as a special delicacy, but they will not eat birds (including chickens), eggs, or foxes. When a man dies, half his wealth (normally young bulls, and cows deficient in milk) is sacrificed, and neighbours and visitors are quick to join such a feast. A cow (or in lesser cases a sheep) will also be sacrificed when disease or sickness claims a Qarawi; its blood is sprinkled over the patient's shoulders and chest.

We passed through scrub and thornbushes, the upland undulating high above the plain. Camels grazed in poor pasture unsuited to cattle. Between Madinat al-Haqq and Shinhaib very little traffic disturbed the silent wilderness. Zig village's new mosque appeared through very low cloud, then we came upon a sign 'Beware of the Cow'. Hairiti was signed 1 km off left, and we saw tomatoes and citrus fruit being tended at the Agricultural Research Station; a white single-decker bus cruised past us on its way to Thumrait. Beyond the left sign to Nahiz Heights, for a spectacular wide view of the Salalah plain, we made another steep descent, as-Sa'adah off left, then Haqif off right, and five kilometres farther on a huge new cattle market just before reaching the Royal Stables.

The evening I spent walking on the soft silver-white sands of Salalah. Coconut palm leaves rattled and threatened like rain on parched earth, but still no rain fell.

Over breakfast of papaya, Danish pastries and honey, and a pot of coffee, I discussed earthquakes with a United Nations expert investigating occurrences throughout Oman. Damage by earthquake has been recorded in Qalhat and Sur as recently as 1983; the epicentre was probably Makran, now a part of West Pakistan, but one should recall the Omani possession of Gwadur in Makran, sold, as Hugh Boustead writes in *The Wind of Morning* (1971), by Sultan Sa'id bin Taimur to Pakistan for the sum of three million pounds.

With Jum'a Ramadan Nasib at the wheel, we headed north to Thumrait, known to the British as 'Midway', which controls the great highway of 1982 north to Muscat and south to Salalah, and the tracks westward to South Yemen (officially known as the People's Democratic Republic of Yemen, ruled from Aden) and eastward to the oilfields such as Dhahaban and Rahab, Marmul and Amal, running parallel with the shore and the Kuria Muria Islands beyond it. From Amal a graded road south brings you to Shalim (report to Wali's Office for permission to continue southward) and eventually to the fishing village of Shuwaimiah, on a magnificent road built by Petroleum Development (Oman) Ltd., after a hair-raising descent from the cliffs to the shore.

Thumrait itself is occupied by two military camps, the village containing a few administrative offices and a small *suq* with supplies for the *badu* over a large area, who use Thumrait as a market town. Baby blankets come from Spain, a huge variety of clothes from Korea and Taiwan, and cartridge belts from Pakistan.

On our way across the high plateau towards Salalah the unforeseen happened: it actually began to rain. Muscat was marked 1020 km distant on the signpost, and I wondered whether rains were falling there too (I learned shortly afterwards that floods had swept away cars in Ruwi's Wadi Kabir!)

Frankincense tree near al-'Uyun, with Jum'a Ramadan Nasib of the 'Awamir

Visibility remained poor as we entered the jabal again along Wadi Harit, and as far as the descent to Haqif. We were on our way to al-'Uyun ('The Springs'), in the zone where the best frankincense is grown. The finest source on the ancient (and modern) trade is *Frankincense and Myrrh* (1981) by Nigel Groom, who has provided conclusive evidence that the ancient frankincense-growing region extended beyond Dhufar westward as far as Wadi Hajr in Hadramaut, where it has been located, albeit sparsely, in recent times. Frankincense has been identified with the tree *Boswellia sacra* and, while this is true of Arabia, in Somalia it is collected from *Boswellia carteri* and *Boswellia frereana* and possibly also from *Boswellia bhaudajiana* (though this last-named has not been located again following its classification in 1860). Inferior qualities come from *Boswellia papyrifera* (East Africa, Sudan, and Ethiopia), from the Indian *Boswellia serrata*, and from a Socotran species of the same tree. Contrary to accepted belief, it does not grow only at an altitude greater than 2,000 feet, though again the quality of gum from low-lying trees is poor.

Myrrh still grows throughout south and south-west Arabia, including 'Asir in Saudi Arabia, Somalia, and Ethiopia. It comes from the different species of the genus *Commiphora*, especially from C. *myrrha*.

The trade in frankincense and myrrh was controlled by South Arabians from the sixth century B.C. to Greece, when incense was first used as a substitute for sacrifice, and from the second century B.C. to Rome, where it was used to propitiate the gods and more commonly to reduce the smells of urban living. The camel was domesticated in south Arabia by the second millennium B.C., but it would not have been brought into significant use as a pack animal for the overland incense to the north before the first millennium. The Queen of Sheba was probably not the ruler of the South Arabian kingdom of Saba, but more likely chief of a northern Arabian tribe of Sabaeans carrying goods on the east-west route between Palestine and the Gulf.

Herodotus states that a flourishing trade existed in 500 B.C., a fact corroborated by archaeological evidence from Eilath (sherds of the period inscribed with South Arabian script). Theophrastus (c. 295 B.C.) published reports from reconnaissance vessels dispatched by Alexander the Great of incense trees growing in South Arabia. Gradual desiccation has affected all Arabian vegetation over the last two thousand years, causing the decline in frankincense trees as in so much else. Its height was probably during the first two hundred years A.D., and its decline was hastened by the spread of Christianity coupled with the destruction or change of use of pagan temples in the Mediterranean, the decline of the Roman Empire generally, and internecine strife between the various South Arabian kings and tribal chiefs. Using Pliny's figures as a guide, the output in the first century A.D. consisted of up to 60 tons of myrrh. A triangle of iron, or *manqif*, shaves strips of bark from the tree trunk, and from the resultant wounds *luban* oozes. After this ooze has hardened, it is collected in a palm-leaf basket *(zanbil)* or a woven milking-bowl *(satmah)*. Trees are not tapped every year, but rested for one or two years. Ownership of the trees and collection of the resin is in the hands of the Bait Kathir and to a smaller extent the Mahra.

Then as now the crop would be harvested from April to June, stored until September or November (whenever the monsoon finished), then shipped to Qana in Hadhramaut, and carried overland to Shabwah together with the crops from Somalia and Socotra. From Shabwah the principal route led west through Timna and Marib, but others would have been dictated by political and climatic factors from time to time. Major staging posts northward might have included Najran, Tathlith, Bishah, Yathrib (now called al-Madinah al-Munawwarah), Hijra (now called Mada'in Salih), Petra and possibly Gaza or Alexandria. Some writers have suggested that a direct overland route ran from Dhufar to Gerrha, but Nigel Groom discounts this possibility. The location of Strabo's 'Gerrha' has incidentally been identified by Daniel T. Potts (*Proceedings* of the Seminar for Arabian Studies, vol. 14, 1984) as Jubail (*port* of Gerrha) and Thaj (*city* of Gerrha).

Today it is still possible to buy incense-burners in the *suq* at Salalah (and above all at Ha'il in Saudi Arabia, where the best I have seen are still made), and even a good bag of the incense can be purchased for a modest sum. Here, if anywhere, is the perfect souvenir of a stay in Dhufar.

In the village of al-'Uyun I spoke to 'Ali Salim 'Ali, who welcomed me, a perfect stranger, with soft drinks and cakes. All of the 500–600 inhabitants of al-'Uyun are Mahri-speaking Ghuwas tribespeople; their school, opened in 1985 with two Egyptian teachers and one from Jordan, has instruction solely in Arabic, the *lingua franca*. After the

Al-'Uyun. New village, with 'Ali Salim 'Ali of the Ghuwas

asphalt road came to al-'Uyun, radio and television followed. 'Ali ushered me into his little shanty shop, where I noted batteries, pens and erasers, feeding-bottles, biscuits and sweets, scissors, a freezer for cold drinks, soap, tissues, towels and the local washing-powder. Crowds of children surrounded our little group. Adults came to shake my hand, exchanging conventional greetings in Arabic; their habit of kissing cheeks reminded me of southern Europe. I felt so easy and relaxed in al-'Uyun that I did not want to leave, suggesting to the children that they showed me the springs their village is named for. Six of them raced off ahead of me, and eventually we found pools, down a steep incline, in a limestone hollow. The more uninhibited boys swam like brown eels in the deep waters, shrieking with excitement.

On the road to Taitam I gave a lift to Sa'id Zailub of the Samhan tribe. Mists ebbed and flowed over this romantic, even legendary heartland of the frankincense trade. Yes, Sa'id Zailub owned frankincense trees, and camels and goats. Round-faced white-bearded, tough and sinewy as a frankincense tree, he was dressed in a camouflage jacket, and a green skirt with an embroidered hem, with plastic sandals of the 'flip-flop' variety. He was the only civilian Dhufari I met who carried binoculars, of particular value for discerning strayed livestock. Taitam is yet another village of the interior with a sparkling new government compound, with a mosque and a school, a dispensary, a police post, new market, and the Deputy Wali's office. Nearby is a Firqat post, for the Home Guard.

From Taitam you can take a poor graded road 14 km to 'Aruqum (not marked on any map I have seen), or a better graded road to Qaftat, recommendable for marvellous views in all directions across

556

Sa'id Zailub of the Samhan, near Taitam

the jabal and, when the drifting mist parted for a while, down to the plain. The government had just provided new drinking-troughs for animals, with pumped water to ensure perpetual supplies, and cows, lazily flicking ears and tail against flies, drank deep from their permanent reservoir. A Flying Doctor service, operated from Salalah, offers emergency aid throughout the far-flung mountain regions, and I heard from a resident of Qaftat that his wife had been taken away for an urgent operation the day before.

On the way downhill we stopped at the Mosque of Nabi Ayub ('The Prophet Job'), with its old tomb, new mosque, and family picnic kiosks, the whole walled and gated. Peace coupled with superb mountain views: the silence of Job's Tomb is broken by birdsong and the rushing wind. And then rain began to fall again, to the delight of Ahmad and Jum'a. 'May Allah be praised!' they chorussed fervently, as the black road glittered shinily ahead. Past the little village of Ghadu we took a turning off left to 'Ain Jarzis, between little shacks with corrugated iron roofs, and cows and camels grazing together, a kind of Arabian Switzerland. The cistern of 'Ain Jarzis, teeming with small fish, is fed by a spring from the limestone mountains surrounded by rocks and trees in a semi-circle like an auditorium. A small latched gate prevents animals from roaming inside. Prayer-hour having arrived, Jum'a and Ahmad performed their religious duties, facing Makkah, then we set off for the road west towards Yemen, the first stop being Raisut. The town is built up on the 200-foot promontory called Ras Raisut, and in the natural bay, where a private Quwwat al-Hudud (Frontier Force) beach offers excellent boating, surfing and sailing facilities. Al-Muhit is a new restaurant complex opened by the

Ministry of Commerce and Industry on the beach near Raisut. Raisut itself is the site of a major new cement factory, and an animal fodder plant.

The fine new blacktop road now swings inland a little way, with an even better view of the limestone mountains on the right.

Glorious empty sandy beaches stretch as far as the eye can see in both directions. Boats are shored up; quartz and mica in the pure sands gleam alluringly. A sign reminds us that it is 'Strictly forbidden to practise fishing or removing any shellfishes or any other marine resources from the sea in all coasts of the Southern Region without a special permit from the Fisheries Department in Salalah.'

Shearwaters and terns encircled a lagoon inshore at Mughsail, an idyllic haven with fishing boats yawning empty-oared, askew on the fine sands. Family kiosks have been erected near the shore, to enable women and children to go to the beach in suitable privacy. A goatherd sat among his eighty animals, half-dormant, half-alert. We took tea at a little restaurant called al-Mughsail, but the place has a military significance that never quite allows one to relax entirely.

A new graded road continues its weaving way more or less parallel with the sealine to the Shahab Asbatib junction, south to Rakhyut, then west to coastal Dalkut; west towards the Yemeni border; or north in a great sweeping curve via Hairun and Mudai to Thumrait.

From Salalah you should explore if possible two picnic areas. 'Ain Arzat or Razat is not a single spring, as its name suggest, but several. Do not touch the water, 'contaminated with bilharzia' as a sign proclaims. Landscaped gardens produce all the plants of which Dhufar is capable, especially the glorious pink oleanders which remind me of the first years of my married life, our garden exploding like fireworks with the radiance of oleander. Cicadas conjured up long Italian summer evenings; small birds darted too quickly to be identified, as if taunting us with their alacrity.

The single spring called 'Ain Humran, halfway between Razat and Tubruq, lies 7 km off the main road to Taqah. Coconut palms have been protected by tall nets and the lush oleanders danced in the gusty wind below them. Unobtrusive parking places are hidden away, so that nature below Jabal Qara blossoms and burgeons without unsightly interference from machines. Attractive hedges have been cut by a topiarist to form slope-roofed green 'cabins' linking at the eaves.

INDEX

ARABIA PAST AND PRESENT

ANNALS OF OMAN
Sirhan ibn Sirhan

ARABIA IN EARLY MAPS
G.R. Tibbetts

ARABIAN GULF INTELLIGENCE
comp. R.H. Thomas

ARABIAN PERSONALITIES OF THE
EARLY TWENTIETH CENTURY
introd. R.L. Bidwell

DIARY OF A JOURNEY ACROSS ARABIA
G.F. Sadleir

THE GOLD-MINES OF MIDIAN
Richard Burton

HA'IL: OASIS CITY OF SAUDI ARABIA
Philip Ward

HEJAZ BEFORE WORLD WAR I
D.G. Hogarth

HISTORY OF SEYD SAID
Vincenzo Maurizi

KING HUSAIN & THE KINGDOM OF HEJAZ
Randall Baker

THE LAND OF MIDIAN (REVISITED)
Richard Burton

MONUMENTS OF SOUTH ARABIA
Brian Doe

OMANI PROVERBS
A.S.G. Jayakar

SOJOURN WITH THE GRAND SHARIF OF MAKKAH
Charles Didier

TRAVELS IN ARABIA (1845 & 1848)
Yrjö Wallin

TRAVELS IN OMAN
Philip Ward